T0206004

Non-equilibrium Statistical Physics with Application to Disordered Systems

Manuel Osvaldo Cáceres

Non-equilibrium Statistical Physics with Application to Disordered Systems

 Springer

Manuel Osvaldo Cáceres
Centro Atómico Bariloche and Instituto
  Balseiro
Comisión Nacional de Energía Atómica
  and Universidad Nacional de Cuyo,
  and Comisión Nacional de Investigaciones
  Científicas y Técnicas
San Carlos de Bariloche
Rio Negro, Argentina

Original Spanish edition published by Reverté, Barcelona, 2002

ISBN 978-3-319-84681-1   ISBN 978-3-319-51553-3   (eBook)
DOI 10.1007/978-3-319-51553-3

Printed on acid-free paper

This Springer imprint is published by Springer Nature
The registered company is Springer International Publishing AG
The registered company address is: Gewerbestrasse 11, 6330 Cham, Switzerland

. . .for my parents and my family. . .

"As if it were some practical purpose, geometricians always talk about squaring, extending, adding, when in fact science is grown for the sole purpose of knowing."
PLATON (República, Libro VII, 527)
"Scientists study Nature not because of its usefulness but for the joy they find in its beauty. If Nature were not beautiful it would not merit our studying it and life itself would not be worth our efforts. I am not referring to the superficial aspects of beauty such as those which only concern external qualities or appearance; not because I hold them in contempt, which would be far from my intention, but because those bear no relation to science. I rather mean that deeper beauty which comes from the harmonious order of all parts to which only pure intelligence is susceptible."
HENRI POINCARÉ

Dedicated to the memory of
Nico G. van Kampen
whose influence on my permanent work runs deeper than I can know.

Preface to the First English Edition

The first edition of this book appeared 13 years ago in Spanish, published by Reverté S.A. As compared with that Spanish edition, the first eight chapters have continued with the same topic, while the contents have been revised and expanded. In particular, advanced exercises with their solutions have been incorporated at the end of each chapter, Appendix I is new and introduces an approach to quantum open systems, and Chap. 9 is new and has been included, in the English edition, with the purpose of relating the stochastic approach with the important study of the relaxation from steady states. This topic brings the opportunity to develop, with some detail, the theory of first passage time in physical problems.

I should like to use this occasion to thank Prof. Dr. V. Grunfeld for the critical revision of the English translation.

Foreword

This text is the result of several courses in nonequilibrium statistics, stochastic processes, stochastic differential equations, anomalous diffusion, and disorder, which I have been giving during the last 25 years at Instituto Balseiro, Centro Atómico Bariloche (Argentina). This book is aimed at university students of physics, chemistry, mathematics, science in general, and engineering. Readers are expected to have a prior knowledge of mathematics and elements of physics from a fourth-year university course. However, less well-known concepts of physics and mathematics are developed not only in sections and special exercises throughout the whole text but also in appendices. Some concepts of quantum mechanics, especially those which first-year students are still not acquainted with, are briefly presented in Appendices F, G, and I and in guided exercises, according to their needs.

Innovations

The physical-mathematical motivation is the main aspect throughout this text. Academic issues regarding probability theory and stochastic processes are presented, as well as new pedagogical aspects in the presentation of the nonequilibrium statistics theory, stochastic differential equations, and disorder. Possible representations for stochastic processes are detailed, and a functional theory is presented for solving linear differential equations with arbitrary noises. In Chap. 4, I talk about the irreversibility problem in particular, and, in this context, we discuss the Fokker-Planck dynamic; the relaxation theory of nonstationary time-periodic Markovian systems is also presented. In Chap. 6, I introduce a presentation of transport phenomena in finite and infinite lattices. In Chap. 7, the anomalous diffusion theme is generally presented. In Chap. 8, bases are given in order to establish the existing relationship between the microscopic aspects of linear response theory and the calculation of the diffusion coefficient in amorphous systems. In Chap. 9, a review on fluctuations around metastable and unstable points is given, and Kramers' activation rate and Suzuki's scaling time are presented. A generalized scaling theory to study the lifetime from arbitrary nonlinear unstable points is presented.

Applications

Different applications and exercises are almost homogeneously found throughout the whole text. In Chap. 2, we introduce, as an application of the theory of random variables, the theory of fluctuations around thermodynamic equilibrium, originally developed by Einstein and later in grater detail by Callen and Landau. In Chap. 3, several physical applications are given, from the stochastic processes' theory to the study of the relaxation in the solid-state area, and also the study of stochastic differential equations and its relationship with the Fokker-Planck equation by means of Stratonovich stochastic differential equations. In Chap. 4, we present general aspects regarding the concept of irreversibility by Onsager and the theory of temporal fluctuation (first fluctuation-dissipation theorem). Several applications of the fluctuation-dissipation theorem to simple, mechanical, electrical, and magnetic systems are also presented in this chapter. In Chap. 5, we will talk about general aspects of the linear response theory developed by Green and Callen, and we will also talk about the (second) fluctuation-dissipation theorem using a magnetic system to introduce an intuitive presentation of it. Other fundamental theorems regarding the linear response theory are also deduced, and some applications in solid state are presented. In Chap. 6, the theory of diffusive transport in ordered media is presented. Emphasis is particularly laid on the analysis of discrete and continuous time Markovian random walks and, in general, on master equations with applications on the study of finite systems with special (absorbing, reflective, and periodic) boundary conditions; finally, we briefly present the statistics problem of the first passage random time through a given boundary. In Chap. 7, we present two alternative and complementary techniques to confront the problem of diffusion in (amorphous) disordered media. The first is based on the effective medium approximation, while the second is based on the non-Markovian random walk theory. Emphasis is laid on the calculation of the diffusion coefficient in disordered media, the displacement variance analysis as a function of time and its scaling laws (universal or not), which depend on the kind of disorder. Finally, the super-diffusion problem and the analysis of diffusion with inner states are presented. Chapter 8 deals with certain quantum aspects of the transport and irreversibility problem. We particularly discuss in detail the formulation of Kubo regarding the analysis of the linear response from a microscopic point of view (third fluctuation-dissipation theorem) and the calculation of the electric conductivity (Green-Kubo formula). Also, we broadly discuss the formula of Scher and Lax for the calculation (within the classical limit) of the electric conductivity in (nonmetallic) disordered materials. Some examples and applications in a Lorentz' gas are presented. Finally, we discuss the relationship between anomalous diffusion and certain characteristics of fractal geometry. In Chap. 9, a review on fluctuations around metastable and unstable points is given. Emphasis is placed on establishing the connection between Kramers' activation time and the theory of the first passage time in stochastic process. Suzuki's scaling theory—for the lifetime from an unstable point—is presented and generalized to study lifetime in critical points, as well as for non-Markovian process.

How This Course Is Designed

This textbook might be useful for the introduction to the study of stochastic processes and its applications in physics, engineering, chemistry, and biology. In this case, Chaps. 1 and 3 constitute the core of a course on random variables, stochastic processes and their relationship with stochastic differential equations. Chapter 2 serves as a presentation to the theory of Einstein that deals with fluctuations around thermodynamic equilibrium, while Chaps. 4 and 5 end the course of nonequilibrium statistics with the analysis of irreversibility in the context of Fokker-Plank equation and the linear response theory. Also Chap. 9 can be included in this introduction study of stochastic process, with the aim of applying the theory of the first passage time in order to tackle the study of the lifetime at metastable and unstable points.

We can also design a course of introduction to the study of anomalous diffusion in disordered, or amorphous, media and its relationship with the calculation of transport coefficients in the context of the linear response theory, which can be studied independently of Chaps. 2, 3, and 4. In this course, Chaps. 6 and 7 give a detailed presentation of the anomalous diffusion problem. Chapter 8 is focused on the microscopic presentation of Kubo's formula for the calculation of electric conductivity. Appendix G.1 particularly presents a alternative demonstration of Kubo's formula (or third theorem), which, from a pedagogical point of view, is easier than that originally introduced by Kubo. This is because, in this new presentation, we use elementary concepts of the time-dependent perturbation theory of quantum mechanics instead of the algebra of superoperators (Liouville-Neumann operator). Appendix I presents a review on quantum open systems with an application to the quantum random walk model.

In general, there are different options regarding exercises throughout the whole text. In particular, we have exercises labeled as "optional," which should be skipped at first reading. On the other hand, sections and chapters indicated with an asterisk are more advanced topics which should later require a second reading. There are also advanced exercises with their solutions presented at the end of each chapter. Appendices from A to H are written with the aim of presenting for the sake of completeness certain physical-mathematical aspects regarding some topics discussed throughout the text. Finally, those sections labeled as "excursus" are specialized comments for those readers who wish to know more about a certain topic.

The history of science shows that the interest toward noise (fluctuations) has varied according to its perception. During the nineteenth century, noise was considered as "annoying," not only in theoretical physics but also in experimental physics. In the beginning of the twentieth century, the study of fluctuations surrounding equilibrium and its symmetries gave origin to the linear response theory, which includes the majestic and ground-breaking works done by Onsager (fluctuation-dissipation), while in the last decades of that same century, noise became essential for understanding self-organized structures out of equilibrium (synergetic). Simultaneously, in the last three decades, disorder (spatial noise) has also occupied a fundamental role in the comprehension of anomalous transport problem.

In the last 40 years, nonequilibrium statistics has made huge progress in the complex understanding of fluctuations and mesoscopic phenomena induced by noise. Nowadays, fluctuation concepts besides equilibrium, stochastic dynamics, noise-induced phase transition, stochastic resonance, chaotic regime, anomalous transport, disorder, and fractal geometry, among others, are being used more and more in basic subjects of exact sciences. It is thus necessary to introduce these basic elements to students in order to prepare them for bigger transformations that they may surely deal with when facing a unified statistical theory of nonequilibrium. This text is expected to give a general idea in order to pave the way for readers toward understanding nonequilibrium statistics and its applications to anomalous transport (e.g., localization).

San Carlos de Bariloche, Rio Negro, Argentina   Manuel O. Cáceres
2017

Acknowledgments

I am pleased to express my gratitude to students, colleagues, and friends, who in one way or another have collaborated in the preparation of this book. Many of their names appear in the references I have used throughout this text. I would like to thank the director of the Instituto Balseiro for logistical support to realize this effort, and in particular the staff of English Academic Corps and the Department of Graphics. My thanks also to the National Atomic Energy Commission, and the Centro Atómico Bariloche and Instituto Balseiro, National University of Cuyo, which for over more than 30 years provided the essential habitat for training and scientific research.

List of Notations and Symbols

List of Notations

rv	random variable
si	statistical independent
F-P	Fokker-Planck
FPTD	first passage time distribution
MFPT	mean first passage time
ME	master equation
pdf	probability distribution function
mv	mean value
sp	stochastic process
sde	stochastic differential equation
Prob.	probability
sirv	statistical independent random variable
GWN	Gaussian white noise
RW	random walk
DOS	density of states
CTRW	continuous time random walk

List of Symbols

$\mathcal{N}$	natural numbers
$\mathcal{Z}$	integer numbers
$\mathcal{O}(1)$	of order one
$\mathcal{R}_e$	real value
$\mathcal{C}$	complex value
$\mathcal{I}_m$	imaginary value
$\Theta(x)$	step function

$\langle\langle\cdots\rangle\rangle$	cumulant
$\langle\cdots\rangle$	mean value
$\delta_{m,n}$	Kronecker symbol
$\delta^{(n)}(x)$	n-th derivative of the delta function
$W(t)$	Wiener process
$dW(t)$	differential of a Wiener process
$\delta(x)$	Dirac delta function

Contents

Chapter 1
Elements of Probability Theory

1.1 Introduction to Random Variables

A crucial, if not the main role of mathematical statistics today is helping us to elucidate Nature's mysteries through detailed models or by providing estimations of most likely values of relevant quantities in real life situations. No mathematical discipline matches Probability Theory in its range of applications which span the whole spectrum all the way from Physics and Chemistry through Biology and Economics. We can no longer disregard the dynamic, random, nonlinear aspects of our Universe, and we are all aware of the huge differences when these nonequilibrium processes are taken into account versus results obtained in stationary (equilibrium) regimes. Traditional treatises on statistical mechanics have focused almost exclusively on equilibrium processes and stationary regimes [1]. It is only recently that texts in statistical mechanics introduce advanced probability concepts such as the theory of stochastic processes and deal with nonequilibrium and nonlinear phenomena. The theory of stochastic processes plays a key role in the dynamic description of modern statistical treatises: the dynamic and probability concepts are the basic elements for building the statistical theory of nonequilibrium [2–4].

Probability theory can be presented in an axiomatic or frequency context; and in the following sections we will briefly describe both presentations. On the other hand, in Chap. 3 we will see that stochastic processes describe the temporal evolution of statistics.

© Springer International Publishing AG 2017
M.O. Cáceres, *Non-equilibrium Statistical Physics with Application to Disordered Systems*, DOI 10.1007/978-3-319-51553-3_1

1.2   Axiomatic Scheme*

In this formulation a random *event* (one which may or may not happen) is considered mathematically as a member of a *sample space*. As an example take the variables for the position of the center of mass, the polarization or the angular momentum of all the H_2O molecules in a glass of water, as *elements* of the sample space S. Then, an event A could be to find 250 molecules with the same polarization in a particular volume ΔV "around" a given point $\mathbf{r}$ in space. A simpler situation—that is, when we determine an expected value[1] for an arbitrary event A—occurs when there is an exhaustive list of the possible elementary events within the sample space S. For example, an ordinary fair die will show any of its six equally likely faces— numbered from 1 to 6—when rolling it. In this case, the elementary sample space is $S = \{1, 2, 3, 4, 5, 6\}$, and a specific event A could be, for example, to obtain an even number when rolling the die. It is clear that the probability of this event taking place will be $P(A) = 1/2$, since there are only 3 possibilities. The axiomatic theory of probability is formulated in such a general way that it encompasses situations such as that previously mentioned and even more complex ones [5].

In the axiomatic scheme, the numbers $P(A), P(B), P(C), \cdots$ give the probabilities of *all* the possible $A, B, C \cdots$ events (outcomes) in a given experiment. Denoting the sample space S, all its elements meet the condition $A \in S, B \in S, C \in S \cdots$. Then, in an axiomatic presentation, the probability is the number (unique) assigned to any event that belongs to the sample space. In addition, the number $P(A)$ assigned to an arbitrary event A follows the axioms

$$P(A) \geq 0 \tag{1.1}$$

$$P(S) = 1.$$

On the other hand, we know from the theory of sets that

- if $w \in A \cup B \Longrightarrow w$ is an element of A or of B
- if $w \in A \cap B \Longrightarrow w$ is an element that is in A and in B
- $\tilde{A}$ indicates the complement of A, that is: $A \cup \tilde{A} = S, A \cap \tilde{A} = \emptyset, \tilde{\emptyset} = S$ ($\emptyset$ is the empty set)
- $A_i \cap A_j = \emptyset \Longrightarrow$ are disjoint sets.

So we conclude in the axiomatic theory of probability that

- $P(A \cup B) = $ Probability that the event A or the event B take place, or that both occur at the same time
- $P(A \cap B) = $ Probability that both events happen (Joint Probability)

---

[1]It is often called *expectation* of A.

- if $A_1, A_2, A_3, \cdots$ are mutually exclusive events[2]
  (disjoint) $\implies P(\bigcup_{n=1}^{\infty} A_n) = \sum_{n=1}^{\infty} P(A_n)$.
  So, in order to apply the axiomatic theory, we have to determine the sample space S and assign the probability P regarding that space, that is, the axioms (1.1).

Example. In the microcanonical quantum *ensemble,*[3] the sample space consists of the different energy levels, all of which are assumed to have equal probabilities. This is called the equal a priori probability postulate, see Appendix A.

Guided Exercise. Show that for every $A \in S$ we have $0 \leq P(A) \leq 1$. Since A and $\tilde{A}$ are mutually exclusive and $\tilde{A} \cup A = S$, from (1.1) it follows that $P(\tilde{A} \cup A) = P(\tilde{A}) + P(A) = P(S) = 1$. Hence $0 \leq P(A) = 1 - P(\tilde{A}) \leq 1$.

Exercise. If $A_1, A_2, A_3, \cdots, A_N$ are mutually exclusive events and also $A_1 \cup A_2 \cup A_3 \cup \cdots \cup A_N = S$, that is, its union represents the set of all the possible events, show that $\sum_{n=1}^{N} P(A_n) = 1$.

Guided Exercise. Show that an event that *cannot happen* has null probability. Given that the complement of S is precisely the empty set $\emptyset$ or nonevent (which cannot happen), we can conclude that $\emptyset \cup S = S$. Moreover, given that $\emptyset, S$ are mutually exclusive, we have $P(\emptyset \cup S) = P(\emptyset) + P(S) = P(S)$, from which we can conclude that $P(\emptyset) = 0$.

Guided Exercise. Show that if two events A, B are mutually exclusive, then their joint probability is null. Using that $P(A \cap B) \implies$ Joint Probability $\equiv P(A; B)$, and the fact that $A \cap B = \emptyset$ and $P(\emptyset) = 0$, we can deduce that $P(A; B) = 0$.

Optional Exercise. If $B_1, B_2, B_3, \cdots, B_N$ are *all* the possible mutually exclusive events, that is: $\bigcup_{n=1}^{N} B_n = S$, show that $\sum_{n=1}^{N} P(A; B_n) = P(A)$. First, we can perceive that $A \cap \left(\bigcup_{n=1}^{N} B_n \right) = A \cap S = A$. Hence, since $\{B_1, \cdots, B_N\}$ are *all* the mutually exclusive events, we can deduce that

$$P\left(A \cap \left(\bigcup_{n=1}^{N} B_n\right)\right) = P\left(\bigcup_{n=1}^{N} A \cap B_n\right) = \sum_{n=1}^{N} P(A \cap B_n),$$

where we have used that $\{A \cap B_1, A \cap B_2, \cdots, A \cap B_N\}$ is also a mutually exclusive set. On the other hand, from

$$P\left(A \cap \left(\bigcup_{n=1}^{N} B_n\right)\right) = P(A \cap S),$$

our statement is proved.

---

[2] Here we are assuming that we have a *numerable* collection of subsets A_n (events). This is a delicate aspect of the axiomatic theory that will not be discussed in the present book; see, for example, [5, 6].

[3] The term *ensemble* is used in many languages with a "collective" meaning.

Example. Take a regular six-face die (equiprobable $P(A_i) = 1/6$, $i = 1, \cdots 6$, where A_i represents each of the elementary events), then $S = \{1, 2, 3, 4, 5, 6\}$. Let us consider the following subsets (or events $A, B, \cdots$): $\emptyset, S, \{1, 3, 5\}, \{2, 4, 6\}$. Then a positive number $P(A), P(B), \cdots$ can be assigned to each of the events previously mentioned.

1.2.1   Conditional Probability

The conditional probability of the event A, knowing with certainty that another event B occurred is commonly denoted as $P(A \mid B)$ and it is given by the quotient of two probabilities:

$$P(A \mid B) = \frac{P(A; B)}{P(B)}, \tag{1.2}$$

where $P(A; B)$ is the joint probability. Obviously, this makes sense *only if* the event $B \neq \emptyset$. That is: *condition B* has to be a possible event.

Optional Exercise. Show that $P(A \mid B)$ is a probability and that it meets all the conditions previously stated.

Optional Exercise. If $A_1, A_2, A_3, \cdots, A_N$ are all the possible mutually exclusive events, show that $\sum_{n=1}^{N} P(A_n \mid B) = 1$.

Guided Exercise. Consider a box C_1 with 3 blue marbles and 1 white marble, and another box C_2 with 2 blue marbles and 1 white marble. One of the boxes is chosen randomly and then 1 marble is extracted from it. Let us calculate what is the probability of extracting a blue marble. Obviously, the result of the experiment can be any of the $4 + 3$ marbles, that is $S = \{a_1, a_2, a_3, a_4, a_5, b_1, b_2\}$. We can represent the problem in the following schematic way

	box C_1	box C_2
event A	a_1, a_2, a_3	a_4, a_5
event B	b_1	b_2

Hence we can define the following subsets

$$C_1 = \{\text{marbles from box 1}\} = \{a_1, a_2, a_3, b_1\}$$

$$C_2 = \{\text{marbles from box 2}\} = \{a_4, a_5, b_2\}$$

$$A = \{\text{blue marbles}\} = \{a_1, a_2, a_3, a_4, a_5\}$$

$$B = \{\text{white marbles}\} = \{b_1, b_2\}.$$

Besides, we assume equal probability of choosing either box, that is: $P(C_1) = P(C_2) = 1/2$. Imagine we choose box 1. As all the marbles have the same probability of being extracted, the conditional probability of getting a blue marble from box 1 is $P(A \mid C_1) = 3/4$. On the other hand, if we choose box 2, the conditional probability of getting a blue marble from box 2 is $P(A \mid C_2) = 2/3$. Hence, using (1.2) to write the joint probability $P(A; C_i)$ and summing over the events C_i, we can deduce that the probability of extracting a blue marble is:

$$P(A) = P(A \mid C_1)P(C_1) + P(A \mid C_2)P(C_2) = \frac{3}{4} \cdot \frac{1}{2} + \frac{2}{3} \cdot \frac{1}{2} = \frac{17}{24}.$$

1.2.2 Bayes' Theorem

Given that $P(A \cap B) \Longrightarrow P(A; B)$ and using (1.2), it follows that[4]

$$P(A; B) = P(A \mid B)P(B)$$

$$P(A; B) = P(B \mid A)P(A),$$

where probabilities $P(B \mid A)$ (that B happens given A) and $P(A \mid B)$ (that given B, A happens) are inverses to each other

$$P(A \mid B) = \frac{P(B \mid A)P(A)}{P(B)}.$$

1.2.3 Statistical Independence

In the probability theory context, events are defined as independent if the conditional probability over any event (or events) has no influence on the calculation of such probability. That is: $P(A \mid B_1, \cdots, B_n) = P(A)$, from which we can deduce that the joint probability $P(A \cap B_1 \cap \cdots \cap B_n)$ can be expressed as

$$P(A; B_1; \cdots ; B_n) = P(A)P(B_1) \cdots P(B_n)$$

Optional Exercises. Let A, B be subsets of a sample space $\mathcal{S}$. If $P(A)$ and $P(B)$ are the respective occurrence probabilities of events A and B, use Venn diagrams to find a geometrical interpretation of the following results:

---

[4]Note that the joint probability meets: $P(A; B) = P(B; A)$. When we introduce the concept of stochastic process we use this same property, but in that case we will be involving events at different instants of time (see Chap. 3).

- $P(A \cup B) = P(A) + P(B) - P(A \cap B)$. But if the events are exclusive: $A \cap B = \emptyset$, we can observe that
- $P(A \cup B) = P(A) + P(B)$.

  We define events to be independent only if $P(A \cap B) = P(A)P(B)$.

- Show that *exclusive* events and *independent* events are not synonymous.
- Calculate the conditional probability $P(A \mid B)$ supposing that A, B are independent events.

1.2.4   Random Variable

When the possible events $A_1, A_2, \cdots$ of a sample space $\mathcal{S}$ are real numbers, such events are clearly mutually exclusive ones. Hence, it is feasible to interpret those numbers[5] as the possible values of a random variable (**rv**). That is, an **rv** X is a function from the sample space $\mathcal{S}$ in a state space (in this case, that of real numbers); in other words, $X : \mathcal{S} \to \mathcal{R}_e$.

One of the advantages of introducing the concept of **rv** resides in the simplification of the use of different functions of **rv**, for example: e^{iX}, X^n, etc. When algebraic equations with constant coefficients are studied, and these coefficients are **rv**, different expectation values or probability distributions will be associated with their roots. The study of differential equations in the presence of **rv** and the modeling of transport in amorphous systems are some of the important fields of application of the elementary concept of **rv**. Making use of this concept, it is possible to calculate the mean value (**mv**) of the *formal* solution given as a function of the **rv** involved in the problem.

If the possible values of the **rv** X are a countable set of numbers $x_1, x_2, \cdots$, we say that the **rv** X is discrete. But if the possible values of the **rv** X are the numbers within the interval $[a, b]$, we say that the **rv** X is continuous. In general, we will call the sample space of a **rv** X its domain, support,[6] etc. and we will denote it as $\mathcal{D}_X$. Hence, in the context of axiomatic probability theory, each event of the **rv** X, that is, the numbers $x_1, x_2, x_3, \cdots$, will be associated with a probability $P(x_1), P(x_2), P(x_3), \cdots$; and, with the previous analysis, we can state that

$$\sum_n P(x_n) = 1. \tag{1.3}$$

---

[5]The symbols $\mathcal{R}_e$ and $\mathcal{C}$ will represent the real numbers and the complex numbers, respectively.

[6]In mathematics, the support of a function is the set of points where the function is not zero-valued.

In case that only *one event* turns to be certain, for example x_p, and none of the rest can happen, then

$$P(x_n) = \delta_{n,p},$$

where $\delta_{n,p}$ is the Kronecker delta and is defined by $\delta_{n,p} = 1$ if $n = p$, and $\delta_{n,p} = 0$ if $n \neq p$.

When the **rv** X is continuous and, for example, defined within an interval (open or closed) $\mathcal{D}_X$, new difficulties appear in the theory, as formula (1.3) cannot be applied.[7] Hence, it is necessary to introduce a probability $P(x_j)dx_j$, which tends to zero when the differential element dx_j tends to zero. In this case, the event that contains only one isolated number, for example x_0, has probability zero. When the **rv** X is continuous, the normalization condition (1.3), will adopt the generalized form

$$\int_{\mathcal{D}_X} P(x)\, dx = 1. \tag{1.4}$$

Hence, the probability that the **rv** X, with domain $\mathcal{D}_X = [a, b]$, adopts a value lower than the number χ (where $a \leq \chi \leq b$) is given by[8]

$$\text{Prob.}[X \leq \chi] = \int_a^\chi P(x)\, dx.$$

Exercise. Show that Prob.$[X \leq \chi]$ is an increasing function of χ, bounded by 1, and $\frac{d}{d\chi}$Prob.$[X \leq \chi]$ is the probability density $P(\chi)$. In some occasions (when necessary) we will denote this probability density as $P_X(\chi)$, but the abridged notation will be used in most cases.

If the function Prob.$[X \leq \chi]$ is at some point discontinuous, the probability density $P(\chi)$ may not exist.[9] If the **rv** X is continuous and has only the certain event x_p, it is necessary to introduce a new notation. For this kind of probability distribution we will denote

$$P_X(\chi) = \delta(\chi - x_p),$$

---

[7]The fact that the sample space is not countable—as happens with the real numbers within a closed interval—introduces difficulties that can only be solved introducing the concept of probability density. Mathematicians do this defining the event $A(x_j, dx_j)$ which means the set given by $[x_j \leq x \leq x_j + dx_j]$.

[8]From now on we will use the names: *probability distribution* or *probability density* indistinctly to characterize the function $P_X(\chi)$. On the other hand, we call function: Prob.$[X \leq \chi]$ the cumulative distribution, see note (11).

[9]However, the singularity introduced by any kind of discontinuity cannot be as abrupt as the Dirac delta (see the next section).

where $\delta(x - x_p)$ is the Dirac delta function; this density will be briefly defined in the next section. Its meaning makes sense only under the integral sign, that is:

$$f(x_p) = \int \delta(x - x_p)f(x)\,dx,$$

and complies with the normalization property (1.4).

Optional Exercise. Show that if X is a discrete **rv**, its probability distribution can be *formally* described as a continuous **rv** using the Dirac delta function , as:

$$P(x) = \sum_n P(x_n)\delta(x - x_n). \tag{1.5}$$

1.3   Frequency Scheme

Probability is a measure used to quantify our *expectation* [7]. Hence, if we want to detect an event A—after conducting a number n of ideally identical experiments—$P(A)$ will be its probability if $nP(A)$ is the value of the number of times the event A occurs.

1.3.1   Probability Density

The basis of every statistical theory is centered on the concept of probability. According to the *frequency* theory, if A is an event that occurs m times in an experiment carried out n times, in the limit $n \to \infty$ we will have

$$m/n \to P(A),$$

where $P(A)$ is the of occurrence (probability) of event A. It is not possible to determine in a rigorous mathematical way how large n should be.

Given a random variable (**rv**), it has associated certain statistical properties that characterize it.[10] One of the simplest characterizations of an **rv** X is its mean value (**mv**). If $x_1, x_2, x_3, \ldots$ are the "outcomes" in a given experiment, the **mv** of the **rv** will be:

---

[10]The axiomatic theory of probability gives a more abstract treatment than the *frequency* scheme. In the *axiomatic* scheme the real axis x is replaced by the event space $\mathcal{S} \equiv \mathcal{D}_X$, the differential interval $x + dx$ by a subset $A \subset \mathcal{S}$ belonging to a family of subsets closed under union and intersection (σ-algebra) [5, 6]. The distribution function assigns a nonnegative number $P(A)$ to each subset of the algebra such that $P(\mathcal{S}) = 1$, where P is called probability. Furthermore, any other set $f(A)$ will be also an **rv**.

$$\langle X \rangle = \lim_{n \to \infty} \frac{x_1 + x_2 + \cdots + x_n}{n}, \tag{1.6}$$

Here, the numbers $x_1, x_2, x_3, \ldots$ represent the possible events of the domain $\mathcal{D}_X$ of the **rv** X. That is, $\langle X \rangle$ is the limit of the arithmetic mean of the values *obtained* when this limit increases indefinitely (which we assume to exist from now on). Other characteristics of an **rv** can also be defined as in the case of the **mv** of more complicated functions, for example: $f(X) = X^2$, $g(X) = \exp(ikX)$, etc. Let us consider, in particular, under which conditions in the outcomes the inequality $X < \chi$ holds; that is, we would like to know the

$$\text{Prob.}[X < \chi].$$

This probability can be calculated considering the **mv** of the following function of the **rv** X

$$\epsilon = \Theta(\chi - X), \quad \text{where } \Theta(z) \begin{cases} = 1 \text{ if } z > 0 \\ = 0 \text{ if } z \leq 0, \end{cases} \tag{1.7}$$

hence, if $\epsilon_1, \epsilon_2, \epsilon_3, \ldots$ are the values obtained in the experiment, the **mv** of $\Theta(\chi - X)$ will be:

$$\langle \Theta(\chi - X) \rangle = \lim_{n \to \infty} [\epsilon_1 + \epsilon_2 + \cdots + \epsilon_n]/n = \text{Prob.}[X < \chi] \tag{1.8}$$

Bearing in mind the definition of the step function $\Theta(z)$, we see that $\text{Prob.}[X < \chi]$ is precisely the quotient m/n, where m is the number of valid inequalities $x_i < \chi$, ($1 \leq i \leq n$), and n is the number of samples or "tosses." That is,

$$\text{Prob.}[X < \chi] = \lim_{n \to \infty} \frac{m}{n}. \tag{1.9}$$

If we analyze $\langle \Theta(\chi - X) \rangle$ as a function of the parameter χ, we will be able to define the cumulative distribution function[11] $F_X(\chi)$ of the **rv** X in the form

$$F_X(\chi) = \langle \Theta(\chi - X) \rangle. \tag{1.10}$$

So its derivative is the so-called *probability density function*[12] (pdf)

$$\frac{dF_X(\chi)}{d\chi} = P_X(\chi) \Rightarrow F_X(\chi) = \int^{\chi} P_X(\chi') \, d\chi'. \tag{1.11}$$

---

[11]In probability theory, the cumulative distribution function of a real-valued random variable X is sometimes called distribution function, but we will prefer to call it the *cumulative* distribution function.

[12]In probability theory, a probability density function (pdf), or density of a continuous random variable, is a function that describes the relative likelihood for this random variable to take on a given value.

Using (1.10) in (1.11) and differentiating under the integral sign we get an alternative definition for the probability density

$$P_X(\chi) = \langle \delta(X - \chi) \rangle, \quad \text{as} \quad \delta(z) = \frac{d\Theta(z)}{dz}, \tag{1.12}$$

where $\delta(z)$ is the Dirac delta function. In what follows we shall be more interested in the probability density (or pdf) $P_X(\chi)$ rather than in the calculation of the cumulative function $F_X(\chi)$.

Note Regarding the Dirac Delta Function

Those readers who are not acquainted with this distribution can think about it as a singular mathematical object that appears in the Fourier transform of a piecewise and absolutely integrable continuous function in $[-\infty, \infty]$ (that is, smooth functions $f(x)$). The Fourier theorem states that

$$f(x) = \frac{1}{2\pi} \int_{-\infty}^{\infty} du \int_{-\infty}^{\infty} dz \, f(z) \exp(iu(z - x)).$$

If we exchange the order of integration, we can define the *object* $\delta(z - x)$ (which diverges at $z = x$). This object can *only* be considered under the sign of integration. Moreover, if we multiply it by an arbitrary function $f(z)$, its integral results in the same function $f(z)$ evaluated at x, that is,

$$f(x) = \int_{-\infty}^{\infty} dz \, f(z) \, \delta(z - x),$$

from which follows the definition of the Dirac delta function (δ–Dirac)

$$\delta(z - x) = \frac{1}{2\pi} \int_{-\infty}^{\infty} du \, \exp(iu(z - x)). \tag{1.13}$$

1.3.2 Properties of the Probability Density $P_X(\chi)$

1. It is defined as positive [this follows immediately from the definition (1.12)].
2. It is normalized [it is trivially shown given that $\langle 1 \rangle = 1$].
3. $P_X(\chi)$ *completely* characterizes the **rv** X. The pdf $P_X(\chi)$ allows us to calculate any *moment*:

$$\langle X^m \rangle \equiv \int \chi^m P_X(\chi) \, d\chi \quad \text{for any} \quad m = 1, 2, 3 \cdots.$$

In particular, the centered second moment $\langle (X - \langle X \rangle)^2 \rangle$ also called dispersion or variance is of particular interest as it gives an idea of the width of the distribution $P_X(\chi)$.

4. $P_X(\chi)$ allows us to calculate the **mv** of *any function* of the **rv** X. Given that $f(x) = \int f(z)\delta(x - z)\, dz$, we can immediately see that

$$\langle f(X) \rangle = \left\langle \int f(z)\delta(X - z)\, dz \right\rangle \tag{1.14}$$

$$= \int f(z)\, \langle \delta(X - z) \rangle\; dz = \int f(z)P_X(z)\, dz.$$

If the **rv** X is discrete, its probability can be incorporated within the same formalism of that of a probability density by the use of the Dirac delta (see Eq. (1.5)). From now on we shall always assume that the **rv** X is continuous, unless explicitly stating the opposite. On the other hand, we will on occasions omit redundant notations and we will simply write $P(x)$ to indicate the probability density $P_X(\chi)$ of the **rv** X.

1.4 Characteristic Function

We can alternatively characterize an **rv** X by means of Fourier's transform of the probability density, that is: from the **mv** of the function $\exp(ikX)$. Most of the time, knowing this **mv** makes statistical analysis easier. We define

$$G_X(k) \equiv \langle \exp(ikX) \rangle = \int_{\mathcal{D}_X} \exp(ikx)\, P_X(x)\, dx, \tag{1.15}$$

hence

$$P_X(x) = \frac{1}{2\pi} \int_{-\infty}^{+\infty} \exp(-ikx)\, G_X(k)\, dk, \tag{1.16}$$

where $\mathcal{D}_X$ is the sample space (domain) of the **rv** X, and $G_X(k)$ is the *characteristic function*. Knowing it makes possible the calculation of all the moments of the **rv** X. If these moments exist, we can expand $G_X(k)$ in Taylor series around $k = 0$, that is:

$$\langle \exp(ikX) \rangle = \sum_{m=0}^{\infty} \frac{(ik)^m}{m!} M_m \tag{1.17}$$

$$M_m \equiv \langle X^m \rangle = \frac{1}{i^m} \frac{d^m}{dk^m} G_X(k) \big|_{k=0}.$$

Note The function $G_X(k)$ is well defined for every $k \in \mathcal{R}_e$; it is continuous and satisfies $G_X(0) = 1$, $|G_X(k)| \leq 1$ (see advanced exercises 1.15.9). We will, on occasions, omit redundant notations and we will simply write $G(k)$ to indicate the characteristic function of the **rv** X.

Guided Exercise. If the sample space of the **rv** S are the integer numbers[13] $\mathcal{D}_S = \mathcal{Z}$, using that

$$P(s) = \int_{-\infty}^{\infty} e^{-iks} \sum_{s' \in \mathcal{Z}} e^{iks'} P(s') \frac{dk}{2\pi},$$

and noting that (compare with the definition of Kronecker delta)

$$\frac{1}{2\pi} \int_0^{2\pi} e^{-ik(s-s')} \, dk = \begin{cases} 1 \text{ if } s = s' \\ 0 \text{ if } s \neq s', \end{cases}$$

show that the inverse Fourier transform is given by

$$P(s) = \frac{1}{2\pi} \int_{-\pi}^{\pi} e^{-iks} G(k) \, dk = \frac{1}{2\pi} \int_0^{2\pi} e^{-iks} G(k) \, dk.$$

Excursus. An interesting result that enhances the importance of the characteristic function is established by the following theorem[14]: if an arbitrary function $G(k)$ meets the requirements (1–3), then a sample space $\mathcal{D}_\Omega$, a probability distribution $P_\Omega(\omega)$ over $\mathcal{D}_\Omega$ and a random variable Ω exist, and such $G(k) = \langle \exp(ik\Omega) \rangle$ is its characteristic function [8].

1. $G(0) = 1$.
2. $G(k)$ is continuous for every $k \in \mathcal{R}_e$.
3. If for any set of N complex numbers $\lambda_1, \lambda_2, \cdots, \lambda_N$, and real numbers $k_1, k_2, \cdots, k_N$ the following applies: $\sum_{ij} \lambda_i^* \lambda_j G(k_i - k_j) \geq 0$.

As a corollary to Bochner's Theorem, we can prove that if $G(k)$ satisfies (1–3) and is analytical in the neighborhood of $k = 0$, the expansion in Taylor series of $G(k)$ converges and its coefficients are related to the moments of the **rv** Ω.

Exercise. Let the Poisson probability be:

$$P_N(n) = \frac{\lambda^n}{n!} \exp(-\lambda); \quad n \in [0, 1, 2 \cdots], \ \lambda \in (0, \infty),$$

---

[13]In what follows, we will use the symbol $\mathcal{Z}$ to denote integer number and the symbol $\mathcal{N}$ to denote natural numbers whenever these do not cause confusion.

[14]S. Bochner; also see F. Riesz and B. St.-Nagy, *Functional Analysis*, New York, Frederick Ungar Publishing (1955).

where $P_N(n)$ gives the probability that the **rv** N takes a natural value n (or zero). Show that its characteristic function is given by

$$G_N(k) = \exp\left[-\lambda(1 - \exp(ik))\right].$$

Calculate all the moments of the **rv** N.

Exercise. Let the binomial probability be:

$$P_B(n) = \frac{M!}{n!\,(M-n)!}\, p^n\,(1-p)^{M-n}; \quad n \in [0, 1, \cdots, M], M \in \mathcal{N}, p \in [0, 1].$$

In this experiment there are two possible outcomes ($\mathcal{A}$ and $\mathcal{B}$) with intrinsic probabilities $P(\mathcal{A}) = p$ and $P(\mathcal{B}) = 1 - p$ respectively; hence, if we carry out M independent trials and we ask ourselves how we can obtain n events $\mathcal{A}$ [in any way], $P_B(n)$ is the expectation given. Find its corresponding characteristic function:

$$G_B(k) = (p\,\exp(ik) + (1-p)\,)^M.$$

Exercise. Consider the Gaussian probability distribution:

$$P_X(x) = \frac{1}{\sqrt{2\pi\sigma^2}}\exp\left(\frac{-(x-\mu)^2}{2\sigma^2}\right); \quad x \in [-\infty, +\infty], \{\mu, \sigma\} \in \mathcal{R}_e,$$

where $P_X(x)dx$ gives the probability of the **rv** X being in the differential interval $[x, x + dx]$. Note that the domain of the **rv** X is the continuum of values of the real line. Find its corresponding characteristic function:

$$G_X(k) = \exp\left(ik\mu - \frac{\sigma^2}{2}k^2\right).$$

Calculate all the moments of the **rv** X. In the case $\mu = 0$ and $\sigma = 1$ this pdf is called *the normal distribution*.

Note that, in general, if the probability (or pdf) is symmetric, then $G(k) \in \mathcal{R}_e$.

Exercise. Using the characteristic function of the binomial probability and invoking a suitable limit, obtain the Poisson probability where $\lambda = Mp$.

Exercise. Show that the characteristic function of the gamma probability function[15]

$$P_X(x) = \frac{c^b}{\Gamma(b)}x^{b-1}\,e^{-cx}; \quad x \in [0, \infty], \{b, c\} > 0,$$

---

[15]Here $\Gamma(b)$ is the gamma function of argument b, see for example [5].

is

$$G_X(k) = \frac{c^b}{(c - ik)^b}.$$

Note that, in this case, the **rv** X has a strictly nonnegative support; on the other hand, the case $b = 1$ corresponds to the exponential distribution. From $G_X(k)$, show that the nth moment of the **rv** X is

$$\langle X^n \rangle = \frac{b\,(b+1)\cdots(b+n-1)}{c^n}.$$

Show that the dispersion is given by $\sigma_X^2 \equiv \left\langle (X - \langle X \rangle)^2 \right\rangle = b/c^2$.

Exercise. Calculate the normalization of a Gaussian distribution with nonnegative support and most probable value x_p.

1.4.1 The Simplest Random Walk*

Consider the sum of several statistically independent[16] and equally distributed **rv** X_i

$$Y_r = X_1 + X_2 + \cdots + X_r,$$

where each of the variables X_i can take the value ± 1 with probability p, q, respectively ($p + q = 1$). The characteristic function of the **rv** Y_r will be given by

$$G_{Y_r}(k) = \left\langle e^{ikY_r} \right\rangle_{P(Y_r)} = \left\langle e^{ikX_1} \right\rangle_{P(X_1)} \cdots \left\langle e^{ikX_r} \right\rangle_{P(X_r)}$$
$$= \left\langle e^{ikX_i} \right\rangle_{P(X_i)}^r = \left(p\, e^{ik} + q\, e^{-ik} \right)^r.$$

Defining $Z \equiv e^{ik}$ and using Newton's binomial expansion we have

$$(A + B)^r = \sum_{n=0}^{r} \frac{r!}{n!(r-n)!} A^n B^{r-n}.$$

Then

$$G_{Y_r}(k) = \sum_{n=0}^{r} \frac{r!}{n!(r-n)!} p^n q^{r-n} Z^{2n-r}.$$

---

[16]In Sect. 1.2.3 the concept of statistical independence is presented.

Defining a new dummy variable $y = 2n - r$, we can rewrite the previous sum as:

$$G_{Y_r}(k) = \sum_{y=-r}^{y=r} \frac{r!}{\left(\frac{r+y}{2}\right)! \left(\frac{r-y}{2}\right)!} p^{(r+y)/2} q^{(r-y)/2} Z^y. \qquad (1.18)$$

From the equality $Y_r = \sum_{i=1}^{r} X_i$ we can deduce that the allowed values of the **rv** Y_r may be: $y = 0, \pm 1, \pm 2, \cdots \pm r$. Hence, we can immediately observe from (1.18) that

$$G_{Y_r}(k) = \sum_{y=-r}^{y=r} \mathcal{P}_r(y) e^{iky} \equiv \left\langle e^{iky} \right\rangle_{\mathcal{P}_r(y)},$$

where

$$\mathcal{P}_r(y) = \frac{r!}{\left(\frac{r+y}{2}\right)! \left(\frac{r-y}{2}\right)!} p^{(r+y)/2} q^{(r-y)/2}. \qquad (1.19)$$

Therefore, we can conclude that $\mathcal{P}_r(y)$ is the probability of the **rv** Y_r taking the value y when the sum of the number of random variables is r. In other words: $\mathcal{P}_r(y)$ is the probability of a "drunkard" being at $y \in [-r, r]$ after r random steps forwards and backwards along a straight line; see Fig. 1.1.

Fig. 1.1 The random value y corresponds to the distance from the starting position. The number of variables r is the number of steps taken by the "walker." The figure shows the representation of four possible random walks

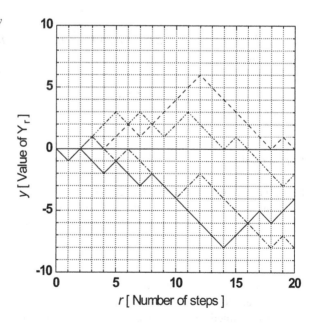

Optional Exercise. From (1.19) prove that with $p = q = 1/2$ and lattice parameter a, in the limit of large "times" and small lattice parameter $a \ll 1$

$$\mathcal{P}_r(y) \rightarrow 1/\sqrt{4\pi dt} \exp\left(-x^2/4dt\right),$$

where $x = ya$, $t = r\tau$. Hence, when $a \rightarrow 0$, $\tau \rightarrow 0$ we have that $\langle x(t)^2 \rangle = 2dt$, where $d = a^2/2\tau$ is the diffusion coefficient. (Hint: use the Stirling approximation for the factorial function: $\Gamma(X + 1) = X! \simeq X^{X+1/2} e^{-X} \sqrt{2\pi}$).

Exercise (**Characteristic Function of a Geometric Probability**). Consider now the case of a random walk where the steps $|s - s'|$ are of arbitrary length but distributed with a decreasing probability for larger jumps, it is useful to bear in mind the following probability defined over the integers:

$$P(s - s') = \mathcal{N}\,(1 - \gamma)\left(\gamma^{|s-s'|} - \delta_{s,s'}\right); \quad \gamma \in (0, 1), |s - s'| = 0, 1, 2, 3, \cdots.$$

Note that for this probability null (zero length) steps are not allowed. Show that the normalization constant is $\mathcal{N} = 1/2\gamma$ and that the characteristic function is

$$G(k) = \frac{1 - \gamma}{\gamma}\left(\frac{1 - \gamma \cos k}{1 - 2\gamma \cos k + \gamma^2} - 1\right). \tag{1.20}$$

In Chap. 6 we will present a thorough analysis of the different kinds of random walks that may appear, depending on the different possible jumps or steps $|s - s'|$.

1.4.2 Examples in Which $G(k)$ Cannot Be Expanded in Taylor Series*

Here are some examples in which the characteristic function cannot be expanded in Taylor series around $k = 0$.

Lorentz pdf

Let X be a (continuous) **rv** characterized by the Lorentzian distribution, whose moments are not defined. From the probability density

$$P(x) = \frac{\gamma/\pi}{[(x - a)^2 + \gamma^2]}, \quad x \in [-\infty, +\infty], a \in \mathcal{R}_e, \gamma > 0,$$

we obtain the characteristic function $G(k) = \langle \exp(ikX) \rangle = \exp(ika - |k|\gamma)$. We note here that $G(k)$ is not differentiable at $k = 0$; that is, its moments are not rigorously defined. However, we can say that the distribution $P(x)$ will be typically centered around the value a and will have a "width" characterized by γ.

Weierstrass Probability

Let S be an **rv** (discrete) characterized by the Weierstrass probability, which we will generally refer to as the Lévy-like probability

$$P(s) = \frac{a-1}{2a} \sum_{n=0}^{\infty} 1/a^n \left(\delta_{s,b^n} + \delta_{s,-b^n} \right), \quad s \in \mathcal{D}_S, \ a > 1, \ b > 1.$$

That is: $1/a$ is the probability of s taking the value $\mp b$; $(1/a)^2$ is the probability of s taking the value $\mp b^2$, and so on. Hence, when $1/a$ is small, the probabilities of taking larger values are lower. In the context of a random walk, this probability leads us to a very interesting structure of "jumps" in the form of clusters in real space (that is, there is not a unique characteristic length for the size of the "jumps" but rather a peculiar self-similarity at many different scales forming "gaps" or "clusters" of points). In nonequilibrium statistical mechanics, this jump distribution leads us to a series of critical phenomena that can be studied in the context of Markov chains [9] (or random walks, see Chap. 6). We can immediately see from $P(s)$ that its second moment is divergent if $b^2 > a$, as

$$\langle s^2 \rangle = \sum_{\mathcal{D}_S} s^2 P(s) = \frac{a-1}{a} \sum_{n=0}^{\infty} (b^2/a)^n.$$

Note that the characteristic function of the Weierstrass probability is well defined for every $k \in \mathcal{R}_e$, and it is continuous over the entire real axis[17]

$$G(k) = \sum_{\mathcal{D}_S} \exp(iks)P(s) = \frac{a-1}{a} \sum_{n=0}^{\infty} 1/a^n \cos(b^n k). \tag{1.21}$$

Although it is a continuous function, we note from its second derivative expression:

$$\frac{d^2 G(k)}{dk^2} = -\frac{a-1}{a} \sum_{j=0}^{\infty} (-1)^j/(2j)! \, k^{2j} \sum_{n=0}^{\infty} (b^{2j+2}/a)^n,$$

that $d^2 G(k)/dk^2 \,|_{k=0} \to \infty$, if $b^2/a > 1$ (for example, consider the term $j = 0$ of the sum). From here, we can deduce that the second moment of the **rv** s is not defined but divergent. The mathematical structure of the Weierstrass function $G(k)$ has extremely complicated properties as a function of k; for example, $G(k)$ has no derivative at any point if $0 < \ln a / \ln b < 1$.[18]

---

[17]This function was presented by K. Weierstrass in 1872 and was later on published in Mathematische Werke. II, 71–74, Berlin, Meyer & Muller (1895).

[18]It is available in: B.D. Hughes, E.W. Montroll and M.F. Shlesinger, J. Stat. Phys. **30**, 273, (1983).

Excursus (**Clusters of Visited Points**). The characteristic function (1.21) can also be written in the form

$$G(k) = \frac{a-1}{a} \cos(k) + \frac{a-1}{a^2} \sum_{m=0}^{\infty} \frac{1}{a^m} \cos(b^{m+1}k)$$

$$= \frac{a-1}{a} \cos(k) + \frac{1}{a} G(bk).$$

On the other hand, if we decompose $G(k)$ into two terms, $G(k) = G_S(k) + G_R(k)$, we can prove by direct substitution in $G(k) = \frac{a-1}{a}\cos(k) + \frac{1}{a}G(bk)$ that the (regular) analytic part is

$$G_R(k) = 1 + \frac{a-1}{a} \sum_{m=1}^{\infty} \frac{(-k^2)^m}{(2m)![1 - b^{2m}/a]}.$$

Hence, the singular contribution $G_S(k)$ is a *nonanalytic* function in k that satisfies the scale relation $G_S(k) = \frac{1}{a}G_S(bk)$. From this relation we can prove that the singular part is of the form $G_S(k) = |\ k\ |^{\mu}\ Q_{\mu}(k)$, where $\mu = \ln a/\ln b$ and the function $Q_{\mu}(k)$ is periodic in $\ln b$; that is, $Q_{\mu}(\ln k) = Q_{\mu}(\ln k + \ln b)$. Hence, one can prove that when $k \to 0$, the behavior of the Fourier variable k is dominated by $\propto |\ k\ |^{\mu}$, with $0 < \mu < 2$ if $b^2 > a$, where the exponent μ characterizes a certain fractal dimension of the set (clusters) of "visited" points in a lattice (that is, the instantaneous positions in a random walk). Note that there will be approximately a jumps of length b^m for each jump of length b^{m+1}. This is the basic idea whereby μ can be associated with the fractal dimension in the sense of Mandelbrot [10]. In Sect. 6.2.3 of Chap. 6, a numerical simulation of this phenomenon is shown. A simple analysis of the concept of fractal dimension is presented in Appendix H.

1.4.3 Characteristic Function in a Toroidal Lattice*

The simplest case of a toroidal lattice [9] with integer sites $\{s\}$ is a $1D$ circular lattice (ring where $s \in \mathcal{D}_S \equiv [1, \cdots, N]$). From the characteristic function definition, when the sample space is $\mathcal{D}_S$, we conclude that

$$G(k) = \sum_{s=1}^{N} \exp(iks)\, P(s). \tag{1.22}$$

On the other hand, using the periodicity condition of probability (that is, $P(s) = P(s + N)$), we can observe that the Fourier conjugate variable is discretized according to $kN = 2\pi\nu$, where $\nu \in \mathcal{D}_S$. Then, we conclude that the inverse Fourier transform (discrete) takes the form:

$$P(s) = \frac{1}{N} \sum_{v=1}^{v=N} G(k = \frac{2\pi v}{N}) \exp(\frac{-i2\pi vs}{N}). \tag{1.23}$$

This formula can be interpreted as the application of a linear transformation $\mathcal{Q}$ on a vector space of dimension N, where the matrix elements of the operator $\mathcal{Q}$ are given by

$$\mathcal{Q}_{vs} = \frac{1}{N} \exp(\frac{-i2\pi vs}{N}) \quad \text{with} \quad \{v,s\} \in \mathcal{D}_S.$$

That is, using the abridged notation $P_s \equiv P(s)$, $G_v \equiv G(k = \frac{2\pi v}{N})$, the expression (1.23) is written in the form

$$P_s = \sum_v \mathcal{Q}_{sv} G_v,$$

hence, its inverse relation will be $G_s = \sum_v \mathcal{Q}_{sv}^{-1} P_v$, where $\mathcal{Q}^{-1}$ is the inverse matrix of $\mathcal{Q}$.

Another alternative way of proving the formula (1.23) is by using the method of images [11] in order to find the probability in a *finite* lattice with periodic boundary conditions in $\mathcal{D}_S$ [9].

Guided Exercise. Study the moments $\langle s^n \rangle$ when the probability $P(s)$ is uniform in the ring (assuming there are N sites). From the definition of the characteristic function $G(k)$, the inverse transform gives

$$P(s) = \frac{1}{2\pi} \int_{-\infty}^{\infty} e^{-ik} G(k) dk.$$

Imposing the condition $P(s) = P(s+N) \Rightarrow k = 2\pi v/N$, $v = 1, 2, \cdots, N$. So the inverse is the finite dimensional Fourier transform (1.23). Now, if $P(s) = 1/N$ we get

$$G(k) = \frac{\mathcal{A}^{N+1} - \mathcal{A}}{N(\mathcal{A}-1)}, \ \mathcal{A} = e^{ik}, \ k = \frac{2\pi v}{N}.$$

Check that

$$\lim_{k \to 0} G(k) \to 1,$$

$$\frac{dG}{d(ik)}\Big|_{k=0} \to \frac{1}{2}(1+N) = \langle s \rangle$$

$$\frac{d^2G}{d(ik)^2}\Big|_{k=0} \to \frac{1}{6}(1+3N+2N^2) = \langle s^2 \rangle.$$

Guided Exercise (**Poisson Sum**). If $\mathcal{P}(s)$ represents the probability in the infinite lattice, let us show that the probability $P(s)$ in the toroidal lattice $1D$ is given by the sum:

$$P(s) = \sum_{l=-\infty}^{l=+\infty} \mathcal{P}(s + lN). \tag{1.24}$$

Note that the periodicity condition $P(s) = P(s + N)$ is fulfilled. On the other hand, from this expression and the use of the Poisson sum [5]:

$$\sum_{q=-\infty}^{q=+\infty} \exp(i2\pi mq) = \sum_{q=-\infty}^{q=+\infty} \delta(m + q), \tag{1.25}$$

we can also obtain (1.23). Given that $\mathcal{P}(s)$ is defined on the infinite lattice, we can deduce that its characteristic function is $G(k) = \sum_{s=-\infty}^{\infty} \exp(iks)\mathcal{P}(s)$; hence, the inverse transform is given by

$$\mathcal{P}(s + lN) = \frac{1}{2\pi} \int_0^{2\pi} dk\, G(k) \exp\left[-ik\,(s + lN)\right].$$

Introducing this expression in (1.24), it follows that

$$P(s) = \frac{1}{2\pi} \int_0^{2\pi} dk\, G(k) \sum_{l=-\infty}^{l=+\infty} \exp\left[-ik\,(s + lN)\right],$$

now, using the Poisson sum (1.25), we obtain

$$P(s) = \frac{1}{N} \int_0^{2\pi} dk\, G(k) \exp(-iks) \sum_{l=-\infty}^{l=+\infty} \delta(k + \frac{2\pi l}{N}),$$

but given that the integration is only over the range $[0, 2\pi]$, we can immediately see that the only contributions come from the terms with $l = \{1, \cdots, N\}$, we thus get the expected expression (1.23). The extension of this formula to nD toroidal lattice is analogous [9].

Optional Exercise. From (1.23) find the formula for $P(s)$ when the lattice is infinite (that is, $N \to \infty$). Show that, in this case, the sum (1.23) becomes an integration over dk in the first Brillouin zone: $k \in [-\pi, \pi]$.

1.4.4 Function of Characteristic Function

Of particular interest is the calculation of the probability distribution of the sum of a countable set of random variables statistically independent[19] of each other (**sirv**).

Let r be a random positive integer characterized by the probability P_r, and $\{X_j\}$ a set of r **sirv** with the same probability distribution function (pdf) $P(X_j)$. Then, the sum

$$Y = X_1 + X_2 + \cdots + X_r \quad \text{with} \quad r = 0, 1, 2, 3 \cdots ,$$

is a new random variable, since each of the variables X_j as well as the number r are **rv**. The characteristic function of the **rv** Y is given by

$$G_Y(k) \equiv \left\langle e^{ikY} \right\rangle$$

$$= \sum_{r=0}^{\infty} P_r \int \cdots \int dX_1 \cdots dX_r \prod_{j=1}^{r} P(X_j) \exp\left[ik(X_1 + \cdots + X_r)\right],$$

where we have used the fact that X_j are **sirv** with equal distribution,[20] then $G_{X_j}(k)$ is the same for each of the **rv** X_j. Therefore, the characteristic function of the **rv** Y is given by

$$G_Y(k) = \sum_{r=0}^{\infty} P_r \left[\int dX\, P(X) \exp(ikX) \right]^r .$$

If we use the concept of the generating function (for the random variable r over $\mathcal{N}$), i.e.:

$$f_r[Z] = \sum_{r=0}^{\infty} P_r Z^r ,$$

we can write $G_Y(k)$ in the form

$$G_Y(k) = f_r[G_X(k)] , \qquad (1.26)$$

which is the expression we are looking for. Note that in general the generating function and the characteristic function are related by $f_r\left[Z = e^{ik}\right] = \sum_{r=0}^{\infty} P_r e^{ikr} = \left\langle e^{ikr} \right\rangle_{P_r} \equiv G_r(k)$.

---

[19]See Sect. 1.2.3.

[20]Here we have used a short notation, because each of the **rv** X_j has a pdf $P_{X_j}(x)$, but we have simplified the notation as $P(X_j)$.

Exercise. What kind of moments can easily be calculated using the generating function $f_r[Z]$ of an **rv** over the positive integers?

Exercise (**Random Walk**). Suppose that $Y = X_1 + X_2 + \cdots + X_r$, but that r is not random; that is, its probability is 1 for a specific value of r, let say $r = t$. In this case $P_r = \delta_{r,t}$, then the generating function of P_r is $f_r[Z] = \sum_{r=0}^{\infty} P_r Z^r = Z^t$. Using (1.26) prove (1.19).

Exercise (**Sum of Gaussian Random Variables**). Suppose that X_j is a Gaussian **rv** of mean zero and variance 1. Calculate the characteristic function of the sum of variables X_j assuming that the number of terms is characterized by a Poisson probability $P_r = \exp(-\lambda)\lambda^r/p!$. In this case the generating function of P_r will be

$$f_r(Z) = \exp\left(-\lambda\left[1 - Z\right]\right),$$

and, given that $G_X(k) = \langle e^{ikX} \rangle = e^{-k^2/2}$, we finally obtain that

$$G_Y(k) = f_r[Z]|_{Z=G_X(k)} = \exp\left(-\lambda\left[1 - \exp\left(-k^2/2\right)\right]\right).$$

Show that the **mv** of the **rv** $Y = \sum_{j=0}^{r} X_j$ is zero while for the second moment we have

$$\langle Y^2 \rangle = \lambda.$$

Optional Exercise (**Sum of Geometrically Distributed Random Variables**). Suppose that each of the X_j is a geometrically distributed **rv** (in (1.20) we gave its corresponding characteristic function). Calculate the characteristic function of the sum $Y = \sum_{j=0}^{r} X_j$ assuming that the number of terms r is random with a Poisson probability. Calculate the dispersion $\langle Y^2 \rangle$. How does this dispersion differ from that of the conventional model of a random walk?

1.5   Cumulant Expansion

Alternatively we can characterize an **rv** from its cumulants. These quantities are related to the moments through:

$$\ln G(k) = \sum_{n=1}^{\infty} \frac{(ik)^n}{n!} K_n. \tag{1.27}$$

It is possible to connect K_n with M_n in a simple way. Thus, for example, by calculating a "simple" determinant, we have that

$$K_n = (-1)^{n-1} \det \mathbf{A},$$

where $\mathbf{A}$ is given by the $n \times n$ matrix

$$
\mathbf{A} = \begin{pmatrix}
M_1 & 1 & 0 & 0 & 0 & \cdots \\
M_2 & M_1 & 1 & 0 & 0 & \cdots \\
M_3 & M_2 & \binom{2}{1}M_1 & 1 & 0 & \cdots \\
M_4 & M_3 & \binom{3}{1}M_2 & \binom{3}{2}M_1 & 1 & \cdots \\
M_5 & M_4 & \binom{4}{1}M_3 & \binom{4}{2}M_2 & \binom{4}{3}M_1 & \cdots \\
\cdot & \cdot & \cdot & \cdot & \cdot & \cdots \\
\cdot & \cdot & \cdot & \cdot & \cdot & \cdots \\
\cdot & \cdot & \cdot & \cdot & \cdot & \cdots
\end{pmatrix}.
$$

Here $\binom{p}{q} = p!/((p-q)!\,q!)$ are binomial coefficients. It is also possible to obtain an inverse relationship between M_n and K_n, which is left as an exercise for the reader.

Excursus. From a more general definition, introducing many **rv**s, it is possible to relate diagrammatically M_n with K_n; see Sect. 1.10. Expansion in cumulants is useful when we are interested in calculating average values, and these can be obtained by perturbations around the correlation length of the **rv**s involved in the problem [12], see advanced exercise 1.15.5.

Exercise. Show that the cumulants of an **rv** X are not the centered moments. For example, the first four cumulants are given by

$$K_1 = M_1$$

$$K_2 = M_2 - M_1^2 = \left\langle (X - M_1)^2 \right\rangle$$

$$K_3 = M_3 - 3M_1 M_2 + 2M_1^3 = \left\langle (X - M_1)^3 \right\rangle$$

$$K_4 = M_4 - 3M_2^2 - 4M_1 M_3 + 12M_1^2 M_2 - 6M_1^4 \neq \left\langle (X - M_1)^4 \right\rangle$$

Optional Exercise. Outline schematically the calculation of M_n as a function of the cumulants.

Exercise. Show that the cumulants of a Poisson probability are all equal.

Exercise. Show that all **rv** X that have a null variance will be deterministic (i.e., singular or delta distributed).

Excursus. There is a very interesting result that says: if the cumulants K_n are null for all $n \geq N \in \mathcal{N}$, then they are null for every $n > 2$ and the **rv** is Gaussian [J. Marcinkiewicz, Mathematisches Zeitschrift **44**, 612, (1939)].

1.6 Central Limit Theorem

We now seek to characterize the probability distribution of an infinite sum of random variables equally distributed. Consider, for example, the following sum

$$X = X_1 + X_2 + \cdots + X_N, \tag{1.28}$$

where it is assumed that each of the **rv** X_i has a well-defined first and second moment

$$\langle X_i \rangle \equiv M_1^{(i)} < \infty, \quad \langle X_i^2 \rangle \equiv M_2^{(i)} < \infty.$$

Here we have used the same notation as in the preceding paragraphs, the superscript (i) indicates the i-th **rv**. If the **rv** X_i are statistically independent of each other,[21] from the sum (1.28) we deduce that

$$\sigma^2 \equiv \left\langle (X - \langle X \rangle)^2 \right\rangle = \sigma_1^2 + \sigma_2^2 + \cdots + \sigma_N^2, \tag{1.29}$$

where $\sigma_i^2 \equiv K_2^{(i)}$ is the second cumulant of the **rv** X_i.

The central limit theorem states that when $N \to \infty$

$$P_X(x) \to \frac{1}{\sigma\sqrt{2\pi}} \exp\left[-(x - M)^2 / 2\sigma^2 \right], \tag{1.30}$$

here $M \equiv \langle X \rangle = M_1^{(1)} + M_1^{(2)} + \cdots + M_1^{(N)}$. To prove this theorem we take (without loss of generality) $M_1^{(i)} = 0$ for all **rv** X_i. Defining a new **rv** Z in the form

$$Z = \frac{X_1 + X_2 + \cdots + X_N}{\sigma}, \tag{1.31}$$

using statistical independence and the fact that each and every **rv** has the same distribution, we get that

$$\sigma^2 = N\sigma_i^2. \tag{1.32}$$

Then, the characteristic function of the **rv** Z is given by the expression

$$G_Z(k) = \langle \exp(ikZ) \rangle = \langle \exp(ik(X_1 + X_2 + \cdots + X_N)/\sigma) \rangle \tag{1.33}$$

$$= \left(\left\langle \exp\left(ik\frac{X_i}{\sqrt{N}\sigma_i} \right) \right\rangle \right)^N.$$

---

[21] See Sect. 1.2.3.

In the limit of $N \to \infty$ we can expand the function $\exp\left(ik\dfrac{X_i}{\sqrt{N}\sigma_i}\right)$, that is under the sign of **mv**, and then we keep only the dominant term:

$$G_Z(k) = \left(1 + \left(i\frac{k}{\sqrt{N}\sigma_i}\right)^2 M_2^{(i)}/2 + \mathcal{O}(N^{-3/2})\right)^N \simeq \left(1 - \frac{k^2}{2N}\right)^N. \qquad (1.34)$$

In this last expression we have used the fact that again $\sigma_i^2 \equiv M_2^{(i)}$; then, using the known asymptotic formula: $\lim_{N\to\infty}\left(1 - \frac{\Omega}{N}\right)^N = e^{-\Omega}$, we get that

$$\lim_{N\to\infty} G_Z(k) \to \exp\left(-\frac{k^2}{2}\right), \qquad (1.35)$$

whereby the theorem is proved.

Excursus (**Gaussian Law of Attraction**). There is a more refined version of the previous theorem, which proves that the restriction $\langle X_i^2 \rangle \equiv M_2^{(i)} < \infty$ is only a sufficient condition [6]. Thus, in the context of the central limit theorem, we can prove the existence of a wider family of distributions which is within the *basin of Gaussian attraction*. For example, if asymptotically $P(X_i) \sim |X_i|^{-3}$ for $X_i \to \infty$ we obtain $\langle X_i^2 \rangle = \infty$; nevertheless, $P(X_i)$ meets the generalized version of the central limit theorem; as well as other distributions decaying faster.

Exercise. From the characteristic function (1.35) obtain the probability distribution (1.30).

Guided Exercise (**Gaussian Approximation**). Use the central limit theorem to find the probability distribution of the sum of several **rv** X_i equally distributed, with uniform distribution in the interval $[0, T]$. For the case of only two **rvs**, graph your result and compare the approximation $P_{\text{Gauss}}(X)$ from (1.30) with the exact result that can be obtained from the transformation $X = X_1 + X_2$. Since $P(X_i) = 1/T$ it follows that

$$\langle X_i \rangle = \frac{T}{2}; \quad \langle X_i^2 \rangle = \frac{T^2}{3},$$

therefore $M = \langle X \rangle = T$ and $\sigma^2 = \sigma_1^2 + \sigma_2^2 = T^2/6$. Introducing these quantities in (1.30) it must hold that

$$P_{\text{Gauss}}(X) = \sqrt{\frac{3}{\pi T^2}} \exp\left(\frac{-3(X - T)^2}{T^2}\right). \qquad (1.36)$$

Fig. 1.2 Comparison of the pdf for the sum of two **sirv** uniformly distributed in [0, 1] against the Gaussian approximation by using the Central Limit Theorem equation (1.36)

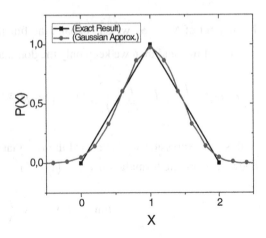

Furthermore, the exact expression for $P(X)$ can be obtained from its characteristic function:

$$G_X(k) = \langle \exp ik (X_1 + X_2) \rangle = \frac{-\left(e^{ikT} - 1\right)^2}{k^2 T^2}.$$

Fourier inversion of $G_X(k)$ yields the distribution $P(X)$; however, this calculation is not trivial, that is why we give the result here:

$$P(X) = \frac{1}{2\pi} \int_{-\infty}^{\infty} \frac{\left(e^{ikT} - 1\right)^2}{-k^2 T^2} \exp(-ikX) \, dk \qquad (1.37)$$

$$= \frac{X}{T^2}\Theta(X)\Theta(T - X) + \frac{2T - X}{T^2}\Theta(X - T)\Theta(2T - X),$$

where $\Theta(X)$ is the step function. Alternatively, and as it will be seen shortly in Sect. 1.13, $P(X)$ can be obtained very simply using the law of transformation for the sum of random variables. In Fig. 1.2 the comparison of the exact result (1.37) against the Gaussian approximation for two **rvs** is plotted (for $T = 1$).

1.7 Transformation of Random Variables

Let us see how the probability density $P_X(x)$ of the **rv** X changes under the transformation law: $Y = g(X)$. The function $g(X)$ is assumed known and its inverse, in general, may not be unique.

If the law is a monotonic transformation ($x = h(y)$, $h = g^{-1}$), the transformation law for the probability density is simple and is given by the Jacobian of the transformation:

$$P_Y(y) = P_X(h(y)) \; | \, dx/dy \, | \, . \tag{1.38}$$

In the general case we proceed as follows: first notice from (1.12) that we can write

$$P_Y(y) = \langle \delta \, (Y - y) \rangle = \langle \delta \, (g(X) - y) \rangle . \tag{1.39}$$

Then, using Eq. (1.14) we obtain

$$P_Y(y) = \int \delta \, (g(X) - y) \, P_X(x) \, dx, \tag{1.40}$$

or, simplifying redundant notation,

$$P(y) = \int \delta \, (g(x) - y) \, P(x) \, dx. \tag{1.41}$$

This equation is more general than (1.38) because, it does not require $g(x)$ to be one to one in the region where $P(x)$ is nonzero.

Finally, using the properties of the Dirac delta function we can integrate (1.41). Then if $x = h_j(y)$, where h_j is one of the inverse transformations,[22] from expression (1.41) we obtain

$$P(y) = \sum_{j=1}^{r} P(h_j(y)) \; | \, dx/dy \, |_{x=h_j(y)}, \tag{1.42}$$

which is the generalization of equation (1.38) for the case when there are r inverse functions $h_j(y)$.

Example. Let $P_V(v)$ be Boltzmann's probability distribution for the speed of a free particle. We calculate the probability distribution of energy $P(E)$. Since $E = \frac{1}{2}mv^2 \equiv g(v)$, there are two roots

$$v_{1,2}(E) = g_{1,2}^{-1}(E) = \pm\sqrt{2E/m},$$

---

[22]For those readers who are unfamiliar with the Dirac-δ function we summarize here one of its fundamental properties:

$$\delta \, (g(x) - y) = \sum_{n} \delta \, (x - x_n) \, \frac{1}{| \, g'(x) \, |_{x=x_n}},$$

where $x_n = g_n^{-1}(y)$ is the n-th root (simple) of equation $g(x) = y$.

and $|\,g'(v)\,|_{v=v_j} = |\,mv_j\,| = +\sqrt{2mE}$. On the other hand, since

$$P_V(v) = \sqrt{m/2\pi k_B T}\exp\left(-\frac{mv^2}{2k_B T}\right),$$

and using $P(E) = \sum_{j=1,2} P_V(g_j^{-1}(E))\,|\,g'(v)\,|_{v_j}^{-1}$ we get

$$P(E) = 2\sqrt{m/2\pi k_B T}\exp\left(-\frac{E}{k_B T}\right)\frac{1}{\sqrt{2mE}} = \frac{1}{\sqrt{\pi E k_B T}}\exp\left(-\frac{E}{k_B T}\right),$$

where $\mathcal{D}_E = [0,\infty]$.

Example. Consider t_e defined by the transformation $t_e = (A\Omega)^{-2/3}$ where Ω is a Gaussian **rv** with zero mean and second moment $1/3$. We want to calculate the probability distribution of the random time t_e, where $\mathcal{D}_{t_e} = [0,\infty]$. In this case the inverse transformation is not unique:

$$h_j(t_e) = \frac{\pm 1}{A\sqrt{t_e^3}}.$$

Then, from (1.42) we have

$$P(t_e) = \sum_{j=1}^{2} P_\Omega(h_j(t_e))\left|\frac{-3}{2A}\,t_e^{-5/2}\right|,$$

and since the distribution of the **rv** Ω is

$$P_\Omega(\omega) = \frac{1}{\sqrt{2\pi/3}}\exp\left(\frac{-\omega^2}{2/3}\right),$$

it follows that

$$P(t_e) = \frac{3^{3/2}}{t_e^{5/2}A\sqrt{2\pi}}\exp\left(\frac{-3}{2}\left(\frac{1}{At_e^{3/2}}\right)^2\right)\quad\text{with}\quad A > 0.$$

Excursus. Note that from the above example, asymptotically, $P(t_e \gg A^{-2/3}) \sim t_e^{-5/2}$. This type of probability distribution, with a power law, is often called a long-tail pdf ($\mathcal{O}(t_e^{-5/2})$), and it appears in the study of instabilities induced by noise (in Chap. 3 the concept of noise or stochastic process is defined); i.e., in the analysis of the escape time from a certain "domain" [13]. There exists a relationship between the **rv** t_e and the random instant of the "first passage time" to a particular location (the escape time problem); see Chap. 9.

1.8 Fluctuations Expansion

There are circumstances in which it is desirable to introduce some approximation to evaluate the **mv** of a function of random variable. For example, let X be an **rv** characterized by a probability distribution $P_X(x)$. Suppose we want to know $\langle f(X) \rangle$, where $f(X)$ is a nonlinear function. If fluctuations of the **rv** X, around its **mv**, are small, we may assume that the centered moments about X are also small, and then an expansion in a Taylor series is convenient

$$\langle f(X) \rangle = \sum_{n=0}^{\infty} \frac{1}{n!} \left[\frac{d^n f}{dX^n} \right]_{X=\langle X \rangle} \langle (X - \langle X \rangle)^n \rangle . \tag{1.43}$$

Example. Those cases in which the sum (1.43) ends for a certain value n^* are particularly useful. Assume that $f(X) = X^3$. From (1.43) it is easy to see that the following polynomial is obtained

$$\langle X^3 \rangle = \langle X \rangle^3 + 3 \langle X \rangle \left\langle (X - \langle X \rangle)^2 \right\rangle + \left\langle (X - \langle X \rangle)^3 \right\rangle . \tag{1.44}$$

If $\langle X \rangle = 0$, (1.44) is an identity. Note that (1.44) is valid whatever the distribution $P_X(x)$.

Exercise. In (1.44) consider fluctuations up to the second order and show that

$$\langle X^3 \rangle \simeq 3 \langle X \rangle \langle X^2 \rangle - 2 \langle X \rangle^3 + \mathcal{O}\left(\left\langle (X - \langle X \rangle)^3 \right\rangle \right) . \tag{1.45}$$

This result shows that it is possible to write the following *operational* approach to the **rv** X^3

$$X^3 \simeq \langle X \rangle^3 + 3X \langle X^2 \rangle - 3 \langle X \rangle^3 + \mathcal{O}\left(\left\langle (X - \langle X \rangle)^3 \right\rangle \right) . \tag{1.46}$$

If $\langle X \rangle = 0$, we get the approximation $X^3 \simeq 3X \langle X^2 \rangle$, which is often called Gaussian approximation.

Exercise. Consider the previous exercise, but now with $\langle X \rangle \neq 0$. Using the operational approximation (1.46) show that the third and fourth cumulants are identically zero; hence the approach (1.46) is called Gaussian approximation , see also advanced exercise 1.15.4.

Optional Exercise. Study higher order cumulants considering the operational approach (1.46).

1.9   Many Random Variables (Correlations)

Suppose now that we have a set of **rv** $(X_1, X_2, \cdots, X_n) \equiv \{X_j\}$. The set $\{X_j\}$ will be fully characterized if we know the n-dimensional joint pdf $P_n(\{X_j\})$, that is to say[23]:

$$P_{X_1, X_2, \cdots, X_n}(x_1; x_2; \cdots; x_n) = \langle \delta(X_1 - x_1)\delta(X_2 - x_2) \cdots \delta(X_n - x_n) \rangle . \qquad (1.47)$$

The **mv** of a function of the set $\{X_j\}$ is calculated in the usual manner (similar to the one-dimensional case) using the n-dimensional joint pdf $P_n(\{X_j\})$.

In particular, the correlation between two **rv**s is defined as

$$\langle\langle X_1 X_2 \rangle\rangle = \langle X_1 X_2 \rangle - \langle X_1 \rangle \langle X_2 \rangle ,$$

that is, the second cumulant between the two random variables.

1.9.1   Statistical Independence

If $\{X_j\}$ is a set of n **rv**s statistically independent (**si**) of one another, then $P_n(\{X_j\})$ is reduced to a product of n probability densities. Here, one notes immediately that the second cumulant vanishes identically, since

$$\langle X_1 X_2 \rangle = \int \int X_1 X_2 \, P_2(X_1; X_2) \, dX_1 \, dX_2, \qquad (1.48)$$

and if X_1, X_2 are **sirv**,

$$\langle X_1 X_2 \rangle = \int X_1 P_1(X_1) \, dX_1 \int X_2 \, P_1(X_2) \, dX_2 \qquad (1.49)$$

$$= \langle X_1 \rangle \langle X_2 \rangle .$$

Exercise. Let X be a **rv** uniformly distributed in the interval $[0, L]$; consider the one-parameter transformation

$$Y_a = \ln(X/a),$$

this law fully characterizes the **rv** Y_a. Calculate the Prob.$[Y_a \in (0, 1)]$. Show that the **mv** $\langle Y_1 Y_2 \rangle$ is

$$\langle Y_1 Y_2 \rangle = 2 + \ln(L/2)\ln(L) - \ln(L/2) - \ln(L).$$

Generalize this result for the calculation of the correlation $\langle\langle Y_1 Y_{1+\alpha} \rangle\rangle$.

---

[23]In general, in the case of many variables we use the shorthand notation $P_{X_1, X_2, \cdots, X_n}$ $(x_1; x_2; \cdots; x_n) = P_n(x_1; X_2; \cdots; X_n) = P_n(\{X_j\})$

Guided Exercise (**Constraining the Correlation Function**). Let Ω and Θ be two **rvs** with joint probability distribution $P(\Omega; \Theta)$. Let us show that the following inequality is satisfied

$$| \int \int (\Omega\Theta - \langle\Omega\rangle \langle\Theta\rangle) \, P(\Omega; \Theta) \, d\Omega \, d\Theta \, |^2 \leq \langle\langle\Theta^2\rangle\rangle \langle\langle\Omega^2\rangle\rangle.$$

We can also write this inequality in a more convenient form

$$\frac{| \langle\langle\Theta\Omega\rangle\rangle |}{\sqrt{\langle\langle\Theta^2\rangle\rangle \langle\langle\Omega^2\rangle\rangle}} \leq 1.$$

To prove the latter relationship we use that $\left\langle [a (\Omega - \langle\Omega\rangle) + (\Theta - \langle\Theta\rangle)]^2 \right\rangle \geq 0$ for all values of $a \in \mathcal{R}_e$. If we expand the square of each of the terms we easily obtain

$$a^2 \left\langle (\Omega - \langle\Omega\rangle)^2 \right\rangle + 2a \left\langle (\Omega - \langle\Omega\rangle)(\Theta - \langle\Theta\rangle) \right\rangle + \left\langle (\Theta - \langle\Theta\rangle)^2 \right\rangle =$$

$$a^2 \left\langle \langle\Omega^2\rangle \right\rangle + 2a \left\langle \langle\Theta\Omega\rangle \right\rangle + \left\langle \langle\Theta^2\rangle \right\rangle \geq 0.$$

On the other hand, this is true $\forall a$, it means: the discriminant of this quadratic equation has to be negative

$$\langle\langle\Theta\Omega\rangle\rangle^2 - \langle\langle\Omega^2\rangle\rangle \langle\langle\Theta^2\rangle\rangle < 0.$$

Then our result follows. Note that the equal sign holds when

$$\langle\langle\Theta\Omega\rangle\rangle = \langle\langle\Omega^2\rangle\rangle = \langle\langle\Theta^2\rangle\rangle = 0.$$

Exercise. Let Ω and Θ be two **rvs** with joint probability distribution $P(\Omega; \Theta)$. Show that

$$\langle\Theta\Omega\rangle^2 \leq \langle\Theta^2\rangle \langle\Omega^2\rangle.$$

Usually this inequality is called the cosine inequality, and its proof is similar to the one in the previous exercise. Hint: use the positivity of $\left\langle (a\Omega - \Theta)^2 \right\rangle$ for all $a \in \mathcal{R}_e$.

1.9.2 Marginal Probability Density

Let the set of two **rvs** $\{X_1, X_2\}$ be characterized by the joint pdf $P_2(X_1; X_2)$. Then, we define the marginal probability density of the **rv** X_1 as:

$$P_1(X_1) = \int_{\mathcal{D}_{X_2}} P_2(X_1; X_2) \, dX_2. \tag{1.50}$$

Obviously, (1.50) can be generalized to an arbitrary number of **rv**, where integration will be $(n-1)$-dimensional. In general, a s-dimensional marginal probability density is defined in complete analogy:

$$P_s(X_1; \cdots ; X_s) = \int_{\mathcal{D}_{X_{s+1}}} \cdots \int_{\mathcal{D}_{X_n}} P_n(X_1; \cdots ; X_s; X_{s+1}; \cdots ; X_n) \, dX_{s+1} \cdots dX_n.$$

1.9.3   *Conditional Probability Density*

Sometimes it is also necessary to work with a different probability density: *the conditional probability*. Suppose we divide the set of n **rv** $\{X_j\}$ in two groups,

$$\{X_1, \cdots, X_s\} \quad \text{and} \quad \{X_{s+1}, \cdots, X_n\},$$

and the second group is characterized with default values $\{X_{s+1}, \cdots, X_n\}$; then, under these conditions, the remaining **rv** will be characterized by the conditional probability

$$P(X_1; \cdots ; X_s \mid X_{s+1}, \cdots, X_n).$$

This probability density contains more information than the joint probability density $P_s(X_1; \cdots ; X_s)$. The relationship between this pdf and the n-dimensional joint probability density is given by the Bayes rule (see Sect. 1.2.2):

$$P(X_1; \cdots ; X_s \mid X_{s+1}, \cdots, X_n) = \frac{P_n(X_1; \cdots ; X_s; X_{s+1}; \cdots ; X_n)}{P_{n-s}(X_{s+1}; \cdots ; X_n)}. \tag{1.51}$$

Note that in these equations we have omitted any redundant notation. On the other hand, one can easily see that the denominator of the second member of Eq. (1.51) is none other than the marginal pdf of the **rv** $(X_{s+1}, \cdots, X_n)$.

Guided Exercise. Consider the joint probability distribution of two **rv**s with domain $[-a, a]$, given by

$$P_2(X_1; X_2) = \pi^{-1} \, \delta(X_1^2 + X_2^2 - a^2).$$

This probability distribution is a certainty, it is "deterministic" for the radius vector $\vec{X} = (X_1, X_2)$. We calculate now the marginal and conditional distributions. From the definition of conditional probability we note that if X_1, X_2 were **sirv**, then the conditional probability $P(X_2 \mid X_1)$ would be a function independent of X_1. To calculate this probability distribution, first we need to know the marginal distribution $P_1(X_1)$:

$$P_1(X_1) = \int_{-a}^{a} P_2(X_1; X_2) \, dX_2 = \int_{-a}^{a} \pi^{-1} \delta(X_1^2 + X_2^2 - a^2) \, dX_2$$

$$= \int_{-a}^{a} \delta(X_2 - \sqrt{a^2 - X_1^2}) \frac{dX_2/\pi}{|2\sqrt{a^2 - X_1^2}|}$$

$$+ \int_{-a}^{a} \delta(X_2 + \sqrt{a^2 - X_1^2}) \frac{dX_2/\pi}{|-2\sqrt{a^2 - X_1^2}|}$$

$$= \frac{1/\pi}{\sqrt{a^2 - X_1^2}}$$

Note that both terms contribute the same amount (remember that $X_2 \in [-a, a]$). It is immediately verified that this marginal distribution, $P_1(X_1)$, is properly normalized in the range $[-a, a]$. From (1.51) it follows that the conditional probability is given by

$$P(X_2 \mid X_1) = \frac{\pi^{-1} \delta(X_1^2 + X_2^2 - a^2)}{P(X_1)} = \delta(X_1^2 + X_2^2 - a^2)\sqrt{a^2 - X_1^2},$$

or equivalently

$$P(X_2 \mid X_1) = \frac{1}{2}\left(\delta(X_2 - \sqrt{a^2 - X_1^2}) + \delta(X_2 + \sqrt{a^2 - X_1^2})\right).$$

From this expression it follows trivially that X_1, X_2 are not **si**.

Exercise (**Conditional Moments**). Let X_1, X_2 be two **rvs** characterized by $P(X_1; X_2)$ in a given domain $\mathcal{D}_X$. Try to calculate conditional moments

$$\langle (X_1 - X_2)^m \rangle,$$

where the **rv** X_2 takes a fixed value X'. That is, for the first moment we have

$$\left\langle X_1 - X' \right\rangle = \int_{\mathcal{D}_X} \left(X_1 - X'\right) P(X_1 \mid X') \, dX_1.$$

Check that all conditional moments are obtained easily defining a conditional characteristic function, i.e.:

$$G_{X_1}(k, X') = \int_{\mathcal{D}_X} e^{ik(X_1 - X')} P(X_1 \mid X') \, dX_1,$$

where

$$P(X_1 \mid X') = P(X_1; X') \Big/ P(X').$$

1.10   Multidimensional Characteristic Function

The characteristic function of a multidimensional probability is defined in the same way as in (1.15). Let $\{X_l\}$ be a set of n rv. Then

$$G_n(\{k_l\}) = \langle \exp(ik_1 X_1 + ik_2 X_2 + \cdots + ik_n X_n) \rangle, \tag{1.52}$$

and its Taylor series expansion in n-dimensions, gives the expression for the moments

$$G_n(\{k_l\}) = \sum_0^\infty \frac{(ik_1)^{m_1}(ik_2)^{m_2}\cdots(ik_n)^{m_n}}{m_1!\,m_2!\,\cdots m_n!} \langle X_1^{m_1}\ldots X_n^{m_n}\rangle. \tag{1.53}$$

Exercise. Show that the cumulants expansion is given by

$$\ln[G_n(\{k_l\})] = \sum_0^\infty{}' \frac{(ik_1)^{m_1}(ik_2)^{m_2}\cdots(ik_n)^{m_n}}{m_1!\,m_2!\,\cdots m_n!} \langle\langle X_1^{m_1}\ldots X_n^{m_n}\rangle\rangle. \tag{1.54}$$

In the sum above the term where all the m_l are zero is absent. Then, from (1.54) it follows that

$$\langle\langle X_1^{m_1}\ldots X_n^{m_n}\rangle\rangle = \left(\frac{\partial}{\partial ik_1}\right)^{m_1}\left(\frac{\partial}{\partial ik_2}\right)^{m_2}\cdots\left(\frac{\partial}{\partial ik_n}\right)^{m_n}\ln[G_n(\{k_l\})]\Bigg|_{\{k_l\}=0}.$$

Example. Let us show that if at least one X_j is statistically independent of the remaining $n-1$ variables, then all cumulants

$$\left\langle\!\left\langle X_1^{m_1}\cdots, X_j^{m_j},\cdots X_n^{m_n}\right\rangle\!\right\rangle$$

vanish identically for $m_j \neq 0$. If X_j is the statistically independent one, then

$$P_n(\{X_l\}) \equiv P_n(X_1; X_2; \cdots; X_n) = P_1(X_j)P_{n-1}(X_1; \cdots; X_{j-1}; X_{j+1}; \cdots; X_n), \tag{1.55}$$

from which it follows that

$$G_n(\{k_l\}) \equiv G_n(k_1, k_2, \cdots, k_n) = G_1(k_j)G_{n-1}(k_1, \cdots, k_{j-1}, k_{j+1}, \cdots, k_n). \tag{1.56}$$

Then, in general,

$$\left\langle\!\left\langle X_1^{m_1}, \cdots, X_j^{m_j}, \cdots, X_n^{m_n} \right\rangle\!\right\rangle =$$

$$= \left(\frac{\partial}{\partial i k_1}\right)^{m_1} \cdots \left(\frac{\partial}{\partial i k_j}\right)^{m_j} \cdots \left(\frac{\partial}{\partial i k_n}\right)^{m_n} \left[\ln[G_1(k_j)] + \ln[G_{n-1}(\{k_l\})]\right]\Big|_{\{k_l\}=0} = 0. \tag{1.57}$$

1.10.1 Cumulant Diagrams (Several Variables)

We know that a second cumulant is related to its lower order moments:

$$K_2 = M_2 - M_1^2.$$

Then, for two **rvs** we have the following generalization:

$$\langle\langle X_1 X_2 \rangle\rangle = \langle X_1 X_2 \rangle - \langle X_1 \rangle \langle X_2 \rangle.$$

This relationship can be represented diagrammatically easily, see Fig. 1.3a. In the case of the third cumulant we have

$$\langle\langle X_1 X_2 X_3 \rangle\rangle = \quad \langle X_1 X_2 X_3 \rangle - \{\langle\langle X_1 X_2 \rangle\rangle \langle X_3 \rangle + \langle\langle X_1 X_3 \rangle\rangle \langle X_2 \rangle$$
$$+ \langle\langle X_2 X_3 \rangle\rangle \langle X_1 \rangle\} - \{\langle X_1 \rangle \langle X_2 \rangle \langle X_3 \rangle\}.$$

In the same way a diagram for this third cumulant can be drawn, see Fig. 1.3b.

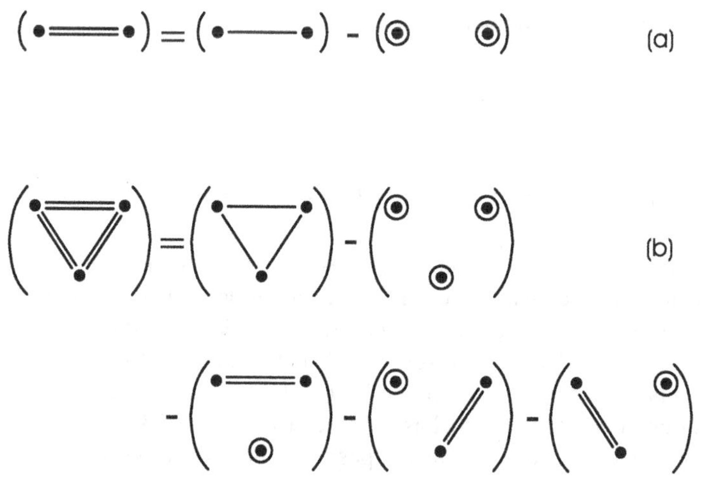

Fig. 1.3 Cumulant diagrams representative of second (**a**) and third order (**b**)

Building Rules

In general we have the following rules for the construction of all diagrams [14]:

1. For every random variable X_i a vertex • is introduced in the graph.
2. Each double bond indicates a cumulant of order of the number of vertices involved.
3. Each single bond indicates a moment of the order of the number of vertices involved.
4. Each isolated vertex (locked) indicates the first moment of the vertex.
5. The cumulant of order n is constructed by subtracting from the moment of order n all combinations (graphs) possible with cumulants of lower orders to exhaust all possibilities.

Exercise. Draw the diagrams corresponding to the calculation of other higher cumulants. In particular show that there are 37 diagrams needed for calculating $\langle\langle X_1 X_2 X_3 X_4 X_5\rangle\rangle$.

Guided Exercise (**Correlation Between Random Variables**). Let X be an **rv** uniformly distributed in the range $[0, 2\pi]$ and let $Y_t(X)$ be the transformation

$$Y_t(X) = \sin[t + X].$$

Then Y_t defines a one-parameter family of **rv** (of course, not uniformly distributed over $[0, 2\pi]$). It is easy to notice that the odd moments $\langle Y_1 Y_2 \cdots Y_{2m+1}\rangle$ vanish identically and, on the other hand, even moments are invariant under continuous translation $t \to t + \tau$ for all τ. For the correlation of two "points" we get:

$$\langle Y_{t_1} Y_{t_2}\rangle \equiv \langle Y_{t_1}(X) Y_{t_2}(X)\rangle = \frac{1}{2\pi} \int_0^{2\pi} \sin[t_1 + X] \sin[t_2 + X]\, dX = \frac{1}{2}\cos(t_2 - t_1),$$

and for the fourth order moment:

$$\langle Y_{t_1} Y_{t_2} Y_{t_3} Y_{t_4}\rangle = \frac{1}{2\pi} \int_0^{2\pi} \sin[t_1 + X] \sin[t_2 + X] \sin[t_3 + X] \sin[t_4 + X]\, dX =$$

$$\frac{1}{8} \left(\cos(t_1 + t_2 - t_3 - t_4) + \cos(t_1 + t_3 - t_2 - t_4) + \cos(t_1 + t_4 - t_2 - t_3)\right).$$

It is easy to get, analogously, a general expression for the moment $\langle Y_1 Y_2 \cdots Y_{2m}\rangle$ and its cumulants. The fact that the correlation functions depend only on the differences (symmetrized) of parameters t_j will be critical in defining stationary stochastic processes. Note also that these correlations do not asymptotically decay to zero even for large separations from their "points," $|t_j - t_m| \to \infty$; this fact is important in the definition of non-ergodic stochastic processes [see Chap. 3]. Finally, note that a correlation function need not always be positive. Interpret the meaning of a negative correlation.

1.11 Terwiel's Cumulants*

Another class of cumulants, extremely important and useful in perturbation theory of differential equations, are the Terwiel cumulants [15]. To define them we need to think about the **mv** of a function as the result of applying a projection operator on it[24] $\mathcal{P} = \mathcal{P}^2$. Then, if X is an **rv** characterized by the probability distribution $P_X(x)$ on $\mathcal{D}$, we define the following projection operators $\mathcal{P}$ and $\mathcal{Q} \equiv \mathbf{1} - \mathcal{P}$:

$$\mathcal{P}f(X) \equiv \langle f(X) \rangle, \tag{1.58}$$

and

$$\mathcal{Q}f(X) = [\mathbf{1} - \mathcal{P}]f(X) \equiv f(X) - \langle f(X) \rangle, \tag{1.59}$$

where, of course $\langle f(X) \rangle = \int_{\mathcal{D}} f(x)P_X(x)\,dx$, it is easy to see that $\mathcal{P}\mathcal{Q}f(X) = \mathcal{Q}\mathcal{P}f(X) = 0$.

If $\{X_j\}$ is a set of n **rv** characterized by the joint distribution $P_n(\{X_j\})$, the Terwiel cumulant of such variables is defined according to:

$$\langle X_1 \cdots X_n \rangle_T = \mathcal{P}X_1\,(\mathbf{1} - \mathcal{P})\,X_2\,(\mathbf{1} - \mathcal{P})\,X_3 \cdots (\mathbf{1} - \mathcal{P})\,X_n. \tag{1.60}$$

Note that this cumulant preserves the order of the n **rv** X_j, $\{j = 1, \cdots n\}$. In general, it may be that these variables represent random operators, so that their order is very important because they may not commute. It is interesting to establish a relationship between Terwiel's cumulants and the usual moments. To this end, we present the following relationship

$$\langle X_1 \cdots X_n \rangle_T = \sum_{j=0}^{n-1}(-1)^j \sum_{1 \le p_1 < p_2 \cdots < p_j \le n} \langle X_1 \cdots X_{p_1} \rangle \langle X_{p_1+1} \cdots X_{p_2} \rangle \cdots \langle X_{p_j+1} \cdots X_n \rangle. \tag{1.61}$$

As an example, let us write here some nontrivial Terwiel's cumulants

$$\langle X_1 X_2 X_3 \rangle_T = \langle X_1 X_2 X_3 \rangle - \langle X_1 \rangle \langle X_2 X_3 \rangle - \langle X_1 X_2 \rangle \langle X_3 \rangle + \langle X_1 \rangle \langle X_2 \rangle \langle X_3 \rangle$$

$$\langle X_1 X_2 X_3 X_4 \rangle_T = \langle X_1 X_2 X_3 X_4 \rangle - \langle X_1 \rangle \langle X_2 X_3 X_4 \rangle - \langle X_1 X_2 \rangle \langle X_3 X_4 \rangle - \langle X_1 X_2 X_3 \rangle \langle X_4 \rangle$$

$$+ \langle X_1 \rangle \langle X_2 \rangle \langle X_3 X_4 \rangle + \langle X_1 X_2 \rangle \langle X_3 \rangle \langle X_4 \rangle + \langle X_1 \rangle \langle X_2 X_3 \rangle \langle X_4 \rangle - \langle X_1 \rangle \langle X_2 \rangle \langle X_3 \rangle \langle X_4 \rangle. \tag{1.62}$$

---

[24]Let a vector space be $\mathbf{V} = \mathbf{S} \oplus \mathbf{W}$, every $z \in \mathbf{V}$ can be expressed univocally as $z = x + y$ where $x \in \mathbf{S}$ and $y \in \mathbf{W}$. The projection operator $\mathcal{P} : \mathbf{V} \to \mathbf{V}$ defined by $\mathcal{P}\,z = x$ is called projector $\mathbf{V}$ on $\mathbf{S}$ parallel to $\mathbf{W}$. Then, $\mathcal{P}$ is projector iff $\mathcal{P} = \mathcal{P}^2$; also in a matrix representation (finite vector space) it is easy to see that this type of matrices $\mathcal{P}$ can only have eigenvalues: $\lambda_0 = 0$ and $\lambda_1 = 1$ (possibly multiple).

Exercise. Show explicitly the difference between a "normal" cumulant of order three and the Terwiel cumulant $\langle X_1 X_2 X_3 \rangle_T$.

Note that if all **rv** X_j are statistically independent, then it is easy to see that the Terwiel cumulant vanishes identically. Other interesting properties are set in the following exercises. (Hint: use the fact that $\mathcal{P} = \mathcal{P}^2$).

Exercise. Show that Terwiel's cumulants have the "partition" property. That is, if the set of n **rv** is ordered such that the first s **rv** form a set of variables independent of the remaining $n - s$, Terwiel's cumulant vanishes identically. (Hint: use the fact that $\mathcal{P}_{1,\cdots s, s+1, \cdots, n} = \mathcal{P}_{1,\cdots,s} \mathcal{P}_{s+1,\cdots,n}$).

$$\langle X_1 \cdots X_s X_{s+1} \cdots X_n \rangle_T = 0. \tag{1.63}$$

Exercise. Show that if the first moments are null, then the expression for the Terwiel cumulant $\langle X_1 \cdots X_j \cdots X_n \rangle_T$ is greatly simplified.

Excursus. Consider the following sum of fourth order Terwiel's cumulants

$$\sum_{j,k,l,m \,=-\infty}^{\infty} \langle X_j X_k X_l X_m \rangle_T \,,$$

where $\{j, k, l, m\} \in \mathbb{Z}$, and analyze the case when all the first moments are zero. Consider the situation when the sum excludes all equal contiguous indices. If for each value $\{j, k, l, m\}$ the **rv** is statistically independent of each other, study the class of Terwiel's cumulants

$$\langle X_j X_k X_l X_m \rangle_T \,,$$

which will give a nonzero contribution to the mentioned sum. Try to draw "diagrams" to characterize these contributions [16].

1.12   Gaussian Distribution (Several Variables)

From the definition (1.54) it follows that if all cumulants, except those which

$$m_1 + m_2 + \cdots + m_n \leq 2, \tag{1.64}$$

vanish identically, we shall have a n-dimensional Gaussian distribution. Then, from the definition of characteristic function it follows that[25]

---

[25]The factor $1/2$ in front of the unrestricted sum is to avoid adding twice.

$$G_n(\{k_p\}) = \exp\left(\sum_{j=1}^{n} ia_j k_j + \frac{1}{2}\sum_{j,l=1}^{n} \sigma_{jl}\,(ik_j)(ik_l)\right). \tag{1.65}$$

Then, the first moments are obtained in the form

$$\langle X_j\rangle = \left(\frac{\partial}{\partial ik_j}\right)[G_n(\{k_p\})]\Big|_{\{k_p\}=0} = a_j \tag{1.66}$$

$$\langle X_j X_l\rangle = \left(\frac{\partial}{\partial ik_j}\right)\left(\frac{\partial}{\partial ik_l}\right)[G_n(\{k_p\})]\Big|_{\{k_p\}=0} = \sigma_{jl} + a_j a_l, \tag{1.67}$$

whereby the variance matrix (covariance if $j \neq l$) is characterized by

$$\langle(X_j - \langle X_j\rangle)(X_l - \langle X_l\rangle)\rangle = \sigma_{jl}. \tag{1.68}$$

The pdf is given by the inverse Fourier transform of $G_n(\{k_p\})$, that is to say:

$$P_n(\{X_p\}) \equiv P_n(X_1;\cdots;X_n) \tag{1.69}$$

$$= (2\pi)^{-n}\int_{-\infty}^{+\infty}\cdots\int_{-\infty}^{\infty}\exp\left(\sum_{j=1}^{n} ia_j k_j + \frac{1}{2}\sum_{j,l}^{n}\sigma_{jl}\,(ik_j)(ik_l)\right)$$

$$\times \exp\left(-i\sum_{j=1}^{n} X_j k_j\right)\Pi_{j=1}^{n}\,dk_j.$$

As σ_{lj} is the covariance matrix (*positive definite*[26]) and therefore it is symmetrical, it follows that

$$\sigma_{lj}^{-1} = \sigma_{jl}^{-1} \quad \text{(there is an inverse)}$$

$$\sigma_{lj}^{1/2} = \sigma_{jl}^{1/2} \quad (\sqrt{\sigma}\text{ exists}) \tag{1.70}$$

$$\sigma_{lj}^{-1/2} = \sigma_{jl}^{-1/2} \quad ((\sqrt{\sigma})^{-1}\text{ exists}).$$

---

[26]In general, a matrix $a \in C$ is *semipositive definite* if the associated quadratic form is *semipositive definite*. That is, if the elements a_{jl} meet the conditions $a_{ll} \geq 0$ and $a_{ll} \cdot a_{jj} - |a_{lj}|^2 \geq 0$, $\forall j, l$. This ensures that the eigenvalues of the matrix a will be $\lambda_i \geq 0$. It is said that a matrix is positive definite if it complies strictly greater than zero condition. See also Sects. 4.2 and 6.1.1.

Introducing the change of variable

$$\alpha_j = \sum_{l=1}^{n} \left[\sigma_{jl}^{1/2} k_l + i\sigma_{jl}^{-1/2} (X_l - a_l) \right] \tag{1.71}$$

in the integrand (1.69), the term of the form $e^{[\cdots]}$ can be rewritten noting that

$$[\cdots] = \frac{-1}{2} \sum_j \alpha_j \alpha_j - \frac{1}{2} \sum_{j,l} \sigma_{jl}^{-1} (X_j - a_j)(X_l - a_l). \tag{1.72}$$

That is, we have "diagonalized" the matrix in the new coordinate system $\{\alpha_p\}$. The Jacobian of the transformation is given by the expression[27]

$$\frac{dk_1 dk_2 \cdots dk_n}{d\alpha_1 d\alpha_2 \cdots d\alpha_n} = \left(\frac{d\alpha_1 d\alpha_2 \cdots d\alpha_n}{dk_1 dk_2 \cdots dk_n} \right)^{-1} = \left[\det \left(\sigma^{1/2} \right) \right]^{-1}, \tag{1.73}$$

whereby the integration is simply reduced to

$$\int \cdots \int \exp \left(\frac{-1}{2} \sum_{j=1}^{n} \alpha_j \alpha_j \right) \Pi_{j=1}^{n} \, d\alpha_j = \left(\int_{-\infty}^{\infty} \exp(-\alpha^2/2) \, d\alpha \right)^n = (2\pi)^{n/2}. \tag{1.74}$$

Finally, the joint probability distribution is expressed in the form:

$$P_n(X_1; \cdots ; X_n) = (2\pi)^{-n/2} \left[\det \sigma^{-1} \right]^{1/2} \exp \left(-\frac{1}{2} \sum_{j,l} \left(\sigma^{-1} \right)_{jl} (X_j - a_j)(X_l - a_l) \right), \tag{1.75}$$

which is the multidimensional Gaussian distribution we were looking for.

1.12.1 Gaussian Distribution with Zero Odd Moments

Consider n **rv** $Y_j = X_j - a_j$, where X_j are Gaussian with mean value a_j; then the characteristic function of the new variables Y_j is given by[28]

$$G_Y(\{k_p\}) \equiv \langle \exp (ik_1 Y_1 + \cdots + ik_n Y_n) \rangle \tag{1.76}$$

$$= \langle \exp (ik_1 X_1 + \cdots + ik_n X_n) \exp (-ik_1 a_1 - \cdots - ik_n a_n) \rangle$$

---

[27]In general, we use the term "diagonalized" to indicate that we have reduced the matrix to diagonal form.

[28]Note that here we emphasize the type of **rv**, therefore we denote $G_Y(\{k_p\})$ to signify the characteristic function of the set $\{Y_j\}$, i.e.: $G_{Y_1;Y_2;\cdots,Y_n}(\{k_p\}) \equiv G_n(\{k_p\})$.

$$= \exp\left(-ik_1 a_1 - \cdots - ik_n a_n\right) G_X(\{k_p\})$$

$$= \exp\left(\frac{-1}{2} \sum_{j,l}^{n} \sigma_{jl}\, k_j k_l\right),$$

because $G_X(\{k_p\})$ is given by (1.65). From which we see that for $m \in \mathcal{N}$

$$\langle Y_1 Y_2 \cdots Y_{2m+1}\rangle = 0. \tag{1.77}$$

Exercise. Show that a linear transformation

$$X_l' = \sum_m \beta_{lm} X_m + B_l,$$

preserves the Gaussian structure of the probability distribution.

1.12.2 Novikov's Theorem*

Consider r **rv** $(X_1, \cdots, X_r) \equiv \{X_p\}$ to be Gaussian of mean zero (generalization to the case of $a_j \neq 0$ is straightforward), then it holds that[29]

$$\langle X_l f(\{X_p\})\rangle = \sum_j \langle X_l X_j\rangle \left\langle \frac{\partial f(\{X_p\})}{\partial X_j}\right\rangle. \tag{1.78}$$

To prove (1.78) note that if the joint distribution is Gaussian

$$P_r(X_1; \cdots; X_r) = (2\pi)^{-r/2} \left[\det \sigma_{jl}\right]^{-1/2} \exp\left(-\frac{1}{2}\sum_{j,l}^{r} \sigma_{jl}^{-1}\, X_j X_l\right), \tag{1.79}$$

it holds that $\langle X_j X_l\rangle = \sigma_{jl}$; then from (1.78) we get

$$\langle X_l f(\{X_p\})\rangle = \sum_j \sigma_{lj} \left\langle \frac{\partial f(\{X_p\})}{\partial X_j}\right\rangle, \tag{1.80}$$

---

[29]In the continuous case (infinite dimensional) we have the generalization:

$$\langle X(r) f[X(\bullet)]\rangle = \int_{\mathcal{D}} \langle X(r) X(r')\rangle \left\langle \frac{\delta f[X(\bullet)]}{\delta X(r')}\right\rangle dr',$$

where $\mathcal{D}$ means that the integral extends over the region (including all points) in which the functions are defined. In Chap. 3 we define in detail the meaning of the object $\dfrac{\delta f(X(\bullet))}{\delta X(r')}$, which can be regarded as the generalization of a derivative, see also advanced exercise 1.15.10.

and multiplying (1.80) by σ^{-1} we get:

$$\sum_l \sigma_{ml}^{-1} \langle X_l f(\{X_p\}) \rangle = \sum_l \sum_j \sigma_{ml}^{-1} \sigma_{lj} \left\langle \frac{\partial f(\{X_p\})}{\partial X_j} \right\rangle = \left\langle \frac{\partial f(\{X_p\})}{\partial X_m} \right\rangle .$$

On the other hand,

$$\sum_l \sigma_{ml}^{-1} \langle X_l f(\{X_p\}) \rangle =$$

$$\frac{(\det \sigma)^{-1/2}}{(2\pi)^{r/2}} \int_{-\infty}^{\infty} \cdots \int_{-\infty}^{\infty} \Pi_{n=1}^r \, dX_n \, f(\{X_p\}) \sum_l \sigma_{ml}^{-1} X_l \exp\left(-\tfrac{1}{2} \sum_{j,k} \sigma_{jk}^{-1} X_j X_k\right) .$$

This integral can be written in matrix notation in the form

$$\sum_l \sigma_{ml}^{-1} \langle X_l f(\{X_p\}) \rangle =$$

$$\frac{-(\det \sigma)^{-1/2}}{(2\pi)^{r/2}} \int_{-\infty}^{\infty} \cdots \int_{-\infty}^{\infty} \Pi_{n=1}^r \, dX_n \, f(\{X_p\}) \frac{\partial}{\partial X_m} \exp\left(-\tfrac{1}{2} X \cdot \sigma^{-1} \cdot X\right) ,$$

then integrating by parts, we get

$$\sum_l \sigma_{ml}^{-1} \langle X_l f(\{X_p\}) \rangle = \left\langle \frac{\partial f(\{X_p\})}{\partial X_m} \right\rangle , \qquad (1.81)$$

which completes the proof.

Note This theorem provides for rules (diagrams) to calculate all the moments of functions of r Gaussian variables.[30]

Exercise. Show that if $\{X_p\}$ are r Gaussian **rv** with **mv** $\langle X_p \rangle \neq 0$, Novikov's theorem (1.78) generalizes as follows:

$$\langle X_l f(\{X_p\}) \rangle = \sum_j \langle\langle X_l X_j \rangle\rangle \left\langle \frac{\partial f(\{X_p\})}{\partial X_j} \right\rangle + \langle X_l \rangle \langle f(\{X_p\}) \rangle ,$$

where $\langle\langle X_l X_j \rangle\rangle$ is the second cumulant.

Exercise. Show that if $\{X_p\}$ are r Gaussian **rv** with **mv** $\langle X_p \rangle = 0$, then all even moments are easily obtained from the formula:

$$\langle X_1 \cdots X_{2m} \rangle = \sum_{\text{symmetrized}} \langle X_i X_j \rangle \langle X_l X_p \rangle \cdots \langle X_k X_q \rangle \quad \text{with} \quad m \in \mathcal{N} .$$

For example: $\langle X_1 X_2 X_3 X_4 \rangle = \langle X_1 X_2 \rangle \langle X_3 X_4 \rangle + \langle X_1 X_3 \rangle \langle X_2 X_4 \rangle + \langle X_1 X_4 \rangle \langle X_2 X_3 \rangle .$

---

[30]K. Furutsu, J Res. NBS, D67, 303 (1963); E.A. Novikov, Sov. Phys. JETP, 20, 1290 (1965). See also related ideas in L. Onsager, Phys. Rev. 38, 2265, (1931).

Exercise. Let X be a Gaussian **rv** with mean zero and variance 1 (normal distribution). Show that for $a \gg 1$ the **mv** of $1/(a + X^2)$ can be approximated by

$$\left\langle \frac{1}{a + X^2} \right\rangle \approx \frac{1}{a + 1} + \mathcal{O}(1/a^3),$$

while for the limit $a \to 0$ the **mv** does not exist.

1.13   Transformation of Densities in n-Dimensions

In the study of thermodynamic fluctuations, in both equilibrium and nonequilibrium dissipative systems, the mathematical problem can be formulated in terms of a set of random variables. This raises the question how the joint probability density of n **rv** $P_n(\{X_j\})$ is modified when a coordinate transformation is applied. Suppose we know the transformation laws:

$$Y_1 = g_1(X_1, X_2, \cdots, X_n) \tag{1.82}$$
$$Y_2 = g_2(X_1, X_2, \cdots, X_n)$$
$$\cdots$$
$$Y_n = g_n(X_1, X_2, \cdots, X_n),$$

here $g_1, \cdots, g_n$ are taken to be continuous functions. Assuming that the system (1.82) is invertible (i.e., $x_1 = h_1(y_1, y_2, \ldots, y_n)$, etc., and functions $h_1, \cdots, h_n$ are all single-valued) $P_n(\{Y_j\})$ is transformed according

$$P_{Y_1, Y_2, \ldots, Y_n}(y_1; y_2; \ldots; y_n) = P_{X_1, X_2, \cdots, X_n}(h_1; h_2; \ldots; h_n) |D_n|, \tag{1.83}$$

where D_n is the Jacobian of the transformation $X \to Y$. We see, then, that (1.83) is the multidimensional generalization (1.38).

Excursus. Let $A(\mathbf{r})$ be a vector field in R^3 characterized by a Gaussian probability distribution. Let $B(\mathbf{r}) = \mathbf{Rotor} A(\mathbf{r})$ be an associated field. What can be said about the probability distribution of the random field $B(\mathbf{r})$? A collection of interesting results obtained from the theorem of transformation of **rv** is presented in: D.T. Gillespie, Am. J. Phys. **51**, 520, (1983).

Exercise (**Gaussian Random Variables**). Let X_1, X_2 be two **sirv** and uniformly distributed in the interval $[0, 1]$. Consider the nonlinear transformation:

$$G_1 = \sqrt{-2 \ln(X_1)} \cos(2\pi X_2), \quad G_2 = \sqrt{-2 \ln(X_1)} \sin(2\pi X_2).$$

Show that the joint distribution $P_2(G_1; G_2)$ corresponds to a Gaussian probability density of two **sirv** with zero mean and variance 1.

Exercise (**Ellipsoidal Random Variables**). Let X_1, X_2 be two **sirv** and uniformly distributed in the interval $[0, 1]$. Consider the nonlinear transformation [17]:

$$E_1 = X_1 \cos(2\pi X_2), \quad E_2 = X_1 \sin(2\pi X_2 + \beta). \tag{1.84}$$

From (1.84) all the cumulants of the random variables E_j can be calculated as a function of the correlation parameter β. When $\beta \gtrless 0$ the ellipsoidal correlations are negative or positive, respectively; and for $\beta = 0$ the correlation is isotropic. Discuss how would the pdf look in the plane (E_1, E_2).

Exercise. Let b be an **rv** uniformly distributed in the range $[0, 1]$. Consider the complex random number $Z = (1 + i)b$. What is the probability distribution associated with the complex number $Z = Z_1 + iZ_2$? Calculate the moments $\langle Z^n \rangle$ using the joint probability distribution $P_2(Z_1, Z_2)$ and compare the results with those obtained from direct calculation: $\langle Z^n \rangle = (1 + i)^n \langle b^n \rangle$ [18].

Guided Exercise. Let X_1, X_2 be two **rvs** characterized by the joint probability distribution $P_2(X_1, X_2)$, with support $\mathcal{D}_{X_1}$ and $\mathcal{D}_{X_2}$. We calculate here the probability distribution of the sum $s = X_1 + X_2$. Taking

$$Y_1 = g_1(X_1, X_2) = X_1$$
$$Y_2 = g_2(X_1, X_2) = X_1 + X_2,$$

and its inverse relationship:

$$X_1 = h_1(Y_1) = Y_1$$
$$X_2 = h_2(Y_1, Y_2) = Y_2 - Y_1,$$

which involves the following Jacobian

$$|D_2| = \begin{vmatrix} \dfrac{\partial h_1}{\partial Y_1} & 0 \\ \dfrac{\partial h_2}{\partial Y_1} & \dfrac{\partial h_2}{\partial Y_2} \end{vmatrix} = \left| \dfrac{\partial h_2}{\partial Y_2} \right|.$$

Then, from (1.83) we get that[31]

$$P_Y(Y_1; Y_2) = P_X(Y_1; h_2(Y_1, Y_2)) \left| \dfrac{\partial h_2}{\partial Y_2} \right|.$$

---

[31] Note that here it is useful to emphasize the random variable involved, which is why we use the notation: $P_Y(Y_1; \cdots ; Y_r)$.

Then the (marginal) probability density of the sum is given (with $Y_2 \equiv s = X_1 + X_2$) by

$$P(s) = \int_\mathcal{D} P_Y(Y_1; Y_2)\, dY_1 = \int_\mathcal{D} P_X(Y_1; Y_2 - Y_1)\, dY_1$$

$$= \int_\mathcal{D} P_X(X_1; s - X_1)\, dX_1.$$

In general, the limits of integration should be taken with care according to the law of transformation and domains $\mathcal{D}_{X_1}$ and $\mathcal{D}_{X_2}$.

Example. Consider the situation when two **sirv** have uniform distribution on the interval $\mathcal{D}_X = [0, T]$. In this case the distribution of the sum $s = X_1 + X_2$ is given by

$$P(s) = \int_{\mathcal{D}_{X_1}} P_X(X_1; s - X_1)\, dX_1$$

$$= \frac{1}{T^2} \int_0^T \Theta(X_1)\Theta(T - X_1)\Theta(s - X_1)\Theta(T - s + X_1)\, dX_1,$$

where $\Theta(z)$ is the step function introduced in (1.7). Hence, we note that there are two different situations: if $s < T$, we get that

$$P(s) = \frac{1}{T^2} \int_0^s dX_1 = \frac{s}{T^2},$$

while if $s > T$, it follows that

$$P(s) = \frac{1}{T^2} \int_{s-T}^T dX_1 = \frac{2T - s}{T^2}.$$

On the other hand, the sum is bounded by $s < 2T$. Show that the Fourier transform of $P(s)$ is given by

$$G_s(k) = \int_0^{2T} e^{iks} P(s)\, ds = \frac{-\left(e^{ikT} - 1\right)^2}{k^2 T^2}$$

and compare it with the guided exercise of the Sect. 1.6.

Exercise. Show that the (marginal) probability density of the difference, product and the quotient of the **rv** X_1, X_2, are given by

$$P_1(X_1 - X_2) = P(d) = \int_{\mathcal{D}_{X_1}} P_X(X_1; X_1 - d)\, dX_1$$

$$P_1(X_1 \cdot X_2) = P(p) = \int_{\mathcal{D}_{X_1}} P_X(X_1; p/X_1) \, \frac{1}{|X_1|} \, dX_1$$

$$P_1(X_2/X_1) = P(c) = \int_{\mathcal{D}_{X_1}} P_X(X_1; cX_1) \, |X_1| \, dX_1.$$

Guided Exercise. Consider the case in which domains $\mathcal{D}_{X_1}$ and $\mathcal{D}_{X_2}$ are nonnegative semi-infinite, i.e.: $X_i \in [0, \infty]$. Show that the probability distribution of the sum $P(s)$ takes the form:

$$P(s) = \int_0^s P_X(Y_1; s - Y_1) \, dY_1. \tag{1.85}$$

Hint: to get this result, simply use the previous example considering that $P_X(X_1, X_2) = 0$ if $X_1 < 0$ and/or $X_2 < 0$, then you can write:

$$P(s) = \int_{\mathcal{D}_{Y_1}} P_X(Y_1; s - Y_1) \, \Theta(Y_1) \, \Theta(s - Y_1) \, dY_1,$$

from which (1.85) follows.

Exercise. Consider the case in which domains $\mathcal{D}_{X_1}$ and $\mathcal{D}_{X_2}$ are nonnegative semi-infinite. Show that the probability distribution of the difference $d = X_1 - X_2$ of two **sirv** takes the form:

$$P(d) = \int_d^{\infty} P_{X_1}(Y_1) \, P_{X_2}(Y_1 - d) \, dY_1.$$

Exercise. Consider two **sirv** with monotonically decreasing probability distributions of the form $P(X_i) = \alpha_i \exp(-\alpha_i X_i)$, with supports $\mathcal{D}_{X_1}$ and $\mathcal{D}_{X_2}$ semi-infinite nonnegative. Prove that the probability distribution of the sum $s = X_1 + X_2$ has a maximum, in $s^* \in (0, \infty)$, and it is given by the expression:

$$P(s) = \int_0^s P_{X_1}(Y_1) \, P_{X_2}(s - Y_1) \, dY_1 = \frac{\alpha_1 \alpha_2}{\alpha_2 - \alpha_1} \left(e^{-\alpha_1 s} - e^{-\alpha_2 s} \right).$$

1.14 Random Perturbation Theory*

We now show how to calculate the **mv** of a time-dependent function when this function is given in terms of a linear evolution with random elements. Particular emphasis is put on the fact that the character of the random elements has an arbitrary distribution. If the time is continuous or discrete the way to solve the problem involves two different techniques for transforming the evolution equation into an algebraic problem.

1.14.1   Continuous Time Evolution

Consider the differential equation

$$\frac{d}{dt}F(t) = (H + B)F(t), \ t \geq 0$$

where H and B are constant in time and B has a random character. Using the Laplace transform[32] we can turn this differential equation into an algebraic equation for the unknown function $F(u)$, that is to say

$$uF(u) - F(0) = (H + B)F(u). \tag{1.86}$$

In particular, it is interesting to study this equation when $F(0)$ is not random. If we use the concept of projector (see Sect. 1.11) we can find a general expression for the **mv** of $F(u)$. First apply $\mathcal{P}$ to Eq. (1.86) and obtain[33]:

$$u\mathcal{P}F - F(0) = H\mathcal{P}F + \mathcal{P}BF = H\mathcal{P}F + \mathcal{P}B\mathcal{P}F + \mathcal{P}B\mathcal{Q}F. \tag{1.87}$$

then apply $\mathcal{Q}$ to Eq. (1.86) and obtain:

$$u\mathcal{Q}F = H\mathcal{Q}F + \mathcal{Q}BF = H\mathcal{Q}F + \mathcal{Q}B\mathcal{P}F + \mathcal{Q}B\mathcal{Q}F. \tag{1.88}$$

The formal solution of (1.88) is $\mathcal{Q}F = (u - H)^{-1}[\mathcal{Q}B\mathcal{P}F + \mathcal{Q}B\mathcal{Q}F]$. We solve (1.88) by iteration and get

$$\mathcal{Q}F = \sum_{k=1}^{\infty}(G\mathcal{Q}B)^k \mathcal{P}F \quad \text{where} \quad G \equiv (u - H)^{-1}.$$

If we introduce this expression in (1.87), we finally get a "closed" equation for the **mv** of $F(u)$, that is to say:

$$u\mathcal{P}F - F(0) = H\mathcal{P}F + \mathcal{P}B\sum_{k=0}^{\infty}(G\mathcal{Q}B)^k \mathcal{P}F. \tag{1.89}$$

---

[32]The Laplace transform is defined by the integral $f(u) = \mathcal{L}[f(t)] \equiv \int_0^{\infty} f(t)e^{-ut}dt$. Note that with the argument u we denote the Laplace transform of the function $f(t)$. For analytic functions $f(u)$ the inverse transform exists and is given by the integral $f(t) = \frac{1}{2\pi i}\int_{\gamma-i\infty}^{\gamma+i\infty} e^{+ut}f(u)du$. Where γ is a real constant that exceeds the real part of all the singularities of $f(u)$ for $u \in \mathcal{C}$. In addition, note that $\mathcal{L}[df(t)/dt] \equiv uf(u) - f(t = 0)$.

[33]For simplicity in the notation we write $F \equiv F(u)$.

Exercise. Show that this equation can be rewritten in the form:

$$u \langle F \rangle - F(0) = H \langle F \rangle + \left\langle \sum_{k=0}^{\infty} \{BG(1 - \mathcal{P})\}^k B \right\rangle \langle F \rangle.$$

From this series expansion it is possible to observe the structure of Terwiel's cumulants [15]. This equation is the starting point for formulating a random perturbation theory for calculating the mean value of the function $F(t)$, whatever the probability distribution associated with the amount B is. In Chaps. 6 and 7 it will be seen that the study of transport in disordered media is related to this example, that is to say: a differential equation with random coefficients.

Exercise. From (1.89) show that if in the series $\sum_{k=0}^{\infty} (G\mathcal{Q}B)^k$ we do not take into account all the terms with $k \geq 1$, the solution in the "temporal" representation is given by

$$\langle F(t) \rangle = F(0) \, \exp \left[(H + \langle B \rangle) \, t \right] \quad \text{with} \quad t \geq 0.$$

On the other hand, if B has a Gaussian character, Novikov's theorem is very useful and allows alternative perturbations for calculating $\mathcal{P}F$. This theorem gives rise to one of the possible methods for studying the problem of thermal fluctuations in dynamical systems [2], analysis of thermal fluctuations will be dealt with in Chaps. 3 and 4. Note that in the case where B is simultaneously a function of time and of an **rv**, the problem gives rise to the study of stochastic disturbances [12], this topic will be presented in detail in Chap. 3. Finally, it should be noted that the expression (1.89) is also valid in the case where A and B are both operators.

1.14.2 Discrete Time Evolution

Consider the linear matrix dynamics (recurrence relation) written in the form:

$$X_{n+1} = \mathbf{M} \cdot X_n, \; \mathbf{M} \in \mathcal{R}_e^{m \times m}, \; n = 0, 1, 2, \cdots \tag{1.90}$$

where X_n is a state vector of dimension m at time step n. The recurrence relation (1.90) can be solved using a generating function technique. We define the generating function (vector) $G(z)$ associated with the state vector X_n by

$$G(z) = \sum_{n=0}^{\infty} z^n X_n, \tag{1.91}$$

this is a z-transform,[34] and then using (1.90)

$$\mathbf{M} \cdot G(z) = \sum_{n=0}^{\infty} z^n X_{n+1} = \frac{1}{z} \sum_{n=0}^{\infty} z^{n+1} X_{n+1}$$

$$= \frac{1}{z} (G(z) - X_0) .$$

From this equation, introducing the identity matrix $\mathbf{1}$, we can solve

$$G(z) = [\mathbf{1} - z\,\mathbf{M}]^{-1} \cdot X_0 \equiv \mathbf{G}(z) \cdot X_0,$$

where we have defined the associated Green function matrix:

$$\mathbf{G}(z) = [\mathbf{1} - z\,\mathbf{M}]^{-1} . \qquad (1.92)$$

Now consider the case when (1.90) has a random contribution, i.e., we split $\mathbf{M}$ in the form $\mathbf{M} \equiv \mathbf{H} + \mathbf{B}$, where $\mathbf{H}$ is nonrandom and $\mathbf{B}$ is random. We can write a random equation for the Green's function (1.92) as:

$$\frac{1}{z} (\mathbf{G}(z) - \mathbf{1}) = (\mathbf{H} + \mathbf{B}) \cdot \mathbf{G}(z), \qquad (1.93)$$

As before, the average of $\mathbf{G}(z)$ can formally be carried out introducing the projector operator $\mathcal{P}$ that averages over the random matrix $\mathbf{B}$, and its complement projector is $\mathcal{Q} \equiv (1 - \mathcal{P})$, i.e.:

$$\langle \mathbf{G}(z) \rangle = \mathcal{P}\mathbf{G}(z), \qquad \mathbf{G}(z) = \mathcal{P}\mathbf{G}(z) + \mathcal{Q}\mathbf{G}(z).$$

Using this projector technique a closed *exact* evolution equation can be obtained. Applying the operator $\mathcal{P}$ to (1.93) we obtain

$$\frac{1}{z} [\mathcal{P}\mathbf{G}(z) - \mathbf{1}] = \mathbf{H}\mathcal{P}\mathbf{G}(z) + \mathcal{P}\mathbf{B}\mathcal{P}\mathbf{G}(z) + \mathcal{P}\mathbf{B}\mathcal{Q}\mathbf{G}(z). \qquad (1.94)$$

Also, applying the operator $\mathcal{Q}$ to (1.93) we obtain

$$\frac{1}{z} \mathcal{Q}\mathbf{G}(z) = \mathbf{H}\mathcal{Q}\mathbf{G}(z) + \mathcal{Q}\mathbf{B}\mathcal{P}\mathbf{G}(z) + \mathcal{Q}\mathbf{B}\mathcal{Q}\mathbf{G}(z). \qquad (1.95)$$

---

[34]The z-transform is defined by the sum $f(z) = \sum_{n=0}^{\infty} z^n f_n$. Note that with the argument z we denote the z-transform of the function f_n. The inversion of the z-transform is given by the integral: $f(t) = \frac{1}{2\pi i} \oint z^{-n-1} f(z) dz$, and the integral is over the unit circle.

A formal solution of (1.95) can be obtained using the *nonrandom* Green matrix:

$$\mathbf{G}^0 \equiv \left[\frac{1}{z}\mathbf{1} - \mathbf{H}\right]^{-1}. \tag{1.96}$$

Applying $\mathbf{G}^0(z)$ to (1.95) and using the definition given in (1.96), we get

$$\mathcal{Q}\mathbf{G}(z) = \mathbf{G}^0\left[\mathcal{Q}\mathbf{B}\mathcal{P}\mathbf{G}(z) + \mathcal{Q}\mathbf{B}\mathcal{Q}\mathbf{G}(z)\right]. \tag{1.97}$$

This equation can be solved iteratively for $\mathcal{Q}\mathbf{G}(z)$,

$$\mathcal{Q}\mathbf{G}(z) = \sum_{k=1}^{\infty}\left[\mathbf{G}^0\mathcal{Q}\mathbf{B}\right]^k \mathcal{P}\mathbf{G}(z). \tag{1.98}$$

Substituting this formal solution in (1.94) we find a closed *exact* equation for the average of the Green's function $\mathcal{P}\mathbf{G}(z)$,

$$\mathcal{P}\mathbf{G}(z) - \mathbf{1} = z\left[\mathbf{H}\mathcal{P}\mathbf{G}(z) + \mathcal{P}\mathbf{B}\mathcal{P}\mathbf{G}(z) + \mathcal{P}\mathbf{B}\sum_{k=1}^{\infty}\left[\mathbf{G}^0\mathcal{Q}\mathbf{B}\right]^k \mathcal{P}\mathbf{G}(z)\right]. \tag{1.99}$$

This equation can be rewritten in a more friendly form

$$\langle\mathbf{G}(z)\rangle = \left[\mathbf{1} - z\left(\mathbf{H} + \left\langle\sum_{k=0}^{\infty}\left[\mathbf{B}\mathbf{G}^0\mathcal{Q}\right]^k \mathbf{B}\right\rangle\right)\right]^{-1}. \tag{1.100}$$

Here we can see the nontrivial Terwiel structure of the *mean* Green function (this is a consequence of its evolution in discrete time).

Note from (1.96) that $\mathbf{G}^0$ is a function of z; if we can make an expansion of $\langle G(z)\rangle$ in powers of z, noting that: $\langle G(z)\rangle \equiv \langle\mathbf{G}(z)\rangle \cdot X_0$, and from (1.91) we can obtain $\langle X_n\rangle$ as the coefficient of z^n.

Equation (1.100) is the starting point for studying random maps in a perturbative way, in particular this result has been used to predict the extinction in random population dynamics [19].

Excursus. Note the difference between the random model (1.90), with the case of the multiplication of random matrices:

$$\mathbf{P}^{(j)} = \overbrace{\mathbf{M} \cdot \mathbf{M} \cdot \mathbf{M} \cdots \mathbf{M}}^{\text{j-times}},$$

this last problem can be associated with (1.90) when the *disorder* is time dependent (dynamical disorder), i.e., when the random matrix $\mathbf{M} = \mathbf{M}(n)$ is a function of the discrete time n. In this case, the recurrence relation would be

$$X_{n+1} = \mathbf{M}(n) \cdot X_n, \ \mathbf{M}(n) \in \mathcal{R}_e^{m \times m}, \ n = 0, 1, 2, \cdots.$$

This model can also be worked out using the technique of Terwiel's cumulants.

1.15   Additional Exercises with Their Solutions

1.15.1   Circular Probability Density

Consider an **rv** X to be characterized by the pdf $P(X) = \left(2/\pi a^2\right)\sqrt{a^2 - X^2}$ for $X \in [-a, a]$. Show that this **rv** cannot be *approximated*, in any limit, by a Gaussian one. Hint: the simplest way to tackle this problem is to calculate its characteristic function

$$G(k) = \int_{-a}^{a} P(X) e^{ikX} dX = \frac{2}{ka} J_1(ka),$$

where $J_1(z)$ is the Bessel function of order 1. Then expanding the Bessel function, around $k = 0$, it is simple to calculate moments (by symmetry odd moments are zero):

$$\langle X^2 \rangle = a^2/4, \ \langle X^4 \rangle = a^4/8, \ \text{etc.}$$

Also the cumulants can be calculated expanding $\ln G(k)$, then we get

$$\langle\langle X^1 \rangle\rangle = \langle\langle X^3 \rangle\rangle = 0, \ \langle\langle X^2 \rangle\rangle = a^2/4, \ \langle\langle X^4 \rangle\rangle = -a^4/16, \ \text{etc.}$$

1.15.2   Trapezoidal Characteristic Function

Examine the question whether there exists an **rv** X whose characteristic function $G_X(k)$ fulfills: $G_X(k) \in \mathcal{R}_e$; $\forall k \in \mathcal{R}_e$; $G_X(0) = 1$, and has the *form* of a trapezoid with a base of length $2B$ and a top of length $2A$ with $B > A$. By using the Bochner theorem it is possible to see that the nonnegative definite condition is fulfilled only if $A = 0$ or $B \to \infty$. Therefore, the *form* of the characteristic function cannot be a trapezoid. Another alternative to see whether there exists an **rv** X associated with this characteristic function, is to calculate explicitly the inverse Fourier transform of

$G_X(k)$, and see if this function is positive and normalized. The associated pdf $P_X(x)$ should be given by the integral

$$P_X(x) = \frac{1}{2\pi} \int_{-\infty}^{\infty} e^{-ikx} G_X(k)\, dk, \tag{1.101}$$

where

$$G_X(k) = \begin{cases} \frac{B+k}{B-A}, & \text{if } -B \le k \le -A \\[2ex] 1, & \text{if } -A \le k \le A \\[2ex] \frac{B-k}{B-A}, & \text{if } A \le k \le B \\[2ex] 0, & \text{elsewhere} \end{cases} \tag{1.102}$$

Introducing (1.102) in the integral (1.101) we get

$$P_X(x) = \frac{1}{\pi} \frac{\cos(Bx) - \cos(Ax)}{x^2}. \tag{1.103}$$

From this result it follows that the pdf is normalized (as expected, $G_X(0) = 1$). Nevertheless, the condition that it be positive for all x cannot be satisfied unless A or B is zero. But because we require $B > A$ the only solution is $A = 0$ (triangle *form* for $G_X(k)$), or $B \to \infty$ (*constant* function for $G_X(k)$). These last conclusions can easily be drawn using trigonometric relations and noting that $P_X(x)$ can also be written in the form:

$$P_X(x) = \frac{2}{\pi} \frac{\sin\left(\frac{(B-A)}{2}x\right)\sin\left(\frac{(B+A)}{2}x\right)}{(B-A)x^2}.$$

Therefore, when $G_X(k)$ is a triangle $\Rightarrow P_X(x) = \frac{2}{\pi}\frac{\sin\left(\frac{B}{2}x\right)^2}{Bx^2}$, and when $B \to \infty$ we get $P_X(x) = \delta(x)$.

1.15.3 Using Novikov's Theorem

Let X be a Gaussian **rv** with mean value zero and variance 1. Show that for $a \gg 1$ the **mv** of $1/(a + X^2)$ can be approximated by

$$\left\langle \frac{1}{a + X^2} \right\rangle \approx \frac{1}{a+1} + \mathcal{O}(1/a^3).$$

From the identity

$$1 = \left\langle \frac{(a + X^2)}{(a + X^2)} \right\rangle = a \left\langle \frac{1}{(a + X^2)} \right\rangle + \left\langle \frac{X^2}{(a + X^2)} \right\rangle, \tag{1.104}$$

and using Novikov's theorem in the second term we have

$$\left\langle \frac{X^2}{(a + X^2)} \right\rangle = \left\langle X \frac{X}{(a + X^2)} \right\rangle = \langle X^2 \rangle \left\langle \partial_X \frac{X}{(a + X^2)} \right\rangle \tag{1.105}$$

$$= \langle X^2 \rangle \left\{ \left\langle \frac{1}{(a + X^2)} \right\rangle - 2 \left\langle \frac{X^2}{(a + X^2)^2} \right\rangle \right\},$$

thus, we get

$$1 = a \left\langle \frac{1}{(a + X^2)} \right\rangle + \langle X^2 \rangle \left\{ \left\langle \frac{1}{(a + X^2)} \right\rangle - 2 \left\langle \frac{X^2}{(a + X^2)^2} \right\rangle \right\}. \tag{1.106}$$

Now because X is normal we get $\langle X^2 \rangle = 1$ so from (1.106) we have

$$1 = (a + 1) \left\langle \frac{1}{(a + X^2)} \right\rangle - 2 \left\langle \left(\frac{X}{(a + X^2)} \right)^2 \right\rangle. \tag{1.107}$$

From which we can solve the required quantity as

$$\left\langle \frac{1}{(a + X^2)} \right\rangle = \frac{1}{(a + 1)} + \frac{2}{(a + 1)} \left\langle \left(\frac{X}{(a + X^2)} \right)^2 \right\rangle, \tag{1.108}$$

thus, taking the limit of $a \gg 1$ we can write

$$\lim_{a \gg 1} \left\langle \frac{1}{(a + X^2)} \right\rangle \rightarrow \frac{1}{(a + 1)} + \frac{2}{(a + 1)} \left\langle \frac{1}{a^2} \left(\frac{X}{(1 + (\frac{x}{a})^2)} \right)^2 \right\rangle \tag{1.109}$$

$$= \frac{1}{(a + 1)} + \frac{2}{(a + 1)} \left\langle \frac{1}{a^2} \left(X \left[1 - \left(\frac{X}{a} \right)^2 + \left(\frac{X}{a} \right)^4 - \cdots \right] \right)^2 \right\rangle$$

$$\simeq \frac{1}{(a + 1)} + \frac{2}{a^3} + \cdots,$$

which is what we wanted to prove.

In the opposite limit $a \to 0$, from (1.108) we are led to an absurd result because $\left\langle \dfrac{1}{X^2} \right\rangle$ is not defined

$$\left\{ \left\langle \frac{1}{X^2} \right\rangle = 1 + 2 \left\langle \frac{1}{X^2} \right\rangle \right\} \;\Rightarrow\; -1 = \left\langle \frac{1}{X^2} \right\rangle. \tag{1.110}$$

1.15.4 Gaussian Operational Approximation

Let ϕ be an **rv** characterized by a symmetric pdf $P(\phi)$ over a given $\mathcal{D}_\phi$. Consider the function $g(\phi) = b\phi^4$, and justify the use of the Gaussian operational approximation:

$$g(\phi) \simeq 3b\phi^2 \left\langle \phi^2 \right\rangle. \tag{1.111}$$

In general from the Taylor series around $\langle \phi \rangle = M_1$, see Sect. 1.8, we get

$$\langle g(\phi) \rangle = bM_1^4 + 6bM_1^2 \left\langle (\phi - M_1)^2 \right\rangle + 4bM_1 \left\langle (\phi - M_1)^3 \right\rangle + b \left\langle (\phi - M_1)^4 \right\rangle. \tag{1.112}$$

This is an exact expansion and gives the starting point to study $\langle g(\phi) \rangle$ as a function of the fluctuations around the **mv** M_1. In the symmetrical case odd moments vanish, then we get the identity $\langle g(\phi) \rangle = bM_4$.

From the definition of the fourth cumulant we have

$$\left\langle\!\left\langle \phi^4 \right\rangle\!\right\rangle = M_4 - 3M_2^2 - 4M_1M_3 + 12M_1^2M_2 - 6M_1^4 = M_4 - 3M_2^2, \tag{1.113}$$

therefore, using (1.113) to eliminate M_4 in the identity $\langle g(\phi) \rangle = bM_4$ we get

$$\langle g(\phi) \rangle = b \left\{ \left\langle\!\left\langle \phi^4 \right\rangle\!\right\rangle + 3M_2^2 \right\}.$$

Assuming a Gaussian pdf, the fourth cumulant is zero then we can approximate $\langle g(\phi) \rangle \simeq 3bM_2^2$. Finally, after some algebra, we get the operational approximation (1.111).

1.15.5 Cumulant Diagrams

In Sect. 1.10.1 we have presented a general procedure for studying correlation functions. Show that the four-points correlation function $\langle\!\langle X_1X_2X_3X_4 \rangle\!\rangle$ can easily be calculated using the diagrams from Fig. 1.4. Also show that in the limit $X_1 \to X_2 \to X_3 \to X_4 = X$ the expression reduces to the formula $\langle\!\langle X^4 \rangle\!\rangle = K_4$ given in Sect. 1.5. Note that even in the case when the pdf $P(X)$ is symmetric the quantity

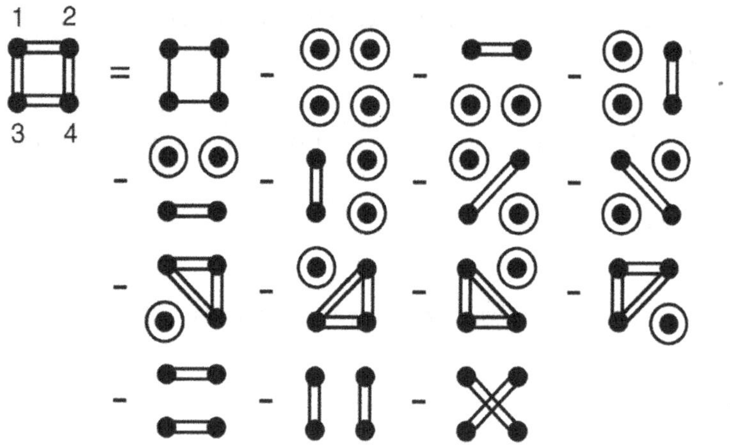

Fig. 1.4 Diagrammatic representation of the cumulant $\langle\langle X_1 X_2 X_3 X_4\rangle\rangle$

K_4 can be positive or negative, see for example the circular pdf in exercise 1.15.1. Information about the shape of a pdf $P(X)$ sometimes is referred by the "kurtosis" $C_4 \equiv K_4/(K_2)^2$, find a lower bound for C_4.[35]

1.15.6 Addition of rv with Different Supports

Let X_1 and X_2 be two **rvs** characterized by the joint pdf $P_{X_1,X_2}(x_1, x_2)$, assume that the two **rvs** lie over different domains: $x_1 \in \mathcal{D}_{X_1}$ and $x_2 \in \mathcal{D}_{X_2}$. The pdf of the sum of these **rv** is

$$P(s) = \int_{\mathcal{D}} P_{X_1,X_2}(y_1, s - y_1) dy_1. \tag{1.114}$$

Here the integrand has to be taken with care. To have an idea of the method we adopt

$$x_1 \in \mathcal{D}_{X_1} = (\frac{1}{2}L, L)$$

$$x_2 \in \mathcal{D}_{X_2} = (0, L),$$

using these domains in (1.114) we get

$$P(s) = \int_{\frac{1}{2}L}^{L} P_{X_1,X_2}(y_1, s - y_1) \Theta(s - y_1) \Theta(L - (s - y_1)) dy_1.$$

---

[35]More inequalities of this types are given by A.A. Dubkov and Malakhov, Radiophys Quantum Electron. (USA) **19**, 833 (1977).

Now from the fact that $s \in (\frac{1}{2}L, 2L)$ we can take the integral over different domains:
if $s \in (\frac{1}{2}L, L)$

$$P(s) = \int_{\frac{1}{2}L}^{L} P_{X_1,X_2}(y_1, s - y_1) \, \Theta \, (s - y_1) \, dy_1 = \int_{\frac{1}{2}L}^{s} P_{X_1,X_2}(y_1, s - y_1) dy_1,$$

if $s \in (L, \frac{3}{2}L)$

$$P(s) = \int_{\frac{1}{2}L}^{L} P_{X_1,X_2}(y_1, s - y_1) dy_1,$$

if $s \in (\frac{3}{2}L, 2L)$

$$P(s) = \int_{\frac{1}{2}L}^{L} P_{X_1,X_2}(y_1, s - y_1) \, \Theta \, (L - (s - y_1)) \, dy_1 = \int_{s-L}^{L} P_{X_1,X_2}(y_1, s - y_1) dy_1.$$

Collecting all these results we get the pdf we were looking for. In the particular case when the **rvs** are **si** and their corresponding distributions are uniform the result is

$$P(s) = \begin{cases} \frac{2}{L^2} \left(s - \frac{L}{2} \right), & \text{if} \quad s \in (\frac{1}{2}L, L) \\[2mm] \frac{1}{L}, & \text{if} \quad s \in (L, \frac{3}{2}L) \\[2mm] \frac{2}{L^2} (2L - s), & \text{if} \quad s \in (\frac{3}{2}L, 2L) \end{cases}$$

1.15.7   Phase Diffusion (Periodic Oscillations)

Consider a particle whose velocity follows the law: $V(t) = A\omega \cos(\omega t + \phi)$. If we now assume that the phase ϕ is random this law defines a stochastic process (see Chap. 3). If the phase is uniformly distributed in 2π the (1-time) pdf can easily be calculated from the formula

$$P_1(V) = \langle \delta \, (V - V(t)) \rangle = \int_0^{2\pi} \delta \, (V - A\omega \cos(\omega t + \phi)) \, \frac{d\phi}{2\pi}$$

$$= \frac{1}{\pi} \frac{1}{\sqrt{(A\omega)^2 - V^2}}, \quad V \in [-A\omega, A\omega],$$

here we have used the trigonometric relation $\cos^{-1}[x] = \sin^{-1}\left[\sqrt{1-x^2}\right]$, $x \in [0, 1]$, to work out the Jacobian. What would be the conclusion if the pdf $P(\phi)$ were not uniform?

1.15.8   Random Matrices

Consider two real random matrices A, B of dimension $n \times n$. Suppose we want to calculate the average $\langle e^A B \rangle$ for the case when the matrices are not **si**. In order to perform this calculation we first expand the exponential inside the **mv**, then

$$
\langle e^A B \rangle = \left\langle \left(1 + A + \frac{1}{2!}A^2 + \frac{1}{3!}A^3 + \frac{1}{4!}A^4 + \cdots\right) B \right\rangle
$$

$$
= \langle B \rangle + \langle AB \rangle + \frac{1}{2!}\langle A^2 B \rangle + \frac{1}{3!}\langle A^3 B \rangle + \frac{1}{4!}\langle A^4 B \rangle + \cdots .
$$

Now if we add *zero*, in a suitable way to this last expression, we get

$$
\langle e^A B \rangle = \langle B \rangle + \langle AB \rangle + \frac{1}{2!}\langle A^2 B \rangle + \frac{1}{3!}\langle A^3 B \rangle + \frac{1}{4!}\langle A^4 B \rangle + \cdots
$$

$$
+ \left[\langle A \rangle \langle B \rangle - \langle A \rangle \langle B \rangle\right] + \frac{1}{2!}\left[\langle A^2 \rangle \langle B \rangle - \langle A^2 \rangle \langle B \rangle\right]
$$

$$
+ \frac{1}{3!}\left[\langle A^3 \rangle \langle B \rangle - \langle A^3 \rangle \langle B \rangle\right] + \frac{1}{4!}\left[\langle A^4 \rangle \langle B \rangle - \langle A^4 \rangle \langle B \rangle\right] + \cdots .
$$

This expression can be reorganized in the form

$$
\langle e^A B \rangle = \left[1 + \langle A \rangle + \frac{1}{2!}\langle A^2 \rangle + \frac{1}{3!}\langle A^3 \rangle + \frac{1}{4!}\langle A^4 \rangle + \cdots\right] \langle B \rangle
$$

$$
+ \left[\langle AB \rangle - \langle A \rangle \langle B \rangle\right] + \frac{1}{2!}\left[\langle A^2 B \rangle - \langle A^2 \rangle \langle B \rangle\right]
$$

$$
+ \frac{1}{3!}\left[\langle A^3 B \rangle - \langle A^3 \rangle \langle B \rangle\right] + \frac{1}{4!}\left[\langle A^4 B \rangle - \langle A^4 \rangle \langle B \rangle\right] + \cdots .
$$

Therefore, we end with the identity

$$
\langle e^A B \rangle = \langle e^A \rangle \langle B \rangle + \langle\langle AB \rangle\rangle + \frac{1}{2!}\langle\langle A^2 B \rangle\rangle + \frac{1}{3!}\langle\langle A^3 B \rangle\rangle + \frac{1}{4!}\langle\langle A^4 B \rangle\rangle + \cdots .
$$

As an application of this result consider the calculation of $\langle e^{\alpha X + \beta Y} Y \rangle$ where X, Y are correlated **rvs** and α, β are real numbers.

1.15.9 Properties of the Characteristic Function

Let X be an **rv** with pdf $P_X(x)$ defined on the domain $\mathcal{D}_X$. Show that: (a) the characteristic function $G_X(k) = \langle e^{ikX} \rangle$ always exists as an ordinary function and implies knowledge of the pdf. (b) $G_X(k)$ is continuous on the real axis k, even when $P_X(x)$ could have discontinuities. (c) $G_X(k) = G_X(-k)^*$. (d) $G_X(k)$ is nonnegative definite.

(a) Follows because $G_X(k)$ is the Fourier transform of the absolutely integrable function $P_X(x)$. The strict equivalence between the knowledge of all moments and the pdf is obtained under additional analyticity conditions of the characteristic function in the neighborhood of the origin. However, it is worth noting that the characteristic function exists when the moments do not. It is also obvious that

$$|G_X(k)| \le \int \left| e^{ikx} P_X(x) \right| dx = \int P_X(x)\, dx = G_X(0) = 1.$$

(b) Follows noting that

$$|G_X(k+\delta) - G_X(k)| \le \int \left| e^{i(k+\delta)x} - e^{ikx} \right| P_X(x)dx$$

$$= 2 \int \left| \sin \frac{x\delta}{2} \right| P_X(x)dx,$$

then $\lim_{\delta \to 0} |G_X(k+\delta) - G_X(k)| \to 0$.
(c) Follows because $P_X(x) \in \mathcal{R}_e$.
(d) To prove this fact we need to show that for any set of N real numbers $\{k_1, k_2, \cdots, k_N\}$ and complex numbers $\{\lambda_1, \lambda_2, \cdots, \lambda_N\}$ it follows that

$$\sum_{i,j}^{N} \lambda_i \lambda_j^* G(k_i - k_j) \ge 0. \tag{1.115}$$

The condition (1.115) follows using $\left\langle \left| \sum_{j=1}^{N} \lambda_j e^{ixk_j} \right|^2 \right\rangle \ge 0$, and noting that

$$\left\langle \left| \sum_{j=1}^{N} \lambda_j e^{ixk_j} \right|^2 \right\rangle = \sum_{i,j}^{N} \lambda_i \lambda_j^* \int_{\mathcal{D}_X} e^{i(k_j - k_i)x} P_X(x)\, dx$$

$$= \sum_{i,j}^{N} \lambda_i \lambda_j^* G(k_i - k_j).$$

1.15.10 Infinite Degrees of Freedom

If the set of random variables $\{X_i\} = X_1, X_2, \cdots, X_n$, with its n-th joint pdf $P(\{X_i\}) \equiv P_n(X_1; X_2; \cdots; X_n)$ on the domain $\mathcal{D}_X$, were thought of as a set with a continuous index t, the joint pdf $P(\{X_t\})$ will turn out to be a joint pdf of infinite order for all t, or a probability functional, in which $\{X_t\} \equiv \{X(t)\}$ stands for the infinite set of all $X's$ for all times t. However the explicit form of $P(\{X_t\})$ is rarely known. Nevertheless, the concept of probability functional is sometimes useful (see Chap. 3). For example the **mv** of any function of $\{X_t\}$ can be written in the form

$$\langle f(\{X_t\}) \rangle = \int f(\{X_t\}) P(\{X_t\}) d\{X_t\},$$

where $d\{X_t\}$ is a short notation for $\prod_{\text{all } t} dX(t)$. Therefore, the n-th dimensional characteristic function can be generalized to a characteristic functional as

$$G_X(k_1, k_2, \cdots, k_n) = \left\langle \exp\left(i \sum_{j=1}^n k_j X_j\right)\right\rangle$$

$$\rightarrow \int \exp\left(i \int k(t) X(t) dt\right) P(\{X_t\}) d\{X_t\} = G_X([k(\bullet)]).$$

How can the moments $\langle X(t_1)^{m_1} \cdots X(t_j)^{m_j}\rangle$ be calculated from the characteristic functional $G_X([k(\bullet)])$? Hint: see Sect. 2.3.1 and Chap. 3.

Remarks (Infinite Dimensional Gaussian Case) For the Gaussian case consider Novikov's theorem when $f[X(\bullet)] = X(t')$ with $t' \in [0, t]$

$$\langle X(t) f[X(\bullet)]\rangle = \int_{\mathcal{D}} \langle X(t)X(s)\rangle \left\langle \frac{\delta f[X(\bullet)]}{\delta X(s)}\right\rangle ds$$

$$= \int_0^t \langle X(t)X(s)\rangle \delta(t' - s) ds$$

$$= \langle X(t)X(t')\rangle,$$

here taking the appropriate limit we can make sure that $t' \subset \mathcal{D}$. For an arbitrary functional $f[X(s)]$ with $s \in \mathcal{D} = [0, t]$ and for the singular case (white-noise): $\langle X(t)X(s)\rangle = \delta(t - s)$, we get

$$\langle X(t) f[X(\bullet)]\rangle = \int_0^t \delta(t - s) \left\langle \frac{\delta f[X(\bullet)]}{\delta X(s)}\right\rangle ds,$$

but note that here we have to set a convention for the value of $\int_0^t \delta(t - s) ds$.

References

1. L.D. Landau, E.M. Lifshitz, *Física Estadística* (Ed. Reverté, S.A. Barcelona, 1969)
2. N.G. van Kampen, *Stochastic Process in Physics and Chemistry*, 2nd edn. (Amsterdam, North-Holland, 1992)
3. R. Kubo, M. Toda, N. Hashitsume, *Statistical Physics II: Nonequilibrium Statistical Mechanics* (Springer, Berlin, 1985)
4. L.E. Reichl, *A Modern Course in Statistical Physics*, 2nd edn. (Edward Arnold Publishers Ltd., Austin, 1992)
5. A. Papoulis, *Probability Random Variables and Stochastic Process*, 3rd edn. (McGraw-Hill, New York, 1991)
6. W. Feller, *An Introduction to Probability Theory and its Applications*, 2nd edn. (Wiley, New York, 1971)
7. R.L. Stratonovich, *Topics in the Theory of Random Noise*, vols. 1, 2 (Gordon and Breach Science Publishers, New York, 1963)
8. S. Bochner, *Lectures on Fourier Integrals* (Princeton, NJ, Princeton University Press, 1959); R.R. Goldberg, *Fourier Transform* (Cambridge University Press, Cambridge, 1961)
9. E.W. Montroll, B.J. West, *Fluctuation Phenomena*, ed. by E.W. Montroll, J.L. Lebowitz (North-Holland, Amsterdam, 1987)
10. M.F. Shlesinger, B.D. Hughes, Physica **A 109**, 597 (1981)
11. S. Chandrasekhar, Rev. Mod. Phys. **15**, 1 (1943)
12. N.G. van Kampen, Phys. Rep. **24 C**, 171 (1976)
13. M.O. Cáceres, C.E. Budde, G.J. Sibona, J. Phys. A Math. Gen. **28**, 3877 (1995); idem **28**, 7391 (1995); M.O. Cáceres, M.A. Fuentes, C.E. Budde **30**, 2287 (1997); M.O. Cáceres, M.A. Fuentes, **32**, 3209 (1999)
14. M.O. Cáceres, A.A. Budini, J. Phys. A Math. Gen. **30**, 8427 (1997)
15. R.H. Terwiel, Physica **A 74**, 248 (1974)
16. E. Hernández-García, M.A. Rodríguez, L. Pesquera, M. San Miguel, Phys. Rev. **B 42**, 10653 (1990)
17. R.R. Pool, M.O. Cáceres, Phys. Rev. E **82**, 0352031 (2010)
18. M.O. Cáceres, D. Strier, E.R. Reyes, Rev. Mex. de Física **45**, 608 (1999)
19. M.O. Cáceres, I. Caceres-Saez, J. Math. Biol. **63**, 519 (2011); M.O. Cáceres, I. Caceres-Saez, Ecol. Mod. **251**, 312 (2013)

Chapter 2
Fluctuations Close to Thermodynamic Equilibrium

2.1 Spatial Correlations (Einstein's Distribution)

As an application of the ideas discussed in the previous chapter, we present here the study of the fluctuations around thermodynamic equilibrium. Inspired by Boltzmann's entropy concept (as the logarithm of the statistical weight of a macroscopic state; see Appendix A.1), Einstein in 1910 introduced [1] an alternative method for analyzing thermodynamic fluctuations [2, 3]. This method is based on the following assumptions:

- Fluctuations of thermodynamic variables are characterized by an exponential probability density in the entropy variation

$$P(\mathbf{X}) \propto \exp\left(\Delta S(\mathbf{X})/k_B\right), \tag{2.1}$$

  where $\Delta S(\mathbf{X})$ is the entropy variation of the isolated system,[1] formally considered as a function of variations $\mathbf{X} \equiv \{X_i\}$ of extensive variables $\{x_i\}$.
- For an isolated system, entropy has its maximum value at equilibrium (i.e., we have $\mathbf{X} = 0$). Then, for small deviations from thermodynamic equilibrium, we can expand $\Delta S(\mathbf{X})$ to second order, $\mathcal{O}(\mathbf{X}^2)$, in their small fluctuations.

In general, here $\mathbf{X}$ denotes a vector of arbitrary dimension, in consonance with the number of thermodynamic variables needed to characterize the system.

In a groundbreaking work, Einstein developed a straightforward method for calculating the probability distribution of the deviations of state variables close to equilibrium[2]

---

[1]The definition of an isolated system is presented in Appendix A.

[2]Kubo [4] demonstrates—in a very elegant way—that the probability distribution $P(\mathbf{X})$ can be obtained from the principle of equal probability a priori. In fact, if the variations of some of the thermodynamic variables are not small, the expression $P(\mathbf{X}) \propto \exp \frac{1}{k_B}$ $\{S(E, N, V, x^* + X) - S(E, N, V, x^*)\}$ remains valid regardless of the magnitude of the deviation

© Springer International Publishing AG 2017
M.O. Cáceres, *Non-equilibrium Statistical Physics with Application to Disordered Systems*, DOI 10.1007/978-3-319-51553-3_2

2.1.1 Gaussian Approximation

Consider an isolated system with total entropy S_T, total energy U_T, and number of particles N_T in a box of volume V_T. If we divide the volume V_T into m elementary cells Δr_i, in *thermal equilibrium* the volume, entropy, internal energy, and the number of particles in each cell i will be $V_i^o, S_i^o, U_i^o, N_i^o$, respectively. Note that we are here considering that each unit cell Δr_i is a subsystem[3] which is in contact with the rest of the volume, and that "rest" behaves like a "heat bath". On the other hand, at *equilibrium*, intensive variables: pressure, temperature, and chemical potential (P^o, T^o, μ^o) must be the same in all cells. Since there is a finite number of particles, the thermodynamic variables fluctuate spontaneously around their equilibrium values in each unit cell Δr_i. These fluctuations should be such that the quantities $V_T = \sum_i V_i$, $U_T = \sum_i U_i$ and $N_T = \sum_i N_i$ remain constant. On the other hand, since the system is isolated and in a state of thermodynamic equilibrium, these fluctuations must make S_T decrease, since otherwise the system would not be stable. Assuming that fluctuations are small we can expand the local entropy of each cell Δr_i in Taylor series around its value at thermodynamic equilibrium. Then the spontaneous change $\Delta S_T = \sum_{i=1}^{m}(S_i - S_i^o)$ will be, to second order in the variations $\mathbf{X} = \{\Delta V_i, \Delta U_i, \Delta N_i\}$:

$$\Delta S_T = \frac{1}{2} \sum_{i=1}^{m} \Delta(\partial S_i/\partial U_i)_{V,N} \, \Delta U_i + \Delta(\partial S_i/\partial V_i)_{U,N} \, \Delta V_i + \Delta(\partial S_i/\partial N_i)_{V,U} \, \Delta N_i$$

$$+ \, \mathcal{O}(\Delta^2), \tag{2.2}$$

here $\mathcal{O}(\Delta^2)$ means larger than the second order.

Note that the first variations are canceled by the hypothesis of local equilibrium in each unit cell. The total variation ΔS_T depends on the configuration of deviations from equilibrium, that is, on the random variables $\{\Delta U_i, \Delta V_i, \Delta N_i\}$. These **rv**, in turn, obey the constraint that V_T, U_T, N_T are constants. Here $\Delta(\cdots)$ represents a second derivative that expresses a variation evaluated at equilibrium. As an example, we write here the first term of (2.2)

$$\Delta(\partial S_i/\partial U_i)_{V,N} = \left[\, (\partial^2 S_i/\partial U_i^2) \right]_{V,N}^o \, \Delta U_i + [\partial/\partial V_i \, (\partial S_i/\partial U_i)_{V,N}]_{U,N}^o \, \Delta V_i$$

$$+ [\partial/\partial N_i \, (\partial S_i/\partial U_i)_{V,N}]_{U,V}^o \, \Delta N_i.$$

$\mathbf{X}$; then it is observed that the Gaussian distribution is confined only to small fluctuations. Here x^* represents the *most probable* value, which coincides with the maximum of $S(E, N, V, x)$ for $x = x^*$.

[3]Each elementary cell Δr_i is assumed to be small but large enough to ensure local equilibrium. That is, each elementary cell is an open subsystem.

Using intensive parameters in the entropy representation (local definition of intensive variables), i.e.:

$$(\partial S_i/\partial U_i)_{V,N} = \left(\frac{1}{T}\right)_i, \quad (\partial S_i/\partial V_i)_{U,N} = \left(\frac{P}{T}\right)_i, \quad (\partial S_i/\partial N_i)_{V,U} = \left(\frac{-\mu}{T}\right)_i,$$

we can write the expression (2.2) in the form

$$\Delta S_T = \frac{1}{2}\sum_{i=1}^{m}[\Delta(1/T)_i\,\Delta U_i + \Delta(P/T)_i\,\Delta V_i - \Delta(\mu/T)_i\,\Delta N_i] + \mathcal{O}(\Delta^2). \quad (2.3)$$

Given that each elementary cell is in local equilibrium (using the first law of thermodynamics, see Appendix A.2) we conclude that

$$\Delta U_i = T_o\,\Delta S_i - P_o\,\Delta V_i + \mu_o\,\Delta N_i,$$

and after substituting

$$\Delta(1/T)_i = -\left(\frac{1}{T^2}\right)^o\,\Delta T_i$$

$$\Delta(P/T)_i = -\left(\frac{P}{T^2}\right)^o\,\Delta T_i + \left(\frac{1}{T}\right)^o\,\Delta P_i$$

$$\Delta(\mu/T)_i = -\left(\frac{\mu}{T^2}\right)^o\,\Delta T_i + \left(\frac{1}{T}\right)^o\,\Delta\mu_i,$$

in (2.3), we get the expression for the variation of entropy of the closed and adiabatically isolated system

$$\Delta S_T \simeq \frac{1}{2T_o}\sum_{i=1}^{m}[-\Delta T_i\,\Delta S_i + \Delta P_i\,\Delta V_i - \Delta\mu_i\,\Delta N_i]. \quad (2.4)$$

This expression gives, in general, the (total) variation of the entropy due to local fluctuations in each unit cell Δr_i. We can study ΔS_T in terms of any combination of independent variables that we choose for each thermodynamic problem of interest. Note that in this expression we are assuming that in the system there is only one "type" of particle.

Guided Exercise. Here we present in broad terms, how to calculate the entropy variation ΔS_T using thermodynamic variables such as pressure, temperature, and number of particles. From the general formula (2.4) we note that we need certain differential expressions (variations μ_i, S_i, V_i) as functions of chosen thermodynamic variables: T, P, N. Namely,

$$\Delta \mu_i = \left(\frac{\partial \mu_i}{\partial T_i} \right)^o_{P,N} \Delta T_i + \left(\frac{\partial \mu_i}{\partial P_i} \right)^o_{T,N} \Delta P_i + \left(\frac{\partial \mu_i}{\partial N_i} \right)^o_{P,T} \Delta N_i \qquad (2.5)$$

$$\Delta S_i = \left(\frac{\partial S_i}{\partial T_i} \right)^o_{P,N} \Delta T_i + \left(\frac{\partial S_i}{\partial P_i} \right)^o_{T,N} \Delta P_i + \left(\frac{\partial S_i}{\partial N_i} \right)^o_{P,T} \Delta N_i$$

$$\Delta V_i = \left(\frac{\partial V_i}{\partial T_i} \right)^o_{P,N} \Delta T_i + \left(\frac{\partial V_i}{\partial P_i} \right)^o_{T,N} \Delta P_i + \left(\frac{\partial V_i}{\partial N_i} \right)^o_{P,T} \Delta N_i.$$

Note that we have assumed independence between cells. Substituting these expressions into (2.4) and using Maxwell's relations[4]:

$$\left(\frac{\partial S}{\partial P} \right)_{T,N} = - \left(\frac{\partial V}{\partial T} \right)_{P,N}$$

$$\left(\frac{\partial S}{\partial N} \right)_{P,T} = - \left(\frac{\partial \mu}{\partial T} \right)_{P,N}$$

$$\left(\frac{\partial V}{\partial N} \right)_{P,T} = \left(\frac{\partial \mu}{\partial P} \right)_{T,N}.$$

We get:

$$\Delta S_T = \frac{1}{2T_o} \sum_{i=1}^m \left[- \frac{C_p}{T_o} (\Delta T_i)^2 + 2 \left(\frac{\partial V_i}{\partial T_i} \right)^o_{P,N} \Delta P_i \, \Delta T_i \right.$$

$$\left. + \left(\frac{\partial V_i}{\partial P_i} \right)^o_{T,N} (\Delta P_i)^2 - \left(\frac{\partial \mu_i}{\partial N_i} \right)^o_{P,T} (\Delta N_i)^2 \right], \qquad (2.6)$$

where we have used the definition of heat capacity at constant pressure, $C_p = T \left(\frac{\partial S}{\partial T} \right)_{P,N}$. This expression shows that pressure and temperature fluctuations are correlated. Prove that the correlation is given by

$$\langle \Delta P_i \, \Delta T_i \rangle = \frac{k_B}{\text{Det}} \frac{1}{T_o} \left(\frac{\partial V}{\partial T} \right)^o_{P,N}$$

where

$$\text{Det} \equiv \frac{-1}{T_o^2} \left[\left(\left(\frac{\partial V}{\partial T} \right)^o_{P,N} \right)^2 + \frac{C_p}{T_o} \left(\frac{\partial V}{\partial P} \right)^o_{T,N} \right],$$

---

[4] A detailed description of these relationships can be found in the book of Reichl [3], pág. 38.

and quadratic fluctuations are given by

$$\langle \Delta P_i^2 \rangle = \frac{k_B}{\text{Det}} \frac{C_p}{T_o^2}, \quad \langle \Delta T_i^2 \rangle = \frac{-k_B}{\text{Det}} \frac{1}{T_o} \left(\frac{\partial V}{\partial P} \right)_{T,N}^o.$$

From the expression for the total variation ΔS_T, (2.6), we note that the variations of different cells are uncorrelated [for example, spatial correlations $\langle \Delta P_i \, \Delta T_j \rangle$ are zero]. This is because there is no coupling between different cells [as seen in (2.5)]. Note that indices have been occasionally omitted for simplicity.

Optional Exercise. Using relations

$$C_p - C_V = -T \frac{\left(\left(\frac{\partial V}{\partial T} \right)_{P,N} \right)^2}{\left(\frac{\partial V}{\partial P} \right)_{T,N}}, \quad \left(\frac{\partial V}{\partial P} \right)_{S,N} = \frac{C_V}{C_p} \left(\frac{\partial V}{\partial P} \right)_{T,N},$$

and considering that heat capacities meet that $C_p > C_V > 0$, prove from the previous exercise that $\text{Det} = \frac{-C_V}{T_o^3} \left(\frac{\partial V}{\partial P} \right)_{T,N}^o > 0$, and show that fluctuations can be rewritten in the form:

$$\langle \Delta P_i^2 \rangle = -k_B T_o \left(\frac{\partial P}{\partial V} \right)_{S,N}^o, \quad \langle \Delta T_i^2 \rangle = \frac{k_B T_o^2}{C_V}.$$

Furthermore, considering the Jacobian of the transformation,[5] it is easy to verify that

$$\frac{\left(\frac{\partial V}{\partial T} \right)_{P,N}}{\left(\frac{\partial V}{\partial P} \right)_{T,N}} = -\left(\frac{\partial P}{\partial T} \right)_{V,N}.$$

Using this result and the expression for the correlation $\langle \Delta P_i \, \Delta T_i \rangle$, show that

$$\langle \Delta P_i \, \Delta T_i \rangle = \frac{k_B T_o^2}{C_V} \left(\frac{\partial P}{\partial T} \right)_{V,N}^o.$$

---

[5] For example, if we write the partial derivative in the form $\left(\frac{\partial V}{\partial T} \right)_P = \frac{\partial(V,P)}{\partial(T,P)}$, where $\frac{\partial(u,v)}{\partial(x,y)}$

$$= \begin{vmatrix} \frac{\partial u}{\partial x} & \frac{\partial u}{\partial y} \\ \frac{\partial v}{\partial x} & \frac{\partial v}{\partial y} \end{vmatrix},$$ it is simple to verify that $\frac{\partial(u,v)}{\partial(x,y)} = -\frac{\partial(v,u)}{\partial(x,y)}$, etc.

In general, in this order of approximation, the entropy variation is characterized by a *quadratic form*, i.e.:

$$\Delta S_T / k_B = -\frac{1}{2} (\mathbf{X} \cdot \mathbf{g} \cdot \mathbf{X}) \equiv -\frac{1}{2} \sum_{i,j=1}^{n} g_{ij} X_i X_j, \qquad (2.7)$$

then the probability $P(\{X_l\}) \propto e^{\Delta S_T / k_B}$ is a Gaussian distribution in n-dimensions (see Chap. 1). Moments and all correlations of the vector $\mathbf{X}$ can be calculated from the characteristic function $G(\mathbf{K}) \equiv G(\{K_l\})$:

$$G(\mathbf{K}) = \int \cdots \int \prod_{j=1}^{n} dX_j \sqrt{\frac{|\mathbf{g}|}{(2\pi)^n}} \exp\left(-\frac{1}{2}(\mathbf{X} \cdot \mathbf{g} \cdot \mathbf{X}) + i\mathbf{K} \cdot \mathbf{X}\right) \qquad (2.8)$$

$$= \exp\left(\frac{-1}{2} \mathbf{K} \cdot \mathbf{g}^{-1} \cdot \mathbf{K}\right),$$

leading to

$$\langle X_i X_j \rangle = \frac{1}{i^2} \left(\partial^2 / \partial K_i \partial K_j\right) G(\mathbf{K}) \Big|_{\{K_l\}=0} = (\mathbf{g}^{-1})_{ij}, \langle X_j^2 \rangle = (\mathbf{g}^{-1})_{jj} > 0, \text{ etc.}$$

Note that $\mathbf{g}$ is a *positive definite* matrix.

Exercise (**Symmetries**). Following the summation convention for repeated indices, assume that Einstein's probability distribution (2.1) is characterized by (2.7). Then that distribution can be written in the abbreviated form $P(\{z_l\}) = \mathcal{N} \exp(-\frac{1}{2} g_{ij} z_i z_j)$; that is, $-\partial^2 S / \partial z_i \partial z_j = k_B \, g_{ij}$. Now consider "thermodynamic forces" defined by the relationship $Z_i = -\partial S / \partial z_i = k_B \, g_{ik} z_k$. Show that z_i, Z_j are **sirv** iff $i \neq j$. Hint: use that $\langle z_i Z_j \rangle = k_B \, \delta_{i,j}$; here the symbol $\delta_{i,j}$ represents the Kronecker delta [see guided exercise of Sect. 1.4]. This result is crucial in the proof of Onsager's Theorem on symmetries of dissipative coefficients [2, 5].

Excursus. For a *power law* type probability distribution, it is also possible to calculate in closed form a generalization of the mean values $\langle z_i Z_j \rangle$; however in this case it is not possible to ensure statistical independence between z_i and Z_j [6].

Let us now try to simplify the above problem. Suppose fluctuations are confined within elementary cell Δr_i, but $\Delta N_i = 0, \forall i$; that is, in this system the number of particles is constant in each unit cell. If we take $\Delta T_i, \Delta V_i$ as independent variables from (2.4), using

$$\Delta P_i = \left(\frac{\partial P_i}{\partial T_i}\right)^o_{V,N} \Delta T_i + \left(\frac{\partial P_i}{\partial V_i}\right)^o_{T,N} \Delta V_i$$

$$\Delta S_i = \left(\frac{\partial S_i}{\partial T_i}\right)^o_{V,N} \Delta T_i + \left(\frac{\partial S_i}{\partial V_i}\right)^o_{T,N} \Delta V_i,$$

and from Maxwell's relations,[6] we get:

$$\Delta S_T = -\frac{1}{2T_o} \sum_{i=1}^{m} \left[\frac{C_V}{T_o} (\Delta T_i)^2 + \frac{1}{\chi_T V} (\Delta V_i)^2 \right]. \tag{2.9}$$

Here $\chi_T = -1/\left(V\left(\frac{\partial P}{\partial V}\right)_T\right)$ is the isothermal compressibility and $C_V = T\left(\frac{\partial S}{\partial T}\right)_{V,N}$ the heat capacity at constant volume. The appearance of a fluctuation $\mathbf{X} \equiv \{\Delta T_i, \Delta V_i\}$ (in a macroscopic equilibrium state) can be studied from Einstein's postulate; i.e., it is characterized by the probability $P(\mathbf{X}) = \mathcal{N} \exp\left(\Delta S_T(\mathbf{X})/k_B\right)$, where the constant $\mathcal{N}$ is determined by normalization: $\int \cdots \int P(\mathbf{X}) \, d\mathbf{X} = 1$. In the present case $\mathbf{X}$ is a vector of $2m$-dimensions, because there are m elementary cells. Since $P(\mathbf{X})$ is a multidimensional Gaussian distribution, characterized by a *positive definite* quadratic form, from (2.9) and the fact that the absolute temperature is positive, we conclude that the constants $\{C_V, \chi_T\}$ must be positive too. Also from (2.9) we note that fluctuations $\{\Delta T_i, \Delta V_i\}$ are independent of each other. The positivity of C_V and χ_T is an example of the famous principle of Le Chatelier: *"If a system is in thermodynamic equilibrium, then any spontaneous change in its parameters must produce a process that tends to restore equilibrium to the system."*

In our above example, with $\Delta N_i = 0$ and m elementary cells Δr_i, and introducing shorthand notation in (2.9),

$$\{X_1, \cdots, X_m, X_{m+1}, \cdots, X_{2m}\} \equiv \{\Delta T_1, \cdots, \Delta T_m, \Delta V_1, \cdots, \Delta V_m\},$$

the matrix $\mathbf{g}$ is given by

$$\mathbf{g} = \begin{cases} C_V/\left(k_B T_o^2\right) & \text{if } i = j & \text{with } 1 \le i \le m \\ 1/(\chi_T k_B T_o V) & \text{if } i = j & \text{with } m+1 \le i \le 2m \\ 0 & \text{elsewhere } (i,j). \end{cases} \tag{2.10}$$

From which we obtain:

$$\langle \Delta V_i^2 \rangle = \chi_T k_B T_o V \quad \text{(density fluctuations)} \tag{2.11}$$

$$\langle \Delta T_i^2 \rangle = k_B T_o^2 / C_V \quad \text{(temperature fluctuations)}$$

$$\langle \Delta V_i \rangle = \langle \Delta T_i \rangle = 0 \quad \text{(the most probable thermodynamic state)}$$

$$\langle \Delta V_i \, \Delta T_i \rangle = 0 \quad (\Delta V_i, \Delta T_i \text{ sirv}).$$

---

[6]In this case the useful relationship is $\left(\frac{\partial S}{\partial V}\right)_{T,N} = \left(\frac{\partial P}{\partial T}\right)_{V,N}$.

As in previous cases, see for example (2.5), in this model there is no coupling between different cells i, j (the matrix $\mathbf{g}$ is diagonal). This fact leads effectively to $\langle \Delta V_i \Delta V_j \rangle = 0$, etc. We note further that fluctuations in temperature and volume (density) are **si**. If we use $\{P, S\}$ as thermodynamic variables to describe the system, we can prove that pressure and entropy fluctuations are **si**; however, other combinations of thermodynamic variables may not be a set of **sirv**. The variance of fluctuations in the volume is proportional to the isothermal compressibility, and since χ_T increases near a phase transition, it can be expected that near the critical point density fluctuations will be very large. Then, near the critical point the Gaussian approximation is no longer useful (the quadratic form vanishes at the critical point). A description of fluctuations near the critical point should be studied in the context of a "scaling" theory[7] [3], or higher order approximations [2]; contexts that are beyond the scope of this introduction.

2.2 Minimum Work*

This section describes the fluctuations of a subsystem I that is small compared to the rest of the system (which we will call subsystem II). Both subsystems, I and II, make up the total system considered. Then, it is possible to prove that $\Delta S_T / k_B = -W_{\min} / k_B T_o$, where $W_{\min}$ is the minimum reversible work needed to take the subsystem I from an equilibrium state to another under the action of the rest of the system (subsystem II), and T_o is the equilibrium temperature (in addition, since system I is much smaller than system II, T_o is also the temperature for subsystem II).

First we note that for Einstein's distribution,

$$P(\mathbf{X}) \propto \exp(\Delta S_T(\mathbf{X})/k_B),$$

the variation ΔS_T need not be small, but it is only when fluctuations are small that it is possible to approximate ΔS_T to second order in the deviations. From this we conclude that $P(\mathbf{X})$ will be Gaussian. In this section we restrict ourselves to the case of small fluctuations. However, in many circumstances it is more convenient to estimate $W_{\min}$ than to calculate ΔS_T, that is why it is of interest to establish their equivalence.

It is possible to associate a thermodynamic concept to $W_{\min}$. Let S_T be the total entropy of the system I + II. Since the total system is closed,[8] its volume remains

---

[7]The first to introduce the idea of a "scaling" theory was B. Widom [J. Chem. Phys. 43, 3898 (1965)], who stressed that the thermodynamic functions are *homogeneous functions* of the distance (or separation) from the critical point. The study of the free energy in terms of a regular part plus a singular part (the homogeneous function) is similar to the analysis of the characteristic function, that we presented in the excursus of Sect. 1.4.2, when we studied self-similar structures in a random walk.

[8]Closed systems are defined in Appendix A.

Fig. 2.1 Representation of
the total entropy S_T as a
function of the total energy
E_T

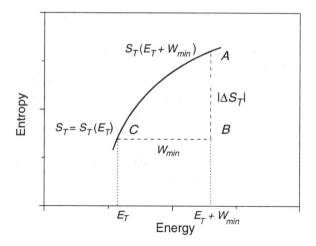

constant and therefore, the entropy $S_T = S_T(E_T)$ is a function only of the total
energy of system I + II. Suppose a given body (subsystem I) is not in equilibrium
with the environment (there has been a change in the subsystem I), but the whole
system has the same energy; then the value of the total entropy $S_T(E_T)$ differs by the
amount $\Delta S_T < 0$ (negative, because the system is closed and adiabatically isolated,
see Fig. 2.1).

For example, if C is a state of equilibrium (between I and II) with Entropy S_T
and energy E_T, and B is a nonequilibrium state (between I and II) with energy[9]
$E_T + W_{min}$, then lines AB (vertical) and CB (horizontal) of the Fig. 2.1 represent,
respectively, the variation of the total entropy and reversible work delivered by
the "work source" to the body to bring it to the state B. Note, of course, that the
equilibrium state C does not correspond to the equilibrium state A. Then, since the
body (small subsystem I) introduces only small changes in the energy and entropy,
we can approximate $| \Delta S_T | / \Delta E \sim dS/dE$, and thus

$$dS_T(E_T)/dE_T = -\Delta S_T/W_{min},$$

but, since $S_T(E_T)$ represents the equilibrium curve , $dS(E_T)/dE_T = 1/T_o$. Then it
follows that

$$\Delta S_T/k_B = -W_{min}/k_B T_o, \tag{2.12}$$

---

[9]To consider the minimum work W_{min} we have imagined that within the environment, in addition
to the body, there is a "work source" thermally insulated from the body in question and the
environment.

which is what we wanted to prove.[10] The expression (2.12) gives the total change in the entropy of a closed system: *body + environment*, according to a certain minimum work delivered to the temperature of the medium. Let us see how that minimal work is calculated.

If we consider the total system (I + II): a body (which may or may not be mechanical) plus its immediate environment, is possible to study the probability distribution of fluctuations of the macroscopic *mechanical* variables of the body. During a fluctuation the body could exchange heat and work with the environment. To calculate the work W_{min} imagine that within the environment, in addition to the body, there is a thermally insulated "reversible work source". Then (from the first law of thermodynamics) the total variation of energy ΔU of the body is due to three effects: (1) work done on the body by the *work source*, W; (2) heat gained from the environment, $-T_o \, \Delta S_o$; and (3) work done by the environment on the body,[11] $P_o \, \Delta V_o$; that is to say:

$$\Delta U = W + P_o \, \Delta V_o - T_o \, \Delta S_o. \tag{2.13}$$

Since the volume of the total system is constant it means that $\Delta V_o = -\Delta V$, and since for an irreversible process the total entropy will increase: $\Delta S_o + \Delta S \geq 0$. Note that the entropy of the external source will not change because the "work source" is isothermal. Then $\Delta S_o \geq -\Delta S$, from which, and using (2.13), it follows that

$$\Delta U \leq W - P_o \, \Delta V + T_o \, \Delta S,$$

the equal sign being valid for reversible processes. Then, the minimum work to be done on the body to vary its energy by ΔU is bounded by the amount [2]

$$W \geq \Delta U + P_o \, \Delta V - T_o \, \Delta S. \tag{2.14}$$

That is, since P_o and T_o are constant,

$$W_{min} = \Delta \left(U + P_o V - T_o S \right) > 0. \tag{2.15}$$

This is equivalent to saying that the maximum work done by the body on the "work source" is equal to

$$W_{max} = -\Delta \left(U + P_o V - T_o S \right).$$

---

[10]This analysis is based on Landau's proof [2]. There is an alternative one by Kubo [4] in terms of the minimum reversible work $W_{min} = \int_C^B dU - T \, dS$, where U and S are, respectively, the energy and entropy of the total system (I + II) and T is the intermediate temperature between the states C and B.

[11]The subscript notation o refers to the "environment."

Then, substituting (2.12) and (2.15) in (2.1) leads to the following expression for Einstein's probability distribution:

$$P(\mathbf{X}) \propto \exp\left(\frac{-1}{k_B T_o}[\Delta U + P_o\,\Delta V - T_o\,\Delta S]\right), \tag{2.16}$$

where $\mathbf{X} \equiv \{\Delta S, \Delta V, \Delta U\}$ are variations of entropy, volume, and energy of the body (or subsystem I), P_o and T_o are equilibrium mean values of pressure and temperature of the body.

2.2.1 The Thermodynamic Potential Φ

Alternatively, from the expression

$$P(\mathbf{X}) \propto \exp\left(\Delta S_T(\mathbf{X})/k_B\right) \propto \exp(-\Delta\Phi/k_B T_o), \tag{2.17}$$

where $\Delta\Phi$ is the variation of the thermodynamic potential

$$\Delta\Phi = \{\Delta U + P_o\,\Delta V - T_o\,\Delta S\}, \tag{2.18}$$

we can see that P_o is the reservoir pressure and V the volume of the subsystem considered (the body). Since the two subsystems are in contact the choice of P and T for the reservoir (subsystem II) will specify the state of the body (subsystem I), and the variations $\Delta V, \Delta U, \Delta S$ will be the same for both.

2.2.2 Fluctuations in Terms of $\Delta P, \Delta V, \Delta T, \Delta S$

From expression (2.17), we can also express this probability in terms of variations $\Delta P, \Delta V, \Delta T, \Delta S$. For this purpose consider ΔU depending on variations ΔV and ΔS:

$$\begin{aligned}
\Delta U = {}& \left(\frac{\partial U}{\partial S}\right)^o_V \Delta S + \left(\frac{\partial U}{\partial V}\right)^o_S \Delta V \\
& + \frac{1}{2}\left\{\left(\frac{\partial^2 U}{\partial S^2}\right)^o_V \Delta S^2 + 2\left(\frac{\partial U}{\partial S\partial V}\right)^o \Delta S\,\Delta V + \left(\frac{\partial^2 U}{\partial V^2}\right)^o_S \Delta V^2\right\} \\
& + \mathcal{O}(\Delta^2),
\end{aligned} \tag{2.19}$$

where $\mathcal{O}(\Delta^2)$ indicates orders higher than the second. Substituting

$$\left(\frac{\partial U}{\partial S}\right)_V = T \quad \text{and} \quad \left(\frac{\partial U}{\partial V}\right)_S = -P$$

in (2.19), we get reordering terms

$$\Delta U \simeq T_o \, \Delta S - P_o \, \Delta V + \tag{2.20}$$

$$+ \frac{1}{2}\left[\left(\frac{\partial^2 U}{\partial S^2}\right)_V^o \Delta S + \left(\frac{\partial U}{\partial S \partial V}\right)^o \Delta V\right] \Delta S$$

$$+ \frac{1}{2}\left[\left(\frac{\partial U}{\partial S \partial V}\right)^o \Delta S + \left(\frac{\partial^2 U}{\partial V^2}\right)_S^o \Delta V\right] \Delta V.$$

On the other hand, from the expressions for T and P it follows that

$$\Delta T = \Delta\left(\frac{\partial U}{\partial S}\right) = \left(\frac{\partial^2 U}{\partial S^2}\right)^o \Delta S + \left(\frac{\partial U}{\partial S \partial V}\right)^o \Delta V \tag{2.21}$$

$$-\Delta P = \Delta\left(\frac{\partial U}{\partial V}\right) = \left(\frac{\partial U}{\partial S \partial V}\right)^o \Delta S + \left(\frac{\partial^2 U}{\partial V^2}\right)^o \Delta V.$$

Then, substituting (2.21) in (2.20) and the latter in (2.18) the desired expression for the probability is finally obtained as:

$$P(\mathbf{X}) \propto \exp(-\Delta\Phi/k_B T_o) \propto \exp\left(\frac{-1}{2k_B T_o}[\Delta T \, \Delta S - \Delta P \, \Delta V]\right), \tag{2.22}$$

where T_o is the equilibrium temperature of the environment. As mentioned earlier (2.4), from this expression it is easy to find the fluctuations for any given pair of thermodynamic variables.

Exercise (**Variables** V; T). Using (2.22) show that volume and temperature are **sirv** [compare with (2.11)]. Show also that

$$\langle \Delta V^2 \rangle = k_B T_o V \chi_T \quad \text{and} \quad \langle \Delta T^2 \rangle = k_B T_o^2 / C_V,$$

where χ_T is the isothermal compressibility and C_V the heat capacity at constant volume.

Exercise (**Variables** P; S). Use (2.22) to show that pressure and entropy are **sirv**, and that

$$\langle \Delta P^2 \rangle = -k_B T_o \left(\frac{\partial P}{\partial V}\right)_S^o \quad \text{and} \quad \langle \Delta S^2 \rangle = k_B C_p,$$

where C_P is the heat capacity at constant pressure.

Excursus (**Fluctuations in the Number** N). We did not consider possible fluctuations in N for the small body when deriving expression (2.16). We shall do so now, using the expression for minimal work:

$$W_{\min} = \Delta U + P_o \Delta V - T_o \Delta S - \mu_o \Delta N > 0, \qquad (2.23)$$

where μ_o is the chemical potential of the environment and N the number of molecules in the small body (*body + environment system* contains a constant number of molecules). Show the generalized expression (2.22) for the probability distribution is

$$P(\mathbf{X}) \propto \exp\left(\frac{-1}{2k_B T_o} [\Delta T \, \Delta S - \Delta P \, \Delta V + \Delta \mu \, \Delta N] \right); \qquad (2.24)$$

compare with (2.4). Show that fluctuations in the number N are given by

$$\langle \Delta N^2 \rangle = k_B T_o / \left(\frac{\partial \mu}{\partial N} \right)_{P,T}^o.$$

This result is consistent with that obtained calculating $\langle \Delta N^2 \rangle$, using the grand canonical distribution $(T - \mu)$ of equilibrium statistical mechanics (see Appendix F):

$$\mathcal{P}(N, l) = \frac{\exp\left(-\beta \left[E_{N,l} - \mu N \right] \right)}{\sum_{N,l} \exp\left(-\beta \left[E_{N,l} - \mu N \right] \right)} \quad \text{with} \quad \beta = 1 / k_B T.$$

2.3 Fluctuations in Mechanical Variables

Consider an arbitrary *mechanical object* plus its environment as a closed system.[12] A pendulum at room temperature and atmospheric pressure, a free particle at rest relative to the environment, a tight rope, etc. are examples of such systems. To account for fluctuations in the mechanical variables of the body in question (the angle of deviation from the vertical, the momentum of a particle, the transverse displacement in a rope, etc.) we need to calculate $W_{\min}$ bearing in mind the mechanical variations of the object (we assume that the environment is large enough

---

[12]The properties of a body such as the values of its thermodynamic quantities, or probability distributions for the coordinates and velocities of the particles, etc. do not depend on the fact that we consider the body as isolated and placed in an imaginary thermostat (see Chap. 7 in Landau and Lifshitz [2]).

so that its temperature and pressure, T_o, P_o, are constants). The minimum work we have to deliver to the body for a reversible change to take place will depend on the object and the process in question.

Example. Let a free particle of mass M be at rest relative to the environment. Consider the probability that a thermal fluctuation imparts a momentum Mv to the particle. In this case W_{min} will simply be kinetic energy, $W_{min} = \frac{1}{2}Mv^2$. So

$$P(v) \propto \exp\left(-\frac{W_{min}}{k_B T_o}\right) = \exp\left(\frac{-Mv^2}{2k_B T_o}\right),$$

which is the Maxwell–Boltzmann distribution for the velocity of a free particle of mass M in a heat bath at temperature T_o.

Example. Now calculate the mean square value of the vertical deflection of a physical pendulum with small oscillations. In this case

$$W_{min} = \int_0^\theta Mgl \sin\theta' \, d\theta' = -Mgl\,(\cos\theta - 1),$$

is the mechanical work done against the force of gravity. Then, for small oscillations we can approximate $W_{min} \simeq \frac{1}{2}Mgl\theta^2$, from which we get

$$P(\theta) \propto \exp\left(\frac{-Mgl\theta^2}{2k_B T_o}\right).$$

After normalizing this probability distribution we find immediately that the mean square fluctuations of the angle are characterized by

$$\langle \theta^2 \rangle = \frac{k_B T_o}{Mgl}.$$

In general, the intensity of fluctuations of a mechanical variable is always proportional to the temperature of the thermal bath (and inversely proportional to the mass of the object). This does not happen for the mean square fluctuations of variables having purely thermodynamic character, such as entropy or temperature.

2.3.1 Fluctuations in a Tight Rope

As discussed earlier, the occurrence of a fluctuation is proportional to $e^{\Delta S_T/k_B}$, i.e., it is characterized by the variation in the total entropy of the *isolated* system. If the system consists of a *mechanical body + the environment*, the variation ΔS_T can be calculated in terms of the minimum work required to carry out, reversibly, the variation of the thermodynamic quantities of the body. Consider a strained string of length L inside an environment at temperature T_o. Suppose we model the homogeneous string as a continuous sequence of "springs and masses" under tension at rest. To describe and analyze possible deformations, consider a "dummy" deformation such that only a single point x of the rope is displaced upward by y from its resting position. Then W_{min} is simply the mechanical work done against some tension Υ (which is a function of displacement). The minimum work required to achieve this configuration (two straight lines), and considering $y(x) \approx 0$ (small amplitudes) can be approximated by $\Upsilon(\Delta l_1 + \Delta l_2)$, where Δl_1 and Δl_2 are the "stretching" of the rope due to the single deformation $y(x) \neq 0$ at point x. Then we can determine the amplitude of the mean square fluctuation of *this transverse* displacement in the rope, induced by the temperature of its environment.

Let y be the vertical displacement of the point x on the rope at rest. For small variations ($y \equiv y(x) \approx 0$), W_{min} is proportional to the elongation. That is, the minimum work is given by

$$W_{min} = \Upsilon(\Delta l_1 + \Delta l_2),$$

where $\Delta l_1 = \sqrt{x^2 + y^2} - x$ and $\Delta l_2 = \sqrt{(L-x)^2 + y^2} - (L-x)$. Then, since $y \ll x$, W_{min} can be approximated by

$$W_{min} \simeq \frac{\Upsilon}{2} \left(\frac{y^2}{x} + \frac{y^2}{(L-x)} \right),$$

from which it follows that

$$P(y) \propto \exp\left(\frac{-W_{min}}{k_B T_o} \right) = \exp\left(\frac{-y^2}{2k_B T_o} \frac{\Upsilon L}{x(L-x)} \right).$$

Therefore the mean square deviation of the vertical displacement is a function of the horizontal location of the point x along the rope, and is given by

$$\langle y^2 \rangle = k_B T_o \frac{x(L-x)}{\Upsilon L}. \tag{2.25}$$

Similarly, in order to find the correlation between two arbitrary transverse displacements $\{y_1 \equiv y(x_1), y_2 \equiv y(x_2)\}$, separated by a distance $|x_1 - x_2|$, we need the joint probability distribution of the two **rv** $\{y_1, y_2\}$. This distribution can be calculated considering the minimum work needed to produce a configuration of two small vertical displacements y_1 and y_2 at x_1, x_2, respectively, see Fig. 2.2. In this case[13] we get

$$W_{min} = \Upsilon(\Delta l_1 + \Delta l_2 + \Delta l_3).$$

---

[13]This small deformation is just bringing the tight rope into a configuration with three straight lines.

Fig. 2.2 Representation of
the instantaneous state of
deformation $y(x)$ of a tight
rope. The three *straight lines*
represent the approximation
for this state of deformation

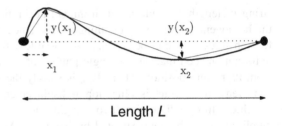

Assuming $x_2 > x_1$ and that $y_1 \ll x_1$ we get

$$\Delta l_1 = \sqrt{x_1^2 + y_1^2} - x_1 \simeq \frac{y_1^2}{2\,x_1},$$

also considering $|y_1 - y_2| \ll |x_1 - x_2|$, we find that

$$\Delta l_2 = \sqrt{(x_1 - x_2)^2 + (y_1 - y_2)^2} - (x_2 - x_1) \simeq \frac{(y_1 - y_2)^2}{2\,(x_2 - x_1)}.$$

Finally, since $y_2 \ll |L - x_2|$ we get

$$\Delta l_3 = \sqrt{(L - x_2)^2 + y_2^2} - (L - x_2) \simeq \frac{y_2^2}{2\,(L - x_2)},$$

whence

$$W_{min} = \frac{\Upsilon}{2} \left[\frac{\left(y_1^2 x_2 (L - x_2) + y_2^2 (L - x_1) x_1 - 2 y_1 y_2 (L - x_2) x_1 \right)}{x_1 (x_2 - x_1)(L - x_2)} \right].$$

In this expression it is easy to recognize the associated quadratic form, i.e.:

$$\frac{W_{min}}{k_B T_o} = \frac{1}{2} \vec{y} \cdot \mathbf{g} \cdot \vec{y},$$

where $\vec{y} \equiv (y_1, y_2)$, then $\mathbf{g}^{-1}$ is the correlations matrix given by

$$\mathbf{g}^{-1} = \frac{x_1 (x_2 - x_1)(L - x_2) T_o}{\Upsilon L} \begin{pmatrix} \frac{L - x_1}{(L - x_2)(x_2 - x_1)} & \frac{1}{(x_2 - x_1)} \\ \frac{1}{(x_2 - x_1)} & \frac{x_2}{x_1 (x_2 - x_1)} \end{pmatrix}.$$

Leading immediately to the second moments and the correlation function

$$\langle y_1^2 \rangle = \frac{x_1 (L - x_1) k_B T_o}{\Upsilon L}$$

$$\langle y_2^2 \rangle = \frac{x_2 (L - x_2) k_B T_o}{\Upsilon L}$$

$$\langle y_1 y_2 \rangle = \frac{x_1 (L - x_2) k_B T_o}{\Upsilon L}. \tag{2.26}$$

Note that the correlation function $\langle y_1 y_2 \rangle$ is nonstationary,[14] and it depends explicitly on the parameter L. In the next chapter we will see the consequence of this fact in the context of the theory of stochastic processes.

Guided Exercise (**Strained Rope**). Let $y(x)$ be a continuous function well behaved in the range $[0, L]$ representing the instantaneous state space of "deformation" of a string under tension with its ends fixed, then $y(0) = y(L) = 0$; see Fig. 2.2. We can expand $y(x)$ in Fourier series as

$$y(x) = \sqrt{\frac{2}{L}} \sum_{n=1}^{\infty} A_n \sin(n\pi \frac{x}{L}). \tag{2.27}$$

From the expression for the energy deformation of a tight string,[15] we obtain:

$$E_d = \frac{\Upsilon}{2} \int_0^L \left(\frac{dy(x)}{dx} \right)^2 dx$$

$$= \frac{\Upsilon}{2} \sum_{n=1}^{\infty} A_n^2 \left(\frac{n\pi}{L} \right)^2.$$

But in our model the deformations are haphazard, and therefore the corresponding numbers A_n will have random values. Therefore, taking

$$P(A; A_2; \cdots ; A_n \cdots) = \mathcal{N} \exp \left(\frac{-E_d}{k_B T_o} \right)$$

as our probability distribution, where $\mathcal{N}$ is the normalization constant, it is readily seen that the correlation between two arbitrary **rvs** A_n and A_m is temperature dependent, and is given by

$$\langle A_m A_n \rangle = \delta_{mn} \frac{k_B T_o}{\Upsilon} \Big/ \left(\frac{n\pi}{L} \right)^2. \tag{2.28}$$

---

[14]For a stationary correlation it will only depend on the difference $|x_1 - x_2|$, see Chap. 3.
[15]See advanced exercise 2.5.2.

From (2.28) and (2.27) we see that A_m are statistically independent Gaussian random variables, that the mean quadratic fluctuation $\langle y(x)^2 \rangle$ is given by (2.25), and in general the correlation $\langle y(x_1)y(x_2) \rangle$ by (2.26). In the next chapter we will see that $y(x)$ is a "spatial" non-Markov stochastic process.

2.4 Temporal Correlations

Consider a system or a small part of it in thermal equilibrium. Over its time evolution the macroscopic fluctuations X_j vary at different times. Therefore one may ask whether there are any time-dependent correlations in the fluctuations, that is we want to study moments of the form $\langle X_i(0) X_j(\tau) \rangle$. A temporal correlation function between different components X_j is defined following the procedure of the previous chapter. Here briefly we show (in the case where $X(\tau)$ is a scalar) that we can define a temporal correlation in the form:

$$\langle\langle X(0) X(\tau) \rangle\rangle = \int \int X'X \, P(X', \tau; X, 0) \, dX' \, dX - \langle X(0) \rangle \langle X(\tau) \rangle . \qquad (2.29)$$

Here $P(X', \tau; X, 0)$ is a joint pdf for two **rv**s, which can also be written in a more illuminating way in the form $P(X(\tau); X(0))$ meaning that the **rv** $X(\tau)$ takes the value X' and the **rv** $X(0)$ takes the value X. Their generalization is analogous in the vectorial case. Note that, inadvertently, we have introduced the concept of stochastic process, which will be developed in detail in the next chapter. A stochastic process $X(\tau)$ can be considered as a new set of **rv**s which is characterized with a continuous index τ. Then it should be noted that the joint probability distribution (for two different times) that appears in (2.29) can be written, in the equilibrium state, as follows:

$$P_{eq}(X, \tau; X, 0) = P(X, \tau \mid X, 0) \, P_{eq}(X), \qquad (2.30)$$

which is, again, Bayes' rule. Then, near thermodynamic equilibrium, $P_{eq}(X)$ is Einstein's probability distribution (2.1). From (2.30) we see that now the problem focuses on the study of the object $P(X, \tau \mid X, 0)$, which is one of the central points in the analysis of stochastic processes. In particular Markov processes play a fundamental role through their simplicity, in the description of the evolution of statistical objects. The calculation of time correlation functions is not restricted solely to situations around thermodynamic equilibrium, the definition (2.29) is valid even in situations far from equilibrium, where Einstein's stationary distribution may no longer be applied. It is worth noting that correlations around equilibrium satisfy certain symmetries. In particular, it will be seen (in Chap. 4) that

$$\langle\langle X_i(0) X_j(\tau) \rangle\rangle_{eq} = \langle\langle X_j(0) X_i(\tau) \rangle\rangle_{eq} . \qquad (2.31)$$

This fact is related to the famous Onsager's Theorem [5] on the symmetries of the transport coefficients. All these points will be studied in detail in the following chapters.

Excursus (**Space Phase**). The symmetry (2.31) is proved by identifying the joint probability $P_{eq}\left(X_i(\tau); X_j(0)\right)$ with a marginal probability distribution of the microscopic Liouville's flow of probability (in phase space), and using the invariance under time reversal of the Hamiltonian system [7]. Onsager's analysis is the starting point for establishing the relationship between correlations in thermodynamic equilibrium and dissipation in the presence of small perturbations (Fluctuation-Dissipation theorem [8]).

2.5 Additional Exercises with Their Solutions

2.5.1 Time-Dependent Correlation in a Stochastic Toy Model

Here we present a simple situation to calculate a time-correlation function using the distribution of equilibrium fluctuations. Many physical systems undergo harmonic motion for small deviations around a steady state. Then, for the purpose of studying a time-dependent correlation we take the harmonic oscillator model. Consider the 1-dimensional Newton's equation of motion for a particle of mass m and Hooke's constant Υ

$$\frac{d^2x}{dt^2} = -\omega^2 x, \quad \omega^2 = \Upsilon/m. \tag{2.32}$$

The general solution of (2.32) is

$$x(t) = x_0 \cos(\omega t) + \frac{v_0}{\omega} \sin(\omega t), \tag{2.33}$$

here x_0 and v_0 are initial conditions (position and velocity respectively) of the particle. Assuming an equilibrium probability distribution $\propto e^{-W_{min}/k_B T}$ for the initial energy (variation) we can write

$$P(x_0; v_0) \propto \exp\left\{-\left[\frac{m}{2}v_0^2 + \frac{\Upsilon}{2}x_0^2\right]/k_B T\right\}. \tag{2.34}$$

This distribution says that x_0 and v_0 are **si** Gaussian random variables, then it follows immediately that $\langle x_0\rangle = \langle v_0\rangle = 0$, $\langle x_0^2\rangle = k_B T/\Upsilon$ and $\langle v_0^2\rangle = k_B T/m$. Using (2.34) and (2.33) we can calculate any time-dependent correlation or cumulant. Show that $\langle x(t)\rangle = \langle v(t)\rangle = 0$, $\langle x(t_1) v(t_2)\rangle = 0$ and

$$\langle x(t_1) x(t_2)\rangle = \frac{k_B T}{\Upsilon} \cos(t_1 - t_2).$$

Fig. 2.3 Representation of a
small static deformation of
the string, then
$\theta \simeq \delta y / \delta x \simeq dy/dx$

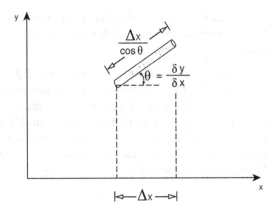

This toy model gives a nondecreasing time-periodic correlation function for the instantaneous position (small variation $x(t)$). The fact that $\langle x(t_1) x(t_2) \rangle$ does not vanish for $|t_1 - t_2| \to \infty$ is a consequence of the oversimplified dynamics (2.32). In the next chapter we will discuss, in detail, how this can be improved.

2.5.2 Energy of a String Deformation

Here we calculate the total energy for a small *static* deformation in a rope of length L.

$$E_d = \int_0^L \frac{\Upsilon}{2} \left(\frac{dy}{dx} \right)^2 dx. \tag{2.35}$$

To derive this formula, consider in Fig. 2.3 an infinitesimal element of length Δx of a string which is located at point x. The potential ΔE energy of this element of the string is the negative of the work ΔW carried out by the tension Υ in stretching it from its unstretched length Δx to its stretched length $\Delta x / \cos \theta$. Hence its potential energy is

$$\Delta E = -\Delta W = -\Upsilon \left[\Delta x - \frac{\Delta x}{\cos \theta} \right] = \Upsilon \Delta x \left[\frac{1 - \cos \theta}{\cos \theta} \right] \simeq \Upsilon \Delta x \left(\frac{1}{2} \theta^2 \right),$$

here we have used a small angle approximation for $\cos \theta$. Substituting $\theta = dy/dx$ in this expression and adding all the contributions for each element Δx of the string, and taking the limit as the length of these elements tends to zero we get (2.35), which is the expression for minimum work, that we have used in Sect. 2.3.1.

Exercise. Using similar arguments consider the kinetic energy, and calculate its contributions to the energy of deformation (2.35). If at time t the element Δx has a vertical displacement y, its velocity at that instant is $\partial y / \partial t$. Then the kinetic energy of this element is

$$\Delta K = \frac{1}{2}\mu\Delta x \left(\frac{\partial y}{\partial t}\right)^2,$$

here μ is the mass per unit length of the string. Then the total energy associated with small amplitude waves on a string of length L is

$$E_w = \frac{1}{2}\int_0^L \left[\mu\left(\frac{\partial y}{\partial t}\right)^2 + \Upsilon\left(\frac{dy}{dx}\right)^2\right] dx. \qquad (2.36)$$

Show that the kinetic energy of each element of the rope is equal to the potential energy of that element. Hint: since the *space-time* variation of the displacement y is given in terms of the variables $(x \pm ct)$ it follows that

$$\frac{\partial y}{\partial x} = \pm\frac{1}{c}\frac{\partial y}{\partial t},$$

then we get

$$\left(\frac{\partial y}{\partial x}\right)^2 = \left(\frac{1}{c}\frac{\partial y}{\partial t}\right)^2 = \frac{\mu}{\Upsilon}\left(\frac{\partial y}{\partial t}\right)^2,$$

where we have used that the velocity of the wave is given by $c^2 = \Upsilon/\mu$. Comparing each term in (2.36) it follows that at each point x the kinetic energy associated with the wave is numerically equal to its potential energy.

Excursus. Using the instantaneous energy of deformation (2.36) we can define an equilibrium probability distribution $\propto e^{-E_w/k_BT}$ for each amplitude A_q of an expansion in wave modes $e^{i(x\pm ct)q}$. Then, it is possible to represent a (free) random field $u(x,t)$ and calculate any thermal space-time correlation function for a wave model [7].

References

1. A. Einstein, Theorie der Opaleszenz von homogenen Flüssigkeiten und Flüssigkeitsgemischen in der Nähe des kritischen Zustandes. Ann Phys. **33**, 1275 (1910)
2. L.D. Landau, E.M. Lifshitz, *Física Estadística* (Reverté S.A., Barcelona, 1969)
3. L.E. Reichl, *A Modern Course in Statistical Physics*, 2nd edn. (Edward Arnold, Austin, 1992)
4. R. Kubo, *Statistical Mechanics: An Advanced Course with Problems and Solutions* (North-Holland, Amsterdam, 1993)
5. L. Onsager, Reciprocal relations in irreversible processes. I. Phys. Rev. **37**, 405 (1931); Reciprocal relations in irreversible processes. II. Phys. Rev. **38**, 2265 (1931)

6. M.O. Cáceres, Irreversible thermodynamics in the framework of Tsallis entropy. Phys. A **218**, 471 (1995)
7. N.G. van Kampen, *Stochastic Process in Physics and Chemistry*, 2nd edn. (North-Holland, Amsterdam, 1992)
8. R. Kubo, M. Toda, N. Hashitsume, *Statistical Physics II: Nonequilibrium Statistical Mechanics* (Springer, Berlin, 1985)

Chapter 3
Elements of Stochastic Processes

3.1 Introduction

The study of the temporal evolution of probability is one of the central themes of modern statistics. Both the analysis of relaxation and the instability of a dynamic system can be placed in direct correspondence with the study of trajectories $X(t)$ of **rv** that evolve over time; hence the importance of defining the concept of stochastic process [1–3].

3.1.1 Time-Dependent Random Variable

A time-dependent random variable or stochastic process (**sp**) is a random function $X_\Omega(t)$ of a real argument[1] (continuous or discrete) $t \in [t_i, t_f]$ and a **rv** Ω on $\mathcal{D}_\Omega$. In what follows we will assume that the **sp** is a scalar, so its generalization to vectors or matrices is obvious. Then, we can interpret an **sp** as a function f of two variables: t (the time[2]) and Ω (the **rv**).

$$X_\Omega(t) = f(t, \Omega). \tag{3.1}$$

---

[1]The generalization to d arguments introduces the concept of stochastic field.

[2]In general, if $X_\Omega(z) = f(z, \Omega)$, where Ω is a **rv** and $z \in \mathcal{R}_e$ is any ordinary variable, the relation $X_\Omega(z)$ also makes it possible to define an **sp** (for example, a spatial **sp**). That is, z is not restricted to time, other choices are possible to define an **sp** correctly. However, the fact that z is the time variable makes possible a very didactic presentation of the concept of **sp** [as counterexample see guided exercise in Sect. 2.3.1].

© Springer International Publishing AG 2017
M.O. Cáceres, *Non-equilibrium Statistical Physics with Application to Disordered Systems*, DOI 10.1007/978-3-319-51553-3_3

There are two "representations" to define an **sp**:

(I) Representation in ensemble of realizations. For each value ω from the domain $\mathcal{D}_\Omega$ of the **rv** Ω, $X_\omega(t) = f(t, \omega)$ is an ordinary function of time, commonly called random function or *realization* of the **sp** $X_\Omega(t)$. Then, characterizing an **sp** is equivalent to giving the ensemble of all realizations[3] $\{X_\omega(t)\}$, $\forall \omega \in \mathcal{D}_\Omega$.

(II) Representation in a countable set of random variables (m-dimensional representation). For any fixed time t', $X_\Omega(t')$ is a new **rv** [actually $X_\Omega(t')$ is a transformation law of the **rv** Ω into $X_\Omega(t')$]. Then, for an arbitrary set of times $\{t_i\}$, the **sp** $X_\Omega(t)$ fully characterizes the set of **rvs** $\{X_\Omega(t_i)\}$.

Note In general, for any t_i, these **rvs** will not have to be **si**. On the other hand, the domain of **rv** $X_\Omega(t_i)$ can be continuous or discrete; also Ω may represent one or more **rvs**.

3.1.2   The Characteristic Functional (Ensemble Representation)

In representation (I), we can calculate mean values on the realizations (thus obtaining ordinary functions t) simply averaging with the probability distribution[4] $P_\Omega(\omega)$, that is:

$$\langle X_\Omega(t) \rangle = \int_{\mathcal{D}_\Omega} X_\omega(t) \, P_\Omega(\omega) \, d\omega. \tag{3.2}$$

This expression is valid for all $t \in [t_i, t_f]$. In general, moments at different arbitrary times are calculated according to the formula:

$$\langle X_\Omega(t_1) X_\Omega(t_2) \cdots X_\Omega(t_n) \rangle = \int_{\mathcal{D}_\Omega} X_\omega(t_1) X_\omega(t_2) \cdots X_\omega(t_n) \, P_\Omega(\omega) \, d\omega. \tag{3.3}$$

Then, we can define a characteristic functional to generate moments of the **sp** $X_\Omega(t)$, with $t \in [t_i, t_f]$, in the form

$$G([k]) \equiv \left\langle \exp\left[i \int_{t_i}^{t_f} k(t) X_\Omega(t) dt \right] \right\rangle_{P_\Omega(\omega)} \tag{3.4}$$

$$= \sum_{m=0}^{\infty} \frac{i^m}{m!} \int_{t_i}^{t_f} \cdots \int_{t_i}^{t_f} k(t_1) \cdots k(t_m) \, \langle X_\Omega(t_1) \cdots X_\Omega(t_m) \rangle \prod_{j=1}^{m} dt_j.$$

---

[3]Sometimes, a realization of the process will be denoted simply in the form $X(t)$.

[4]Generally we assume that the **rv** will be continuous (then we use indistinctly the name pdf or probability distribution to characterize it). For the discrete case, we have shown that the two presentations are equivalent [see Eq. (1.5) in Sect. 1.2.4].

This notation emphasizes that $G([k])$ depends on the whole "test" function $k(t)$, and not only on the value that $k(t)$ takes at a particular time t' (note the similarity to the definition of Fourier's transform). The convergence of the integral (3.4) is assured, because the type of test functions $k(t)$ that are allowed, are those which vanish for large enough $|t|$. From (3.4) it follows that all moments are obtained by differentiation[5] of the functional $G([k])$, thus this functional completely characterizes the **sp** $X_\Omega(t)$. That is, given a functional $G([k])$, if we can expand this in the form:

$$G([k]) = 1 + i \int_{t_i}^{t_f} k(t_1) \langle X_\Omega(t_1) \rangle \, dt_1$$

$$+ \frac{i^2}{2!} \int_{t_i}^{t_f} \int_{t_i}^{t_f} k(t_1)k(t_2) \langle X_\Omega(t_1)X_\Omega(t_2) \rangle \, dt_1 \, dt_2 + \cdots, \quad (3.5)$$

we easily recognize all moments $\langle X_\Omega(t_1) \cdots X_\Omega(t_m) \rangle$ of the **sp** $X_\Omega(t)$ as the coefficients of the terms $k(t_1) \cdots k(t_m)$ in the series expansion (3.5).[6]

Alternatively, we can also define a "partition-characteristic" functional which is what generates the cumulants[7] of the **sp** $X_\Omega(t)$

$$\ln G([k]) = \sum_{m=1}^{\infty} \frac{i^m}{m!} \int_{t_i}^{t_f} \cdots \int_{t_i}^{t_f} k(t_1) \cdots k(t_m) \langle\langle X_\Omega(t_1) \cdots X_\Omega(t_m) \rangle\rangle \prod_{j=1}^{m} dt_j.$$
$$(3.6)$$

Example. Note that in the deterministic limit[8] the logarithm of the characteristic function is linear in $k(t)$, because

$$G([k]) \equiv \left\langle \exp\left[i \int_{t_i}^{t_f} k(t)X_\Omega(t) \, dt \right] \right\rangle = \exp\left[i \int_{t_i}^{t_f} k(t_1)X_{\omega_d}(t_1) \, dt_1 \right].$$

This expression is obtained from (3.4) whereas in the aforementioned limit it has

$$\langle X_\Omega(t_1) \cdots X_\Omega(t_m) \rangle \to X_{\omega_d}(t_1) \cdots X_{\omega_d}(t_m);$$

or, from (3.6) noting that

$$\langle\langle X_\Omega(t_1) \cdots X_\Omega(t_m) \rangle\rangle \to 0 \quad \text{if} \quad m \neq 1,$$

---

[5]In the next optional exercise, the definition of functional derivative is presented in a "simple" way, see also advanced exercise 1.15.10. In Sect. 3.10.2, guidelines to understand the concept of functional derivative as a "discrete derivative" are given.

[6]Note that n-dimensional integrals in (3.5) must be well defined for the moment $\langle X_\Omega(t_1) \cdots X_\Omega(t_n) \rangle$ involved.

[7]On the usual definition of cumulant, see Sect. 1.10. Depending on the type of **rv** we will, generally, define another class of correlation (cumulant), and therefore its generating function will be different. See, for example, [1, 3].

[8]In that limit, $P_\Omega(\omega) \to \delta(\omega - \omega_d)$.

and that the only nonzero cumulant is

$$\langle\langle X_\Omega(t_1)\rangle\rangle \rightarrow X_{\omega_d}(t_1).$$

Exercise. Calculate the characteristic functional[9] $G_X([k])$ for the **sp** $X_\Omega(t) = \psi(t)\Omega$, $t \in [t_i, t_f]$. In general, if the **rv** Ω is characterized by $P_\Omega(\omega)$ on $\mathcal{D}_\Omega$, show that the functional $G_X([k])$ can be written in the form:

$$G_X([k]) = G_\Omega\left(\int_{t_i}^{t_f} k(t)\psi(t)\,dt\right), \quad t \in [t_i, t_f],$$

where $G_\Omega(q) = \langle e^{iq\Omega}\rangle$ is the characteristic function of the **rv** Ω. Show that when Ω ceases to be a **rv** and behaves as a (deterministic) parameter ω_d, the characteristic functional takes a structure in the argument of the exponential, which is *linear*

$$G_X([k]) = \exp\left[i\,\omega_d\int_{t_i}^{t_f} k(t)\psi(t)\,dt\right], \quad t \in [t_i, t_f].$$

Exercise. Consider the **sp** of the previous exercise, but now characterize the probability of the **rv** Ω with a Poisson probability and calculate the characteristic functional using an expansion in the cumulants of Ω.

*Exercise (***Poisson Noise***[10]).* Consider the **sp** $\psi(t) = \sum_{\sigma=1}^{n} A\delta(t-t_\sigma)$, where the random times $\{t_\sigma\}$ are uniformly distributed in the interval $[0, T]$, and the number of times n (for a given realization of the **sp**) also is a **rv** and it is characterized by a Poisson probability $P_n = e^{-\lambda}\lambda^n/n!$, where $n = 0, 1, 2, \cdots$. Show that the characteristic functional of the **sp** $\psi(t)$ is

$$G_\psi([k]) = \exp\left[\int_0^T \lambda e^{iAk(s)}\frac{ds}{T} - \lambda\int_0^T \frac{ds}{T}\right], \quad t \in [0, T].$$

Show that if $T \rightarrow \infty$ and $\lambda \rightarrow \infty$, in such a way that $\dfrac{\lambda}{T} \rightarrow q$, the characteristic functional of the "Poisson noise" adopts a simpler form:

$$G_\psi([k]) = \exp\left[q\int_0^\infty \left(e^{iAk(s)} - 1\right)ds\right], \quad t \in [0, \infty].$$

Exercise. Consider the addition of two **si** Poisson noises of amplitudes $A = \pm 1$, respectively. Show that the characteristic functional of the sum of two Poisson noises $\varphi = \psi_+ + \psi_-$ is given by

---

[9]Sometimes when necessary to emphasize the **sp** $X_\Omega(t)$, we denote the characteristic functional in the form $G_X([k])$.

[10]The term noise or **sp** is equivalent.

$$G_\varphi\left([k]\right) = \exp\left[2q \int_0^\infty \left(\cos k(s) - 1\right)\, ds\right], \quad t \in [0, \infty].$$

Exercise. Consider now the sum of two Poisson noises ψ_1, ψ_2 both **si**, but with random amplitudes $A = \pm 1$ equally likely. Calculate the functional of the noise $\phi = \psi_1 + \psi_2$. Study the difference between the functional $G_\phi\left([k]\right)$ and the previous one $G_\varphi\left([k]\right)$.

Exercise. Consider an **sp** $\psi(t)$ characterized by its functional $G_\psi\left([k]\right)$. Show that all n-time moments and cumulants can be calculated from partial derivatives from the n-time characteristic function. Note that $G_\psi^{(n)}\left(\{k_j, t_j\}\right)$ can be obtained from the functional as:

$$G_\psi^{(n)}\left(\{k_j, t_j\}\right) = G_\psi\left([k(t) = k_1\delta\left(t - t_1\right) + k_2\delta\left(t - t_2\right) + \cdots + k_n\delta\left(t - t_n\right)]\right).$$

Then, for example

$$\langle\langle \psi(t_1)\cdots\psi(t_m)\rangle\rangle = \left.\frac{\partial^n \ln G_\psi^{(n)}\left(\{k_j, t_j\}\right)}{(i)^n\, \partial k_1 \cdots \partial k_n}\right|_{\{k_j=0\}}$$

Optional Exercise. Let $F([\phi])$ be an arbitrary functional. Using the generalized definition of derivative in the functional calculus [4]

$$\frac{\delta F([\phi])}{\delta\phi(\tau)} = \left.\frac{dF\left[\phi(t) + \lambda\,\delta(t - \tau)\right]}{d\lambda}\right|_{\lambda=0},$$

show the following results:

1. If $F([\phi]) = f(\phi(t))$, then:

$$\frac{\delta F[\phi]}{\delta\phi(\tau)} = \frac{\partial f}{\partial\phi}\,\delta(t - \tau).$$

2. If $F[\phi] = f\left(g(\phi)\right)$, then:

$$\frac{\delta F[\phi]}{\delta\phi(\tau)} = \frac{\partial f}{\partial g}\frac{\delta g}{\delta\phi(\tau)}.$$

3. If $F[\phi] = f\left(\phi(t), \dot\phi(t)\right)$, then:

$$\frac{\delta F[\phi]}{\delta\phi(\tau)} = \frac{\partial f}{\partial\phi}\,\delta(t - \tau) - \frac{\partial f}{\partial\dot\phi}\frac{d}{d\tau}\,\delta(t - \tau).$$

4. If $F[\phi] = \int f(\phi, \dot{\phi}) \, dt$, then:

$$\frac{\delta F[\phi]}{\delta \phi(\tau)} = \frac{\partial f}{\partial \phi(\tau)} - \frac{d}{d\tau} \frac{\partial f}{\partial \dot{\phi}(\tau)}.$$

Exercise (**Delta Correlations**). Consider an **sp** $\psi(t)$ such that all cumulants are given by the expression:

$$\langle\langle \psi(t_1) \cdots \psi(t_m) \rangle\rangle = \Lambda_m \delta(t_1 - t_2)\delta(t_1 - t_3) \cdots \delta(t_1 - t_m), \quad \langle\langle \psi(t_1) \rangle\rangle = \Lambda_1, \quad m \geq 2.$$

that is, all correlations are Dirac-δ form. Show that the characteristic functional is

$$G_\psi([k]) = \exp\left\{ \sum_{m=1}^{\infty} \frac{i^m}{m!} \Lambda_m \int_{t_i}^{t_f} (k(t))^m \, dt \right\}, \quad t \in [t_i, t_f].$$

These **sp** are called white noises. The case when $\Lambda_m = 0$, $\forall m \geq 3$ defines a Gaussian white noise (GWN). Using $\Lambda_m = q$, $\forall m \geq 1$, recover the functional of the Poisson noise.

3.1.3   Kolmogorov's Hierarchy (Multidimensional Representation)

Representation (II) shows that to fully characterize the **sp** $X_\Omega(t)$ we need to specify an infinite set of joint probability densities:

$$\begin{aligned}
&P_1(x_1, t_1) \\
&P_2(x_2, t_2; x_1, t_1) \\
&P_3(x_3, t_3; x_2, t_2; x_1, t_1) \\
&\cdots \text{etc.}
\end{aligned} \tag{3.7}$$

This is the Kolmogorov hierarchy of any **sp**. Note that this hierarchy is "over-complete" because P_i can be obtained marginalizing[11] P_{i+1}. The hierarchy should meet the basic Kolmogorov's conditions:

1. positivity: $P_n \geq 0$, $n \in \mathcal{N}$;
2. permutation symmetry:[12] $(x_k, t_k) \longleftrightarrow (x_j, t_j)$;

---

[11]On the definition of marginal distribution, see Sect. 1.9.2.

[12]It is worthwhile to point out that this symmetry must be understood in the context of intersection of subsets (events), i.e., for example for two events: $A \equiv x(t_1) \in x_1 + dx_1$ and $B \equiv x(t_2) \in x_2 + dx_2$. We should interpret: $P(x_1, t_1; x_2, t_2) = P(A \cap B)$; see Sect. 1.2.

3. marginality[13]: $P_n(x_n, t_n; \cdots ; x_1, t_1) = \int P_{n+1}(x_{n+1}, t_{n+1}; \cdots ; x_1, t_1) \, dx_{n+1}$; and
4. normalization: $\int P_1(x_1, t_1) \, dx_1 = 1$.

Kolmogorov's hierarchy helps us, for example, to calculate the probability

$$P_2(x_1, t_1; x_2, t_2) \, dx_1 \, dx_2$$

that the **sp** $X_\Omega(t)$ has the value $[x_1, x_1 + dx_1]$ at instant t_1 and $[x_2, x_2 + dx_2]$ at instant t_2. Then, for all t_1, the first moment or the **mv** of the **sp** is given by the formula[14]

$$\langle X(t_1) \rangle = \int_{\mathcal{D}_x} x_1 \, P_1(x_1, t_1) \, dx_1. \tag{3.8}$$

And, for example, the correlation for two different times is calculated in terms of the second moment:

$$\langle X(t_1) X(t_2) \rangle = \int_{\mathcal{D}_x} \int_{\mathcal{D}_x} x_1 x_2 \, P_2(x_2, t_2; x_1, t_1) \, dx_1 \, dx_2; \tag{3.9}$$

that is, evaluating

$$\langle \langle X(t_1) X(t_2) \rangle \rangle = \langle X(t_1) X(t_2) \rangle - \langle X(t_1) \rangle \langle X(t_2) \rangle.$$

Comparing both *representations* (I) and (II) we note that

$$P_1(x, t) = \langle \delta (x - X_\Omega(t)) \rangle_{P_\Omega(\omega)}. \tag{3.10}$$

This assertion can easily be confirmed as follows. Representation in ensemble (I) shows that the **mv** over all realizations is

$$\langle X_\Omega(t) \rangle = \int_{\mathcal{D}_\Omega} X_\omega(t) P_\Omega(\omega) \, d\omega.$$

On the other hand, using the multidimensional representation (II), and Eq. (3.10) for an arbitrary fixed time t', we have

$$\left\langle X(t') \right\rangle \equiv \int_{\mathcal{D}_x} x \, P_1(x, t') \, dx = \int_{\mathcal{D}_x} x \left\langle \delta \left(x - X_\Omega(t') \right) \right\rangle_{P_\Omega(\omega)} \, dx$$

$$= \left\langle \int x \delta \left(x - X_\Omega(t') \right) \, dx \right\rangle_{P_\Omega(\omega)} = \left\langle X_\Omega(t') \right\rangle_{P_\Omega(\omega)},$$

proving the equivalence.

---

[13]This means that the lower members of the hierarchy can be obtained from the higher ones, i.e., we are considering that $t_1 < t_2 < \cdots < t_n \in [t_i, t_f]$.

[14]In what follows, we shall always assume the domain of the **sp** to be continuous, unless we specify otherwise.

Similarly, we can also write the joint probability distribution for n-times in the form:

$$P_n(x_n, t_n; \cdots; x_1, t_1) = \langle \delta(x_1 - X_\Omega(t_1)) \cdots \delta(x_n - X_\Omega(t_n)) \rangle_{P_\Omega(\omega)}. \qquad (3.11)$$

Depending on the physical processes that we want to model, it may be more convenient to use one representation than the other. The first representation in ensemble is frequently used in mathematical calculations and perturbations, while the second representation is more reminiscent of the intuitive notion of probability density used in equilibrium statistical mechanics.

Examples.

1. Suppose a (time-dependent) physical system can be modeled by a "multiplicative" **sp** like

$$X_\Omega(t) = \psi(t)\Omega,$$

where $\psi(t)$ is an arbitrary function of time and Ω is an arbitrary **rv** characterized by the pdf $P_\Omega(\omega)$ on $\mathcal{D}_\Omega$. From this, we note that moments of the **sp** are determined by moments of the **rv** Ω

$$\langle X_\Omega(t) \rangle = \psi(t) \langle \Omega \rangle$$
$$\langle X_\Omega(t_1)X_\Omega(t_2) \rangle = \psi(t_1)\psi(t_2) \langle \Omega^2 \rangle$$
$$\cdots \text{ etc.}$$

So, in this particular case, it is clear that it is considerably easier to calculate the statistics of the **sp** using the representation (I), in *ensemble*.

2. Suppose we draw an **sp** $X_\Omega(t)$ from some information that we have about the n-times joint pdf P_n (Kolmogorov's hierarchy). It is clear, then, that the simplest representation of use will be the second if, for example, we know that $P_n(x_n, t_n; \cdots; x_1, t_1)$ factorizes for all n according to

$$P_n(x_n, t_n; \cdots; x_1, t_1) = \prod_{i=1}^{n} P_1(x_i, t_i). \qquad (3.12)$$

That is, the **rv** associated at instant t is statistically independent of the **rv** associated with any other time t'. Moreover, for any fixed t the **sp** establishes a specific transformation $\Omega \to X_\Omega(t)$, which is characterized by the distribution $P_1(x, t)$. Here, $P_1(x, t)\, dx$ gives the relative frequency that the realization is in the range $[x, x + dx]$ at the instant t. Then, it is clear that this **sp** is completely characterized by Eq. (3.12).[15]

---

[15]What happens if we try to write the expansion (3.5) for an **sp** defined from Kolmogorov's hierarchy (3.12).

3.1.4 Overview of the Multidimensional Representation

In representation (II), it is usually convenient to eliminate all superfluous notation (subscript Ω, etc.), so henceforth we denote the **sp** $X(t) \equiv X_\Omega(t)$.

The pdf $P_1(x_1, t_1)$ is positive, and for all $t \in [t_i, t_f]$ it is related to the probability that the **rv** $X(t_1)$ takes a certain value

$$\int_a^b P_1(x_1, t_1)\, dx_1 = \text{Prob.}[a < X(t_1) < b].$$

Note the different notation used for characterizing the **sp** $X(t)$, the **rv** $X(t_1)$ and the realization $X(t)$ [i.e., the random function of time $X_\omega(t)$]. In cases where this does not bring any confusion, we will simplify the notation.

It is noteworthy that $P_1(x, t)$ is necessary but not sufficient to characterize the **sp** $X(t)$. To fully describe an **sp** [i.e., the statistical knowledge of every possible realization] it is necessary to know the n-dimensional joint pdf. That is, the joint probability

$$P_n(x_n, t_n; \cdots ; x_2, t_2; x_1, t_1) \prod_{i=1}^n dx_i,$$

for all n and $\{t_l\}$ arbitrary times.

In the particular case of three-times, the quantity $P_2(x_3, t_3; x_2, t_2; x_1, t_1)$ $dx_1\, dx_2\, dx_3$ gives the probability that the realizations $X(t)$ pass through the windows[16]: $[x_1, x_1 + dx_1]$ at instant t_1, through $[x_2, x_2 + dx_2]$ at instant t_2 and through $[x_3, x_3 + dx_3]$ at instant t_3.

Realizations of a Continuous Stochastic Process

Let $X_\Omega(t)$ represent some **sp** defined for $t \in [0, \infty]$. In Fig. 3.1, we show three possible realizations $X(t)^{(j)}$ starting from $X(0)$ at time t_0. For a fixed time $t = t_1$ the value $X(t_1)^{(j)}$ is a random number. Then, the pdf $P(x_1, t_1)$ can formally be built up from an histogram of many realizations like the ones that appear in this figure. Which would be the realizations used to calculate the two-times joint probability $P(x_1, t_1; x_2, t_2)\, dx_1 dx_2$?

A special case is the second example in the previous section. There the **sp** $X(t)$ was fully characterized by $P_1(x, t)$. This is true only if the set of **rvs** $\{X(t_1) \cdots X(t_n)\}$ is **si** for all times $\{t_1, \cdots, t_n\} \in [t_i, t_f]$. That is, in this case, the n-times joint pdf is reduced to the product of one-time densities. These are the

---

[16]Here (;) is meant in a Boolean sense; that is, $P(A \cap B) \Rightarrow P(A; B)$. Remember that $P(A \cap B) = 0$ only if the events are mutually exclusive, see Sect. 1.2.

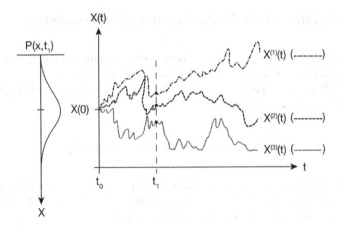

Fig. 3.1 Typical representation of three realizations of the stochastic process $X_\Omega(t)$ starting at t_0, and its associated pdf for a fixed time t_1

uncorrelated **sps**, commonly called *completely random processes*, and they are in direct correspondence with the usual **sirv** in ordinary probability theory.

In what follows, we assume, for the moment, that the **sp** $X(t)$ we are studying has zero **mv**, $\langle X(t) \rangle = 0$. This does not restrict at all our analysis, since it is always possible to consider a new **sp** $X(t) = Y(t) - \langle Y(t) \rangle$. In the representation (II), moments or **mv** of the **sp** $X(t)$ are simply the moments of a multidimensional **rv**. In particular, an n times correlation function is given in terms of the nth moment[17]:

$$\langle X(t_n) \cdots X(t_1) \rangle = \int_{\mathcal{D}_x} \cdots \int_{\mathcal{D}_x} x_1 \cdots x_n \, P_n(x_n, t_n; \cdots ; x_2, t_2; x_1, t_1) \prod_{i=1}^{n} dx_i. \quad (3.13)$$

As already mentioned, the set of n-times joint pdf must meet certain specific conditions. These are the Kolmogorov compatibility conditions, which are merely the generalization of the concept of marginal distributions of the process. A fundamental theorem of Kolmogorov[18] proves that this hierarchy (numerable) of distributions is necessary and sufficient to fully characterize the ensemble of random functions $X_\omega(t)$. This is a nontrivial result, it ensures that we only need to know the probability that the random function $X_\omega(t)$ takes values x_i at *discrete and arbitrary* times t_i, instead of the countless possibilities afforded by the continuous variable t.

---

[17]The generalization of the concept of correlation of n-times is given by the very definition of cumulant of n-times, see Sects. 1.10 and 3.1.2.

[18]This theorem was proved by Kolmogorov in 1933 (published in German). J. M. Kac and Logan have phrased it in a very elegant way on page 14 of the book Fluctuation Phenomena, Eds. E.W. Montroll and J.L. Lebowits, Amsterdam, Elsevier Science Publisher B.V. (1979).

3.1.5 Kolmogorov's Hierarchy from the Ensemble Representation

Kolmogorov's hierarchy can be obtained by quadrature from the characteristic functional of the **sp**. Indeed, in general we can invert the functional $G_X([k])$ introducing a n-dimensional Fourier transform, and thus we can obtain a formal expression for the n-times joint pdf:

$$P_n(x_n, t_n; \cdots ; x_2, t_2; x_1, t_1) = \frac{1}{(2\pi)^n} \int dk_1 \cdots \int dk_n \, e^{-i \sum_{j=1}^n k_j x_j} \tag{3.14}$$

$$\times \, G_X([k(t)])_{k(t) = k_1 \delta(t - t_1) + \cdots + k_n \delta(t - t_n)} \, .$$

Note that this result yields the whole Kolmogorov hierarchy; i.e., the complete characterization of **sp** whatever the kind of process.

Optional Exercise. Suppose the expression for the $\ln G_X([k])$ is cut at the second cumulant. From the formula (3.6), get the probability $P_1(x_1, t_1)$, and the joint pdf of two-times $P_2(x_1, t_1; x_2, t_2)$.

Exercise. Suppose the expression for the n-times pdf is given by (3.12), write the n-times characteristic function $G_X^{(n)}(\{k_1, t_1; \cdots ; k_n, t_n\})$, find what will happen if we try to write the characteristic *functional* $G_X([k])$.

3.2 Conditional Probability

The probability that the random function $X_\omega(t)$ passes through the "window" around x_n at the instant t_n, *having passed* x_{n-1} at the instant t_{n-1}, and x_{n-2} at the instant t_{n-2}, etc., will be given in terms of the *conditional* pdf:

$$P(x_n, t_n \mid x_{n-1}, t_{n-1}; \cdots ; x_i, t_i; \cdots ; x_1, t_1).$$

As in Chap. 1, the relationship between the n-times joint pdf and the conditional pdf is given by Bayes' rule:

$$P(x_n, t_n; x_{n-1}, t_{n-1}; \cdots ; x_i, t_i \mid x_{i-1}, t_{i-1}; \cdots ; x_1, t_1) =$$

$$\frac{P_n(x_n, t_n; x_{n-1}, t_{n-1}; \ldots ; x_i, t_i; x_{i-1}, t_{i-1}; \ldots ; x_1, t_1)}{P_{i-1}(x_{i-1}, t_{i-1}; \ldots ; x_1, t_1)}. \tag{3.15}$$

Here the time set has been ordered: $t_1 < t_2 < \cdots < t_n$. In our notation $(\cdots \mid \cdots)$ means "given that."[19]

---

[19] See Sects. 1.2.1 and 1.9.3.

Exercise. Show, in particular, that the conditional pdf $P(x_2, t_2 \mid x_1, t_1)$ meets the condition: $\lim_{t_2 \to t_1} P(x_2, t_2 \mid x_1, t_1) \to \delta(x_2 - x_1)$.

Optional Exercise. Let $x(t)$ and $p(t)$ represent trajectories of a free particle in phase space. Consider $x(0)$, $p(0)$ as random initial conditions characterized by the pdf $\mathcal{P}(x(0); p(0))$. Calculate:

$$P_1(x_1, p_1, t_1)$$
$$P_2(x_2, p_2, t_2; x_1, p_1, t_1)$$
$$P(x_2, p_2, t_2 \mid x_1, p_1, t_1).$$

Note that here we are introducing an ordered pair $\{x_i, p_i\}$ at every moment t_i.

3.3   Markov's Processes

A particular case of (3.15) occurs when the condition of the "past" is given for a single previous time. For example, $P(x_2, t_2 \mid x_1, t_1)$ is a typical conditional pdf, and often called propagator of the system. A particular class of **sps** are the Markov processes [1–3]. The conditional pdf of these **sps** depends only "on the closest" time. That is, a Markov **sp** does not depend on the complete history of the given conditions at $\{t_j\}$. In a Markov process, the future depends only on the immediately preceding time, namely:

$$P(x_n, t_n; \cdots; x_i, t_i \mid x_{i-1}, t_{i-1}; \cdots; x_1, t_1) = P(x_n, t_n; \cdots; x_i, t_i \mid x_{i-1}, t_{i-1}),$$
(3.16)

here we used $t_n \geq t_{n-1}$. This equation can be taken as the very definition of a Markov process. A Markov **sp** is the immediate and next complication of uncorrelated **sp** (completely random noises). This is, to characterize a Markov **sp**, it is necessary to know *only two* mathematical objects: $P(x_i, t_i \mid x_{i-1}, t_{i-1})$ and $P_1(x, t)$. This structure marks a fundamental difference in the many different applications in which a Markov **sp** is involved. However, this does not mean that "completely random noise" and "white noise" are less important. When different systems are modeled, it is necessary to determine (theoretically) to what extent the physical description in terms of uncorrelated noises is representative of variables that fluctuate in short time scales compared to all time scales of the experiment in question. Markov processes with their more complex structure, arise naturally in many applications of physics, as a solution of differential equations with coefficients that are random δ-correlated functions (i.e., time-dependent coefficients that are white noises[20]).

Exercise. Show that for a Markov **sp** the joint probability density of n-times is completely characterized by the propagator and the probability density of one-time (here, we adopt the convention $t_n \geq t_{n-1} \cdots \geq t_i \geq t_{i-1} \cdots \geq t_1$), then

---

[20]White noises are uncorrelated and also singular **sp**; see, for example, Sect. 3.9.

$$P_n(x_n, t_n; x_{n-1}, t_{n-1}; \cdots; x_i, t_i; x_{i-1}, t_{i-1}; \cdots; x_1, t_1) =$$

$$P(x_n, t_n \mid x_{n-1}, t_{n-1}) \cdots P(x_i, t_i \mid x_{i-1}, t_{i-1}) \cdots P(x_2, t_2 \mid x_1, t_1) P(x_1, t_1). \tag{3.17}$$

Exercise. Show that (3.17) is an alternative definition of a Markov **sp**; i.e., given (3.17) prove (3.16).

Exercise. Prove that the **sp** $X_\Omega(t) = \psi(t)\Omega$, where Ω is any **rv**, is not a Markov **sp**.

Guided Exercise (**Markovian Memory**). Consider the deterministic equation: $\dot{x} = N(x)$. If we write the solution in the form $x(t) = \phi(x_0, t_0; t)$, where (x_0, t_0) is the initial condition of the problem, prove that $x(t)$ can be considered as a (singular) Markov **sp** even when the (deterministic) system has infinite memory. To prove this assertion, we simply use the fact that if an **sp** is Markovian, then it must follow that

$$P(x_n, t_n; x_{n-1}, t_{n-1}; \cdots \mid x_0, t_0) = P(x_n, t_n \mid x_{n-1}, t_{n-1}) \cdots P(x_1, t_1 \mid x_0, t_0).$$

However, from the definition of the two-times joint pdf

$$P(x_2, t_2; x_1, t_1 \mid x_0, t_0) = \langle \delta[x_2 - \phi(x_0, t_0; t_2)] \delta[x_1 - \phi(x_0, t_0; t_1)] \rangle,$$

and the propagator

$$P(x_n, t_n \mid x_{n-1}, t_{n-1}) = \langle \delta[x_n - \phi(x_{n-1}, t_{n-1}; t_n)] \rangle,$$

we conclude that ($\forall\, t_n \geq t_{n-1} \geq \cdots \geq t_0$)

$$\delta[x_2 - \phi(x_0, t_0; t_2)] \delta[x_1 - \phi(x_0, t_0; t_1)] = \delta[x_2 - \phi(x_1, t_1; t_2)] \delta[x_1 - \phi(x_0, t_0; t_1)]. \tag{3.18}$$

Here we have considered that there are no random elements because the system is deterministic, so we can eliminate **mv**; for example: $\langle \delta[x_1 - \phi(x_0, t_0; t_1)] \rangle = \delta[x_1 - \phi(x_0, t_0; t_1)]$, etc. Then, from (3.18) we deduce that $\phi(x_0, t_0; t_2) = \phi(x_1, t_1; t_2)$, which is true because it is a deterministic system where $x_1 = \phi(x_0, t_0; t_1)$. We will reach the same conclusion by studying the n-times joint pdf. That is: a deterministic process has infinite memory but just the immediately preceding condition is sufficient. This latter feature matches the definition of a Markovian **sp**, which uses the fact that only the immediately preceding (conditioned) history is required to characterize the future, however a Markov **sp** introduces a dispersion that starts at the time initial condition. Note, on the other hand, we can also define an **sp** introducing random variables in the deterministic system. For example, the initial value x_0 could be a **rv**, in which case the process will not be, in general, Markovian.

3.3.1 The Chapman-Kolmogorov Equation

The integral Chapman-Kolmogorov equation which satisfies the conditional probability of a Markov **sp** is

$$P(x_n, t_n \mid x_1, t_1) = \int_{\mathcal{D}_x} P(x_n, t_n \mid x_i, t_i) P(x_i, t_i \mid x_1, t_1) \, dx_i, \ \forall t_i \in [t_n, t_1]. \tag{3.19}$$

This equation can also be taken as a definition of a Markov process (necessary condition). Remember that a Markov process is completely characterized by the function $P_1(x, t)$ and the propagator $P(x, t \mid x', t')$. Therefore these two functions cannot be chosen arbitrarily, but they must satisfy the Chapman-Kolmogorov equation (3.19) and the compatibility relationships of the joint hierarchy. That is, two nonnegative functions

$$P_1(x, t) \quad \text{and} \quad P(x, t \mid x', t'),$$

which satisfy the compatibility relationships (3.7) and (3.19) define a unique Markov **sp**. This theorem must not be confused with the characterization (sometimes misunderstood) of the differential equation satisfied by $P_1(x, t)$ (erroneously called Markovian Master Equation),[21] when it does not have a memory kernel).

Exercise (**Proof of 3.19**). Obtain Chapman-Kolmogorov's equation from the fact that the future of a Markov process does not depend on the past but on the present given (3.16).

Exercise (**Lorentz' Distribution**). Consider the following nonnegative functions

$$P_1(x, t) = \frac{t/\pi}{(x^2 + t^2)}$$

$$P(x_2, t_2 \mid x_1, t_1) = \frac{(t_2 - t_1)/\pi}{(x_2 - x_1)^2 + (t_2 - t_1)^2}, \ t_2 \geq t_1,$$

and show that we can define a Markov process with them [Cauchy's *Random Walk*, see also Chap. 6]. Figure 3.2 shows the conditional pdf $P(x_2, t_2 \mid x_1, t_1)$ for $x_1 = 1$ and different values of time $t_2 - t_1$. A discussion on the divergence of the moments of this distribution is presented in Sect. 1.4.2.

Optional Exercise (**Dichotomous Process**). Consider the particular case where the **sp** can only have two values (e.g., $\{\Delta, -\Delta\}$), i.e., the domain $\mathcal{D}_x$ is discrete.

---

[21] A detailed analysis on the Master Equation will be presented in Chap. 6, see also advanced exercise 4.9.3.

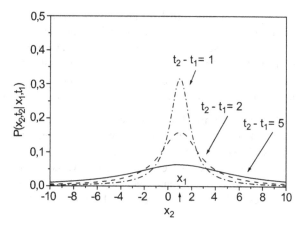

Fig. 3.2 Lorentz probability distribution for three values of time $t_2 - t_1$

As we have already mentioned, we can use the same definitions to characterize processes defined on either a discrete or continuous sample space. So we study here the Chapman-Kolmogorov equation in the case where $\mathcal{D}_x$ is discrete.[22] Consider the one-time probability

$$P_1(x, t) = \frac{\alpha}{\alpha + \gamma} \delta_{x,\Delta} + \frac{\gamma}{\alpha + \gamma} \delta_{x,-\Delta},$$

and the propagator $(t_2 \geq t_1)$

$$P(x_2, t_2 \mid x_1, t_1) = \frac{1}{\alpha + \gamma} \left(\alpha + \gamma e^{-(\alpha+\gamma)(t_2-t_1)} \right) \delta_{x_2,\Delta} \, \delta_{x_1,\Delta}$$

$$+ \frac{\alpha}{\alpha + \gamma} \left(1 - e^{-(\alpha+\gamma)(t_2-t_1)} \right) \delta_{x_2,\Delta} \, \delta_{x_1,-\Delta}$$

$$+ \frac{\gamma}{\alpha + \gamma} \left(1 - e^{-(\alpha+\gamma)(t_2-t_1)} \right) \delta_{x_2,-\Delta} \, \delta_{x_1,\Delta}$$

$$+ \frac{1}{\alpha + \gamma} \left(\gamma + \alpha e^{-(\alpha+\gamma)(t_2-t_1)} \right) \delta_{x_2,-\Delta} \, \delta_{x_1,-\Delta},$$

and show that we can define a Markov process with these functions. This is the dichotomic **sp** or telegrapher process; a matrix representation of the conditional probability is presented in Chap. 6. In particular, in Sect. 6.3.1 we connect the magnitudes α, γ with the physical quantities of the model.[23] Calculate the **mv** $\langle X(t) \rangle$ and the conditional **mv** $\langle X(t) \rangle|_{X(0)=\Delta}$.

---

[22]The integral in (3.19) is replaced by a summation.

[23]In this case, it is easy to visualize a possible realization of the **sp**. Compare with the *discrete-time* dichotomic process, see advanced exercise 6.7.2.

Excursus (**Chapman-Kolmogorov Is Not a Sufficient Condition**). There are stochastic processes verifying the Chapman-Kolmogorov equation without being Markov processes [see, for example, P. Lévy, Comptes Rendus Académie Sciences (París), **228**, 2004, (1949); W. Feller, Ann. Math. Stat. **30**, 1252, (1959)]. From this result, we conclude immediately that the Chapman-Kolmogorov equation is a necessary but not a sufficient condition. See the advanced exercise 3.20.4.

Excursus (**Semi Markov Processes**). It is interesting here to draw attention to the existence of special cases in which the conditional probability may depend on "two" past conditions; however, the **sp** can be considered Markov if states are redefined [see, for example, G.H. Weiss, *Aspects and Applications of the Random Walk*, Amsterdam, North-Holland (1994), pp 6]. See also the advanced exercises 3.20.6; 7.5.3 and Sect. 8.5.1 for a physical application.

3.4   Stationary Processes

In general, an **sp** is stationary if there exists an asymptotic regime in which all n-times joint pdf satisfy, for all τ, an invariance under a continuous temporal translation:

$$P_n(x_n, t_n; x_{n-1}, t_{n-1}; \ldots; x_1, t_1) = P_n(x_n, t_n + \tau; x_{n-1}; t_{n-1} + \tau; \ldots; x_1, t_1 + \tau).$$

Therefore, the one-time pdf must be independent of t if the **sp** is stationary.[24] Then, if the process is stationary, it follows that $P_1(x, t) = P_{st}(x)$.

Exercise. Prove that the correlation function $\langle\langle X(t_1)X(t_2)\rangle\rangle$ of a stationary **sp** only depends on the time difference $|t_1 - t_2|$.

Guided Exercise (**Calculation of the Stationary Probability**). Let us show that the stationary pdf $P_{st}(x)$ can be obtained (if it exists) from the conditional pdf $P(x_1, t_1 \mid x_0, t_0)$. First, note from the definition of the two-time joint pdf and the concept of marginal pdf that

$$P_1(x_1, t_1) = \int P(x_1, t_1 \mid x_0, t_0) \, P_1(x_0, t_0) \, dx_0.$$

But if the process is stationary, it follows that

$$P_{st}(x_1) = \int P(x_1, t_1 - t_0 \mid x_0, 0) \, P_{st}(x_0) \, dx_0, \tag{3.20}$$

---

[24]A typical example of a stationary **sp** is the temporal fluctuations of macroscopic variables around thermodynamic equilibrium, see Chap. 2.

here we see that if it holds, we get that

$$P_{st}(x_1) = \lim_{t_1 \to \infty} P(x_1, t_1 - t_0 \mid x_0, 0) = \lim_{t_0 \to -\infty} P(x_1, t_1 - t_0 \mid x_0, 0), \quad \forall x_0,$$

then, $P_{st}(x_1)$ is a solution of the integral equation (3.20). This result shows that it is possible to obtain, from a suitable limit in the conditional probability, the stationary probability $P_{st}(x)$ of the process. Note that if the **sp** is Markovian and stationary, then the mere knowledge of the conditional probability is sufficient to completely characterize the process.

Example. Let an **sp** be: $\mathbf{T}(t) = X_n$, where $\{X_j\}$ is an infinite set of **sirv** characterized by a unique probability distribution $P(X_j)$, and also

$$n + \Omega < t < n + 1 + \Omega \quad \text{con} \quad n \in \mathcal{Z},$$

where Ω is a random variable with distribution $P_\Omega(\omega)$ on $\mathcal{D}_\Omega = [0, 1]$. A necessary condition for the **sp** $\mathbf{T}(t)$ to be stationary is that its first moment is independent of time. Let us inspect that condition:

$$\langle T(t) \rangle_{P_\Omega(\omega) \prod_j P(X_j)} = \sum_j \langle X_j \rangle \int_0^1 d\omega \, P_\Omega(\omega) \, \Theta(t - j - \omega) \, \Theta(\omega + j + 1 - t), \quad (3.21)$$

where $\Theta(z)$ indicates the step function (see Chap. 1). For a fixed time t, in the sum (3.21) there will exist many X_j that comply with the conditions imposed in the integration by the two step functions: $\Theta(t - j - \omega)$ and $\Theta(\omega + j + 1 - t)$. For example, suppose the integer part of t is m, then we can write $t = m + \epsilon$, where ϵ is an arbitrary number in $[0, 1]$. From (3.21) it is easy to see that there are only two values of j that meet the conditions of the limits of integration. Then, if the condition

$$j = m,$$

is met, it follows that the domain of integration is $[0, \epsilon]$, that is, $\int_0^\epsilon P_\Omega(\omega) \, d\omega$. While from the condition

$$j = m - 1$$

it follows that integration will be $\int_\epsilon^1 P_\Omega(\omega) \, d\omega$. From these results and (3.21) we note that for $t = m + \epsilon$

$$\langle T(t) \rangle_{P_\Omega(\omega) \prod_j P(X_j)} = \langle X_m \rangle \int_0^\epsilon d\omega \, P_\Omega(\omega) + \langle X_{m-1} \rangle \int_\epsilon^1 d\omega \, P_\Omega(\omega),$$

from which it follows, X_j being identically distributed that

$$\langle T(t) \rangle_{P_\Omega(\omega) \prod_j P(X_j)} = \langle X \rangle \int_0^1 d\omega \, P_\Omega(\omega) = \langle X \rangle.$$

Optional Exercise. Show that $\mathbf{T}(t)$ is a stationary **sp**. What is the relevance of the choice of distributions $P_\Omega(\omega)$ and $P(X_j)$?

Exercise. Calculate the **mv** $\mathbf{T}(t)$ when the variables X_j are statistically independent, but not equally distributed. Under these conditions is the **sp** $\mathbf{T}(t)$ stationary?

3.5 2π-Periodic Nonstationary Processes*

Of particular interest are the **sp** that asymptotically have a discrete temporal translation invariance. In general a T-invariant nonstationary **sp** (where T is a certain period of time) must satisfy the time translation ($\forall m = 1, 2 \cdots$):

$$P_n(x_n, t_n; \cdots ; x_1, t_1) = P_n(x_n, t_n + mT; \cdots ; x_1, t_1 + mT), \quad \forall n. \tag{3.22}$$

Such cases occur in noisy dynamic systems which are modulated by periodic forces in time. In these processes it is often important to study the asymptotic behavior (long time) of the n-time joint probability distributions and their correlation functions[25] [3, 5, 6].

Exercise. Show that a 2π-periodic **sp** cannot be stationary.

As a result of the discrete translational symmetry (3.22), in the asymptotic state, moments, correlations, etc., of the **sp** $\mathbf{X}(t)$ will depend on both absolute time $\{t_l\}$ and their differences. However, the dependence on absolute times will still be periodic $\{t_l \to t_l + mT\}$. Then, if $0 < \tau < T$, the statistics of $X(t + \tau)$ has to be different from the statistics of $X(t)$, i.e.:

$$\langle X(t + \tau) \rangle \neq \langle X(t) \rangle$$

$$\langle X(t_1 + \tau)X(t_2 + \tau) \rangle \neq \langle X(t_1)X(t_2) \rangle$$

$$\cdots\cdots$$

We can say that the difference between the **sp** $\mathbf{X}(t+\tau)$ and $\mathbf{X}(t)$ is a phase. Then, in real situations, when the phase is important, we say that the system is coherent, and the usefulness of a 2π-periodic **sp** is obvious. On the other hand, if the phase is not relevant, we say that the system is incoherent, and we can introduce an average in τ on the 2π-periodic **sp**, i.e., to invoke a random phase diffusion, see advanced exercise 1.15.7.

Excursus. The notion of stochastic resonance, i.e., the amplification of a dynamic response by introducing noise in the system, has its genesis in the nonstationary 2π-periodic **sp**, see [3, vol. 1, p. 139]. A review of various

---

[25] An eigenvalue theory for 2π-periodic Markov processes is presented in Sect. 4.8.

applications of stochastic resonance phenomenon can be seen in: Proc. of the NATO Adv. Res. Workshop: Stochastic Resonance in Physics and Biology, J. Stat. Phys. **70**, N: 1/2, (1993).

Excursus. Prove that the correlation function of a nonstationary 2π-periodic **sp**

$$\langle\langle X(t_1)X(t_2)\rangle\rangle ,$$

depends on both arguments. In particular, show that, in the asymptotic regime, the correlation function depends on $\mid t_1 - t_2 \mid$, and also—in a periodic way—on the time argument. Hint: use Fourier series to represent the **sp** $X(t)$ [3], or use the Kolmogorov eigenvalue theory for 2π-periodic **sp**, see Chap. 4.

Another typical example of nonstationary 2π-periodic **sp** is the study of fluctuations in electrical circuits under *ac* current flows; or in the case of a particle moving in time periodic modulated potential, see advanced exercise 4.9.8.

3.6 Brownian Motion (Wiener Process)

The canonical example of Markov processes is Brownian motion, also called Wiener process. We will denote the Wiener process as $\mathbf{W}(t)$. This Markov **sp** (nonstationary) is characterized by the one-time pdf

$$P_1(w, t = 0) = \delta(w) \tag{3.23}$$

and has as propagator the Gaussian density

$$P(w_2, t_2 \mid w_1, t_1) = \frac{1}{\sqrt{2\pi(t_2 - t_1)}} \exp\left(\frac{-(w_2 - w_1)^2}{2(t_2 - t_1)}\right), \ t_2 \geq t_1. \tag{3.24}$$

In Fig. 3.3 we show the propagator $P(w_2, t_2 \mid w_1, t_1)$ for $w_1 = 1$ and different values of $t_2 - t_1$. Here we can see that the "width" of the Gaussian propagator goes like $\sim 2\sqrt{t_2 - t_1}$; this is not the case for the Lorentz propagator, shown in Fig. 3.2.

Exercise. Show, using marginal densities, that the one-time pdf for $t \neq 0$ is given by

$$P_1(w, t) = \frac{1}{\sqrt{2\pi t}} \exp\left(\frac{-w^2}{2t}\right). \tag{3.25}$$

From (3.25), we note that for a fixed time t, $W(t)$ is a Gaussian variable with mean zero and variance t. The **sp** $\mathbf{W}(t)$ will be completely characterized if the n-time joint probability distribution is known. Furthermore, using (3.17), for $t_n > t_{n-1}$, we get that

Fig. 3.3 Wiener propagator for three values of times $t_2 - t_1$. The *dotted lines* correspond to Cauchy's random walk (Lorentz pdf) for the same times. The half-width for the Gaussian pdf [for case $t_2 - t_1 = \sqrt{5} \simeq 2.23$] is shown with a *double arrow*

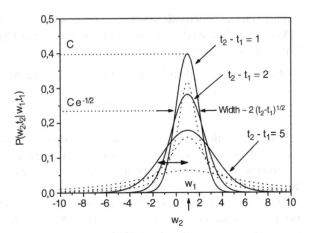

$$P_n(w_n, t_n; w_{n-1}, t_{n-1}; \ldots; w_1, t_1) = \frac{1}{\sqrt{2\pi(t_n - t_{n-1})}} \exp\left(\frac{-(w_n - w_{n-1})^2}{2(t_n - t_{n-1})}\right)$$

$$\times \cdots \frac{1}{\sqrt{2\pi(t_2 - t_1)}} \exp\left(\frac{-(w_2 - w_1)^2}{2(t_2 - t_1)}\right) \frac{1}{\sqrt{2\pi t_1}} \exp\left(\frac{-w_1^2}{2t_1}\right),$$

(3.26)

from which it follows that the Brownian motion is a Gaussian process because

$$\{W(t_1), \ldots, W(t_n)\}$$

are jointly distributed Gaussian **rv**. A random sampling of the **sp** $W(t)$ (for the set $\{t_n\} \in [t_i, t_f]$) will have a statistical weight given by the joint probability distribution (3.26).

Exercise. The time constraint $t_2 \geq t_1$, that appears in the *conditional* pdf is necessary for the correct handling of a Markov **sp**. How can the permutation symmetry $(x_i, t_i) \leftrightarrow (x_j, t_j)$ in the hierarchy of joint distributions be checked?

Excursus. In the limit $\Delta t = |t_j - t_{j-1}| \to 0$ the n-time joint pdf tends to a functional probability distribution. This was the original idea of the functional integral introduced by Wiener in 1921; which then led to the Feynman path integral. It can be shown that all Markov **sp** can completely be characterized by suitable integrals over the random path [7].

Exercise. Show that moments of the Wiener process are given by

$$\langle W(t)^{2n+1}\rangle = 0$$

$$\langle W(t)^{2n}\rangle = \frac{(2n)!}{2^n\, n!} t^n$$

(3.27)

$$\langle W(t)W(s)\rangle = \min(t, s).$$

Hint: when integrated over $dW\,dW'$ use the change of variables $z = W - W'$, and $y = W$.

Optional Exercise. Use Novikov's Theorem (see Chap. 1) to calculate the fourth order moment:

$$\langle W(t_1) W(t_2) W(t_3) W(t_4) \rangle .$$

3.6.1 Increment of the Wiener Process*

Another very important **sp** which appears when differential equations with Gaussian stochastic disturbances are studied (related to the Brownian motion) is the **sp** of increments of a Wiener process:

$$\Delta \mathbf{W}(t) = \mathbf{W}(t + \Delta) - \mathbf{W}(t), \text{ for all value of the parameter } \Delta. \qquad (3.28)$$

The **sp** of the increments of the Wiener process, for an arbitrary fixed time t, is a Gaussian variable with mean zero and variance Δ. Let us see this

$$\begin{aligned}
\langle \Delta W(t)^2 \rangle &= \langle (W(t + \Delta) - W(t))^2 \rangle \\
&= \langle W(t + \Delta)^2 \rangle - 2 \langle W(t + \Delta) \cdot W(t) \rangle + \langle W(t)^2 \rangle \qquad (3.29) \\
&= (t + \Delta) - 2t + t = \Delta.
\end{aligned}$$

The fact that the realizations of a Brownian motion are highly irregular (rough) at different time-scales (magnification factors of the "graphic" $W(t)$) shows that there are some statistically self-similar properties in the register of W versus t.

Exercise. Show that the correlation of two disjoint Wiener increments for times $\{t, s\}$, such that $|t - s| > \Delta$, vanishes identically. Then, since $\Delta \mathbf{W}(t)$ is a Gaussian process with zero mean and $\langle \Delta W(t) \Delta W(s) \rangle = 0$, it follows that Wiener increments are **si** for disjoint times. Remember that **si** is a stronger condition than uncorrelated, but in the Gaussian case it is equivalent.

Excursus. Mandelbrot in 1968 introduced a generalized Wiener process with the name: *The fractional Brownian motion* (fBm)[26] [8]. This modification of Wiener's process enables us to model statistically self-similar (to be precise *self-affine*) objects with different "roughness." In Appendix H, we present the concept of self-similarity, and in Sect. H.2 the definition of the fBm process is given.

---

[26]It is called *fractional Brownian motion* as the variance of **sp** *fBm* is not linear in time but has a fractional exponent.

The tremendous irregularity in the realizations of the Wiener **sp** leads us to conclude that these realizations are not differentiable anywhere. The fact that Wiener increments at disjoint time intervals are **sirv** is another typical characteristic of trajectories of the Brownian motion (or Wiener process). By the same token, the time derivative of the Wiener process is not defined in the ordinary context of an **sp**; this is so because $dW(t)/dt$ is not well defined as random function for all t, in the same way that the Dirac delta function is not in the space of usual functions either.

3.7 Increments of an Arbitrary Stochastic Process*

It is possible to prove that an **sp** with statistically independent increments is a Markov process, however the reverse is not true, as shown by the counterexample of the Ornstein-Uhlenbeck process; see Sect. 3.12.1. To prove this assertion, we note that if $\mathbf{Y}(t)$ is the process under study, the increments (to arbitrary times t_j) can be denoted as follows:

$$\Delta_k = Y(t_k) - Y(t_{k-1}) \tag{3.30}$$

$$\Delta_{k-1} = Y(t_{k-1}) - Y(t_{k-2})$$

$$\cdots$$

$$\Delta_2 = Y(t_2) - Y(t_1),$$

where Δ_k are **rv** and times t_j are sorted increasingly $t_1 < t_2 \cdots < t_{k-1} < t_k$. Using the representation (II) for the **sp**, we can write the random variables $Y(t_j)$ as follows:

$$Y(t_k) = Y(t_{k-1}) + \Delta_k \tag{3.31}$$

$$Y(t_{k-2}) = Y(t_{k-1}) - \Delta_{k-1}$$

$$Y(t_{k-3}) = Y(t_{k-1}) - \Delta_{k-1} - \Delta_{k-2}$$

$$\cdots$$

$$Y(t_1) = Y(t_{k-1}) - \Delta_{k-1} - \cdots - \Delta_2.$$

Note that these equations are only linear transformations among random variables. Consequently, if we set the condition $Y(t_{k-1})$, the conditional probability[27]

$$P(Y(t_k) \mid Y(t_{k-1}))$$

---

[27]Note that we are here using the simplified notation: $P(y_k, t_k \mid y_{k-1}, t_{k-1}) \equiv P(Y(t_k) \mid Y(t_{k-1}))$.

will be determined by the probability distribution $\mathcal{W}(\Delta_k)$. But if we set two conditions: $Y(t_{k-1})$ and $Y(t_{k-2})$, from the second equation (3.31) it follows that Δ_{k-1} will also be fixed. However, since Δ_k, Δ_{k-1} are **sirv**, it follows that $\mathcal{W}(\Delta_k \mid \Delta_{k-1}) = \mathcal{W}(\Delta_k)$. Therefore, the fact that $Y(t_{k-2})$ is fixed does not change the distribution of $Y(t_k)$ conditioned for two times, because Δ_{k-1} does not alter or modify the conditional distribution Δ_k; that is:

$$P(Y(t_k) \mid Y(t_{k-1}), Y(t_{k-2})) = P(Y(t_k) \mid Y(t_{k-1})).$$

This is the first of the conditions that must be met for a process to be Markovian. In general from (3.31) and the fact that $\{\Delta_j\}$ are **sirv**,[28] we conclude that

$$P(y(t_k) \mid y(t_{k-1}), y(t_{k-2}), \cdots, y(t_1)) = P(y(t_k) \mid y(t_{k-1})),$$

which is the necessary and sufficient condition for a Markov process.

3.8 Convergence Criteria*

In probability theory, there is no single way to define the concept of limit[29] of a succession of **rvs** [2, 9, 10]. It is therefore necessary to consider the various definitions of convergence to the limit, in particular the limit in *mean square* is one of the most frequently used in statistical physics. We define here four convergence criteria.

Consider a probability space[30] $\mathcal{S}$ and a sequence of **rv** $\{\Omega_n\}$, with $n \in \mathcal{N}$, defined on $\mathcal{S}$. Then we denote that the limit of this sequence, as $n \to \infty$, is[31]

$$\Omega = \lim_{n \to \infty} \Omega_n. \tag{3.32}$$

Here Ω is, in some sense, the **rv** to which the sequence of **rvs** $\{\Omega_n\}$ converges. The various criteria for the convergence arise when one considers the different possibilities for defining probabilities on $\mathcal{S}$.

---

[28] And then $\mathcal{W}(\Delta_k \mid \Delta_{k-1}, \Delta_{k-2}, \cdots, \Delta_2) = \mathcal{W}(\Delta_k)$.

[29] Recall that a sequence of numbers N_n tends to limit N if, given $\epsilon > 0$, we can find a number n_0 such that $\mid N_n - N \mid < \epsilon$ for all $n > n_0$.

[30] In a mathematical context the probability space (or the probability triple) consists of the sample space, the field of all events and a probability measure on it. This probability is a function from the field of events onto the interval $[0, 1]$, that is the probability is defined for events and not *only* for the elementary outcomes of the sample space.

[31] For a specific measurement, this sequence of numbers might or might not converge. This fact shows that the notion of convergence of **rvs** might have several interpretations.

3.8.1 Definitions

1. Convergence with certainty (probability 1): Ω_n converges with Prob.1 to Ω for
   all $\omega \in S$ (except a set with zero probability) if

$$\Omega(\omega) = \lim_{n \to \infty} \Omega_n(\omega), \qquad (3.33)$$

   and we write[32]

$$ac - \lim_{n \to \infty} \Omega_n \to \Omega.$$

2. Convergence in mean: Ω_n converges to Ω in mean square if

$$\lim_{n \to \infty} \int_S (\Omega_n(\omega) - \Omega(\omega))^2 P_\Omega(\omega)\, d\omega \equiv \lim_{n \to \infty} \left\langle (\Omega_n - \Omega)^2 \right\rangle = 0, \qquad (3.34)$$

   and we write

$$ms - \lim_{n \to \infty} \Omega_n \to \Omega.$$

3. Convergence in probability: Ω_n converges in probability to Ω if the probability
   of deviation tends to zero, that is, if for all $\epsilon > 0$ it is met that

$$\lim_{n \to \infty} \text{Prob.} \left(|\Omega_n - \Omega| > \epsilon \right) = 0,$$

   that is:

$$\lim_{n \to \infty} \int_S \Theta \left[|\Omega_n(\omega) - \Omega(\omega)| - \epsilon \right] P_\Omega(\omega)\, d\omega = 0, \qquad (3.35)$$

   where $\Theta[z]$ is the step function. Then, we write

$$\text{Prob.} \lim_{n \to \infty} \Omega_n \to \Omega.$$

4. Convergence in distribution. This is an even weaker criterion, we say that Ω_n
   converges in distribution to Ω if for any continuous and bounded function $f(X)$,
   it holds that

$$\lim_{n \to \infty} \langle f[\Omega_n] \rangle = \langle f[\Omega] \rangle, \qquad (3.36)$$

---

[32]Note that, in this case, the limit is a number. The sequence consists of all possible events such
that $\Omega(\omega) = \lim_{n \to \infty} \Omega_n(\omega)$.

that is:

$$\lim_{n\to\infty} \int_S [f\,(\Omega_n(\omega)) - f\,(\Omega(\omega))]\,P_\Omega(\omega)\,d\omega = 0.$$

In particular, if we take

$$f(X) = \exp(ikX),$$

we observe that the characteristic functions converge into each other. That is, the probability distribution of Ω_n converges to the probability distribution of Ω. Note that in this case the succession $\Omega_n\,(\omega)$ need not to converge $\forall\omega$.

Note. The following relationships between different convergence criteria can be proved

$$ac - \lim_{n\to\infty} \Omega_n \to \Omega \implies \text{Prob.} \lim_{n\to\infty} \Omega_n \to \Omega$$

$$ms - \lim_{n\to\infty} \Omega_n \to \Omega \implies \text{Prob.} \lim_{n\to\infty} \Omega_n \to \Omega$$

$$\text{Prob.} \lim_{n\to\infty} \Omega_n \to \Omega \implies \Omega_n \text{ converges to } \Omega \text{ in distribution.}$$

Exercise. Consider a general **sp** $X(t)$, $\forall t \geq 0$ on $\mathcal{D}_x$. Analyze the convergence in probability of the sequence of **rvs** $X_\Delta \equiv X(t + \Delta)$ to the fixed value $X(t) = z$ for $\Delta \to 0$. That is, write down explicitly the prescription

$$\text{Prob.} \lim_{\Delta\to 0} X_\Delta \to z.$$

Excursus (**Ergodicity**). Consider a stationary **sp** $X(t)$, $\forall t \geq 0$ on $\mathcal{D}_x$ and define the "step" function (for $x \in \mathcal{D}_x$ and $\epsilon \gtrsim 0$)

$$I_x^\epsilon\,[z] = \begin{cases} 1, & \text{if } z \in [x - \epsilon, x + \epsilon] \\ 0, & \text{if } z \notin [x - \epsilon, x + \epsilon] \end{cases}.$$

A *strong* Ergodic theorem establishes that [10]

$$ac - \lim_{T\to\infty} \frac{1}{T} \int_0^T I_x^\epsilon\,[X(t)]\,dt \to \int_{x-\epsilon}^{x+\epsilon} P_{st}\,(x')\,dx',$$

where $P_{st}\,(x)$ is the stationary distribution of the **sp** $X(t)$. That is, the probability $P_{st}\,(x)\,dx$ equals the fraction of time an arbitrary realization $X(t)$ of the **sp** $X(t)$ spends in an infinitesimal neighborhood of x.

3.8.2 Markov Theorem (Ergodicity)

The concept of ergodicity arises when trying to establish the equivalence between the average distribution (statistical) versus time average (sampling). Consider here the situation when the time average of the **sp** $X(t)$ is taken over a set of discrete data, i.e., collecting experimental values $X(t_j)$. Then given a sequence $X(t_j)$, we write the sample mean as

$$X_n = \frac{1}{n} \sum_{j=1}^{n} X(t_j).$$

Clearly, X_n is a **rv** whose values depend on the experimental outcome. The interesting question is whether the succession X_n converges or not to a certain quantity η when $n \to \infty$. The answer is given by the following theorem due to Markov [9]. If the **rvs** $X(t_j)$ are such that the mean η_n of X_n tends to the limit η and its variance σ_n^2 tends to 0 when $n \to \infty$; that is, if in the limit $n \to \infty$, we get:

$$\langle X_n \rangle = \eta_n \to \eta \tag{3.37}$$

$$\left\langle (X_n - \eta_n)^2 \right\rangle = \sigma_n^2 \to 0, \tag{3.38}$$

then

$$\mathrm{ms} - \lim_{n \to \infty} X_n \to \eta.$$

Proof This is based on the use of the triangle inequality

$$| X_n - \eta |^2 \leq (| X_n - \eta_n | + | \eta_n - \eta |)^2 .$$

Indeed, taking **mv** in this inequality, it follows that

$$\left\langle (X_n - \eta)^2 \right\rangle \leq \left\langle (X_n - \eta_n)^2 \right\rangle + | \eta_n - \eta |^2 + 2 \langle | X_n - \eta_n | \rangle | \eta_n - \eta | .$$

Using (3.37) and (3.38) in this expression, we get: $\lim_{n \to \infty} \left\langle (X_n - \eta)^2 \right\rangle \to 0$, which completes the proof.

3.8.3 Continuity of the Realizations

To get an idea of the irregularity of each realization of the Wiener process, first consider the concept of time derivative in mean square[33]:

$$\lim_{\Delta t \to 0} \frac{\Delta W}{\Delta t} \to \lim_{\Delta t \to 0} \sqrt{\left\langle (W(t + \Delta t) - W(t))^2 \right\rangle} \frac{1}{\Delta t}$$

$$= \lim_{\Delta t \to 0} \sqrt{\left\langle (W(t + \Delta t))^2 \right\rangle + \left\langle (W(t))^2 \right\rangle - 2 \left\langle W(t + \Delta t)W(t) \right\rangle} \frac{1}{\Delta t}$$

$$= \lim_{\Delta t \to 0} \frac{\sqrt{\Delta t}}{\Delta t} \to \infty.$$

This divergence shows that the derivative of Wiener's process is not defined in any point of the realization. However, the realizations of the Wiener process are continuous as can be seen from the following result.

We can study the continuity of any Markov **sp** using the Lindeberg criteria. We shall say, with Prob. 1, the realizations are continuous functions of t if for all $\epsilon > 0$ it holds that[34]

$$\lim_{\Delta t \to 0} \frac{1}{\Delta t} \int_{|x-z|>\epsilon} dx\, P(x, t + \Delta t \mid z, t) \to 0, \tag{3.39}$$

uniformly in z, t and Δt. This means that the probability to be outside some small neighborhood around z (at the instant $t + \Delta t$) goes to zero faster than Δt.

Example. Consider the Wiener process and show that indeed the **sp** $W(t)$ has continuous realizations (but not differentiable, as seen above). In this case the propagator is

$$P(w_2, t_2 \mid w_1, t_1) = \frac{1}{\sqrt{2\pi(t_2 - t_1)}} \exp\left(\frac{-(w_2 - w_1)^2}{2(t_2 - t_1)}\right);$$

then

$$\frac{1}{\sqrt{2\pi\Delta t}} \int_{|w_2 - w_1|>\epsilon} dw_2 \exp\left(\frac{-(w_2 - w_1)^2}{2\Delta t}\right) = \frac{2}{\sqrt{2\pi\Delta t}} \int_{\epsilon}^{\infty} dZ \exp\left(\frac{-Z^2}{2\Delta t}\right)$$

$$= \operatorname{erfc}\left(\epsilon/\sqrt{2\Delta t}\right).$$

---

[33]Limit of the (infinitesimal) difference quotient.

[34]This condition makes large displacements improbable; compare with the exercise of Sect. 3.8. This criterion was introduced by Lindeberg in 1936, in the hope that it would be a necessary and sufficient for the continuity of the realizations. See p. 333 in [11].

Applying Lindeberg's criterion and the expansion erfc $(\chi) \simeq e^{-\chi^2}/\chi\sqrt{\pi}$ when $\chi \to \infty$, we get

$$\lim_{\Delta t \to 0} \frac{1}{\Delta t}\text{erfc}\,(\epsilon/\sqrt{2\Delta t}) \simeq \lim_{\Delta t \to 0} \frac{1}{\Delta t}\exp\left[-(\epsilon/\sqrt{2\Delta t})^2\right]\frac{1}{(\epsilon\sqrt{\pi}/\sqrt{2\Delta t})} \to 0,$$

from which it follows, from (3.39), that the realizations are continuous with Prob. 1.

Optional Exercise. Show that a Cauchy random walk has discontinuous realizations. In general, we can define a Lévy **sp** of which the Cauchy random walk is a particular case[35] [11–14].

Optional Exercise. Consider a realization of Cauchy's random walk $X(t)$. Study the asymptotic limit $X(t)/t$ for $t \to \infty$ and compare this result with the case for the Wiener **sp**. Hint: use the convergence criterion in probability.

3.9   Gaussian White Noise

The relationship between Gaussian white noise (GWN)[36] and Brownian motion (Wiener process) is easy to establish. Since noise is Gaussian, we can fully characterize it knowing only the **mv** and its correlation, that is:

$$\langle \xi(t) \rangle = 0 \tag{3.40}$$

$$\langle \xi(t)\xi(s) \rangle = \delta(t - s). \tag{3.41}$$

The Dirac-delta, δ, in the correlation (3.41) indicates the singular character of the white noise; i.e., $\xi(t)$ is not really a "usual" random function of argument t; that is, $\xi(t)$ is not a good **sp** in the ordinary context. However, Eqs. (3.40) and (3.41) completely define the GWN.[37] The Wiener process is related to a Gaussian white noise by the integral equation

$$\int dW(t) = \int \xi(t)\,dt. \tag{3.42}$$

---

[35] See second exercise of Sect. 3.3.1.

[36] A realization of the GWN can be imagined as a sequence of "narrow" pulses with small random amplitudes (positive and negative). In the limit that pulses (with infinitesimal amplitudes) are Dirac-delta functions (infinitely dense in the line of time) the realization becomes a GWN.

[37] In general we use the terms noise or **sp** interchangeably, and we use Greek letters to characterize it when we refer to differential equations.

Often this equation is written as a differential equation,

$$\frac{dW(t)}{dt} = \xi(t), \tag{3.43}$$

but what is meant is (3.42). This differential equation is the simplest yet fundamental stochastic differential equation (**sde**). We note here the presence of a coefficient that is an **sp**, so knowing the statistics properties of $\xi(t)$, it is possible to know the statistics properties of the **sp** $W(t)$. However, due to the singular character of the **sp** $\xi(t)$, there may be some ambiguities in presenting a differential calculus. These difficulties will be discussed in the following sections.

Exercise. Using moments and correlations of the **sp** $\xi(t)$, show that moments and correlations that define the Wiener process can be found using (3.42).

3.9.1 Functional Approach

If we extend the definition of **sp** to functional spaces,[38] we can reinterpret the relationship between the **sp** $W(t)$ and the GWN. Consider the Gaussian white noise as a functional operator $\xi[\bullet]$, whose argument is an arbitrary function $f(t)$. Now, to get a well-defined **rv** $\xi[f]$, we will "need" to use a "test" function $f(t)$, throughout its domain, rather than fix the time t as it is done in an ordinary **sp**; that is:

$$\xi[f] = \int_0^\infty \xi(t)f(t)\,dt. \tag{3.44}$$

Here we have adopted the domain $t \in [0, \infty]$. This expression is correct for any smooth $f(t)$, i.e., bounded and rapidly decreasing at the integration limits. Thus, now the **rv** $\xi[f]$ is well defined, even though some of the realizations $\xi(t)$ are singular objects such the Dirac delta function.

Example. From the expression (3.44) we can prove the equality

$$\langle \xi[f]^2 \rangle = \int_0^\infty f(t)^2\,dt. \tag{3.45}$$

Then, using $\xi(t) = dW(t)/dt$ with the moments of the **sp** $W(t)$, it follows from (3.44) that

---

[38]This time we do not go into the technical aspect of the functional approach, and only give basic "recipes" to properly manage this type of analysis.

$$\langle \xi[f]^2 \rangle = \left\langle \left(\int_0^\infty \xi(t) f(t)\, dt \right)^2 \right\rangle \tag{3.46}$$

$$= \left\langle \left(\int_0^\infty \frac{dW}{dt} f(t)\, dt \right)^2 \right\rangle$$

$$= \left\langle \left(-\int_0^\infty W(t) \frac{df(t)}{dt}\, dt \right)^2 \right\rangle,$$

where we have integrated by parts ($f(t)$ being smooth) used $f(\infty) = 0$, and also $W(0) = 0$ since we are assuming a Wiener **sp** with zero mean. Using $\langle W(t_1) W(t_2) \rangle = \min(t_1, t_2)$ in (3.46), we finally obtain

$$\langle \xi[f]^2 \rangle = \left\langle \int_0^\infty dt_1 \int_0^\infty dt_2\, f'(t_1) f'(t_2) W(t_1) W(t_2) \right\rangle$$

$$= \int_0^\infty dt_1 \int_0^\infty dt_2\, f'(t_1) f'(t_2) \min(t_1, t_2)$$

$$= \int_0^\infty dt_1\, f'(t_1) \left(\int_0^{t_1} dt_2\, f'(t_2) t_2 + t_1 \int_{t_1}^\infty dt_2\, f'(t_2) \right)$$

$$= \int_0^\infty dt_1\, f'(t_1) \left(-\int_0^{t_1} dt_2\, f(t_2) \right)$$

$$= -\left[f(t) \int_0^t f(t_2)\, dt_2 \right]_0^\infty + \int_0^\infty f(t_1)^2\, dt_1 = \int_0^\infty f(t_1)^2\, dt_1,$$

where we integrate by parts twice (fourth and fifth lines). On the other hand, using relationships (3.40) and (3.41), which define the GWN, Eq. (3.45) is immediately verified, namely:

$$\langle \xi[f]^2 \rangle = \left\langle \int_0^\infty \xi(t_1) f(t_1)\, dt_1 \int_0^\infty \xi(t_2) f(t_2)\, dt_2 \right\rangle$$

$$= \int_0^\infty f(t_1)\, dt_1 \int_0^\infty f(t_2) \langle \xi(t_2) \xi(t_1) \rangle\, dt_2$$

$$= \int_0^\infty f(t_1)\, dt_1 \int_0^\infty f(t_2) \delta(t_1 - t_2)\, dt_2$$

$$= \int_0^\infty f(t)^2\, dt.$$

Optional Exercise (**Higher Moments**). Using (3.44) and moments of the Wiener process, calculate $\langle \xi[f]^4 \rangle$; compare this result with that obtained from direct calculation using (3.40) and (3.41) that define the GWN.

3.10 Gaussian Processes

3.10.1 Non-singular Case

An **sp** $X(t)$ is Gaussian if all Kolmogorov's hierarchy is characterized by Gaussian joint probabilities.

Let us analyze $P_n(x_n, t_n; \cdots ; x_1, t_1)$: since we have n **rvs** its joint probability density must be of the form

$$P_n(x_n, t_n; \cdots ; x_1 t_1) = \frac{(\det A)^{1/2}}{(2\pi)^{n/2}} \exp\left[\frac{-1}{2} \sum_{j,l}^{n} (x_j - \langle X(t_j)\rangle) A_{jl} (x_l - \langle X(t_l)\rangle)\right],$$

$$(3.47)$$

where

$$\left(A^{-1}\right)_{jl} \equiv A_{jl}^{-1}(\{t_n\}) \tag{3.48}$$
$$= \langle (X(t_j) - \langle X(t_j)\rangle) (X(t_l) - \langle X(t_l)\rangle)\rangle$$
$$\equiv \langle\langle X(t_j)X(t_l)\rangle\rangle,$$

in complete analogy with the case of n Gaussian variables. In the context of stochastic processes, $\langle\langle X(t_j)X(t_l)\rangle\rangle$ is a correlation function over time. So from formula (3.47), we conclude that the **sp** $X(t)$ is completely characterized knowing its first two cumulants, i.e., $\langle\langle X(t)\rangle\rangle$ and $\langle\langle X(t)X(t')\rangle\rangle$.

3.10.2 The Singular Case (White Correlation)*

Another way to define an **sp** $X(t)$ is by using the characteristic functional[39]:

$$G_X([k]) = \left\langle \exp\left[i \int k(t)X(t)\, dt\right]\right\rangle_{P([X(\bullet)])}.$$

Then, all formulas valid for n-dimensional random variables will be generalized using the prescription:

$$\frac{\partial}{\partial k_j} \rightarrow \frac{\delta}{\delta k(t_j)} \tag{3.49}$$

---

[39]Comparing with the notation of Eq. (3.4), we note that $P([X(\bullet)])$ represents the functional probability of the ensemble of all realizations of the **sp** $X(t)$. See also the advanced exercise 1.15.10.

$$\sum_j x_j \rightarrow \int x(t) \, dt. \tag{3.50}$$

For example, the moments and correlations are given by

$$\langle X(t)^m X(s)^n \rangle = \frac{1}{i^{m+n}} \frac{\delta^m}{\delta k(t)^m} \frac{\delta^n}{\delta k(s)^n} G_X([k]) \bigg|_{k(t)=0}.$$

That is, for the second cumulant we have

$$\langle\langle X(t) X(s) \rangle\rangle = \frac{1}{i^2} \frac{\delta}{\delta k(t)} \frac{\delta}{\delta k(s)} \ln G_X([k]) \bigg|_{k(t)=0}.$$

Guided Exercise. Calculate the characteristic functional of the GWN $\xi(t)$. Since the **sp** is Gaussian with zero mean, we know that the only cumulant that appears is the second one. Using the definition of the characteristic functional in terms of an expansion in cumulants, we note from (3.6) that

$$\ln G_\xi([k]) = \frac{-1}{2!} \int \int k(t_1) k(t_2) \langle\langle \xi(t_1) \xi(t_2) \rangle\rangle \, dt_1 \, dt_2.$$

In the singular case the correlation is white: $\langle\langle \xi(t_1) \xi(t_2) \rangle\rangle = \Gamma_2 \, \delta(t_1 - t_2)$, then we finally obtain

$$G_\xi([k]) = \exp\left[-\frac{\Gamma_2}{2} \int k(t)^2 \, dt \right], \tag{3.51}$$

which is the desired functional.

Optional Exercise. Using $G_\xi[k(t)]$, with $\Gamma_2 = 1$, show that the moment and correlation of the GWN $\xi(t)$ are those given in (3.40) and (3.41).

Excursus. The generating functional of the cumulants of an **sp** $\xi(t)$ (i.e., the functional $\ln G_\xi([k])$) cannot be a polynomial of a degree greater than two, but must be a series to ensure positivity of the probability. [Reference: Marcinkiewicz' theorem; Math Z. 44, 612, (1939); see also: Rajagopal and Sudarshan; Phys. Rev. A 10, 1852 (1974)].

Optional Exercise (**White Noise**). In Sect. 3.1.5 we have shown, in general, the relationship between the joint probability and the characteristic functional $G_\xi([k])$. From the Gaussian characteristic functional, show that, if the correlation is white, it is not feasible to obtain an expression for the probability distribution. This is just a consequence of the singularity of a white noise.

3.11   Spectral Density of Fluctuations (Nonstationary sp)*

Using the characteristic function $G_X(k, t)$ it is possible, in a direct way, to calculate the power spectrum of fluctuations of the **sp** $\mathbf{X}(t)$. Suppose the **sp** $\mathbf{X}(t)$ satisfies—in distribution—the scale relationship (see Appendix H.2)

$$\frac{\mathbf{X}(\Lambda t)}{\Lambda^H} \doteq \mathbf{X}(t), \tag{3.52}$$

where H is a parameter that can be inferred from the invariance of the associated characteristic function

$$G_X(\frac{k}{\Lambda^H}, \Lambda t) = G_X(k, t); \tag{3.53}$$

for example, in the case of Brownian motion it is $H = \frac{1}{2}$.

From the scale relation (3.52), it is possible to estimate the power spectrum as follows [15]. Consider a fixed value for the parameter Λ; then, we can define a new **sp** $\mathbf{Y}(t)$ as:

$$\mathbf{Y}(t, T) = \begin{cases} \mathbf{Y}(t) = \Lambda^{-H} \mathbf{X}(\Lambda t) & \text{if } 0 < t < T \\ 0 & \text{elsewhere.} \end{cases} \tag{3.54}$$

Adopting the notation for the Fourier transforms[40]

$$\begin{array}{ll}
F_X(f, T), \ F_Y(f, T) & \text{Fourier transform of } \mathbf{X}(t, T) \text{ and } \mathbf{Y}(t, T) \\
S_X(f, T), \ S_Y(f, T) & \text{Spectral density of } \mathbf{X}(t, T) \text{ and } \mathbf{Y}(t, T) \qquad (3.55) \\
S_X(f), \ S_Y(f) & \text{Spectral density of } \mathbf{X}(t) \text{ and } \mathbf{Y}(t),
\end{array}$$

then, for example:

$$F_Y(f, T) = \int_{-\infty}^{\infty} \mathbf{Y}(t, T) \exp(-2\pi i f t) \, dt = \int_0^T \mathbf{Y}(t) \exp(-2\pi i f t) \, dt$$

and using that

$$S_Y(f, T) = \frac{1}{T} \langle |\ F_Y(f, T)\ |^2 \rangle, \text{ etc.}$$

---

[40]On the definition of spectral density (power spectrum) see Sect. 5.1. A detailed presentation on this topic can be seen in [16].

We can now calculate $F_Y(f, T)$ using the scaling (3.52), from which it follows that

$$F_Y(f, T) = \int_0^T \mathbf{Y}(t) \exp(-2\pi ift)\, dt \tag{3.56}$$

$$= \frac{1}{\Lambda} \int_0^T \Lambda^{-H} \mathbf{X}(\Lambda t)\, \exp(\frac{-2\pi if\Lambda t}{\Lambda})\Lambda\, dt$$

$$= \int_0^{T\Lambda} \Lambda^{-H-1} \mathbf{X}(t')\, \exp(\frac{-2\pi ift'}{\Lambda})\, dt'.$$

Therefore, from (3.56) it is obtained that

$$F_Y(f, T) = \frac{1}{\Lambda^{H+1}} F_X(\frac{f}{\Lambda}, T\Lambda). \tag{3.57}$$

Then, $S_Y(f, T)$ is given by

$$S_Y(f, T) = \frac{1}{T}\langle | F_Y(f, T) |^2\rangle \tag{3.58}$$

$$= \frac{1}{T}\left(\frac{1}{\Lambda^{H+1}}\right)^2 \langle | F_X(\frac{f}{\Lambda}, T\Lambda) |^2\rangle$$

$$= \frac{1}{\Lambda^{2H+1}} S_X(\frac{f}{\Lambda}, T\Lambda).$$

In the limit $T \to \infty$ the spectral density $S_Y(f)$ satisfies the relation

$$S_Y(f) = \lim_{T\to\infty} S_Y(f, T),$$

then, from (3.58) we conclude that

$$S_Y(f) = \frac{1}{\Lambda^{2H+1}} S_X(\frac{f}{\Lambda}). \tag{3.59}$$

Since $\mathbf{Y}(t)$ is just the **sp** $\mathbf{X}(t)$, their spectral densities must match. To see this, take for example $\Lambda \equiv 1$, from which it follows $S_Y(f) = S_X(f)$. So from the relationship (3.59) we finally obtain

$$S_X(f) = \frac{1}{\Lambda^{2H+1}} S_X(\frac{f}{\Lambda}). \tag{3.60}$$

Now, formally putting $f = 1$ and replacing $1/\Lambda$ by f in (3.60), we get the desired result:

$$S_X(f) \propto \frac{1}{f^{2H+1}}. \tag{3.61}$$

This expression gives the spectrum of fluctuations from a *nonstationary* **sp** that satisfies the scale relation: $\mathbf{X}(\Lambda t) \doteq \Lambda^H \mathbf{X}(t)$.

Example (**Spectrum of the Wiener sp**). The characteristic function of the diffusion process[41] satisfies the invariance condition:

$$G_W(\frac{k}{\Lambda^{1/2}}, \Lambda t) = \exp\left(-\frac{\Lambda t}{2}\left(\frac{k}{\Lambda^{1/2}}\right)^2\right) = G_W(k, t),$$

then the scale relation is

$$\frac{W(\Lambda t)}{\Lambda^{1/2}} \doteq W(t).$$

Thus, using (3.61) with $H = 1/2$, it follows that

$$S_W(f) \propto \frac{1}{f^2},$$

which is the power spectrum of Brownian motion, see also advanced exercise 3.20.8.

Example (**Spectrum of the GWN**). The characteristic function of GWN is singular (not defined), due to the occurrence of a squared Dirac-delta, δ. This fact can be seen from the characteristic functional of the GWN (3.51). That is, $G_\xi(k, t)$ is obtained by evaluating (3.51) with the test function $k(t) = k\delta(s - t)$, i.e.:

$$G_\xi\left([k(t) = k\delta(s - t)]\right) = G_\xi(k, t) = \exp\left[-\frac{\Gamma_2}{2}k^2 \int_0^\infty \delta(s - t)^2\, ds\right]. \qquad (3.62)$$

Despite the occurrence of $\delta(s - t)^2$ in the integrand, we can study "formally" the invariance property of $G_\xi(k, t)$ from (3.62); let us see:

$$G_\xi\left(\frac{k}{\Lambda^{-1/2}}, \Lambda t\right) = \exp\left[-\frac{\Gamma_2}{2}\left(\frac{k}{\Lambda^{-1/2}}\right)^2 \int_0^\infty \delta(s - \Lambda t)^2\, ds\right] \qquad (3.63)$$

$$= \exp\left[-\frac{\Gamma_2}{2}\Lambda k^2 \int_0^\infty \left(\delta(\Lambda s' - \Lambda t)\right)^2 \Lambda\, ds'\right].$$

Using the property of the Dirac-delta: $\delta(\Lambda s' - \Lambda t) = \delta(s' - t)/\Lambda$, in (3.63), we obtain

$$G_\xi\left(\frac{k}{\Lambda^{-1/2}}, \Lambda t\right) = \exp\left[-\frac{\Gamma_2}{2}\Lambda k^2 \int_0^\infty \left(\frac{\delta(s' - t)}{\Lambda}\right)^2 \Lambda\, ds'\right] = G_\xi(k, t),$$

which "formally" means that realizations of this noise verify the scale relation:

$$\frac{\xi(\Lambda t)}{\Lambda^{-1/2}} \doteq \xi(t). \qquad (3.64)$$

---

[41] To find $G_W(k, t)$, take the Fourier transform of Eq. (3.25).

Note that this relationship implies a negative exponent H. Then, using (3.61) with $H = -1/2$, it follows that

$$S_\xi(f) \propto \frac{1}{f^0} = \text{constant};$$

that is, a white power spectrum.

Excursus. Using the scale relation $\mathbf{X}(\Lambda t) \doteq \Lambda^H \mathbf{X}(t)$ it is also possible to study the fractal dimension of the realizations of **sp** $\mathbf{X}(t)$; that is, the self-similarity at different scales of the graph of X versus t. See, for example, Appendix H and references [13, 15, 17].

Note. An interesting result obtained by Wiener and Khinchin establishes the relationship between the power spectrum $S_X(f)$ of a stationary and ergodic **sp** $\mathbf{X}(t)$ with its correlation function $\langle\langle X(t)X(t')\rangle\rangle$. See Chap. 5.

3.12 Markovian and Gaussian Processes

The kind of processes that are simultaneously Markovian and Gaussian have a universality that can be characterized by their correlation function, as evidenced by the following theorem due to Doob [1, 11].

Consider the first necessary condition. Since it is Gaussian, the **sp** $\mathbf{X}(t)$ is characterized only by the cumulants $\langle\langle X(t)\rangle\rangle$ and $\langle\langle X(t)X(t')\rangle\rangle$. For simplicity and without loss of generality, let $\langle\langle X(t)\rangle\rangle = 0$, then the conditional probability

$$P(x_2, t_2 \mid x_1, t_1) = P_2(x_2, t_2; x_1, t_1)/P_1(x_1, t_1),$$

for $t_1 \le t_2$, is given by the formula

$$P(x_2, t_2 \mid x_1 t_1) = \frac{\frac{(\det A)^{1/2}}{2\pi} \exp\left[\frac{-1}{2} \sum_{(j,l)=1}^{2} x_j A_{jl} x_l\right]}{\frac{1}{\sqrt{2\pi\sigma^2(t_1)}} \exp\left[\frac{-1}{2} x_1^2/\sigma^2(t_1)\right]}. \tag{3.65}$$

Here we have used (3.47) to write P_2, P_1 and $(A^{-1})_{jl} = \langle X(t_j)X(t_l)\rangle$; then, we conclude that $\det A^{-1} = \sigma^2(t_1)\sigma^2(t_2) - \langle X(t_1)X(t_2)\rangle^2$; and in the one-time case we have $A^{-1} = \langle X(t)X(t)\rangle = \sigma^2(t)$.

Using

$$A = \begin{pmatrix} \langle X(t_1)X(t_1)\rangle & \langle X(t_1)X(t_2)\rangle \\ \langle X(t_2)X(t_1)\rangle & \langle X(t_2)X(t_2)\rangle \end{pmatrix}^{-1}$$

$$= \frac{1}{\det A_{jl}^{-1}} \begin{pmatrix} \sigma^2(t_2) & -\langle X(t_1)X(t_2)\rangle \\ -\langle X(t_2)X(t_1)\rangle & \sigma^2(t_1) \end{pmatrix},$$

the conditional pdf (3.65) is written as

$$P(x_2, t_2 \mid x_1, t_1) = \frac{\sigma(t_1)}{\sqrt{2\pi}\sqrt{\sigma^2(t_1)\sigma^2(t_2) - \langle X(t_1)X(t_2)\rangle^2}}$$

$$\times \exp\left[\frac{-\left(x_2^2\sigma^2(t_1) - 2x_1x_2\langle X(t_1)X(t_2)\rangle + x_1^2\sigma^2(t_2)\right)}{2\left(\sigma^2(t_1)\sigma^2(t_2) - \langle X(t_1)X(t_2)\rangle^2\right)} + \frac{x_1^2}{2\sigma^2(t_1)}\right].$$

On the other hand, the term in brackets in the exponential can be written as:

$$\exp\left[\frac{-\left(x_2 - \langle X(t_1)X(t_2)\rangle \frac{x_1}{\sigma^2(t_1)}\right)^2}{2\sigma^2(t_2)\left(1 - \frac{\langle X(t_1)X(t_2)\rangle^2}{\sigma^2(t_1)\sigma^2(t_2)}\right)}\right],$$

whence we can rewrite the conditional probability (or the propagator). Defining a normalized correlation or correlator:

$$\rho(t_2, t_1) = \frac{\langle X(t_1)X(t_2)\rangle}{\sigma(t_1)\sigma(t_2)}, \quad |\rho(t_2, t_1)| < 1, \quad t_2 > t_1 \tag{3.66}$$

we get

$$P(x_2, t_2 \mid x_1, t_1) = \frac{1}{\sqrt{2\pi\sigma^2(t_2)\left(1 - \rho(t_2, t_1)^2\right)}} \exp\left[\frac{-\left(x_2 - \frac{\rho(t_2,t_1)\sigma(t_2)}{\sigma(t_1)}x_1\right)^2}{2\sigma^2(t_2)\left(1 - \rho(t_2, t_1)^2\right)}\right].$$

$$\tag{3.67}$$

Exercise. Justify the condition $|\rho(t_2, t_1)| < 1$. Hint: see the constraint presented in Sect. 1.9.1.

Note that if the process is stationary, it must be, for all τ, that

$$\rho(t_2, t_1) = \rho(t_2 - \tau, t_1 - \tau).$$

In addition, it is clear that if the process is Markovian the only time dependence in the propagator (3.67) must be through the times t_1 and t_2, indicating that the correlator $\rho(t_1, t_2)$ cannot depend parametrically on any other past or future time.

Using the conditional probability (3.67), we can calculate any conditioned **mv.**[42] Then, for example, for $t_1 \le t_3$ we have

$$\langle X(t_3) \rangle_{X(t_1)=x_1} = \int x_3 P(x_3, t_3 \mid x_1, t_1) \, dx_3 \qquad (3.68)$$

$$= \frac{\rho(t_3, t_1)\sigma(t_3)}{\sigma(t_1)} x_1.$$

On the other hand, if the process is Markovian the conditional pdf has to satisfy the Chapman-Kolmogorov equation. Then, we can calculate the conditional moment, Eq. (3.68), in a different way: for $t_1 \le t_2 \le t_3$ we have

$$\langle X(t_3) \rangle_{X(t_1)=x_1} = \int x_3 \left[\int P(x_3, t_3 \mid x_2, t_2) P(x_2, t_2 \mid x_1, t_1) \, dx_2 \right] dx_3. \qquad (3.69)$$

Interchanging the order of integration we have

$$\langle X(t_3) \rangle_{X(t_1)=x_1} = \frac{\rho(t_3, t_2)\sigma(t_3)}{\sigma(t_2)} \int x_2 P(x_2, t_2 \mid x_1, t_1) \, dx_2$$

$$= \frac{\rho(t_3, t_2)\sigma(t_3)}{\sigma(t_2)} \frac{\rho(t_2, t_1)\sigma(t_2)}{\sigma(t_1)} x_1.$$

Then, comparing both results we observe

$$\frac{\rho(t_3, t_1)\sigma(t_3)}{\sigma(t_1)} x_1 = \frac{\rho(t_3, t_2)\sigma(t_3)}{\sigma(t_2)} \frac{\rho(t_2, t_1)\sigma(t_2)}{\sigma(t_1)} x_1,$$

which is equivalent to the following necessary condition (Doob's Theorem)

$$\rho(t_3, t_1) = \rho(t_3, t_2)\rho(t_2, t_1), \quad t_3 \ge t_2 \ge t_1. \qquad (3.70)$$

Exercise. If the process is stationary, (3.70) can be written as: $\rho(\tau) = \rho(\mu)\rho(\epsilon)$, with $\tau = \mu + \epsilon$. Evaluate $d\rho(\tau)/d\tau$ and find the solution $\rho(\tau) = e^{-|\rho'(0)|\tau}$, $\forall \tau \ge 0$, fulfilling (3.66).

Guided Exercise (**Sufficient Condition**). Conversely, we see now that (3.70) is a sufficient condition. This example shows how we can understand that if an **sp** is Gaussian and meets (3.70), the **sp** will be Markovian. Consider the conditional pdf with $t_4 > t_3 > t_2 > t_1$

---

[42] See Sect. 1.9.3.

$$P(x_4, t_4 \mid x_3, t_3; x_2, t_2; x_1, t_1) \equiv P(X_4 \mid X_3; X_2; X_1)$$

If we can show that this distribution does not depend on the conditions $\{x_2, t_2; x_1, t_1\}$ and, in general, if $P(x_j, t_j \mid x_{j-1}, t_{j-1}; \cdots; x_1, t_1) = P(x_j, t_j \mid x_{j-1}, t_{j-1})$, we will have proved that the **sp** is Markovian. For the purposes of simplifying the demonstration now use the following shorthand notation

$$\langle X(t_1) X(t_1) \rangle \equiv \sigma_1^2$$

$$\langle X(t_1) X(t_2) \rangle / \sigma_1 \sigma_2 \equiv \rho_{12}.$$

Because the **sp** is Gaussian, X_4 is a Gaussian **rv** as well as the linear combination $T_4 = X_4 - X_3 \rho_{34} \sigma_4 / \sigma_3$. Furthermore, it follows that $\langle T_4 X_3 \rangle = 0$ and that, being Gaussian, T_4 is a **sirv** of X_3. On the other hand, if (3.70) is met, in particular $\rho_{23} \rho_{34} = \rho_{24}$, it is easy to see that $\langle T_4 X_2 \rangle = 0$, and so $\langle T_4 X_1 \rangle = 0$. Then T_4 is a **sirv** of X_3, X_2, X_1. Given the transformation of Gaussian variables $\{X_4, X_3, X_2, X_1\} \to \{T_4, X_3, X_2, X_1\}$, it follows that we can relate $P_4(X_4; X_3; X_2; X_1) \to P_4(T_4; X_3; X_2; X_1)$. Now, because T_4 is a **sirv** of X_3, X_2, X_1, the joint probability verifies $P_4(T_4; X_3; X_2; X_1) = P_1(T_4) P_3(X_3; X_2; X_1)$, from which follows that the three-times conditional pdf satisfies $P(T_4 \mid X_3; X_2; X_1) = P(T_4)$; that is, the three-times conditional pdf $P(X_4 \mid X_3; X_2; X_1)$ is only a function of $\{X_4, X_3\}$, which is what we wanted to show. This same methodology can be used for any conditional probability of n-times, proving the sufficient condition.

Optional Exercise.   Show that the fluctuations of a stretched string (massless) can be modeled by a Gaussian process (stochastic in space). Show that it is not possible to say that the fluctuations are "Markovian." Note also that the process cannot be stationary, because the correlator depends parametrically on the "past" (left end of the rope) and the "distant future" (right). Hint: see Sect. 2.3.1.

Example (**Wiener Process**). Let us characterize a Gaussian and Markovian **sp** $W(t)$ using the moments:

$$\langle W(t) \rangle = 0$$

$$\langle W(t) W(s) \rangle = \min\{t, s\}.$$

Note that this process is not stationary. In this case, the correlator $\rho(t_2, t_1)$ is given (for $t_2 \geq t_1$) by

$$\rho(t_2, t_1) = \frac{\langle W(t_1) W(t_2) \rangle}{\sigma(t_1) \sigma(t_2)}$$

$$= \frac{t_1}{\sqrt{t_1} \sqrt{t_2}} = \frac{\sqrt{t_1}}{\sqrt{t_2}}.$$

We immediately see that Doob's Theorem is satisfied: $\rho(t_3, t_1) = \rho(t_3, t_2)\rho(t_2, t_1)$, whence we can ensure that such a process exists. Then, we calculate the one-time pdf as

$$P_1(w, t) = \frac{1}{\sqrt{2\pi t}} \exp\left(\frac{-w^2}{2t}\right),$$

because $\sigma^2(t) = t$. Furthermore, the conditional pdf, Eq. (3.67), is given by

$$P(w_2, t_2 \mid w_1, t_1) = \frac{1}{\sqrt{2\pi t_2 (1 - t_1/t_2)}} \exp\left(\frac{-\left(w_2 - \frac{\sqrt{t_1/t_2}\sqrt{t_2}}{\sqrt{t_1}} w_1\right)^2}{2t_2(1 - t_1/t_2)}\right) \qquad (3.71)$$

$$= \frac{1}{\sqrt{2\pi(t_2 - t_1)}} \exp\left(-(w_2 - w_1)^2 / 2(t_2 - t_1)\right),$$

that precisely corresponds to the propagator of the Wiener process.

Exercise. Show that the propagator of the Wiener process (with $t_0 = 0$)

$$P(w, t \mid w_0, 0) = \frac{1}{\sqrt{2\pi t\epsilon}} \exp\left(-(w - w_0)^2 / 2\epsilon t\right),$$

satisfies the partial differential equation

$$\frac{\partial}{\partial t} P(w, t \mid w_0, 0) = \frac{\epsilon}{2} \frac{\partial^2}{\partial w^2} P(w, t \mid w_0, 0) \qquad (3.72)$$

and complies with the initial condition $P(w, t \to 0 \mid w_0, 0) \to \delta(w - w_0)$. This is the diffusion equation analyzed by Einstein (1920) in his studies on Brownian motion. Einstein also managed to relate the diffusion coefficient $\epsilon/2 \equiv D$ [that is, from the dispersion of the displacement $\langle W(t)^2 \rangle = 2Dt = \epsilon t$] with the parameter characterizing dissipation in the system. This analysis was the first attempt to establish a relationship between fluctuation and dissipation.

Exercise. Prove that a zero mean Gaussian **sp** (non-singular) is Markovian and stationary only if the correlator is $\rho(t_1, t_2) = \exp[-\gamma \mid t_1 - t_2 \mid]$. Conversely, if a process is Gaussian and stationary, and has second cumulant with an exponential structure $\langle X(t_1)X(t_2) \rangle = \sigma^2(0) \exp[-\gamma \mid t_1 - t_2 \mid]$, then the **sp** $X(t)$ is unique and is called the Ornstein-Uhlenbeck process.

3.12.1   The Ornstein-Uhlenbeck Process

This process is one of the most important models in the study of relaxation of dynamic systems. For example, when it comes to studying a particle with velocity V immersed in a fluid at temperature T, the influence of thermal fluctuations will be present in the stochastic nature of $V(t)$. On the other hand, the viscous fluid exerts a braking (thermalization) action on the particle speed. It is then assumed that, for the stationary state, the **mv** of the velocity of the particle is zero, and the time correlation function of the velocity decays with an exponential law which depends only on the time difference. We know that these two ingredients completely define a Gaussian and stationary **sp** $V(t)$. In particular, because the correlator $\rho(t_1, t_2)$ satisfies Doob's Theorem, we will be able to ensure that the process is also Markovian.

If the first two moments are

$$\langle V(t) \rangle = 0, \quad \langle V(t)V(s) \rangle = \frac{D}{2\gamma} \exp(-\gamma \mid t - s \mid), \qquad (3.73)$$

we conclude that $\sigma^2(t) = \frac{D}{2\gamma} = \text{constant}$ and we observe that $\rho(t_2, t_1) = \exp(-\gamma(t_2 - t_1))$ for all $t_1 \leq t_2$. Then, trivially, Doob's Theorem is satisfied, and then the process will simultaneously be Gaussian and Markovian. Consistently the stationary probability distribution that characterizes the process is given by

$$P_{\text{st}}(v) = \frac{1}{\sqrt{2\pi \left(\frac{D}{2\gamma}\right)}} \exp\left(\frac{-v^2}{2\left(\frac{D}{2\gamma}\right)}\right). \qquad (3.74)$$

For the corresponding propagator, (3.67), we have the expression

$$P(v_2, t_2 \mid v_1, t_1) = \frac{1}{\sqrt{2\pi\left(\frac{D}{2\gamma}\right)\left(1 - \rho(t_2, t_1)^2\right)}} \exp\left(\frac{-\left(v_2 - \frac{\rho(t_2, t_1)\sigma(t_2)}{\sigma(t_1)}v_1\right)^2}{2\left(\frac{D}{2\gamma}\right)\left(1 - \rho(t_1, t_2)^2\right)}\right). \tag{3.75}$$

Exercise. Show that the set of positive functions (3.74) and (3.75) satisfy the Kolmogorov compatibility conditions; on the other hand, (3.75) satisfies the Chapman-Kolmogorov equation. Then (3.74) and (3.75) define a Markov process.

Exercise. Using (3.75), show that the conditional probability $P(v, t \mid v_0, t_0)$ is the solution of the partial differential equation

$$\partial_t P(v, t \mid v_0, t_0) = \left[\frac{\partial}{\partial v}\gamma v + \frac{D}{2}\frac{\partial^2}{\partial v^2}\right] P(v, t \mid v_0, t_0), \qquad (3.76)$$

and satisfies the initial condition

$$\lim_{t \to t_0} P(v, t \mid v_0, t_0) \to \delta(v - v_0).$$

In the coming sections we will see that Eq. (3.76) is associated with the conditional probability distribution obtained from the **sde** (linear) for velocity:

$$\frac{d}{dt}V = -\gamma V + \xi(t), \qquad\qquad (3.77)$$

where γ is a coefficient of friction and $\xi(t)$ is a GWN with zero mean and intensity

$$\langle \xi(t)\xi(s)\rangle = D\,\delta(t - s). \qquad\qquad (3.78)$$

Note, from a simple dimensional analysis, that if $[V] = \mathrm{cm/seg}$, then $[\gamma] = 1/\mathrm{seg}$ and $[D] = \mathrm{cm}^2/\mathrm{seg}^3$.

Exercise (**Green Function**). When dealing with linear equations, the use of the Green function technique is extremely useful. Show that the solution of (3.77) can be found using the Green function $G(t - t')$ that fulfills

$$\left(\frac{d}{dt} + \gamma\right) G(t - t') = \delta(t - t').$$

Then, the solution of (3.77) can be written as

$$V(t) = V^{(0)}(t) + \int_0^\infty G(t - t')\,\xi(t')\,dt', \qquad\qquad (3.79)$$

where

$$\left(\frac{d}{dt} + \gamma\right) V^{(0)}(t) = 0.$$

Show that $G(t - t') = e^{-\gamma(t-t')}\Theta(t - t')$, where $\Theta(t)$ is the step function. Prove by substitution that (3.79) is a solution of (3.77).

Exercise. Compare $P_{\mathrm{st}}(v)$, given in (3.74), with the classical canonical distribution of equilibrium statistical mechanics and show that the following constraint must be met

$$D = 2\gamma \frac{k_B T}{M}, \qquad\qquad (3.80)$$

where M is the mass of the particle. This means that the intensity of the noise fluctuations is related to the coefficient of friction of the fluid. This is a typical relationship between fluctuation and dissipation characteristic of systems that are in contact with a thermal bath and satisfy the principle of detailed balance (see Chap. 4).

Exercise. Integrate formally (3.77) and use (3.78) to show that, asymptotically, the correlation of the velocity

$$\langle V(t)V(s)\rangle = \frac{D}{2\gamma}\exp(-\gamma \mid t-s\mid)$$

corresponds to the stationary state of the stochastic process defined by **sde** (3.77). Hint: for a fixed realization consider $\xi(t)$ as an ordinary function of time, i.e., use (3.79).

Exercise. Suppose the velocity of a particle is well described statistically by the Ornstein-Uhlenbeck process, then, the displacement of its position will be Gaussian and given by

$$X(t) = \int_0^t V(t')\,dt' + X(0). \tag{3.81}$$

Show, using the initial condition $X(0) = 0$, that

$$\langle X(t_1)X(t_2)\rangle = \frac{D}{2\gamma}\left[\frac{2}{\gamma}\min(t_1, t_2) - \frac{1}{\gamma^2} + \frac{1}{\gamma^2}\left(e^{-\gamma t_1} + e^{-\gamma t_2} - e^{-\gamma\mid t_1 - t_2\mid}\right)\right].$$

Why is the **sp** $X(t)$ defined in this way non-Markovian?

Exercise. What conditions must be imposed on friction so that the **sp** $X(t)$ tends asymptotically to be the Wiener process? Show that asymptotically, the diffusion coefficient that characterizes the second moment of displacement $\langle X(t)^2\rangle = D_X t$ is given by

$$D_X = \frac{2k_B T}{M\gamma}.$$

Exercise. Given *cgs* dimension units on the Ornstein-Uhlenbeck **sp** show that $[D_X] = \text{cm}^2/\text{seg}$.

3.13   Einstein Relation

The diffusion coefficient D_X is defined as the proportionality factor between the second moment of displacement of the particle (which diffuses) and the time elapsed from the initial condition.[43] We can, then, generalize this idea by the following definition

$$\frac{d}{dt}\langle(X(t) - X(0))^2\rangle = 2D_X(t). \tag{3.82}$$

---

[43]Some authors explicitly introduce a factor of 2. That is, asymptotically, if diffusion is not anomalous: $\langle X(t)^2\rangle \sim 2D_X t$.

Using the relation (3.81) we have that

$$\left\langle (X(t) - X(0))^2 \right\rangle = \left\langle \int_0^t ds_2 \int_0^t ds_1 \, V(s_1)V(s_2) \right\rangle_0 = \int_0^t ds_2 \int_0^t ds_1 \, \langle V(s_1)V(s_2) \rangle_0 ,$$

(3.83)

where the subscript (0) indicates a **mv** calculated at equilibrium. Then using that

$$\langle V(s_1)V(s_2) \rangle_0 = f(| s_1 - s_2 |),$$

where f is an arbitrary correlation function, we get, after introducing the change of variable $\tau = s_1 - s_2$, the following alternative expression

$$\left\langle (X(t) - X(0))^2 \right\rangle = 2 \int_0^t ds_1 \int_0^{s_1} ds_2 \, f(| s_1 - s_2 |) = 2 \int_0^t ds_1 \int_0^{s_1} f(\tau) \, d\tau.$$

Then, differentiating with respect to time, that is, using (3.82), we get

$$2D_X(t) = 2 \int_0^t f(\tau) \, d\tau.$$

So the diffusion coefficient can be written as a function of velocity correlation[44] in the form:

$$D_X \equiv D_X(\infty) = \int_0^\infty \langle V(\tau)V(0) \rangle_0 \, d\tau.$$

In particular, Einstein [18] found the relationship between the mobility and the diffusion coefficient D_X. This relationship makes it possible to find the electric conductivity as a function of D_X. The mobility μ can be defined as the final velocity of a charged particle per unit of applied force

$$\langle V \rangle = \mu e E,$$

while D_X is the proportionality constant connecting the particle current to the concentration gradient

$$J_p = \frac{-D_X}{2} \frac{\partial n}{\partial X}.$$

Einstein's original idea was to establish that, at equilibrium, the particle current (concentration) cancels the current induced by the electric field E, from which the first fluctuation-dissipation relationship follows:

---

[44] An alternative derivation of this expression can be obtained using linear response theory, often called Green-Kubo formula (see Chap. 5). On the other hand, the Scher-Lax formula, which is the generalization of Einstein's relationship in anomalous transport situations, is presented in Chap. 8.

$$\mu = \frac{D_X}{2k_BT}. \tag{3.84}$$

Guided Exercise. Prove the fluctuation-dissipation relation (3.84). Assuming Gibbs' formula for particle concentration at equilibrium $n(x) \propto e^{-U(x)/k_BT}$, and using Fick's relation for the expression of the current: $J_p = \frac{-D_X}{2} \partial_x n(x)$ we have that

$$J_p \sim \langle V \rangle n(x) \sim \frac{-D_X}{2} \partial_x n(x) \sim \frac{D_X}{2k_BT} n(x)\ \partial_x U(x) \tag{3.85}$$

Using $\langle V \rangle = \mu eE$ and $U(x) = eEx$ in (3.85), we get $\mu eE = \frac{D_X}{2k_BT} eE$ from which Einstein's formula (3.84) follows.

3.14   Generalized Ornstein-Uhlenbeck Process*

When Eq. (3.77) was presented, it was also said that the **sp** $V(t)$ corresponded to the Ornstein-Uhlenbeck process. To prove this fact, we just have to calculate the characteristic functional:

$$G_V([k]) = \left\langle \exp\left[i \int_0^\infty k(t)V(t)\,dt \right] \right\rangle, \tag{3.86}$$

where $V(t)$ is the formal solution ($t \in [0, \infty]$) of (3.77) for one realization of the process (noise) $\xi(t)$. That is to say,

$$V(t) = \exp(-\gamma t) V_0 + \int_0^t \exp(-\gamma(t-s)) \xi(s)\,ds. \tag{3.87}$$

Introducing (3.87) in (3.86) and invoking an expansion in cumulants, we get

$$G_V([k]) = \exp\left(iV_0 \int_0^\infty dt\, k(t)\, e^{-\gamma t} \right) \tag{3.88}$$

$$\times \exp \sum_{m=1}^\infty \frac{i^m}{m!} \prod_{j=1}^m \int_0^\infty dt_j \int_0^{t_j} ds_j \exp(-\gamma(t_j - s_j)) k(t_1) \cdots$$

$$\times \cdots k(t_m) \langle\langle \xi(s_1) \cdots \xi(s_m) \rangle\rangle.$$

If the process $\xi(s)$ is Gaussian, for example white[45] with zero mean, i.e., characterized by (3.78), the cumulant expansion is cut at the second term. Then, the **sp** $V(t)$ is characterized by the characteristic functional:

$$G_V([k]) = \exp\left(iV_0 \int_0^\infty dt\, k(t)\, e^{-\gamma t} \right) \tag{3.89}$$

$$\times \exp \frac{-D}{2} \int_0^\infty dt_1 \int_0^\infty dt_2\, k(t_1)k(t_2) \int_0^{t_1} ds_1 \int_0^{t_2} ds_2\, e^{-\gamma(t_1-s_1)} e^{-\gamma(t_2-s_2)} \delta\,(s_1-s_2)\,.$$

Integrating over variables s_1 and s_2, we can see from the functional (3.89) that, in the asymptotic regime, all moments of **sp** $V(t)$ coincide with those characterized by the Ornstein-Uhlenbeck process, presented in Sect. 3.12.1.

Exercise. Calculate the correlation of the process $V(t)$ from the functional (3.89), and show that cumulants of higher order than the second one are zero.

If the Gaussian process $\xi(s)$ was not white, the functional (3.89) would remain valid as long as $D\delta\,(s_1 - s_2)$ is replaced by the corresponding nonwhite correlation noise

$$\langle\langle \xi(s_1)\xi(s_2)\rangle\rangle\,.$$

An even more interesting problem[46] is that in which $\xi(s)$ is not a Gaussian process. In this case, the formal expression for the characteristic functional of the **sp** $V(t)$ is (3.88). However, this expression is complicated to handle because we need first to know all cumulants of $\xi(s)$ and then perform the corresponding infinite integrations. So stated, this issue is of enormous complexity.

Alternatively, suppose the characteristic functional of the noise $\eta(s)$ is known. In this case, if

$$G_\eta([k]) = \left\langle \exp\left[i \int_0^\infty k(t)\eta(t)\, dt \right] \right\rangle$$

is a given function, the issue is characterizing the **sp** $V(t)$ when its **sde** is

$$\frac{d}{dt}V(t) = -\gamma V + \eta(t). \tag{3.90}$$

Equation (3.90) defines the generalized Ornstein-Uhlenbeck process. To solve this problem, we note that if we write $\eta(t) = \dot{V} + \gamma V$, we can rewrite $G_\eta([k])$, after integrating by parts, in the form:

---

[45]It is usual to say that a process is white if it has a deltiform correlation function. See the last exercise of Sect. 3.1.2.

[46]For analyzing far-from equilibrium models.

$$G_\eta([k]) = e^{-ik_0 V_0} \, G_V \left(\left[-e^{+\gamma t} \frac{d}{dt} e^{-\gamma t} k(t) \right] \right). \tag{3.91}$$

Then, defining the test function $Z(t) = -e^{+\gamma t} \frac{d}{dt} e^{-\gamma t} k(t)$ we can write

$$G_V([Z]) = e^{+ik_0 V_0} \, G_\eta \left(\left[e^{\gamma t} k(0) - \int_0^t e^{\gamma(t-s)} Z(s) \, ds \right] \right). \tag{3.92}$$

Note, from the definition of the test function $Z(t)$, that the constant k_0 is a functional of $Z(t)$ as

$$k_0 \equiv k(0) = \int_0^\infty Z(s) \exp(-\gamma s) \, ds. \tag{3.93}$$

Functional (3.92) is the solution to the problem, enabling us to obtain all cumulants of the generalized process $V(t)$ if the functional of the noise $G_\eta([k])$ is known [19].

Exercise. From the functional (3.92) show that when $\eta(t)$ is Gaussian and white, the known Ornstein-Uhlenbeck process is recovered.

Optional Exercise. Show explicitly that (3.92) is the general expression whatever the noise characteristic functional is.

Excursus (**Strong Non-Markovian Processes**). Here, we want to characterize an **sp** $X(t)$ when its time evolution is ruled by the **sde**:

$$\frac{d}{dt} X(t) = \eta(t), \tag{3.94}$$

whatever the noise characteristic functional $G_\eta([k])$ is. Equation (3.94) is a particular case of (3.90) and defines the generalized Wiener process. When the noise $\eta(t)$ is "white," the process $X(t)$ will be Markovian but, of course, not necessarily Gaussian. In particular, depending on the "range" of the memory of the correlation of the noise $\eta(t)$, it is possible to classify the **sp** $X(t)$ as non-Markovian in a *strong* or *weak* sense, depending on the behavior of its correlation for long enough times [17, 20].

Excursus (**Lévy Noise**). Defining Lévy's noise through the functional:

$$G_\eta([k]) = \left\langle \exp\left[i \int_0^\infty k(t)\eta(t) \, dt \right] \right\rangle = \exp\left(-b \int_0^\infty |k(s)|^\alpha \, ds \right), \quad \alpha \in (0, 2], b > 0,$$

get the Lévy flights from (3.94). In this context, it is also possible to study their generalizations to the nonautonomous case [13]. In Appendix H.2, a short discussion on the Lévy flights is presented.

Optional Exercise (**Nonstationary Gaussian Processes**).   Study the Markovian and Gaussian character of the **sp** $X(t)$ for $t \in [0, \infty]$ when its time evolution is ruled by the **sde**:

$$\frac{d}{dt}X(t) = \psi\,(t)\,\xi(t), \quad \text{with} \quad \xi\,(t) = \text{GWN},$$

where $\psi\,(t) > 0$ and finite for $t \in [0, \infty]$. That is, a realization can be written in the form:

$$X(t) - X(0) = \int_0^t \psi\,(t')\,dW(t'), \quad \text{with} \quad dW(t) \equiv \xi\,(t)\,dt,$$

take for example functions: $\psi\,(t) = e^{-at}$ or a power-law as ruled by Abel's dpf, see advanced exercise 8.8.4. What would be the answer if the process was defined by a realization of the form:

$$X(t) - X(t_0) = \int_{t_0}^t \psi\,(t - t')\,dW(t').$$

3.15   Phase Diffusion*

An interesting application of the generalized Ornstein-Uhlenbeck process is the study of the rotation of a "spherical" molecule. If we imagine that the molecule is a rigid body, the equation of motion is

$$\frac{d}{dt}\vec{L} = \vec{\Upsilon}, \tag{3.95}$$

where $\vec{L}$ is the angular momentum vector, and $\vec{\Upsilon}$ is the sum of all the applied torques on the body. In the presence of fluctuations, the sum of the torques will have a stochastic contribution $\vec{\eta}(t)$ in addition to any external torque $\vec{\Upsilon}_{\text{ext}}(t)$; then, we also have to consider a dissipative term proportional to the angular velocity $\gamma\vec{\Omega}$. Therefore the equation of motion takes the form

$$\frac{d}{dt}\vec{L} + \gamma\vec{\Omega} = \vec{\Upsilon}_{\text{ext}}(t) + \vec{\eta}(t). \tag{3.96}$$

If the rigid body is a flat rotor (a plate), the angular momentum is simply $I\Omega$, where I is the moment of inertia of the plate. In this case the torques will be vectors in the direction of the angular velocity $\vec{\Omega}$. Then, the stochastic equation of motion for the plane rotor is

$$I\frac{d}{dt}\Omega + \gamma\Omega = \Upsilon_{\text{ext}}(t) + \eta(t). \tag{3.97}$$

If we are interested in the response of the system when the external torque $\Upsilon_{\text{ext}}(t)$ is turned off (for example, at $t = 0$), it can be shown in the context of linear response theory that it is only necessary to study the dynamics of the system without external excitation; see Chap. 5. That is, one needs to study (for $t > 0$) correlation functions and the relaxation of the free angular velocity system, i.e.:

$$I\frac{d}{dt}\Omega + \gamma\Omega = \eta(t). \tag{3.98}$$

3.15.1 Dielectric Relaxation

In 1913, Debye introduced the rotor plane model to study the phenomenon of relaxation of a dielectric fluid. He considered an ensemble of molecules all independent of each other and with a permanent dielectric dipole moment $\vec{\mu}$ free to rotate on a fixed axis.

In this case, the external torque will lie between the electric field vector $\vec{E}$ and the permanent dipole moment $\vec{\mu}$, that is $\Upsilon_{\text{ext}}(t) = \mu E(t)\sin\phi$ where ϕ is the initial angle (in the plane) between the vectors $\vec{E}$ and $\vec{\mu}$. If at $t = 0$ we turn off the electric field from (3.98), it will follow that the evolution equation for the phase ϕ is simply

$$I\frac{d^2}{dt^2}\phi + \gamma\frac{d}{dt}\phi = \eta(t), \quad t > 0. \tag{3.99}$$

This equation can be studied in terms of two sps, $\Omega(t)$ and $\phi(t)$, that is using (3.98) in conjunction with

$$\frac{d}{dt}\phi = \Omega(t). \tag{3.100}$$

Then, the study of this physical system is reduced to the study of the Ornstein-Uhlenbeck process [21]. In particular, if the stochastic torque $\eta(t)$ is not Gaussian, the sp $\Omega(t)$ will be a generalized Ornstein-Uhlenbeck's process [17].

The functional of the angular velocity Ω can be obtained from (3.92) comparing (3.98) with (3.90) (for $I = 1$); that is,

$$G_\Omega([Z]) = e^{+ik_0\Omega_0} G_\eta\left(\left[\int_t^\infty e^{\gamma(t-s)}Z(s)\,ds\right]\right), \tag{3.101}$$

where $k_0 = \int_0^\infty Z(s)\exp(-\gamma s)\,ds$. The functional of the phase ϕ is immediately obtained from (3.100) using the functional $G_\Omega([Z])$ as a "noise" functional. Then again, using (3.92) with $V \rightarrow \phi$ and $\gamma = 0$, $\eta \rightarrow \Omega$, and comparing (3.90) with (3.100), it follows that

$$G_\phi([M]) = e^{+iq_0\phi_0} G_\Omega \left(\left[\int_t^\infty M(s) \, ds \right] \right), \tag{3.102}$$

where $q_0 = \int_0^\infty M(s) \, ds$. Introducing (3.101) in (3.102) it follows that the functional of the phase ϕ is written, for any stochastic torque $\eta(t)$, in the form:

$$G_\phi([M]) = e^{+iq_0\phi_0} e^{+ik_0\Omega_0} G_\eta \left(\left[\int_t^\infty e^{\gamma(t-s)} \int_s^\infty M(s') \, ds' \, ds \right] \right). \tag{3.103}$$

Here ϕ_0 is the initial angle and $\Omega_0 = \dot{\phi}_0$ the initial velocity at $t = 0$. The expression (3.103) is exact and valid for any kind of **sp** $\eta(t)$. In particular, if the stochastic torques are GWN using (3.51), the functional of the phase becomes

$$G_\phi([M]) = e^{+iq_0\phi_0} e^{+ik_0\Omega_0} \exp\left[\frac{-\Gamma_2}{2} \int_0^\infty \left(\int_t^\infty e^{\gamma(t-s)} \int_s^\infty M(s') \, ds' \, ds \right) \right.$$
$$\times \left. \left(\int_t^\infty e^{\gamma(t-s)} \int_s^\infty M(s') \, ds' \, ds \right) dt \right].$$

Since we are interested in studying the relaxation of the dipole moment $\mu \langle \cos\phi(t) \rangle$, all correlations can be obtained from trigonometric relationships considering the *cosine* functional and the *sine* functional, for example:

$$\left\langle \cos\left[\int_0^\infty \phi(t) M(t) \, dt \right] \right\rangle = \mathcal{R}_e \left[G_\phi([M]) \right]. \tag{3.104}$$

Example. We can get $\langle \cos\phi(t_1) \rangle$ considering the functional (3.104) with test function $M(t) = \delta(t - t_1)$. In this case, and for white Gaussian torques, we get

$$\langle \cos\phi(t_1) \rangle = \cos\left[\phi_0 + \frac{1}{\gamma}\left(1 - e^{-\gamma t_1} \right) \dot{\phi}_0 \right] \exp\left(\frac{-\Xi(t_1)}{2} \right),$$

where

$$\Xi(t) = \frac{\Gamma_2}{2\gamma^3} \left[2\gamma t - 3 - e^{-2\gamma t} + 4e^{-\gamma t} \right].$$

That is, for longer times, $\gamma t \gg 1$, dipolar relaxation is exponential.

Optional Exercise. Consider the plane rotor in the presence of nonwhite Gaussian torques and characterized by a long-range correlation function. Show that the dipole moment relaxation is not an exponential function. Calculate the dipole-dipole correlation $\langle \cos \phi(t_1) \cos \phi(t_2) \rangle$. Show that the cosine-correlation function can be studied calculating $\langle \cos[\phi(t_1) - \phi(t_2)] \rangle$ using (3.104) evaluated with the test function $M(t) = \delta(t - t_1) - \delta(t - t_2)$. See, for example, [17].

Excursus. Consider the same plane rotor but when stochastic torques are not Gaussian, see for example [13].

3.16   Stochastic Realizations (Eigenfunction Expansions)

It is usually desirable to have semi-analytical approaches for each realization of a stochastic process, especially when it comes to studying stochastic fields. As discussed at the end of this section, a representation in eigenfunctions can be generalized to the case of stochastic fields of $d + 1$ dimension.

Let $\mathbf{Z}(t) \in \mathcal{R}_e$ be an arbitrary **sp** for $t \in [0, T]$, of which only the first two cumulants are known (in principle there is no reason to assume that higher cumulants are zero). Consider the expansion in eigenfunctions of the second cumulant,

$$\int_0^T \langle\langle Z(t)Z(s) \rangle\rangle \phi_j(s) \, ds = \lambda_j \phi_j(t). \tag{3.105}$$

We will assume that the set of eigenfunctions $\{\phi_j(s)\}$ is a complete set of orthonormal functions, that is:

$$\int_0^T \phi_l(s)\phi_j(s) \, ds = \delta_{lj}$$

$$\sum_j \phi_j(t)\phi_j(t') = \delta(t - t')$$

$$\int_0^T \phi_j(s) \, ds < \infty.$$

Then, each realization $Z(t)$ can be represented in terms of eigenfunctions of the integral operator (3.105) [Fredholm operator with a symmetrical[47] kernel $\langle\langle Z(t)Z(s) \rangle\rangle$] as

---

[47]If the kernel of Fredholm's operator is symmetrical, and supposing the Hilbert-Schmidt condition to be valid
$\int_0^T \int_0^T |\langle\langle Z(s)Z(t) \rangle\rangle|^2 \, dt \, ds < \infty$, it is possible to prove the following propositions:

(a) The integral equation has at least one nonzero eigenvalue.
(b) Each eigenvalue has at most a finite degeneration.
(c) There exists a (finite or infinite) complete set of eigenfunctions, with the property that every function $F(t)$, square integrable on $[0, T]$, can be represented in the sense of convergence in the media in the form:

$$Z(t) = \sum_j z_j \phi_j(t),$$

where z_j are obviously random numbers:

$$z_j = \int_0^T Z(s)\phi_j(s)\, ds.$$

Hence, it is observed that their mean values are given by

$$\langle z_j \rangle = \int_0^T \langle Z(s) \rangle\, \phi_j(s)\, ds,$$

while the second cumulant is characterized by the eigenvalue λ_j:

$$\langle z_j^2 \rangle - \langle z_j \rangle^2 = \lambda_j \int_0^T \phi_j(s)^2\, ds = \lambda_j.$$

Here, we have used the normalization: $\int_0^T \phi_j(s)^2\, ds = 1$. We see, then, that the problem has been reduced to characterizing the **rv** z_j. Therefore, if we only know the first two moments of the **sp** $Z(t)$, this amounts to "building" a Gaussian approximation for the realizations $Z(t)$.

Exercise. Prove that z_j, z_l are uncorrelated. Are z_j, z_l **sirv**?

Exercise. The probability distribution $P(z_j)$ is given by the characteristic function $G_{z_j}(k)$ which, in turn, is given in terms of the eigenfunctions $\phi_j(s)$ and all cumulants of the **sp** $Z(t)$ as:

$$G_{z_j}(k) \equiv \langle \exp\left[ikz_j \right]\rangle$$

$$= \langle \exp\left[ik \int_0^T Z(s)\phi_j(s)\, ds \right]\rangle$$

$$= \exp\left[\sum_{m=1}^{\infty} \frac{(ik)^m}{m!} \int_0^T \cdots \int_0^T \phi_j(s_1)\cdots\phi_j(s_m)\, \langle\langle Z(s_1)\cdots Z(s_m)\rangle\rangle \prod_{l=1}^{m} ds_l \right].$$

(d) $F(t) = h(t) + \sum \phi_n(t) z_n$, where $h(t)$ is the function that verifies $\int \langle\langle Z(t)Z(s)\rangle\rangle\, h(t) dt = 0$ and $z_n = \int_0^T \phi_n(t) Z(t) dt$.

(e) The kernel $\langle\langle Z(t)Z(s)\rangle\rangle$ of the integral equation can be expanded en series in the form $\langle\langle Z(t)Z(s)\rangle\rangle = \sum \lambda_n \phi_n(t)\phi_n(s)$, converging in the mean.

Show that $G_{z_j}(k)$ is not a characteristic functional, here the eigenfunctions $\phi_j(s)$ are weighting factors that come from the expansion in cumulants of the **rv** $z_j \equiv \int_0^T Z(s)\phi_j(s)\,ds$.

Exercise. In this Gaussian approximation, show that when $\langle Z(t) \rangle = 0$ the joint pdf of all the **rvs** z_j; that is, the $P(\{z_j\})$ is completely characterized by the bilinear form: $\vec{z} \cdot \mathbf{A} \cdot \vec{z}$, i.e.,

$$P(\{z_j\}) \propto \exp\left[\frac{-\vec{z} \cdot \mathbf{A} \cdot \vec{z}}{2}\right],$$

where the diagonal matrix $\mathbf{A}$ is given by

$$\mathbf{A}^{-1} = \begin{pmatrix} \cdot & 0 & 0 & 0 & 0 \\ \cdot & \lambda_{i-1} & \cdot & 0 & 0 \\ 0 & \cdot & \lambda_i & \cdot & \cdot \\ 0 & 0 & \cdot & \lambda_{i+1} & \cdot \\ 0 & 0 & 0 & 0 & \cdot \end{pmatrix}.$$

In principle, eigenvalues λ_i can be degenerate but it has been assumed that the set of eigenfunctions $\phi_j(s)$ is complete. Note that the advantage of using this representation (Karhumen-Loéve approach) lies in being able to build (Gaussian) analytical approaches for the realizations $Z(t)$ in terms of uncorrelated **rvs** z_j [16].

Example. Here we show how to work out the integral equation (3.105) in a simple particular case. Suppose the correlation function $\langle\langle Z(t_1)Z(t_2)\rangle\rangle$ has the following dependence on the arguments (t_1, t_2) :

$$\langle\langle Z(t_1)Z(t_2)\rangle\rangle = g\left(\min(t_1, t_2)\right) . \tag{3.106}$$

The linear case $g(t) = t$ corresponds to Brownian motion (Wiener process). Substituting (3.106) in (3.105), we get

$$\int_0^{t_1} g(t_2)\phi(t_2)\,dt_2 + g(t_1)\int_{t_1}^T \phi(t_2)\,dt_2 = \lambda\phi(t_1). \tag{3.107}$$

Then, $\phi(t = 0)$ satisfies the relation

$$\lambda\phi(0) = g(0)\int_0^T \phi(t_2)\,dt_2.$$

If the correlation function satisfies $g(0) = 0$, we obtain for the eigenfunctions the following boundary condition:

$$\phi(0) = 0. \tag{3.108}$$

Differentiating equation (3.107) once with respect to t_1, we obtain[48]

$$g(t_1)\phi(t_1) + g'(t_1)\int_{t_1}^{T}\phi(t_2)\,dt_2 - g(t_1)\phi(t_1) = g'(t_1)\int_{t_1}^{T}\phi(t_2)\,dt_2 = \lambda\phi'(t_1).$$

$$(3.109)$$

Then, if $g'(T) \neq \infty$ is satisfied, a second boundary condition for the eigenfunctions follows:

$$\phi'(T) = 0. \qquad (3.110)$$

Differentiating again Eq. (3.109), we obtain

$$g''(t_1)\int_{t_1}^{T}\phi(t_2)\,dt_2 - g'(t_1)\phi(t_1) = \lambda\phi''(t_1). \qquad (3.111)$$

This is the equation we should solve together with the boundary conditions (3.108) and (3.110) to find the eigenfunctions $\phi(t)$.

Exercise. Show that when $g(t)$ is linear, the eigenfunctions are identified by

$$\phi_j(t) = \sqrt{\frac{2}{T}}\sin(\omega_j t) \quad \text{where} \quad \omega_j = \sqrt{\frac{1}{\lambda_j}} = \frac{(2j+1)\,\pi}{2T}, \quad j = 0,1,2,3\cdots$$

Then, the realizations of the Wiener process can be written in the form

$$Z(t) = \sqrt{\frac{2}{T}}\sum_{j=0}^{\infty}z_j\sin(\omega_j t), \quad z_j = \sqrt{\frac{2}{T}}\int_{0}^{T}Z(t)\sin(\omega_j t)\,dt.$$

If $\langle Z(t)\rangle = 0 \Rightarrow \langle z_j\rangle = 0$, and, as mentioned earlier, $\{z_j\}$ are uncorrelated **rvs** with second moment $\langle z_j^2\rangle = \lambda_j = (2T/\,(2j+1)\,\pi)^2$.

Optional Exercise. Study the expansion in eigenfunctions $\phi_j(t)$ for situations where $g(\min(t,t'))$ is nonlinear, see for example, $g(t) = t^{2H}$, $H \in [0,1]$ in Appendix H.

Excursus. Generalize the eigenfunctions expansion for a $(d+1)$ dimensional stochastic fields. Assume a kernel in the integral operator (3.105) of the form

$$\langle\langle Z(\mathbf{r}_1,t_1)Z(\mathbf{r}_2,t_2)\rangle\rangle,$$

where $\mathbf{r}$ is a vector in d dimensions [16].

---

[48]In general, we use the notation: $g'(t) = \frac{d}{dt}g(t)$, $g''(t) = \frac{d^2}{dt^2}g(t)$, etc.

3.17 Stochastic Differential Equations

3.17.1 Langevin Equation

A differential equation in which random coefficients that are Gaussian functions of time with white correlations appear is a Langevin equation. Furthermore, and as is evident, the Langevin equation is only a particular case of **sde**.

The generalized Langevin forces, commonly referred as *white noise forces*, represent dynamic variables that fluctuate in a short time scale compared to the relaxation of the macroscopic state variables describing a system [1]. Moreover, we are assuming that we know, in a statistical sense, the behavior of these variables, that we call "hidden," and that we represent by one or more **sps**. In what follows, we shall discuss in detail the one-dimensional case.[49] Let

$$\frac{dX}{dt} = h(X, t) + g(X, t)\xi(t) \tag{3.112}$$

be the **sde** for the variable X. Here $\xi(t)$ is a δ-correlated Gaussian **sp** with zero mean value, so, it is completely characterized by (3.40), (3.41) and it holds that $W(t) = \int_0^t \xi(\tau) \, d\tau$, where $W(t)$ is a realization of the Wiener **sp**. In (3.112), $h(X, t)$ and $g(X, t)$ are known functions of the unknown variable[50] X and the time t.

Actually, the **sde** (3.112) is not well defined as the integral

$$\int g(X, t') \, dW(t') \tag{3.113}$$

is not even correctly specified. Each particular choice of the integration rule to be used will correspond to a specific stochastic differential calculus. This distinction leads, for example, to the Itô stochastic differential calculus [2] or to the Stratonovich one [3]. In particular, Stratonovich's calculus coincides with the ordinary differential calculus (it has the same rules of differentiation).

If $g(X, t)$ were, at most, a time-dependent function, this difficulty would not appear, because the **mv** of the integral in question would be the **mv** of an ordinary Riemann-Stieltjes integral:

$$\left\langle \sum g(\tau_j)[W(t_j) - W(t_{j-1})] \right\rangle,$$

---

[49] A brief presentation of a generalization of n coupled equations is presented in the last section.

[50] Here, we should simplify the notation and denote $X(t)$ as the stochastic process characterized by the **sde** (3.112).

which converges to $\langle \int g(t')\, dW(t') \rangle$ regardless of the choice of τ_j, where $t_{j-1} < \tau_j < t_j$. We will not discuss this type of analysis.[51] Here we just limit ourselves to using Stratonovich's stochastic differential calculus.

There are many physical justifications for choosing one calculus rather than another. For the particular case when the correlation function $\langle\langle \xi(t)\xi(t') \rangle\rangle$ approaches a deltiform type, it is Stratonovich's calculus which is appropriate. This is a plausible justification for the use of this particular stochastic calculus. Furthermore, Stratonovich's calculus is physically reasonable when Langevin's forces are external, because correlations approach the Dirac delta function. Note that in that case Eq. (3.112), in itself, does not make sense without specifying a stochastic calculus. That is, it is not defined at what time[52] we must evaluate the function $g(X(t'), t')$ in the integrand (3.113).

Guided Exercise. Consider the case when $g(X, t')$ in the integral (3.113) is at most linear in the Wiener function; then, in this case we should evaluate

$$S(t) = \int_0^t g(W)\, dW(t') \rightarrow S_N = \sum_{j=0}^N W(\tau_j)[W(t_j) - W(t_{j-1})]. \qquad (3.114)$$

Introducing: $t_{j-1} < \tau_j < t_j$, and taking the **mv** we obtain

$$\langle S_N \rangle = \left\langle \sum_{j=0}^N [W(\tau_j)W(t_j) - W(\tau_j)W(t_{j-1})] \right\rangle$$

$$= \sum_{j=0}^N [\min(\tau_j, t_j) - \min(\tau_j, t_{j-1})] = \sum_{j=0}^N [\tau_j - t_{j-1}]. \qquad (3.115)$$

Now, if we select a specific α-discretization: $\tau_j = \alpha t_j + (1 - \alpha) t_{j-1}$, from (3.115) we get

$$\langle S_N \rangle = \alpha\, (t_N - t_0)\,, \ \alpha \in (0, 1)\,,$$

which clearly shows the dependence of the result on the specific α-discretization. Show that Stratonovich's calculus ($\alpha = 1/2$) matches the ordinary one, and convince yourself for the general case: $\int_0^t W(t')^{2n+1}\, dW(t')$, $n \in \mathcal{N}$.

---

[51]A discussion on this topic can be found in the text of Gardiner [2], Horsthemke & Lefever [10], and Langouch, Roekaerts & Tirapegui [7].

[52]In the Stratonovich calculus the instant τ_j that is used is just the midpoint.

3.17.2 Wiener's Integrals in the Stratonovich Calculus

For the purposes of illustrating this particular calculus, we present here some examples of specific applications.

Example. Let $W(t)$ be a realization of the Wiener **sp** (with $W(0) = 0$). If we define the Gaussian process[53] $\Omega(t) = \int_0^t W(s)\, ds$, it may be necessary to consider mean values and correlations between the two processes: $\Omega(t)$ and $W(t)$. For example, let us calculate

$$\langle W(t)\Omega(t)\rangle = \left\langle W(t) \int_0^t W(s)\, ds \right\rangle = \int_0^t \langle W(s)W(t)\rangle\, ds.$$

Using the correlation for the Wiener process $\langle W(s)W(t)\rangle = \min\{t, s\}$ we see immediately that

$$\langle W(t)\Omega(t)\rangle = \int_0^t s\, ds = \frac{t^2}{2}.$$

Exercise. Consider the properties of the Wiener process. Show, integrating by parts that

$$\left\langle \int_t^{t+\tau} W(s)\, dW(s) \right\rangle = \frac{\tau}{2}.$$

Example. Using the ordinary differential calculus we show now that

$$\left\langle \int_0^t \Omega(t')\, dW(t') \right\rangle = 0.$$

To do this, first, we integrate by parts (with the usual rules of differentiation); so

$$\int_0^t \Omega(t') \frac{dW(t')}{dt'}\, dt' = \left[\Omega(t')W(t')\right]_0^t - \int_0^t \dot{\Omega}(t')W(t')\, dt'.$$

From the definition of the process $\Omega(t)$ and noting that $W(0) = \Omega(0) = 0$ we observe that

$$\int_0^t \Omega(t')\, dW(t') = \Omega(t)W(t) - \int_0^t W(t')W(t')\, dt',$$

---

[53] Note that we are using a unique (simplified) notation for the **sp** $\Omega(t)$ and its realization $\Omega(t)$.

then, taking the mean value on the realizations of the Wiener process we finally get

$$\left\langle \int_0^t \Omega(t') \, dW(t') \right\rangle = \left\langle \Omega(t)W(t) - \int_0^t W(t')W(t') \, dt' \right\rangle$$

$$= \langle \Omega(t)W(t) \rangle - \int_0^t \langle W(t')W(t') \rangle \, dt'$$

$$= \frac{t^2}{2} - \int_0^t t' \, dt' = 0.$$

Exercise. Show that the correlation of the process $\Omega(t) = \int_0^t W(s) \, ds$ is given by

$$\langle \Omega(t_1)\Omega(t_2) \rangle = \frac{-1}{6} t_1^3 + \frac{t_2 t_1^2}{2} \quad \text{for} \quad t_1 < t_2.$$

That is, the process $\Omega(t)$ is Gaussian but it is not stationary nor Markovian [22]. Why is that?

Optional Exercise. Calculate the two-time characteristic function of the process $\Omega(t) = \int_0^t W(s) \, ds$ and verify the result of the previous exercise. Hint: use

$$\int_0^\infty k(t) \, \Omega(t) \, dt = \int_0^\infty dt \int_0^\infty k(t) \, \Theta(t-s) \, W(s) \, ds,$$

where $\Theta(z)$ is the step function.

Exercise. Consider the non-Gaussian process $\eta(t) = \int_0^t \Omega(s) \, dW(s)$. Show that the correlation of process $\eta(t)$ is given by

$$\langle \eta(t_1)\eta(t_2) \rangle = \frac{1}{12} [\min(t_1, t_2)]^4.$$

That is, process $\eta(t)$ is not stationary. Why, is it not possible from this result to conclude anything about the non-Markovian condition of the process $\eta(t)$?

Exercise. Consider the non-Gaussian process $\Lambda(t) = \int_0^t W^2(s) \, ds$. Show that the second moment of $\Lambda(t)$ is given by [23]

$$\langle \Lambda^2(t) \rangle = \frac{7}{12} t^4.$$

Optional Exercise. Consider the process $\theta(t) = \int_0^t \eta(s) \, ds$ with $\eta(t) = \int_0^t \Omega(s) \, dW(s)$. Show that the second moment of $\theta(t)$ is given by

$$\langle \theta^2(t) \rangle = \frac{1}{180} t^6.$$

Excursus. Consider the set of four coupled processes

$$\dot{W} = \xi(t), \quad \dot{\Omega} = W(t), \quad \dot{\eta} = \Omega(t)\xi(t), \quad \dot{\Xi} = \eta(t),$$

where $\xi(t)$ is a zero mean Gaussian white noise. Does the joint pdf $P(W; \Omega; \eta; \Xi, t)$ describe a four-dimensional Markov process [22].

3.17.3 Stratonovich's Stochastic Differential Equations

Short Times

If we write (3.112) in its integral form

$$X(t+\tau) = X + \int_t^{t+\tau} h(X(t'),t')\, dt' + \int_t^{t+\tau} g(X(t'),t')\xi(t')\, dt', \qquad (3.116)$$

we see that the problematic term is the third one on the right. It is there that we must specify the stochastic differential calculus to be used [24]. In what follows we will always adopt Stratonovich's calculus [3, 14, 25]. Assuming both $h(X(t'),t')$ as $g(X(t'),t')$ can be expanded in Taylor's series[54] around X.

$$h(X(t'),t') \simeq h(X,t') + h'(X,t')(X(t') - X) + \cdots$$

$$(3.117)$$

$$g(X(t'),t') \simeq g(X,t') + g'(X,t')(X(t') - X) + \cdots,$$

we get from (3.116), that the integral solution only depends on the initial condition X and the realizations of the increments of the Wiener process $dW(t) = \xi(t)\, dt$. By this procedure, it is possible to calculate all the moments of $\mathcal{O}(\tau^n)$.

If the **sde** were linear, it would be feasible to obtain the exact solution $X([\xi(t)], t)$ as a functional of the **sp** $\xi(t)$. In this case, we can calculate every moment of **sp** $X(t)$ from statistical knowledge of **sp** $\xi(t)$. Alternatively, it is also possible to calculate the characteristic functional of the process under study [see Sect. 3.14]. In most nonlinear cases, it is not possible to obtain a closed solution of **sde** (3.112); however, one can still obtain a partial differential equation for the propagator of **sp** $X(t)$ (i.e., $P(x, t \mid x', t')$). This evolution equation is called the Fokker-Planck equation and it will be analyzed in detail in the next section.

---

[54]This section $h'(X, t)$ or $h''(X, t)$, etc. will represent the first or second derivative of the function $h(X, t)$ with respect to X.

Optional Exercise. Substituting (3.117) in (3.116), get after repeated iterations

$$X(t+\tau) - X = \int_t^{t+\tau} h(X,t')\,dt' + \int_t^{t+\tau} h'(X,t')\int_t^{t'} h(X,t'')\,dt''\,dt'' \qquad (3.118)$$

$$+ \int_t^{t+\tau} h'(X,t')\int_t^{t'} g(X,t'')\xi(t'')\,dt''\,dt' + \int_t^{t+\tau} g(X,t')\xi(t')\,dt'$$

$$+ \int_t^{t+\tau} g'(X,t')\int_t^{t'} h(X,t'')\xi(t')\,dt''\,dt'$$

$$+ \int_t^{t+\tau} g'(X,t')\int_t^{t'} g(X,t'')\xi(t'')\xi(t')\,dt''\,dt' + \cdots.$$

Exercise. Based on the fundamental theorem of analysis (mean value theorem for continuous functions) show that, in the limit $\tau \to 0$ we obtain

$$\lim_{\tau \to 0} \int_t^{t+\tau} h'(X,t')\int_t^{t'} h(X,t'')\,dt'\,dt'' \to \frac{\tau^2}{2!}h'(X,t)h(X,b) + \frac{\tau^3}{3!}h''(X,t)h(X,b)$$

$$\cdots + \frac{\tau^{n+1}}{(n+1)!}h^n(X,t)h(X,b),$$

where $b \in [t+\tau, t]$. That is, the integrals not containing Langevin forces $\xi(t)$, behave in the limit $\tau \to 0$ of $\mathcal{O}(\tau^2)$.

Guided Exercise. Let us show, using the Wiener differential, that in the limit $\tau \to 0$ we obtain

$$\lim_{\tau \to 0} \left\langle \int_t^{t+\tau} g'(X,t')\int_t^{t'} g(X,t'')\xi(t')\xi(t'')\,dt'\,dt'' \right\rangle \to \frac{\tau}{2}g'(X,t)g(X,t) + \mathcal{O}(\tau^n),$$

$$(3.119)$$

with $n > 1$. To prove this result we first identify $dW(t') = \xi(t')\,dt'$. Then, from the fundamental theorem of analysis we can write:

$$\int_t^{t+\tau} g'(X,t')\,dW(t')\int_t^{t'} g(X,t'')\,dW(t'')$$

$$= g(X,b)\int_t^{t+\tau} g'(X,t')\,dW(t')\left[W(t') - W(t)\right],$$

where $b \in (t, t')$. In the limit $\tau \sim 0$, we can approximate this result in the form:

$$g(X, b) \int_t^{t+\tau} g'(X, t') dW(t') \left[W(t') - W(t) \right] \approx$$

$$\approx g(X, b) g'(X, t) \int_t^{t+\tau} dW(t') \left[W(t') - W(t) \right]$$

$$\approx g(X, b) g'(X, t) \left(\int_t^{t+\tau} W(t') \, dW(t') - W(t) \int_t^{t+\tau} dW(t') \right).$$

If we now take the mean value and use

$$\left\langle W(t) \int_t^{t+\tau} dW(t') \right\rangle = \left\langle W(t) W(t+\tau) - W(t)^2 \right\rangle = 0,$$

it follows that

$$\left\langle \int_t^{t+\tau} g'(X, t') \, dW(t') \int_t^{t'} g(X, t'') \, dW(t'') \right\rangle \cong g(X, b) g'(X, t) \left\langle \int_t^{t+\tau} W(t') \, dW(t') \right\rangle$$

$$\cong g(X, b) g'(X, t) \frac{\tau}{2},$$

and in the limit $\tau \to 0$ this corresponds to the result (3.119), because $b \to t$.

Optional Exercise. Show that the result of the previous exercise can also be obtained using the properties of cumulants of $\xi(t)$, if in addition, we use the interpretation:

$$\int_b^a f(x) \delta(x - a) \, dx = \frac{1}{2} f(a).$$

This definition is in accordance with the interpretation of Stratonovich's differential calculus.

Guided Exercise. Using (3.118) and the statistical properties of **sp** $\xi(t)$, obtain in the limit $\tau \to 0$

$$\langle X(t + \tau) - X \rangle = \left[h(X, t) + \frac{1}{2} g'(X, t) g(X, t) \right] \tau + \mathcal{O}(\tau^2)$$

$$(3.120)$$

$$\left\langle (X(t + \tau) - X)^2 \right\rangle = [g(X, t) g(X, t)] \tau + \mathcal{O}(\tau^2).$$

Note that τ is the main order. Also show that higher moments are $\mathcal{O}(\tau^2)$. This is the reason why a **sde** like (3.112) characterizes, in general, a process $X(t)$ as a Markov one, induced by the GWN $\xi(t)$.

Excursus. In general, for an arbitrary **sde** such as (3.112) it is possible to see that if the noise $\xi(t)$, whichever it is, is δ-correlated, then the process $X(t)$, induced by this noise, will be Markovian.[55]

Optional Exercise. Starting from the **sde** (3.77) and for the case where the noise $\xi(t)$ has an exponential correlation, $\langle \xi^2 \rangle e^{-|t|/\tau_c}$, show that the diffusivity: $\lim_{\tau \to 0} \langle \Delta V(\tau)^2 \rangle / \tau$, is canceled if $\tau_c \neq 0$; from here it is possible to infer that the **sp** $V(t)$ is not Markovian [26].

Optional Exercise. Consider the **sde** (3.112) for the case when $g(x,t) = 1$. A numerical realization of the **sp** $X(t)$ can be built up taking into account the scaling of Wiener's increments: $\Delta W(t_i) \doteq \sqrt{\Delta} \eta_i$, here the increments of time are $t_{i+1} - t_i = \Delta$ and η_i is a normal distributed random number (see advanced exercise 3.20.7). Therefore, integrating formally the **sde** we get

$$X(t_{i+1}) = X(t_i) + h(X(t_i), t_i)\Delta + \sqrt{\Delta}\eta_i$$

Show that an improvement to this iteration can be written in the form[56]

$$X(t_{i+1}) = X(t_i) + \frac{1}{2}\left[h(X(t_i), t_i) + h\left(\hat{X}(t_{i+1}), t_{i+1}\right)\right]\Delta + \sqrt{\Delta}\eta_i$$

$$\hat{X}(t_{i+1}) = X(t_i) + h(X(t_i), t_i)\Delta + \sqrt{\Delta}\eta_i.$$

Write down a generalized expression (in Stratonovich's calculus) when $g(x,t) \neq$ constant.

Singular (Temporal) Perturbations

In general, a **sde** cannot be analyzed for long times by a perturbation expansion of the type (3.118), because this type of analysis involves a singular perturbation. The following "deterministic" example shows such a problem

Consider the differential equation:

$$\frac{d}{dt}X = -\epsilon X, \quad X(0) = 1, \epsilon > 0, t \in [0, \infty] \tag{3.121}$$

---

[55] As a particular case consider the Langevin **sde** $dX = h(X,t)dt + g(X,t)dW(t)$. If this equation satisfies the condition of *existence and uniqueness* theorem, the solution $X(t)$ of the **sde**, for arbitrary initial values, is a Markov process; see [14, p. 146].

[56] Honerkamp J. Stochastische Dynamische Systeme (VCH Verlagsgesellschaft, Weinhein, 1990), Chap. 8. See also Gardiner's book third edition Chap. 10.

The obvious solution is: $X(t) = \exp(-\epsilon\, t)$, and its asymptotic limit $X(t \to \infty) \to 0$ is also immediate. This simple result cannot be obtained by a perturbation theory in time.

Exercise. Show that the perturbative solution of (3.121) is:

$$X(t) = \sum_{n=0}^{\infty} \frac{(-\epsilon\, t)^n}{n!}. \tag{3.122}$$

Obviously, this expression is the expansion of an exponential in series obtained from an iterative perturbation in time. If we did not know this sum and cut (3.122) at order p, we get

$$X_{(p)}(t) \sim \sum_{n=0}^{p} \frac{(-\epsilon\, t)^n}{n!}. \tag{3.123}$$

This expression approximates the correct result only for times $t \ll p/\epsilon$. Moreover, this approach would give an incorrect asymptotic result [$X(t \to \infty) \to$ diverges], this simple fact shows that, in general, time perturbative expansions are singular.

Exercise. Show that for increasing times the approximation (3.123) makes sense only if $\left| \frac{(-\epsilon\, t)^{n-1}}{(n-1)!} \right| > \left| \frac{(-\epsilon\, t)^n}{n!} \right|$.

Optional Exercise. Show that if a perturbation expansion in the Laplace conjugate variable is introduced,[57] this singular phenomenon disappears.

Unfortunately, the Laplace transform is not useful in the type of **sde** (3.112). So, we need some extra hypothesis that allows us to obtain information on **sp** $X(t)$ at all times, knowing only their perturbative moments expansion for short times. This hypothesis is provided by the structure of Markov processes, that is why these Markov **sps** are of importance in nonequilibrium statistical theory. See Sect. 3.18.

Self-Consistent Approach for Small Fluctuations*

If the noise term in a Langevin equation is small, and the deterministic part of the differential equation admits a stationary and stable solution, it is possible to study the corresponding **sde**, on average, in the context of a perturbation theory, where the small parameter is the amplitude of the fluctuations.

---

[57]The Laplace transform of an arbitrary function $f(t)$ is defined as follows: $f(u) \equiv \mathcal{L}_u[f(t)] = \int_0^{\infty} f(t)e^{-ut}\, dt$, where u is an arbitrary number provided the integral converges. The study $f(u)$ in the limit $u \to 0$ gives information about the behavior of $f(t)$ in the regime of long time: $t \to \infty$. In Chap. 7 some properties about the function $f(u)$ will be given.

In Chap. 1, we saw that if $h(X)$ is an arbitrary nonlinear function of X, it is feasible to introduce an expansion in fluctuations to calculate its **mv**:

$$\langle h(X) \rangle = \sum_{n=0}^{\infty} \frac{1}{n!} \left[\frac{d^n h}{dX^n} \right]_{X=\langle X \rangle} \langle (X - \langle X \rangle)^n \rangle . \tag{3.124}$$

Then, in particular conditions and for each special case, it is possible to replace the nonlinear term $h(X)$ in the **sde** by a simpler expression. That is, it is possible to study in a self-consistent way the evolution of $\langle X(t) \rangle$ in a perturbative approach. For example, if the **sde** is

$$\frac{d}{dt} X = h(X) + \sqrt{d}\, \xi(t) , \tag{3.125}$$

where $\xi(t)$ is a GWN with mean value zero. We can take the average value in (3.125) and using (3.124) we obtain:

$$\frac{d}{dt} \langle X \rangle = \sum_{n=0}^{\infty} \frac{1}{n!} \left[\frac{d^n h}{dX^n} \right]_{X=\langle X \rangle} \langle (X - \langle X \rangle)^n \rangle . \tag{3.126}$$

In general, this system is not closed[58] but it can be solved roughly by truncating the series conveniently. On the other hand, if $h(X)$ is a polynomial, a Gaussian approximation can be introduced; then, (3.126) can be solved to that order in a self-consistent way, see also advanced exercise 1.15.4.

Excursus. This self-consistent approach can also be used in models of stochastic partial differential equations, that is, in the analysis of stochastic fluctuating fields (nonhomogeneous systems) [27].

3.18   The Fokker-Planck Equation

The problem of finding the moments and correlations of the **sp** $X(t)$, characterized by **sde** (3.112), is reduced to calculating averages knowing the hierarchy of Kolmogorov's distributions. As already mentioned for a Markov process, the knowledge of the conditional probability (propagator) and of the one-time pdf allows the complete characterization of the **sp**. The relationship between these two objects, $P(x_1, t_1 \mid x_0, t_0)$ and $P_1(x, t)$, is given by the compatibility of Kolmogorov's hierarchy and Bayes' rule (definition of conditional probability). In general, if the **sp** is stationary, we will almost always be interested in asymptotic results (long times). In which case, the stationary pdf (that is independent of time t) is

---

[58] A hierarchy of coupled equations for all the moments can be obtained.

obtained as the asymptotic limit $t \to \infty$ from the propagator.[59] Furthermore, the differential equation governing the propagator of the system is the same as the evolution equation for the one-time pdf. Then, we will try to get this partial differential equation.

There are many methods for obtaining the Fokker-Planck equation from a Langevin **sde**. For example: the method of paths integral [7], expansion in ordinary cumulants and Terwiel's cumulants[60], functional [1, 3, 25] analysis [28], etc. Here in particular, we are going to introduce one of the simplest methods which relies heavily on calculating the characteristic function, as presented in Chap. 1. Furthermore, this method permits direct analysis in cases where the stochastic forces are neither Gaussian nor delta correlated. Of course, in the particular case that the noise does not have delta correlation, the conditional probability (or propagator) will not be sufficient to fully characterize the system, since this **sp** is not Markovian.

In the representation (II) for a given time value, $t \in [t_i, t_f]$, the **sp** $X(t)$ is equivalent to a **rv** characterized by the pdf $P(x, t)$. In particular, if the process takes, with certainty, a given value x' at time t', the distribution of the **rv** (at time t) is given by the conditional pdf $P(x, t \mid x', t')$. This distribution can be studied in terms of the characteristic function (its Fourier transform). Taking $t = t + \Delta t$, $t' = t$ and $x' = X(t)$, the characteristic function is given in terms of the conditional moments:

$$\langle [X(t + \Delta t) - X(t)]^n \rangle ,$$

which are assumed to be known from its corresponding **sde**.[61] Then, in general, the conditional characteristic function[62] is given by

$$G(k, x', t, \Delta t) = \int_{-\infty}^{\infty} \exp\left(ik(x - x')\right) P(x, t + \Delta t \mid x', t) \, dx \qquad (3.127)$$

$$= 1 + \sum_{n=1}^{\infty} (ik)^n M_n(x', t, \Delta t)/n!.$$

Here $M_n(x', t, \Delta t) \equiv \langle [X(t + \Delta t) - X(t)]^n \rangle_{X(t) = x'}$. The notation of *conditional* moments indicates that the solution of the **sde** has been taken with the initial condition $X(t) = x'$. Then, the propagator can be obtained by Fourier inversion of the function (3.127)

---

[59]See second exercise Sect. 3.4.

[60]The definition of both cumulants is presented in Chap. 1. The starting point to go from a nonlinear stochastic differential equation to a linear partial differential equation for the probability density is based on the use of the Liouville equation [for a not necessarily incompressible fluid] and of van Kampen's lemma [1].

[61]For example, in Eq. (3.120) the first two moments for $\tau \to 0$ are presented.

[62]See Sect. 1.9.3.

$$P\left(x, t + \Delta t \mid x', t\right) = \frac{1}{2\pi} \int_{-\infty}^{\infty} \exp\left(-ik(x - x')\right) G(k, x', t, \Delta t) \, dk \qquad (3.128)$$

$$= \frac{1}{2\pi} \int_{-\infty}^{\infty} \exp\left(-ik(x - x')\right) \left[1 + \sum_{n=1}^{\infty} (ik)^n M_n(x', t, \Delta t)/n!\right] dk.$$

Noting that $n \geq 1$, we can use the following Fourier's operational relationship

$$\frac{1}{2\pi} \int_{-\infty}^{\infty} \exp\left(-ik(x - x')\right)(ik)^n \, dk = \left(-\frac{\partial}{\partial x}\right)^n \delta(x - x'), \qquad (3.129)$$

note that this operational relationship has a meaning only under the sign of integration; then, using the properties of the Dirac delta function from (3.128), we obtain

$$P(x, t + \Delta t \mid x', t) = \left[1 + \sum_{n=1}^{\infty} (-\frac{\partial}{\partial x})^n M_n(x, t, \Delta t)/n!\right] \delta(x - x'). \qquad (3.130)$$

If $\Delta t \to 0$, this equation represents the infinitesimal solution of the propagator of the system in terms of the moments of the **sp** $X(t)$. The solution (3.130) also satisfies the usual initial condition from the conditional pdf

$$\lim_{\Delta t \to 0} P(x, t + \Delta t \mid x', t) \to \delta(x - x').$$

The evolution equation for the conditional pdf is obtained by constructing the time derivative from the infinitesimal propagator

$$\lim_{\Delta t \to 0} \left[P(x, t + \Delta t \mid x', t) - P(x, t \mid x', t)\right] / \Delta t = \qquad (3.131)$$

$$= \lim_{\Delta t \to 0} \left[\sum_{n=1}^{\infty} (-\frac{\partial}{\partial x})^n \frac{M_n(x, t, \Delta t)}{\Delta t \, n!}\right] \delta(x - x') \equiv \mathcal{L}(x, t)\delta(x - x').$$

Then, $\mathcal{L}(x, t)$ is a differential operator on the initial condition. This is the partial differential equation we were looking for. So, we can write an equation for the time derivative of the conditional pdf at the initial time:

$$\frac{\partial}{\partial \tau} P(x, \tau \mid x', t')\Big|_{\tau = t'} = \mathcal{L}(x, \tau) P(x, \tau \mid x', t')\Big|_{\tau = t'}. \qquad (3.132)$$

In particular, if the **sde** (3.112) involves a Gaussian white noise, $\xi(t)$, only the first two moments $M_n(x', t, \Delta t)$ are of $\mathcal{O}(\Delta t)$. Then, from (3.131), we get a partial differential equation of second order which is called the Fokker-Planck equation **(F-P)**.

The **F-P** equation represents the generalization of the diffusion equation, as there appears in it a *drift* term proportional to

$$\lim_{\Delta t \to 0} \frac{M_1(x, t, \Delta t)}{\Delta t},$$

as the usual diffusion term proportional to

$$\lim_{\Delta t \to 0} \frac{M_2(x, t, \Delta t)}{2! \Delta t}.$$

In general, these coefficients are not homogeneous in the state variable x. Defining

$$\lim_{\Delta t \to 0} \frac{M_1(x, t, \Delta t)}{\Delta t} \equiv K(x, t) \tag{3.133}$$

$$\lim_{\Delta t \to 0} \frac{M_2(x, t, \Delta t)}{2! \Delta t} \equiv D(x, t), \tag{3.134}$$

we can write the coefficients of the **F-P** equation based on the coefficients from the Langevin equation:

$$dx = h(x, t)dt + g(x, t)dW(t). \tag{3.135}$$

Using the results (3.133), (3.134), and (3.120) we get

$$K(x, t) = h(x, t) + \frac{1}{2} g'(x, t) g(x, t) \tag{3.136}$$

$$D(x, t) = \frac{1}{2} [g(x, t)]^2. \tag{3.137}$$

Then, it follows that the **F-P** differential equation (in Stratonovich's calculus) is written for all t, t' in the form[63]

$$\frac{\partial}{\partial t} P(x, t \mid x', t') = \left[-\frac{\partial}{\partial x} K(x, t) + \frac{\partial^2}{\partial x^2} D(x, t) \right] P(x, t \mid x', t'). \tag{3.138}$$

Note that, integrating on the initial condition, we obtain the same evolution equation for the one-time pdf $P_1(x, t)$.

---

[63]Some authors define $D(x, t) = \lim_{\Delta t \to 0} M_2(x, t, \Delta t)/\Delta t$. Therefore, in Eq. (3.138) a factor $\frac{1}{2}$ appears in front of the diffusion coefficient.

Exercise (**Continuous Markov Processes**). Let $X(t)$ be a (continuous) Markov **sp** characterized for $\Delta t \to 0$ by $M_1(x', t, \Delta t) \sim \mathcal{O}(\Delta t)$, $M_2(x', t, \Delta t) \sim \mathcal{O}(\Delta t)$ and $M_n(x', t, \Delta t) \sim \mathcal{O}(\Delta t^\nu)$ with $\nu > 1$ for all $n \geq 3$, where

$$M_n(x', t, \Delta t) \equiv \langle [X(t + \Delta t) - X(t)]^n \rangle_{X(t)=x'}.$$

Show that the infinitesimal propagator satisfies

$$P(x, t + \Delta t \mid x', t) = \left[1 + \sum_{n=1}^{2} (-\frac{\partial}{\partial x})^n M_n(x, t, \Delta t)/n! \right] P(x, t \mid x', t) + \mathcal{O}(\Delta t^\nu).$$

On the other hand, since the **sp** $X(t)$ is Markovian, the Chapman-Kolmogorov equation holds, i.e.:

$$P(x, t + \Delta t \mid x', t') = \int P(x, t + \Delta t \mid z, t) P(z, t \mid x', t') \, dz; \quad \forall \, t \geq t'.$$

Using (3.128), we can relate $P(x, t + \Delta t \mid z, t)$ to the Fourier transform of $G(k, z, t, \Delta t)$; then, Chapman-Kolmogorov's equation is written, in terms of $M_n(z, t, \Delta t)$, as follows ($\forall \, t \geq t'$):

$$P(x, t + \Delta t \mid x', t') = \int \int e^{-ik(x-z)} \left[1 + \sum_{n=1}^{\infty} (ik)^n \frac{M_n(z, t, \Delta t)}{n!} \right] P(z, t \mid x', t') \frac{dz \, dk}{2\pi}.$$

$$(3.139)$$

This is the Chapman-Kolmogorov equation in differential form.

Exercise (**Proof of 3.138**). Using the behavior of the moments $M_n(z, t, \Delta t)$ (for $\Delta t \to 0$), and the operational Fourier's relation (3.129), obtain from the Chapman-Kolmogorov equation, in its differential form, the **F-P** equation for the Markov **sp** $X(t)$, i.e., (3.138).

Optional Exercise (**Backward F-P Equation**). For time-independent moments $M_n(z, \Delta t)$ write the Chapman-Kolmogorov equation in the form:

$$P(x, t \mid x', t') = \int P(x, t \mid z, t' + \Delta t) P(z, t' + \Delta t \mid x', t') \, dz; \quad \forall \, t \geq t' + \Delta t \geq t'.$$

$$(3.140)$$

Then, using the Taylor series expansion of the δ function:

$$\delta(y - x'') = \delta(x' - x'' + y - x') = \sum_{n=0}^{\infty} \frac{(y - x')^n}{n!} \left(\frac{\partial}{\partial x'} \right)^n \delta(x' - x'')$$

in (3.140) obtain $[P(x,t \mid x',t') - P(x,t \mid x',t' + \Delta t)]/\Delta t \rightarrow -\partial P(x,t \mid x',t')/\partial t'$.
Then, from (3.135) to (3.137), we can write the *backward* **F-P** equation

$$\frac{\partial P(x,t \mid x',t')}{\partial t'} = -\left[K(x') \frac{\partial}{\partial x'} + D(x') \frac{\partial}{\partial x'} \frac{\partial}{\partial x'} \right] P(x,t \mid x',t')$$

$$\equiv -\mathcal{L}^\dagger(x') P(x,t \mid x',t'). \tag{3.141}$$

This is also an equation of motion for $P(x,t \mid x',t')$ where we differentiate with
respect to the initial condition. Both Eqs. (3.138) and (3.141) lead to the same result
for the propagator. The usefulness of the adjoint **F-P** operator will be seen in the
context of extreme densities (see Chaps. 4, 6 and 9) [10, 11].

Exercise. Using (3.136), (3.137), and (3.138) show that the **F-P** equation
associated with the Langevin equation can be written, also, in the form:

$$\frac{\partial}{\partial t} P(x,t \mid x',t') = \left[-\frac{\partial}{\partial x} h(x,t) + \frac{1}{2} \frac{\partial}{\partial x} g(x,t) \frac{\partial}{\partial x} g(x,t) \right] P(x,t \mid x',t'); \quad \forall\, t \geq t'. \tag{3.142}$$

Excursus. If instead of using Stratonovich's differential calculus we had used the
Itô calculus, the **F-P** equation associated with the Markov process $X(t)$ would be

$$\frac{\partial}{\partial t} P(x,t \mid x',t') = \left[-\frac{\partial}{\partial x} h(x,t) + \frac{1}{2} \frac{\partial}{\partial x} \frac{\partial}{\partial x} g(x,t)^2 \right] P(x,t \mid x',t'); \quad \forall\, t \geq t'. \tag{3.143}$$

Hence, we note that for a **sde** with $g = g(t)$ there is no difference between the two
calculus formulations. In the event that g is a function of the state variable, $g = g(x)$,
often this **sde** is said to have *multiplicative noise*.[64]

Optional Exercise. Consider a differential-like equation as in (3.135) under a
transformation $\bar{x} = \phi(x)$; then, the coefficients become

$$\bar{h}(\bar{x}) = h(x) \frac{d\phi}{dx}, \quad \bar{g}(\bar{x}) = g(x) \frac{d\phi}{dx}.$$

Show that the **F-P** (3.142) also transforms with the same coefficients; so in
the Stratonovich sense equations (3.135) and (3.142) are invariant for nonlinear
transformation of the variable x. Nevertheless, in the Itô interpretation the proper
transformation formula for the coefficients of (3.143) are different. A detailed
comment on this fact can be seen in [1, 10, 14].

---

[64]In 1D, this classification is superfluous since it is always possible to make a change of variable
(nonlinear) "to avoid" this multiplicative character. Note that making a nonlinear change of
variable, the ambiguity that exists in the **sde** (3.112) is not eliminated when g is a state function
$g(x)$. That is, in this case it is essential to specify (define) a particular stochastic differential
calculus. See the last exercise in this section.

Excursus (**Itô-Stratonovich Dilemma**). If we compare (3.138) with the **F-P** equation that we would have obtained using Itô's calculus, it is easy to see that the difference is a spurious drift. But since either formulation is valid Eq. (3.135) is just a symbolic *pre-equation* and meaningless until a specific calculus is chosen [24]. Therefore, both Markov **sp** $X(t)$ are different and are correctly defined.

In general, if we know that the **sp** $X(t)$ is Markovian, but not all moments $M_n(x', t, \Delta t)$ with $n \geq 3$ are of $\mathcal{O}(\Delta t^\nu)$ with $\nu > 1$, the evolution equation for the propagator will be

$$\frac{\partial}{\partial t} P(x, t \mid x', t') = \left[\sum_{n=1}^{\infty} (-\frac{\partial}{\partial x})^n \lim_{\Delta t \to 0} \frac{1}{\Delta t} M_n(x, t, \Delta t)/n! \right] P(x, t \mid x', t'); \quad \forall\, t \geq t';$$

this would be the case of having a non-Gaussian white noise $\xi(t)$ in the **sde** (3.112).

Exercise. Show that the propagator (3.71) is the conditional pdf

$$P(w, t \mid w', t'), \quad \forall\, t \geq t',$$

of Wiener's process defined by the **sde** $\dot{W} = \xi(t)$, with $\xi(t)$ a Gaussian white noise, and has an associated **F-P** equation

$$\frac{\partial}{\partial t} P(w, t \mid w', t') = \frac{1}{2} \frac{\partial^2}{\partial w^2} P(w, t \mid w', t'),$$

along with the initial condition $\lim_{t \to t'} P(w, t \mid w', t') \to \delta(w - w')$.

Excursus. In 1906 Smoluchowski[65] introduced a method for studying the spatial fluctuations of a particle that diffuses in the presence of an external force $F(x, t)$. This method is based on the analysis of the diffusion equation

$$\frac{\partial}{\partial t} P(x, t \mid x', t') = \frac{1}{m\gamma} \left[-\frac{\partial}{\partial x} F(x, t) + k_B T \frac{\partial^2}{\partial x^2} \right] P(x, t \mid x', t'); \quad \forall\, t \geq t',$$

$$(3.144)$$

where m is the mass of the particle and γ the damping coefficient of the velocity. This equation is correct if the force is approximately constant along the distance over which the velocity is damped. If this condition is not met, it is more appropriate to consider the joint pdf of the position and velocity of the particle; in this case, the description necessarily leads to a multidimensional **F-P** equation.[66]

In a modern context, Smoluchowski's equation (3.144) is the first systematic correction in the small γ^{-1} parameter (large dissipation). Successive corrections can be studied, for example, for the case of a time-independent force using a singular perturbation expansion, as presented by van Kampen [1].

---

[65]M. von Smoluchowski, Ann. Physik **21**, 756, (1906).

[66]The **F-P** equation in the position and velocity variables is Kramers' equation; see Chap. 4.

Exercise (**Smoluchowski's Equation**). Find the **sde** associated with the **F-P** equation (3.144) and generalize it to arbitrary dimension.

Optional Exercise (**Reflecting Boundary Condition**). Consider that a particle diffuses in the presence of the force of gravity in the $-x$ direction (in one-dimension). Show from the previous exercise, in the *overdamped* approximation, that the equation for the probability density in space is

$$\frac{\partial}{\partial t} P(x, t \mid x', t') = \frac{1}{\gamma} \left[g \frac{\partial}{\partial x} + \frac{k_B T}{m} \frac{\partial^2}{\partial x^2} \right] P(x, t \mid x', t'); \quad \forall t \geq t'.$$

Where, now, we have to solve $P(x, t \mid x', t')$ for all $\{x, x'\} \geq 0$, under the condition of reflection at $x = 0$, i.e.:

$$\left[g + \frac{k_B T}{m} \frac{\partial}{\partial x} \right] P(x, t \mid x', t') = 0 \quad \text{en} \quad x = 0.$$

This is a typical boundary condition of zero current probability, see Sect. 4.7. Show that the time-independent solution is the barometric density:

$$P_{eq}(x) \propto \exp \left(-\frac{mg}{k_B T} x \right) \quad \text{con} \quad x \in [0, \infty].$$

The problem of finding the time-independent solution of the **F-P** equation will be presented in Chap. 4. It should be noted that the study of solutions with special boundary conditions is a nontrivial problem and we leave part of its analysis to Chaps. 4 and 6. Concerning regular boundary conditions, see also the advanced exercise 3.20.14.

Exercise (**Absorbing Boundary Condition**). Consider a particle that diffuses on the positive domain in the presence of an absorbing boundary condition at the origin. Due to the symmetry of the problem, if the initial condition is at $x_0 > 0$, the propagator can be written using the method of images in the space of the realizations.[67] In this form, any realization of the process reaching the origin is excluded from the statistics. So we can subtract a *copy* of the propagator starting from the position $-x_0$ at time $t = 0$, thus:

$$P_{Abs}(x, t \mid x_0, 0) = \frac{1}{\sqrt{2\pi\sigma^2 t}} \left(e^{-(x-x_0)^2/2\sigma^2 t} - e^{-(x+x_0)^2/2\sigma^2 t} \right), \quad x \in [0, \infty]$$

$$= 2 \frac{e^{-x_0^2/2\sigma^2 t}}{\sqrt{2\pi\sigma^2 t}} \left[e^{-x^2/2\sigma^2 t} \sinh\left(2xx_0/2\sigma^2 t\right) \right]. \tag{3.145}$$

---

[67] In computing the conditional pdf $P_{Abs}(x, t \mid x_0, 0)$ we should count the number of distinct sequences of steps which lead to x excluding all sequences which include even a single arrival to $x = 0$. Chandrasekar, Rev. Mod. Phys. 15, pp1, 1943; see also Sect. 6.5.

Prove that this propagator is a solution of

$$\partial_t P(x,t \mid x_0,0) = \frac{\sigma^2}{2}\partial_x^2 P(x,t \mid x_0,0), \quad \forall x \geq 0$$

fulfilling absorbing boundary condition at $x = 0$; i.e., $P(0,t \mid x_0,0) = 0, \forall t \geq 0$, and the initial condition $\lim_{t\to 0} P(x,t \mid x_0,0) \to \delta(x - x_0)$ for $x \geq 0$. Show that the propagator (3.145) is not normalized and the *area*: $A_t = \int_0^\infty P(x,t \mid x_0,0)dx$ is a decreasing function of time bounded from below. Nevertheless the first moment is constant $\langle x(t)\rangle = x_0$ and independent of the diffusion parameter σ^2, but the dispersion is an increasing function of time. Hint: integrate the diffusion equation by parts, proving that $\frac{d}{dt}\langle x^2(t)\rangle > 0$ and show asymptotically for long times that $\langle x^2(t)\rangle \sim \sqrt{t}$.

Exercise. Consider one-dimensional "multiplicative" **sde**:

$$dX/dt = h(X) + g(X)\xi(t). \tag{3.146}$$

If $g(X) \neq 0$, $\forall X \in \mathcal{D}_X$, we can divide (3.146) by $g(X)$. Then, defining a new variable Y by

$$Y = \int^X \frac{dX'}{g(X')} = f(X),$$

if $X = f^{(-1)}(Y)$, where $f^{(-1)}$ represents the inverse function of $Y(X)$. Show that we can rewrite (3.146) in a non-multiplicative form:

$$\frac{dY}{dt} = H\left[f^{(-1)}(Y)\right] + \xi(t),$$

where the new deterministic part is characterized by the function $H[X] = h(X)/g(X)$.

Excursus (**van Kampen's Lemma**). Consider the one-dimensional **sde**:

$$\frac{dX}{dt} = h(X) + g(X,[\xi(t)]), \quad \text{where } \xi(t) \text{ is an arbitrary } \textbf{sp}, \tag{3.147}$$

even when the function $g(X,[\xi(t)])$ cannot be factored as in the previous exercise, there is an interesting result (due to van Kampen) that addresses the problem of calculating the evolution equation for the probability density (given an arbitrary initial condition $X(0)$):

$$\frac{\partial P(x,t)}{\partial t} = -\frac{\partial}{\partial x}\langle [h(X) + g(X,[\xi(t)])]\ \delta(x - X(t))\rangle_{\xi(t)}, \tag{3.148}$$

where $P(x,t) = \langle \delta(x - X(t))\rangle_{\xi(t)}$ and the **vm** is taken over all the realizations of the **sp** $\xi(t)$. Note that if the **sp** $\xi(t)$ is Gaussian, Novikov's Theorem is very useful for calculating the **vm** on the second member of (3.148) (see advanced exercise 3.20.12). However, due to the complexity of the problem,[68] even from this lemma, it is still necessary to use a perturbation treatment to obtain a closed equation for the evolution of $P(x, t)$.

Excursus. Show that an **sp** is continuous and Markovian, if and only if the evolution equation for the conditional pdf is the **F-P** equation [9, 14].

Excursus. Consider the **sde**: $\dot{V}(t) = -\gamma V + \eta(t)$ when $\eta(t)$ is a dichotomous noise; using van Kampen's lemma, find the evolution equation for the one-time pdf, convince yourself that the stationary solution is different from the Maxwell velocity distribution [26].

3.19 The Multidimensional Fokker-Planck Equation*

The **F-P** equation is the starting point for the study of temporal fluctuations around thermal equilibrium, and also for the analysis of temporal fluctuations of macroscopic variables far from thermodynamic equilibrium. If the process was a discrete Markov **sp**, the evolution equation for the conditional probability would be the Master Equation.[69] In Chap. 6 the Master Equation will be discussed in detail in connection to the problem of diffusion in disordered media.

Exercise. Consider n coupled **sde**s in the presence of n white noises $\xi_j(t)$, $(j = 1, \ldots, n)$ Gaussian with mean value zero and correlations $< \xi_j(t)\xi_l(s) > = \delta_{j,l}\ \delta(t-s)$

$$\frac{d}{dt}X_j(t) = h_j(\{X_m\}, t) + g_{jl}(\{X_m\}, t)\xi_l(t), \tag{3.149}$$

where we have used the convention of summing over repeated indices. Generalize (3.136) and (3.137), to prove that

$$K_i(\{x_m\}, t) = h_i(\{x_m\}, t) + \frac{1}{2}g_{kj}(\{x_m\}, t)\frac{\partial}{\partial x_k}g_{ij}(\{x_m\}, t) \tag{3.150}$$

$$D_{ij}(\{x_m\}, t) = \frac{1}{2}g_{ik}(\{x_m\}, t)g_{jk}(\{x_m\}, t). \tag{3.151}$$

Exercise. Show that $D_{ij}(\{x_m\}, t)$ is a symmetric nonnegative definite matrix.

---

[68]If $\xi(t)$ is not a white noise, $X(t)$ will be a *non-Markovian* process. Due to the extent of this problem, this issue will not be discussed here. A good reference where you can see this lemma and its applications is: van Kampen, Phys Rep **24**, 171, (1976).

[69]The physical description of the system (in its mesoscopic dynamics) is given through discrete variations.

Exercise. Using Stratonovich's calculus, show that the **F-P** equation associated with **sde** (3.149) is given by

$$\frac{\partial}{\partial t} P(\{x_m\}, t \mid \{x'_m\}) = \left[-\frac{\partial}{\partial x_i} K_i(\{x_m\}, t) + \frac{\partial}{\partial x_i} \frac{\partial}{\partial x_j} D_{ij}(\{x_m\}, t) \right] P(\{x_m\}, t \mid \{x'_m\}),$$

(3.152)

where the **F-P** coefficients are given by (3.150) and (3.151), the distribution complies with the initial condition $\lim_{t \to 0} P(\{x_m\}, t \mid \{x'_m\}) \to \delta(\{x_m\} - \{x'_m\})$. The analysis and solution of this equation will be discussed in the next chapter.

Excursus (**Change of Variables**). Consider the **F-P** equation (3.152) in the variables x_j, suppose we want to know the corresponding **F-P** equation in some new set of variables: $y_j = f_j(\{x_j\})$. The simplest way to introduce the effect of this change of variables is to work out the associated **sde**; then, a specific stochastic calculus must be invoked. Alternatively, we can use the law of transformation of **rv**. Convince yourself that the desired **F-P** equation follows from the use of the transformation law

$$P(\{y_j\}, t) = P(\{x_j\}, t) \left| \frac{\partial(x_1, x_2, \cdots)}{\partial(y_1, y_2, \cdots)} \right|.$$

Although, the algebra is rather complicated it can sometimes be simplified by suitable symmetries.

Excursus. If D_{ij} is a positive definite matrix, there exists a function f_{in} such that $(D^{-1})_{ij} = f_{ni}f_{nj}$. If f_{in} satisfies the integrability condition:

$$\frac{\partial f_{in}}{\partial x_m} = \frac{\partial f_{im}}{\partial x_n},$$

(3.153)

it is possible to define a change of coordinate: $d\tilde{x} = f_{kn}dx_n$ such that $\tilde{D}_{ij} = \delta_{ij}$. A necessary and sufficient condition for integrability is the existence of a potential ϕ such that

$$f_{in} = \frac{\partial^2 \phi}{\partial x_i \partial x_n},$$

in this way the structure of the **F-P** equation simplifies notably, see L. Garrido, D. Lurié, & M. San Miguel, Phys. Letters A **67**, 243, (1978).

Optional Exercise. Let the following (linear) system of coupled differential equations be

$$\frac{dX_1}{dt} = X_2$$

$$\frac{dX_2}{dt} = \frac{4}{3}\Lambda X_1 - 2k + \sqrt{\epsilon}\xi_2(t),$$

where $X_j \in [-\infty, \infty]$. Consider a pair of Gaussian white noises $\xi_l(t)$, such that $\langle \xi_l(t) \rangle = 0$ and $\langle \xi_l(t)\xi_j(t') \rangle = 2\delta_{l,j}\,\delta(t - t')$. Find the partial differential equation governing the evolution of the conditional pdf $P(\{x_m\}, t \mid \{x'_m\}, t')$. Study the **mv** $\langle X_j(t) \rangle$ in the long times regime and the quadratic relative dispersion $\sqrt{\sigma_{jj}(t)}/\langle X_j(t) \rangle$, for $j = 1, 2$. Show that these dispersions remain constant only if $\Lambda \neq 0$ [29].

Excursus. Now, consider the following system of nonlinear coupled **sde**:

$$\frac{dX_1}{dt} = X_2$$

$$\frac{dX_2}{dt} = a\frac{X_2^2}{X_1} + bX_1 + \sqrt{\epsilon}\xi_2(t).$$

Find the **F-P** equation associated with this set of **sde**s. In general, in the nonlinear case we cannot find the exact solution of the **F-P** equation. However, in the limit of small fluctuations ($\epsilon \sim 0$) we may use stochastic perturbative techniques to study the problem. In particular, the method of van Kampen's Ω-expansion is very useful. Show for this model that, if ϵ is small, the squared relative dispersion remains constant (at long times) only if $b \neq 0$ [30]. A short discussion on the Ω-expansion is presented in the advanced exercises of Chap. 6.

Excursus (**Coupled Logistic Carrying Capacity Model**). The logistic equation has been widely studied in population and ecological modeling. If $N(t)$ denotes a population density, r the intrinsic growth rate, and K the carrying capacity, the logistic equation for the population $N(t)$, from an initial condition $N(0) = N_0$, is (Verhulst's equation)

$$\frac{dN}{dt} = rN\left(1 - \frac{N}{K}\right), \quad N(t) \geq 0. \tag{3.154}$$

Show that solutions of this equation, for all strictly positive N_0, approach the constant value carrying capacity, K, as time, t, tends to infinity. As carrying capacity regarded as a constant is not often realistic many variations of the logistic equation have been introduced. An alternative route is to consider that the carrying capacity is a stochastic state variable coupled to N; therefore, we can model the dynamics of $N(t)$ to be governed by (3.154) coupled to

$$\frac{dK}{dt} = f(N, K) + \sqrt{\epsilon}\,\xi(t), \quad K(t) \geq 0, \tag{3.155}$$

here $\xi(t)$ is a GWN. If the noise intensity ϵ is zero, we recover the familiar coupled logistic carrying capacity model [see for example: J.H.M. Thornley & J. France, Eco. Modelling, 184, 257, 2005; M.O. Cáceres, Phys. Rev. E, 90, 022137 (2014)].

If $f(N, K) = 0$ this *free*-model represents a pure stochastic carrying capacity limit. Show that in this case it is possible to write the realizations of $N(t)$ in terms of the Wiener process $dW(t) = \xi(t)\,dt$ using the propagator (3.145).[70] Write the **F-P** equation for the conditional pdf $P(N, K, t \mid N_0, K_0)$.

Excursus (**Extended Systems**). Consider the situation where the system of interest has a very large number of dynamic variables X_i, $i = 1, 2, \cdots m$ [2]. In particular, in the case when $m \to \infty$ we can consider the continuous limit $X_i \to r$ [this results in a functional **F-P** equation[71]]. This is the case where the dynamic variable of interest (the order parameter) is a scalar field in a space of arbitrary dimension, i.e.: $\phi(r, t)$. The evolution of the dynamic system could, for example, have nonlinear terms, and even diffusion terms of form $\nabla^2 \phi(r, t)$ (or even more complex); if the system also has random coefficients (stochastic), we say that we are in presence of a stochastic partial differential equation. Such equations are often called reaction-diffusion equations [this is the genesis of the scene of complex systems that present spatiotemporal patterns, spatial chaos, spirals, etc., a good introductory reference on this topic can be found in the book: Nicolis G., *Introduction to Nonlinear Science*, Cambridge University Press (1995)]. Some advanced exercises with infinite degrees of freedom are presented in Sect. 1.15.10, also in Chap. 9 an introduction to the stability analysis for stochastic field are presented.

3.19.1   Spherical Brownian Motion

When it comes to understanding an erratic "spherical" movement, it is natural to generalize the concept of Brownian motion in the plane, to the stochastic study of the displacement of a point on the surface of a sphere[72] of constant radius r (for example, $r = 1$). A specific application of the analysis of Brownian motion on the sphere is the study of magnetic relaxation [21]. The starting point for studying magnetic relaxation is established by the phenomenological dynamics of a magnetic vector moment $\mathbf{M}$. According to the treatment of Gilbert-Brown this dynamic evolution is controlled by the deterministic equation

$$\frac{d\mathbf{M}}{dt} = \gamma_0 \mathbf{M} \times \left(\frac{-\partial V}{\partial \mathbf{M}} - \eta \frac{d\mathbf{M}}{dt} \right), \tag{3.156}$$

---

[70] A small-time perturbation expansion (on the realizations) can be introduced noting that population's concentration can be written in the form $N(t) = N_0 e^{rt} (1 + S(t))^{-1}$ where $S(t) = rN_0 \int_0^t \frac{ds}{K(s)}$. Here the carrying capacity $K(s)$ is a Wiener **sp** with absorbing condition at the origin; note that $\langle S(t) \rangle < \infty$, $\forall t \geq 0$.

[71] See for example: J. Garcia-Ojalvo and J.M. Sancho, in *Noise in Spatially Extended Systems*, Berlin, Springer (1999), and reference therein.

[72] A rigorous presentation of the problem of Brownian motion on a differential manifold can be read in the work of van Kampen: J. Stat. Phys. **44**, 1, (1986).

where γ_0 is the ratio between the magnetic moment and angular momentum (gyro-magnetic ratio), η is the damping constant and $V = V(\theta, \phi)$ represents the potential of conservative forces. Then, the vector $\left(\frac{-\partial V}{\partial \mathbf{M}} - \eta \frac{d\mathbf{M}}{dt} \right)$ represents an *effective field* acting on $\mathbf{M}(t)$. Taking into account a Langevin phenomenological description, the pure damping factor is replaced by a term also including fluctuations, i.e.:

$$-\eta \frac{d\mathbf{M}}{dt} \rightarrow -\eta \frac{d\mathbf{M}}{dt} + \vec{\xi}(t),$$

where the vector $\vec{\xi}(t)$ is a zero mean GWN $\langle \xi_j(t) \xi_l(t + \tau) \rangle = \epsilon \, \delta_{j,l} \, \delta(\tau)$. Replacing this in (3.156), a **sde** is obtained, from which it is possible to write the **F-P** equation associated with the probability distribution $P(\theta, \phi, t)$. This probability characterizes the angular direction of the magnetic moment $\mathbf{M}(t)$; that is, angles $\theta(t)$ and $\phi(t)$ characterize the stochastic magnetic vector $\mathbf{M}$ at instant t. Adopting a Cartesian coordinate system: $M_x = M_s \sin \theta \cos \phi$, $M_y = M_s \sin \theta \sin \phi$, and $M_z = M_s \cos \theta$; we can define a *surface* probability density $\mathcal{W}(\theta, \phi, t)$ by the relation:

$$d\mathcal{W}(\theta, \phi, t) = \mathcal{W}(\theta, \phi, t) \sin \theta \, d\theta \, d\phi$$

$$= dP(\theta, \phi, t)$$

$$= P(\theta, \phi, t) \, d\theta \, d\phi,$$

from which the **F-P** equation associated with the density $\mathcal{W}(\theta, \phi, t)$ is given by

$$\frac{\partial \mathcal{W}}{\partial t} = \frac{1}{\sin \theta} \frac{\partial}{\partial \theta} \left\{ \sin \theta \left[\left(h' \frac{\partial V}{\partial \theta} - \frac{g'}{\sin \theta} \frac{\partial V}{\partial \phi} \right) \mathcal{W} + k' \frac{\partial \mathcal{W}}{\partial \theta} \right] \right\} \qquad (3.157)$$

$$+ \frac{1}{\sin \theta} \frac{\partial}{\partial \phi} \left\{ \left(g' \frac{\partial V}{\partial \theta} + \frac{h'}{\sin \theta} \frac{\partial V}{\partial \phi} \right) \mathcal{W} + \frac{k'}{\sin \theta} \frac{\partial \mathcal{W}}{\partial \phi} \right\}.$$

Here, we have used that $\frac{-\partial V}{\partial \mathbf{M}} \equiv \left(\partial V/\partial M_x, \partial V/\partial M_y, \partial V/\partial M_z \right)$, along with a definition of constants

$$g' = \frac{\gamma_0}{(1 + \alpha^2) M_s}; \quad h' = g' \alpha; \quad \alpha = \eta \gamma_0 M_s; \quad k' = \frac{\epsilon}{2} M_s^2 \left[h'^2 + g'^2 \right].$$

Optional Exercise. Using the relationship between polar and Cartesian coordinates in (3.156), and the multidimensional Fokker-Planck theory it is possible to obtain the corresponding operator (3.157). However, by using intuitive concepts of probability currents, this result can also be obtained in a much simpler way [31].

In the next chapter we shall see (under the restriction of detailed balance) that it is possible to link the dissipation coefficient η with the fluctuation parameter ϵ appearing in the **sde**. In this way, it is easy to see how it is possible to obtain the stationary probability distribution corresponding to the equilibrium statistical mechanics from Gilbert-Brown's **sde**; see Sect. 4.6.2.

3.20 Additional Exercises with Their Solutions

3.20.1 Realization of a Campbell Noise

Consider the Campbell noise $\psi\,(t)$ to be defined by the expression:

$$\psi\,(t) = \sum_{\sigma=0}^{n} \frac{A}{\tau} \exp\left(-\left|t - t_\sigma\right|/\tau\right), \quad t \in [0, T]. \tag{3.158}$$

Here A is a random number that can take the values ± 1 with the same probability. The set of values $\{t_\sigma\}$ are uniformly distributed in the interval $[0, T]$ and the number of times n is characterized by the Poisson probability $P_n = \frac{\lambda^n}{n!} e^{-\lambda}$, $\lambda > 0, n \in [0, 1, 2, \cdots]$. Figure 3.4 shows three possible realizations of the noise $\psi\,(t)$. For a fixed time t_1 consider the possibility to plot the histogram $P(\psi_1, t_1)$ associated with all possible realizations of Campbell's noise. In the limit $\tau \to 0$, this noise turns out to be white (see Sect. 3.1.2). What happens with the histogram for the **rv** $\psi\,(t_1)$? Is it possible to plot $P(\psi_1, t_1)$ in the limit $\tau \to 0$? Compare with the histogram of a nonwhite noise, like the one in Sect. 3.1.4.

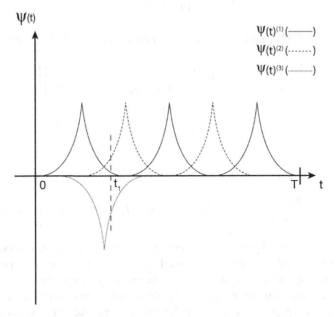

Fig. 3.4 Representation of three realizations of Campbell's noise $\psi\,(t)$. Note that in these realizations: $\psi\,(t)^{(1)}$ has three points t_σ, $\psi\,(t)^{(2)}$ has two points, and $\psi\,(t)^{(3)}$ just one

3.20.2   Gaussian White Noise Limit

Consider Campbell's noise (3.158) in the limit when the width of the *impulses* goes to zero ($\tau \to 0$), the amplitude $A \to \beta C_\sigma$ where C_σ are **sirv** with mean value zero and finite moments $\langle C_\sigma^{2n} \rangle < \infty, \forall n \geq 1$, and the intensity parameter $\beta \to 0$. Then, taking $T \to \infty$, and a large density of Poisson's time points t_σ (i.e., $\lambda \to \infty$), in such a way that $\beta^2 \lambda \to q =$ constant, show that the functional of this process tends to $G_\psi([k(t)]) = \exp\left(-\frac{q\langle C_\sigma^2\rangle}{2} \int_0^\infty k(t)^2 \, dt\right)$, i.e., the white Gaussian functional. Therefore, the process

$$\psi(t) = \beta \sum_{\sigma=0}^n \frac{C_\sigma}{\tau} \exp\left(-|t - t_\sigma|/\tau\right), \quad t \in [0, T],$$

can be used to compute a Gaussian white noise numerically.

3.20.3   On the Realizations of a Continuous Markov Process

Let $\mathbf{X}(t)$ represent an **sp** defined for times $t \in [0, \infty]$. In Fig. 3.5 we show three possible realizations $X^{(j)}(t)$ passing through X_1 at time t_1, and in the little box $[x_2 + dx_2]$ at time t_2. For a fixed time $t = t_2$, the value $X^{(j)}(t_2)$ is a random number; then, the conditional probability $P(x_2, t_2| x_1, t_1; x_0, t_0) \, dx_2$ can formally be built from an histogram considering all possible realizations that pass through the previously mentioned values X_j at times t_j ($t_0 < t_1 < t_2$). Argue which of these realizations will be the one needed to calculate the conditional pdf $P(x_2, t_2| x_1, t_1)$ if the **sp** were a Markov process.

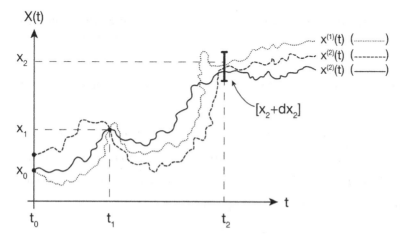

Fig. 3.5 Three possible realizations of a continuous stochastic process

3.20.4 On the Chapman-Kolmogorov Necessary Condition

We have already commented in Sect. 3.3.1 that the Chapman-Kolmogorov (Ch-K) equation is a necessary but not sufficient condition to be a Markov process. First note that the Ch-K equation imposes an integral condition on $P(x, t| x_0, t_0)$ (Ch-K is a matrix multiplication in the case of a discrete **sp**) while the Markov definition imposes restrictions on all elements of the conditional pdf $P(x, t| x', t')$. If the **sp** is Markovian, the n-time joint pdf can be written as:

$$P_n(x_n, t_n; x_{n-1}, t_{n-1}; \cdots ; x_1, t_1) = P(x_n, t_n| x_{n-1}, t_{n-1}) P(x_{n-1}, t_{n-1}| x_{n-2}, t_{n-2}) \cdots$$
$$\times P(x_2, t_2| x_1, t_1) P_1(x_1, t_1).$$

Here we have chosen to order times as: $t_n \geq t_{n-1} \geq \cdots \geq t_1$.

To show that Ch-K is not sufficient, for example, note that from the three-time joint pdf we can get a marginal two-time joint pdf as

$$\int P(x_3, t_3; x_2, t_2| x_1, t_1) P(x_1, t_1) dx_2 = P_2(x_3, t_3; x_1, t_1) = P(x_3, t_3| x_1, t_1) P_1(x_1, t_1),$$
(3.159)

now, if we used the Ch-K equation in the RHD of (3.159) we would get

$$\int P(x_3, t_3; x_2, t_2| x_1, t_1) P(x_1, t_1) dx_2 = \int P(x_3, t_3| x_2, t_2) P(x_2, t_2| x_1, t_1) P_1(x_1, t_1) dx_2.$$
(3.160)

Since $P_1(x, t) > 0$ this last equation can be written in the form

$$\int [P(x_3, t_3; x_i, t_i| x_1, t_1) - P(x_3, t_3| x_i, t_i) P(x_i, t_i| x_1, t_1)] dx_i = 0, \ t_3 > t_i > t_1$$
(3.161)

Nevertheless, Eq. (3.161) does not imply a Markov **sp** because the integrand can be positive or negative. A similar relation can be found, for example, working the four-time joint pdf, and so on.

Find an explicit non-Markov model that satisfies Ch-K and show that in fact the *integrand* in (3.161) is different from zero. A possible model is given by the following hierarchy of functions

$$P_1(x_1, t_1) = \mathcal{A}(x_1, t_1), \text{ (initial distribution at time } t_1)$$

$$P_2(x_2, t_2; x_1, t_1) = \mathcal{A}(x_2, t_2) \mathcal{A}(x_1, t_1), \text{ (\textbf{si} events at adjacent times)}$$

$$P_3(x_3, t_3; x_2, t_2; x_1, t_1) = \mathcal{T}(x_3, t_3| x_1, t_1) \mathcal{A}(x_1, t_1) \mathcal{A}(x_2, t_2), \text{ (}\mathcal{T} \Rightarrow \text{transition)}$$

$$P_4(x_4, t_4; x_3, t_3; x_2, t_2; x_1, t_1) = \mathcal{T}(x_4, t_4| x_2, t_2) \mathcal{A}(x_2, t_2) \mathcal{T}(x_3, t_3| x_1, t_1) \mathcal{A}(x_1, t_1)$$

$$\cdots \quad \text{etc.} \quad \cdots$$

By construction, this model is non-Markovian, there is a correlation between next neighbors nonadjacent times. This **sp** is fully characterized by two functions: the pdf $A(x, t)$, the transition pdf $T(x, t | x', t')$ between *nonadjacent* times, and a ladder of times.

We can see that Kolmogorov's consistency condition is satisfied (for any integer $s < n$)

$$\int \cdots \int P_n(x_n, t_n; x_{n-1}, t_{n-1}; \cdots; x_s, t_s; \cdots; x_1, t_1) \, dx_n \cdots dx_{s+1} = P_s(x_s, t_s; \cdots; x_1, t_1).$$

Note also that:

$$\int P_4(x_4, t_4; x_3, t_3; x_2, t_2; x_1, t_1) \, dx_2 = T(x_3, t_3 | x_1, t_1) \, A(x_1, t_1)$$

$$\times \int T(x_4, t_4 | x_2, t_2) \, A(x_2, t_2) \, dx_2$$

$$\neq P_3(x_4, t_4; x_3, t_3; x_1, t_1),$$

because of the *memory* associated with the origin of time at t_1. By construction, the two-time conditional pdf is clearly non-Markovian

$$P_3(x_3, t_3 | x_2, t_2; x_1, t_1) = \frac{P_3(x_3, t_3; x_2, t_2; x_1, t_1)}{P_2(x_2, t_2; x_1, t_1)}$$

$$= T(x_3, t_3 | x_1, t_1),$$

and the one-time conditional pdf is

$$P(x_3, t_3; x_2, t_2 | x_1, t_1) = \frac{P_3(x_3, t_3; x_2, t_2; x_1, t_1)}{P_1(x_1, t_1)}$$

$$= T(x_3, t_3 | x_1, t_1) \, A(x_2, t_2).$$

Therefore, the integrand in (3.161) is different from zero

$$[P(x_3, t_3; x_i, t_i | x_1, t_1) - P(x_3, t_3 | x_i, t_i) \, P(x_i, t_i | x_1, t_1)]$$

$$= T(x_3, t_3 | x_1, t_1) \, A(x_i, t_i) - A(x_3, t_3) \, A(x_i, t_i)$$

$$= [T(x_3, t_3 | x_1, t_1) - A(x_3, t_3)] \, A(x_i, t_i) \neq 0.$$

If there is no "correlations ladder" all events are **si** and the integrand is zero.

In addition, if we consider "Gaussian" functions in the present non-Markov model

$$\mathcal{T}\left(x, t \mid x', t'\right) = \frac{\exp\left(\frac{-(x-x')^2}{2(t-t')}\right)}{\sqrt{2\pi\,(t-t')}}, \quad t > t'$$

$$\mathcal{A}\left(x, t\right) = \frac{\exp\left(\frac{-x^2}{2t}\right)}{\sqrt{2\pi t}}, \quad t \geq t_1.$$

All the calculations can be done analytically; for example, the fourth moment is $\langle x_1 x_2 x_3 x_4 \rangle = t_1 t_2$ (which is different from the Wiener process).

Show that the fourth cumulant $\langle\langle x_1 x_2 x_3 x_4 \rangle\rangle = 0$ and that Doobs' Theorem does not apply, as expected, because the process is Gaussian but non-Markovian.

3.20.5   Conditional Probability and Bayes' Rule

Consider the **sp** $X_\Omega\left(t\right) = e^{-\Omega t}$ for times $t \in [0, \infty]$, with Ω an **rv** characterized by the pdf $P_\Omega\left(\omega\right)$ in the support $\mathcal{D}_\omega$. Study whether this process is Markovian or not. To do this, we should calculate any conditional probability and show that

$$P\left(x_n, t_n; \cdots; x_s, t_s \mid x_{s-1}, t_{s-1}; \cdots; x_0, t_0\right) = P\left(x_n, t_n; \cdots; x_s, t_s \mid x_{s-1}, t_{s-1}\right).$$

The simplest object in this hierarchy concerns the calculation of

$$P\left(x_3, t_3 \mid x_2, t_2; x_1, t_1\right) \text{ with } t_3 \geq t_2 \geq t_1.$$

This conditional pdf can be calculated using Bayes' rule; then, we first calculate the three-times joint pdf

$$
\begin{aligned}
P\left(x_3, t_3; x_2, t_2; x_1, t_1\right) &= \left\langle\left(x_3 - e^{-\Omega t_3}\right) \delta\left(x_2 - e^{-\Omega t_2}\right) \delta\left(x_1 - e^{-\Omega t_1}\right)\right\rangle \\
&= \int_{\mathcal{D}_\omega} \delta\left(x_3 - e^{-\omega t_3}\right) \delta\left(x_2 - e^{-\omega t_2}\right) \delta\left(x_1 - e^{-\omega t_1}\right) P_\Omega\left(\omega\right) d\omega \\
&= \frac{1}{x_1 t_1} P_\Omega\left(\frac{\ln x_1}{-t_1}\right) \delta\left(x_3 - x_1^{t_3/t_1}\right) \delta\left(x_2 - x_1^{t_2/t_1}\right).
\end{aligned}
$$

On the other hand, the two-times joint pdf

$$
\begin{aligned}
P\left(x_2, t_2; x_1, t_1\right) &= \left\langle\delta\left(x_2 - e^{-\Omega t_2}\right) \delta\left(x_1 - e^{-\Omega t_1}\right)\right\rangle \\
&= \int_{\mathcal{D}_\omega} \delta\left(x_2 - e^{-\omega t_2}\right) \delta\left(x_1 - e^{-\omega t_1}\right) P_\Omega\left(\omega\right) d\omega \\
&= \frac{1}{x_1 t_1} P_\Omega\left(\frac{\ln x_1}{-t_1}\right) \delta\left(x_2 - x_1^{t_2/t_1}\right).
\end{aligned}
$$

Therefore, the conditional pdf that we were looking for is given by

$$P(x_3, t_3 | x_2, t_2; x_1, t_1) = \frac{P(x_3, t_3; x_2, t_2; x_1, t_1)}{P(x_2, t_2; x_1, t_1)}$$

$$= \delta\left(x_3 - x_1^{t_3/t_1}\right).$$

Note that, in all the calculations, we have used the ordering $t_3 \geq t_2 \geq t_1$. This result shows that indeed the conditional pdf depends on the remote time t_1; therefore, the **sp** $X_\Omega(t)$ cannot be Markovian.

Compare this conclusion with the model when $X_\Omega(t)$ is defined by: $X_\Omega(t) = \exp \Omega(t)$, where $\Omega(t) = \int_0^t dW(s)$ is an **sp** ($dW(s)$ is a Wiener's differential).

3.20.6 Second-Order Markov Process

The next stage in complexity of a possible characterization of an **sp** is to consider that the conditional probability depends only on two most recent values, i.e.:

$$P(x_n, t_n; \cdots; x_s, t_s | x_{s-1}, t_{s-1}; x_{s-2}, t_{s-2}; \cdots; x_0, t_0) = \tag{3.162}$$

$$P(x_n, t_n; \cdots; x_s, t_s | x_{s-1}, t_{s-1}; x_{s-2}, t_{s-2}).$$

With the help of the same arguments used in the Markov process prove that, to characterize the whole Kolmogorov's hierarchy, we need to know only three distributions:

$$P_3(x_3, t_3; x_2, t_2; x_1, t_1), \; P_2(x_2, t_2; x_1, t_1), \; P_1(x_1, t_1).$$

Using (3.162) and the marginal operation:

$$P_2(x_2, t_2; x_1, t_1) = \int P_3(x_3, t_3; x_2, t_2; x_1, t_1) \, dx_3.$$

Show that the Chapman-Kolmogorov equation adopts now the generalized structure

$$P(x_4, t_4 | x_3, t_3; x_2, t_2) = \int_{\mathcal{D}_x} P(x_4, t_4 | x_3, t_3; x_2, t_2) P(x_3, t_3 | x_2, t_2; x_1, t_1) \, dx_3,$$

for $t_4 > t_3 > t_2 > t_1$. Build a model and try to find a solution for this equation. Compare it with the usual first-order Markov process.

3.20.7 Scaling Law from the Wiener Process

From the one-time pdf of the Wiener **sp**, it is easy to see the scaling:

$$P\left(\Lambda^H W, \Lambda t\right) = \Lambda^{-H} P(W, t), \; H = \frac{1}{2},$$

note that the factor Λ^{-H} assures that the pdf is correctly normalized, then we can write in distribution: $\mathbf{W}\,(\Lambda t) \doteq \Lambda^{H}\mathbf{W}\,(t)$. This scaling relation shows how to plot a realization of the Wiener process. To realize this fact, consider the sde: $dW/dt = \xi\,(t)$ where $\xi\,(t)$ is a GWN. Integrating this equation, we can write:

$$W\,(t_{i+1}) = W\,(t_i) + \int_{t_i}^{t_{i+1}} \xi\,(s)\,ds. \tag{3.163}$$

Introducing the change of variable $\tau = s - t_i$ and denoting $t_{i+1} - t_i \equiv \Delta$, we get

$$\Delta W\,(t_i) \equiv W\,(t_{i+1}) - W\,(t_i) = \int_0^{\Delta} \xi\,(\tau + t_i)\,d\tau, \tag{3.164}$$

taking $t_i = 0$, using the GWN scaling: $\xi\,(\Lambda t) \doteq \Lambda^{-1/2}\xi\,(t)$ and introducing the change of variable $\tau = s/\Delta$, from (3.164) we finally get

$$\Delta W\,(t_i) = \sqrt{\Delta}W_i(1). \tag{3.165}$$

So, for every discrete time t_i we can reproduce the increment of a Wiener realization by taking *at random* a normal distributed number $W_i(1)$ and multiplying this value by $\sqrt{\Delta}$.

Consider now the *smoother* Gaussian process: $\Omega\,(t) = \int_0^t W\,(s)\,ds$. Prove the scaling in distribution: $\Omega\,(t) \doteq t^{3/2}\Omega\,(1)$, where $\Omega\,(1)$ is a Gaussian number with mean value zero and dispersion $\frac{1}{3}$ [22].

3.20.8 Spectrum of the Wiener Process

Calculate the spectrum of the Wiener process without using scaling theory (Sect. 3.11). A realization of the Wiener process defined on the interval $t \in [0, T]$ can be written using the Karhumen-Loéve eigenfunction theory in the form (see Sect. 3.16)

$$W\,(t, T) = \sqrt{\frac{2}{T}}\sum_{j=0}^{\infty} Z_j \sin\,(\omega_j t)\,, \tag{3.166}$$

where Z_j are **si** Gaussian **rv** characterized by $\langle Z_j \rangle = 0$, $\langle Z_j Z_l \rangle = \delta_{j,l}\left(\frac{2T}{(2j+1)\pi}\right)^2$, and the characteristic frequencies are given by $\omega_j = \left(\frac{2T}{(2j+1)\pi}\right)^{-1}$. The spectrum of the Wiener process $W\,(t)$ is

$$\lim_{T \to \infty} S_W\,(f, T) = \frac{1}{T}\left\langle |F_W\,(f, T)|^2 \right\rangle, \tag{3.167}$$

where $F_W(f, T) = \int_0^T W(t, T)e^{-i2\pi ft}dt$. Introducing the expression (3.166) in (3.167), using that $\langle Z_j Z_l \rangle = \langle Z_j^2 \rangle \delta_{j,l}$, noting that $\lim_{T \to \infty} \omega_j \to 0$ and $\lim_{T \to \infty} \omega_j T \to (2j + 1)\pi/2$, we get, after some algebra,

$$\lim_{T \to \infty} S_W(f, T) = \sum_{j=0}^{\infty} \left[\frac{2}{(2j+1)\pi} \right]^2 \frac{1}{(2\pi f)^2} [1 - \cos \pi (2j + 1)]$$

$$= \frac{8}{\pi^2 (2\pi f)^2} \sum_{j=0}^{\infty} \frac{1}{(2j+1)^2} = \frac{1}{(2\pi f)^2}$$

3.20.9   Time Ordered Correlations

Consider the integral of the process $W(s)$

$$\Omega(t) = \int_0^t W(s)\, ds. \tag{3.168}$$

Here $W(s)$ is a nonstationary stochastic process characterized by all its n-time moments

$$\langle W(s_0) \rangle = W(0) \tag{3.169}$$

$$\langle W(s_0)W(s_1) \cdots W(s_n) \rangle = f(s_0, s_1, \cdots s_n).$$

Where $f(s_0, s_1, \cdots s_n)$ is a given function under the condition $(s_0 \le s_1 \le \cdots \le s_n)$. Suppose that we want to calculate the two-time correlation function

$$\langle \Omega(s_0)\Omega(s_1) \rangle - \langle \Omega(s_0) \rangle \langle \Omega(s_1) \rangle. \tag{3.170}$$

It is obvious that in order to calculate the moment $\langle \Omega(s_0)\Omega(s_1) \rangle$, we have to make a partition in the two-dimensional integral in such a way that it satisfies the condition $(s_0 \le s_1)$. In two dimensions this situation is simple and will lead to the calculation of two integrals inside suitable domains.

Now, suppose that we want to calculate a higher order moment of the form

$$\langle \Omega(t)^3 \rangle = \int_0^t \int_0^t \int_0^t \langle W(s_1)W(s_2)W(s_3) \rangle \, ds_1 ds_2 ds_3. \tag{3.171}$$

Which is the partition in the positive three-dimensional quadrant necessary to do this calculation?

The partition should preserve the ordering of all times appearing in the integrand. To solve this, we can rewrite (3.171) in a slightly different form

$$\int_0^t dt_1 \cdots \left(\int_0^{t_1} dt_2 \cdots + \int_{t_1}^t dt_2 \cdots \right) \left(\int_0^{t_1} dt_3 \cdots + \int_{t_1}^{t_2} dt_3 \cdots + \int_{t_2}^t dt_3 \cdots \right) \cdots .$$

(3.172)

Here, each integral "operates" to the right. Of all the possible combinations, there is one that cannot be fulfilled:

$$\int_0^t dt_1 \int_0^{t_1} dt_2 \int_{t_1}^{t_2} dt_3 \cdots ,$$

and there are two that have to be re-ordered again:

$$\int_0^t dt_1 \int_0^{t_1} dt_2 \int_0^{t_1} dt_3 \cdots , \qquad \int_0^t dt_1 \int_0^{t_1} dt_2 \int_{t_2}^t dt_3 \cdots .$$

(3.173)

Re-ordering these last integrals, there are two that are the same. So from (3.173), we end with three re-ordered integrals. Then, the total number of integrals to be calculated is six. As expected, the number six comes from the time-ordering combinatorial of the integrand $\langle W(s_1)W(s_2)W(s_3)\rangle$. Therefore, the partitions that we were looking for are:

$$\int_0^t dt_1 \int_0^{t_1} dt_2 \int_0^{t_2} dt_3 \cdots ; \quad \int_0^t dt_1 \int_0^{t_1} dt_2 \int_{t_2}^{t_1} dt_3 \cdots ; \quad \int_0^t dt_1 \int_0^{t_1} dt_2 \int_{t_1}^t dt_3 \cdots$$

$$\int_0^t dt_1 \int_{t_1}^t dt_2 \int_0^{t_1} dt_3 \cdots ; \quad \int_0^t dt_1 \int_{t_1}^t dt_2 \int_{t_1}^{t_2} dt_3 \cdots ; \quad \int_0^t dt_1 \int_{t_1}^t dt_2 \int_{t_2}^t dt_3 \cdots .$$

3.20.10   On the Cumulants of Integrated Processes

Consider an **sp** to be defined by the integral of an arbitrary process $\eta(t)$; i.e.,

$$X(t) = \int \eta(s) \, ds. \tag{3.174}$$

Show that, indeed, the following equation is valid for all processes $\eta(t)$, and arbitrary n:

$$\langle\langle X(t_1) \cdots X(t_n)\rangle\rangle = \int \cdots \int ds_1 \cdots ds_n \, \langle\langle \eta(s_1) \cdots \eta(s_n)\rangle\rangle . \tag{3.175}$$

In the case of the second cumulant, the relationship is obvious. Let us prove (3.175) for the third cumulant. In general, using the diagrammatic sketch[73] it follows that

---

[73] See Sect. 1.10.

$$\langle\langle X(t_1)X(t_2)X(t_3)\rangle\rangle = \langle X(t_1)X(t_2)X(t_3)\rangle - \langle\langle X(t_1)X(t_2)\rangle\rangle \langle X(t_3)\rangle$$

$$-\langle\langle X(t_1)X(t_3)\rangle\rangle \langle X(t_2)\rangle - \langle\langle X(t_2)X(t_3)\rangle\rangle \langle X(t_1)\rangle - \langle X(t_1)\rangle \langle X(t_2)\rangle \langle X(t_3)\rangle \,,$$

and substituting (3.174) in this expression we get

$$\langle\langle X(t_1)X(t_2)X(t_3)\rangle\rangle = \int\int\int \{\langle\eta(s_1)\eta(s_2)\eta(s_3)\rangle - \langle\langle\eta(s_1)\eta(s_2)\rangle\rangle \langle\eta(s_3)\rangle$$

$$- \langle\langle\eta(s_1)\eta(s_3)\rangle\rangle \langle\eta(s_2)\rangle - \langle\langle\eta(s_2)\eta(s_3)\rangle\rangle \langle\eta(s_1)\rangle$$

$$- \langle\eta(s_1)\rangle \langle\eta(s_2)\rangle \langle\eta(s_3)\rangle\} \; ds_1 \; ds_2 \; ds_3$$

$$= \int\int\int \langle\langle\eta(s_1)\eta(s_2)\eta(s_3)\rangle\rangle \; ds_1 \; ds_2 \; ds_3,$$

which is the result we wanted to prove. Show that the formula (3.175) is valid for all n.

Example (**Kubo's Oscillator**). We can invoke an expansion in cumulants to average solutions involving stochastic integrals. An example of application is the **sde**

$$dX/dt = -\varepsilon(t)X,$$

where $\varepsilon(t)$ is an arbitrary **sp**. This **sp** $X(t)$ is the so-called Kubo's oscillator [1, 19]. Its solution is given in terms of an integral for each realization in the form: $X(t) = X(0)\exp\left(-\int_0^t \varepsilon(s)\, ds\right)$. Show explicitly that

$$\langle X(t)\rangle = X(0)\exp\left[\sum_{m=1}^{\infty} \frac{1}{m!} \left\langle\!\left\langle \left(-\int_0^t \varepsilon(s)\, ds\right)^m \right\rangle\!\right\rangle\right]$$

$$= X(0)\sum_{m=0}^{\infty} \frac{1}{m!} \left\langle \left(-\int_0^t \varepsilon(s)\, ds\right)^m \right\rangle$$

Exercise (**Lorentz' Force**). In a similar way solve the dynamics of a Lorentz' *random charge* model. Consider a particle with velocity $\vec{V}$ in a magnetic field in the $\hat{z}$ direction. Using Lorentz' force[74] the perpendicular velocity satisfies the stochastic evolution equation:

---

[74]Lorentz' force $\vec{F}$ is exerted on a charged particle q moving with velocity $\vec{V}$ through an electric $\vec{E}$ and magnetic field $\vec{B}$ in the form: $\vec{F} = q\vec{E} + q\vec{V} \times \vec{B}$, in SI units.

$$\dot{V}_x(t) = \varepsilon(t) V_y(t), \quad \dot{V}_y(t) = -\varepsilon(t) V_x(t).$$

Assuming that the random charge is $\varepsilon(t) = q_0 + q(t)$ where $q(t)$ is a GWN with mean value zero, find the **mv** $\langle V_\perp(t) \rangle$ and the correlation function $\langle\langle V_\perp(t) V_\perp(t') \rangle\rangle$.

3.20.11   Dynamic Versus Static Disorder

Kubo's oscillator mentioned in the previous exercise is a minimal model to show the fundamental differences when a deterministic system is perturbed by stochastic fluctuations (dynamic disorder), compared to the case when it is perturbed by a random variable (static disorder). Let the **sde** be

$$\frac{dX}{dt} = (r + \epsilon\xi(t)) X(t), \ \xi(t) = \text{Noise (dynamic disorder)}$$

Consider the normalized variance: $\sqrt{\left\langle X(t)^2 \right\rangle - \langle X(t) \rangle^2}/\langle X(t) \rangle$. Show that when $\xi(t)$ is a GWN with mean value zero, the normalized variance is

$$\frac{\sigma_X(t)}{\langle X(t) \rangle} = \sqrt{\exp(\epsilon^2 t) - 1}.$$

While if the **sde** is perturbed by a **rv**

$$\frac{dX}{dt} = (r + \beta) X(t), \ \beta = \text{rv (static disorder)},$$

the process is non-Markovian and behaves quite differently. If the pdf of the **rv** β is Gaussian: $P(\beta) = \exp(-\beta^2/2\theta^2)/\sqrt{2\pi\theta^2}$, show that in this case the result is

$$\frac{\sigma_X(t)}{\langle X(t) \rangle} = \sqrt{\exp(\theta^2 t^2) - 1}.$$

This difference can be understood if a Gaussian noise has an *infinite* correlation in time. Hint: use a cumulant expansion, note that ϵ and β have different dimensions.

3.20.12   On the van Kampen Lemma

Consider the **sde** $dX/dt = h(X) + \xi(t)$ where $\xi(t)$ is a GWN. Use van Kampen's lemma to recover the **F-P** equation for the conditional pdf $P(x, t \mid x_0, t_0)$ in the limit when the noise is white. Taking $\rho(x, t) = \delta(X(t) - x)$ where $X(t) = X(t, x_0, t_0)$

is the solution of the stochastic flow, we can write van Kampen's lemma (using Novikov's Theorem) as

$$\frac{\partial \langle \rho(x,t) \rangle}{\partial t} = -\frac{\partial}{\partial x} \langle [h(x) + \xi(t)] \rho(x,t) \rangle$$

$$= -\frac{\partial}{\partial x} h(x) \langle \rho(x,t) \rangle - \frac{\partial}{\partial x} \langle \xi(t) \rho(x,t) \rangle$$

$$= -\frac{\partial}{\partial x} h(x) \langle \rho(x,t) \rangle - \frac{\partial}{\partial x} \int_D \langle \xi(t) \xi(s) \rangle \left\langle \frac{\delta \rho(x,t)}{\delta \xi(s)} \right\rangle ds. \qquad (3.176)$$

In general using that $\rho(x,t) = \delta(X(t) - x)$ the functional derivative of $\rho(x,t)$ is

$$\frac{\delta \rho(x,t)}{\delta \xi(s)} = -\frac{\partial \delta(X(t) - x)}{\partial x} \frac{\delta X(t)}{\delta \xi(s)}. \qquad (3.177)$$

For the present stochastic flow, a realization can be written as:

$$X(t) = \int_{t_0}^t h(X(s')) \, ds' + \int_{t_0}^t \xi(s') \, ds', \quad t \geq t_0.$$

Then, we can obtain an integral equation for the unknown *response*: $\frac{\delta X(t)}{\delta \xi(s)}$, $s \leq t$; that is,

$$\frac{\delta X(t)}{\delta \xi(s)} = \int_{t_0}^t \left[\frac{\partial h(X(s'))}{\partial X(s')} \frac{\delta X(s')}{\delta \xi(s)} + \int_{t_0}^t \delta(s' - s) \right] ds', \quad t_0 \leq s \leq t$$

$$= \int_s^t \frac{\partial \dot{X}(s')}{\partial X(s')} \frac{\delta X(s')}{\delta \xi(s)} ds' + \int_{t_0}^t \delta(s' - s) \, ds'.$$

The lower limit in the integral comes from causality: the solution $X(t)$ cannot depend on the stochastic force at a later instant of time. Calling $\frac{\delta X(t)}{\delta \xi(s)} \equiv R(t,s)$ and differentiating with respect to t, we can write an evolution equation for $R(t,s)$

$$\frac{d}{dt} R(t,s) = \frac{\partial \dot{X}(t)}{\partial X(t)} R(t,s) + \delta(t - s).$$

This is a *linear* equation for the (Green) function $R(t,s)$, which can be solved as

$$R(t,s) = \Theta(t - s) \exp \left(\int_s^t \frac{\partial \dot{X}(s')}{\partial X(s')} ds' \right). \qquad (3.178)$$

Note that in the limit $t \to s$ we have to specify a prescription for the value of the step function $\Theta(0)$. Introducing (3.178) in (3.177) and finally in (3.176), we get

$$\frac{\partial P(x,t \mid x_0, t_0)}{\partial t} = -\frac{\partial}{\partial x} h(x) P(x,t \mid x_0, t_0)$$

$$-\frac{\partial}{\partial x} \int_{\mathcal{D}} \langle \xi(t) \xi(s) \rangle \left(-\frac{\partial \delta(X(t) - x)}{\partial x} R(t,s) \right) ds,$$

which is an integro-differential equation. In the Gaussian *color* case:

$$\langle \xi(t) \xi(s) \rangle = e^{-|t-s|/\epsilon}/2\epsilon,$$

a perturbative approach, when $\epsilon \to 0$, can be used.[75]

In the Gaussian singular case $\langle \xi(t) \xi(s) \rangle = \delta(t-s)$, we must choose a convention to follow, in particular taking $\Theta(0) = \frac{1}{2}$ we get the usual **F-P** equation:

$$\frac{\partial P(x,t \mid x_0, t_0)}{\partial t} = -\frac{\partial}{\partial x} h(x) P(x,t \mid x_0, t_0) + \frac{\partial^2}{\partial x^2} \langle \delta(X(t) - x) R(t,t) \rangle$$

$$= -\frac{\partial}{\partial x} h(x) P(x,t \mid x_0, t_0) + \Theta(0) \frac{\partial^2}{\partial x^2} P(x,t \mid x_0, t_0),$$

see also advanced exercise 1.15.10.

3.20.13   *Random Rectifier*

It may be that, in the course of time, we want to know if the value of an **sp** $W(t)$ is larger than a given threshold a. Then, we can consider the following **sp**

$$\psi(t) = \Theta(W(t) - a). \tag{3.179}$$

Here, $\Theta(z)$ is the step function and $W(t)$ is an arbitrary process.[76] A typical realization of the **sp** $\psi(t)$ is shown in Fig. 3.6. Let us calculate the first moment $\langle \psi(t) \rangle$,

$$\langle \psi(t) \rangle = \langle \Theta(W(t) - a) \rangle = \int_{-\infty}^{\infty} \Theta(W - a) P(W,t) \, dW$$

$$= \int_{a}^{\infty} P(W,t) \, dW.$$

---

[75]There are several contributions on this problem: Stratonovich; Sancho and San Miguel; Lindenberg et al.; Grigolini; Hanggi, in *"Noise in nonlinear dynamics systems"*, V1, Eds. F. Moss and P.V.E, McClintock, Cambridge University Press, N.Y. 1989.

[76]Sometimes it may be useful to use the representation of the step function in the form: $\Theta(W(t) - a) = \Theta(W(t)) - \int_0^a \delta(W(t) - x) \, dx$.

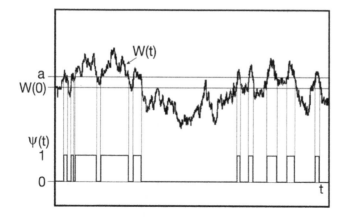

Fig. 3.6 Representation of a realization of the random rectifier process $\psi(t)$

Now consider the case when the one-time **pdf** of the **sp** $W(t)$ is a Gaussian density with arbitrary variance

$$P(W,t) = \frac{\exp\left(-W^2/2\sigma_t^2\right)}{\sqrt{2\pi\sigma_t^2}},$$

in this case, the first moment is

$$\langle \psi(t) \rangle = \frac{1}{2}\left(1 - \text{erf}\left(\frac{a}{\sqrt{2\pi\sigma_t^2}}\right)\right), \tag{3.180}$$

in particular if $W(t)$ is a Wiener process, we know that $\sigma_t^2 = t$. The variance of the **sp** $\psi(t)$ can be calculated in the same way

$$\sigma_\psi^2(t) = \left\langle \psi(t)^2 \right\rangle - \langle \psi(t) \rangle^2$$

$$= \int_a^\infty P(W_1,t)\, dW_1 \left(1 - \int_a^\infty P(W_2,t)\, dW_2\right)$$

$$= \frac{1}{4}\left(1 - \text{erf}\left(\frac{a}{\sqrt{2\pi\sigma_t^2}}\right)^2\right).$$

Form this result, we see that for long times the dispersion saturates to the value $\frac{1}{4}$. Evaluate the correlation function $\langle\langle \psi(t_1)\,\psi(t_2)\rangle\rangle$ when $W(t)$ is the Wiener process. In Fig. 3.7, we have plotted $\langle \psi(t) \rangle$ as a function of time when $a = 1$ and for the general situation when $\sigma_t^2 = t^{2H}$. As can be seen, the asymptotic value $\langle \psi(t \to \infty) \rangle \to \frac{1}{2}$ is delayed for small values of H.

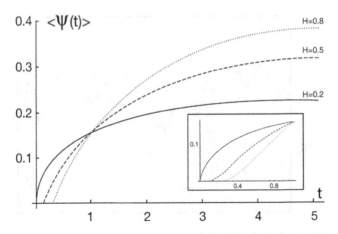

Fig. 3.7 First moment of the process $\psi(t)$, using (3.180) with $a = 1$ and $\sigma_t^2 = t^{2H}$, and for three different values of H. The *inset* shows the same functions for small times

3.20.14 Regular Boundary

Suppose that the one-dimensional state space of the **sp** $X(t)$ is the interval $[b_1, b_2]$, where possibly $|b_1| = \infty$. Furthermore, the process $X(t)$ is characterized by the **sde**:

$$dx = h(x)\,dt + g(x)\,dW(t). \tag{3.181}$$

We assume that $h(x)$ and $g(x)$ are both continuous and differentiable; that is, the same condition which guarantees existence and uniqueness of the Langevin equation. We call b_i an intrinsic boundary of the **sp** $X(t)$ if either $g(b_i) = 0$ or $|h(b_i)| = \infty$. The important question concerning what type of boundary condition (b.c.) can be imposed on a general diffusion process like (3.181) was studied and classified independently by Feller, and Gihman & Skorohod (for a detailed analysis concerning the slightly different way in which it was classified see [10]). From now on, we will follow the work of Horsthemke & Lefever in order to classify the nature of the boundaries. To do this, it is necessary to study the integrability of functions made up of $h(x)$ and $g(x)$. In the Stratonovich calculus, we know that the drift and the diffusion are given by: $K(x) = h(x) + \frac{1}{2}g(x)g'(x)$ and $D(x) = \frac{1}{2}g(x)^2$ respectively (see Sect. 3.18). The character of the boundaries is determined by the behavior of the $K(x)$ and $D(x)$ near the boundaries. In fact, the analytical condition for the boundary b_i can be put in terms of the integrability of functions L_1, L_2 and L_3:

$$L_1(b_i) = \int_{b_i}^{\beta} \phi(x)\,dx \tag{3.182}$$

$$L_2 (b_i) = \int_{b_i}^{\beta} \frac{1}{2D (y)} \left(\int_{b_i}^{y} \phi (x) \, dx \right) \frac{dy}{\phi (y)} \tag{3.183}$$

$$L_3 (b_i) = \int_{b_i}^{\beta} \frac{1}{2D (y)} \frac{dy}{\phi (y)}, \tag{3.184}$$

where β is a point near the boundary b_i, and the function $\phi (x)$ is given by

$$\phi (x) = \exp - \int_{\beta}^{x} \frac{K (z)}{D (z)} \, dz. \tag{3.185}$$

(1) The so-called *natural* boundary corresponds to the case when $L_1 (b_i) = \infty$ (the process never reaches the boundary with a finite probability). Thus b_i is inaccessible and no b.c. can be put on this point.
(2) A boundary is called *inaccessible* but *attracting* if $L_1 (b_i) < \infty$ and $L_2 (b_i) = \infty$ (no b.c. can be put on this point).
(3) An *accessible* boundary corresponds to the case when $L_1 (b_i) < \infty$ and $L_2 (b_i) < \infty$ (the process is not solely defined by the drift and diffusion, but also by the b.c. that has to be prescribed). The prescribed b.c. can be: *absorption, reflection, jump reflection,* or more complicated cases that are combinations of these three). When $L_1 (b_i) < \infty$ and $L_2 (b_i) < \infty$, there are also two subclasses:
(3.1) The boundary acts naturally as an absorbing boundary if $L_3 (b_i) = \infty$ (only absorption or jump reflection are compatible with it and can be imposed on the process).
(3.2) A boundary is called *regular* if also $L_3 (b_i) < \infty$ (in this case any b.c. can be prescribed).

Exercises (**Intrinsic Boundary Points**).

(a) Show that the Ornstein-Ulhenbeck process has natural b.c. at $b_i = \pm \infty$.
(b) Show that an "unstable parabolic" model, that is, the process:

$$\dot{x} = + |\gamma| x + \sqrt{\epsilon} \xi (t),$$

has attracting points at $b_i = \pm \infty$.
(c) Show that the Ornstein-Ulhenbeck process on the domain $[0, \infty]$ has the point $b_1 = 0$ regular.
(d) Show that the Wiener process defined on a finite domain $[b_1, b_2]$ has regular boundaries.
(e) Consider the nonlinear **sde**:

$$\frac{d}{dt} \cos \theta = -\gamma \cos \theta + \sqrt{\epsilon} \xi (t) , \ \gamma > 0, \ \xi (t) = \text{GWN}.$$

This last model may be related to the study of the perpendicular velocity in a Lorentz' force field. Classify the nature of the possible intrinsic points in $[0, 2\pi]$.

Exercise (**On the Itô vs Stratonovich Interpretation**). In a logistic model with a stochastic growth rate:

$$\dot{x} = \left(r + \sqrt{\epsilon}\xi\,(t)\right) x\,(1 - x)\,,$$

the state space of the **sp** is the interval $[0, 1]$ and the points $x = 0$ and 1 are intrinsic boundaries. For example, in the Stratonovich sense, near $x = 0$, we have $K(x) \simeq (r + \epsilon/2)\,x$ and $D\,(x) \simeq \epsilon x^2/2$; then classify all boundary points in $[0, 1]$. For the case when the probability current is zero, it is simple to calculate, for the stochastic logistic model, the stationary pdf of the **F-P**. Show that it depends on the stochastic calculus we may have used.

References

1. N.G. van Kampen, *Stochastic Process in Physics and Chemistry*, 2nd edn. (North-Holland, Amsterdam, 1992)
2. C.W. Gardiner, *Handbook of Stochastic Methods* (Springer, Berlin, 1985)
3. R.L. Stratonovich, *Topics in the Theory of Random Noise*, vols. 1 and 2 (Gordon and Breach, New York, 1963)
4. P. Hangii, *Stochastic Processes Applied to Physics*, ed. by L. Pesquera, M.A. Rodríguez (World Scientific, Singapore, 1985)
5. M.O. Cáceres, A. Becker, L. Kramer, Phys. Rev. A **43**, 6581 (1991); A. Becker, M.O. Cáceres, L. Kramer, Phys. Rev. A **46**, R4463 (1992)
6. M.O. Cáceres, A.M. Lobos, J. Phys. A Math. Gen. **39**, 1547 (2006)
7. F. Langouche, D. Roekaerts, E. Tirapegui, *Functional Integration and Semiclassical Expansions* (D. Reidel, Dordrecht, 1982)
8. B.B. Mandelbrot, *The Fractal Geometry of Nature* (W.H. Freeman, New York, 1983)
9. A. Papoulis, *Probability Random Variables, and Stochastic Processes*, 3rd edn. (McGraw-Hill, New York, 1991)
10. W. Horsthemke, R. Lefever, *Noise-Induced Transitions* (Springer, Berlin, 1984)
11. W. Feller, *An Introduction to Probability Theory and Its Applications*, 2nd edn. (Wiley, New York, 1971)
12. E.W. Montroll, B. West, *Fluctuation Phenomena*, ed. by E.W. Montroll, J.L. Lebowits (Elsevier, Amsterdam, 1979)
13. M.O. Cáceres, J. Phys. A Math. Gen. **32**, 6009 (1999)
14. L. Arnold, *Stochastic Differential Equation* (Wiley, New York, 1974)
15. D. Saupe, *The Science of Fractal Images*, ed. by H.-O. Peitgen, D. Saupe (Springer, Berlin, 1988)
16. L. Mandel, E. Wolf, *Optical Coherence on Quantum Optics* (Cambridge University Press, Cambridge, 1995)
17. M.O. Cáceres, Phys. Rev. E **60**, 5208 (1999)
18. A. Einstein, Ann. Phys. **17**, 549 (1905); **19**, 371 (1906)
19. M.O. Cáceres, A.A Budini, J. Phys. A Math. Gen. **30**, 8427 (1997)
20. M.O. Cáceres, Braz. J. Phys. **29**, 125 (1999)
21. W.T. Coffey, Yu.P. Kalmykov, J.T. Waldron, *The Langevin Equation* (World Scientific, Singapore, 1996)

22. M.O. Cáceres, C.E. Budde, G.J. Sibona, J. Phys. A Math. Gen. **28**, 3877 (1995); M.O. Cáceres, M.A. Fuentes, C.E. Budde, J. Phys. A Math. Gen. **30**, 2287 (1997)
23. P. Colet, F. de Pasquale, M.O. Cáceres, M. San Miguel, Phys. Rev. A **41**, 1901 (1990)
24. N.G. van Kampen, J. Stat. Phys. **24**, 175 (1981)
25. H. Risken, *The Fokker-Planck Equation* (Springer, Berlin, 1984)
26. M.O. Cáceres, Phys. Rev E **67**, 016102 (2003)
27. O. Osenda, C.B. Briozzo, M.O. Cáceres, Phys. Rev. E **54**, 6944 (1996); O. Osenda, C.B. Briozzo, M.O. Cáceres, Phys. Rev. E **57**, 412 (1998)
28. F. Moss, P.V.E. McClintock (eds.), *Noise in Nonlinear Dynamics Sytems*, vol. 1 (Cambridge University Press, Cambridge, 1989)
29. M.O. Cáceres, M.C. Diaz, J.A. Pullin, Phys. Lett. A **123**, 329 (1987)
30. C.B. Briozzo, C.E. Budde, M.O. Cáceres, P.W. Lamberti, Phys. Lett. A **129**, 363 (1988)
31. W.F. Brown, Phys. Rev. **130**, 1677 (1963)

References

Chapter 4
Irreversibility and the Fokker–Planck Equation

A fundamental problem in theoretical physics is that of explaining the origin of macroscopic *irreversibility* (time's arrow), given that the laws of physics are *time-reversible*. Planck was aware of this as early as 1897 and although his work in this area was not transcendental it led him eventually to his quantum hypothesis.[1]

Intuitively irreversibility is a simple concept widely observed in Nature. It is seen both in "closed" systems (ideally assumed not to interact with their environment, as in the case of expansion of a gas into a vacuum) as well as in "open" ones which can exchange energy and/or mass with the outside, a prime example of which are living beings.

In thermodynamics irreversibility may be formulated through the concept of entropy (defined for equilibrium states, see Appendix A). When a system (open or closed) goes from equilibrium state 1 to equilibrium state 2, the entropy change (difference between final and initial entropy) is $\Delta S_{21} = \Delta S_e + \Delta S_i \geq 0$. For a closed system $\Delta S_e = 0$, the entropy variation of the system (internal ΔS_i) is always positive if the process is irreversible, and zero $\Delta S_{21} = \Delta S_i = 0$ if it is reversible (i.e., the second law of thermodynamics). If it is open, ΔS_e can be positive or negative, depending on the type of interaction with the external environment.[2]

From the point of view of irreversible thermodynamics a conceptually clear way of characterizing irreversibility is[3]: we say that a system is irreversible if it is not *invariant* under time reversal $t \rightarrow -t$.

---

[1] A review of the work of Planck can be read at: M. Planck, *Treatise on Thermodynamics*, 3rd ed., New York, Dover (1945).

[2] For example, if the interaction with the environment involves only the transfer of heat ΔQ at temperature T (from the environment to the system), then $\Delta S_e = \Delta Q/T > 0$.

[3] If there is local equilibrium it is possible to write a balance equation involving a density of entropy ρ_s, a flow of entropy J_s (positive or negative) and a (density) source of internal production of entropy $\sigma_s \geq 0$ (which is zero *only* if the process is reversible). That is, the balance equation is $\frac{d}{dt}\rho_s = -\nabla \cdot J_s + \sigma_s$. It is evident that if the process is irreversible the equation is not the same under transformation $t \rightarrow -t$.

© Springer International Publishing AG 2017
M.O. Cáceres, *Non-equilibrium Statistical Physics with Application to Disordered Systems*, DOI 10.1007/978-3-319-51553-3_4

Boltzmann's famous *H* theorem (1870) embodies many of the efforts to solve the paradox of microscopic reversibility versus macroscopic irreversibility [1]. Boltzmann's kinetic theory shows that to solve this paradox it is necessary to introduce probabilistic elements in the dynamic description of the particles. This is the fundamental principle of Boltzmann, the *Stosszahlansatz* (or molecular Chaos). In the remainder of this text we will not go into the problem of irreversibility, or the arrow of time,[4] as it has been called by Prigogine [2], this topic is beyond the scope of this book. Here, when we want to describe irreversible phenomena, we simply do it in terms of stochastic processes [3]. On the other hand, the Fokker–Planck (**F-P**) equation is of particular interest because it gives an appropriate description of the time evolution of classical fluctuations around thermodynamic equilibrium.

4.1   Onsager's Symmetries

Suppose a physical system can be characterized by a set of macroscopic variations of variables $\{X_m\}$ such as energy, momentum, and particle concentration. If the temporal evolution of these macroscopic variables can be described by a continuous Markov **sp**, the evolution equation for the conditional probability $P(\{X_m(t)\} \mid \{X_m(0)\})$ is the **F-P** equation [4, 5]. If the system under study is closed and isolated, the stationary solution $P_{st}(\{X_m\})$ of the **F-P** equation must coincide with Einstein's probability distribution around thermodynamic equilibrium (see Chap. 2).

Using the concept of **sp** associated with the temporal fluctuations of the state variables $\{A_m\}$, Onsager concluded that the spontaneous force resulting from an instantaneous variation $X_j(t)$ can produce a flow and vice versa [1, 6]. The importance of this fact is that the proportionality coefficients are the same even when the physical processes involved are different. That is, the transport coefficients have intrinsic symmetries. The assumptions involved in the proof of Onsager's Theorem are as follows:

1. The mean values of the fluctuations $X_m(t)$ decay over time following an exponential law, analogous to that followed by macroscopic phenomenological laws. That is, the evolution equation is linear[5]:

$$\frac{d}{dt} \langle \mathbf{X}(t) \rangle_{\mathbf{X}(0)} = -\mathbf{M} \cdot \langle \mathbf{X}(t) \rangle_{\mathbf{X}(0)}. \tag{4.1}$$

---

[4]Prigogine and his colleagues recently suggested that the essential condition for understanding the arrow of time is that the microscopic description of the universe involves unstable dynamical systems (i.e., chaotic).

[5]Here we have used the vector notation $\mathbf{X}(t)$ to indicate the components $\{X_m(t)\}$, i.e., all possible instantaneous fluctuations (variations with respect to equilibrium) associated with state variables $\{A_m\}$. Note that in principle there is no reason to assume any symmetry in the matrix $\mathbf{M}$. In the linear approximation around equilibrium this means that the regression of fluctuations is governed by the macroscopic law.

2. Fluctuations around thermodynamic equilibrium are described by Einstein's probability distribution: $P_{eq}(\mathbf{X}) \propto \exp{(+\Delta S_T/k_B)}$, where the variation of the total entropy of the isolated system is given by the bilinear form $\Delta S_T/k_B \equiv \frac{-1}{2}(\mathbf{X} \cdot \mathbf{g} \cdot \mathbf{X})$.
3. Microscopic reversibility.[6]

To prove this theorem we proceed as follows:
First, the concept of generalized forces, we define [produced by spontaneous fluctuations of the state variables $\{A_m\}$]

$$\mathcal{F}_i = -\frac{\partial \Delta S_T}{\partial X_i} = -\frac{\partial S}{\partial A_i} = +k_B \sum_l g_{il} X_l. \tag{4.2}$$

Note that at thermodynamic equilibrium S is maximum, so that $\langle \partial S/\partial A_i \rangle_{eq} = 0$, then $\langle \mathcal{F}_i \rangle_{eq} = 0$.

The concept of generalized current is then introduced [produced by the temporal variation or fluctuation $X_i(t)$]

$$J_i = \frac{dX_i}{dt}, \tag{4.3}$$

from which it is observed that the production of entropy is given by

$$\frac{d\Delta S_T}{dt} = \sum_j \frac{\partial \Delta S_T}{\partial X_j} \frac{dX_j}{dt} = -\sum_j \mathcal{F}_j J_j. \tag{4.4}$$

Then, the problem is reduced to expressing currents J_i in terms of the forces $\mathcal{F}_m$ through a matrix (transport coefficients) that will have some intrinsic symmetry.

The solution of (4.1), for short times is given by the expansion

$$\langle \mathbf{X}(t) \rangle_{\mathbf{X}(0)} = \exp{(-\mathbf{M}t)} \cdot \mathbf{X}(0) \simeq \mathbf{X}(0) - t\mathbf{M} \cdot \mathbf{X}(0) + \mathcal{O}(t^2). \tag{4.5}$$

Then, for $t \simeq 0$ the conditioned mean values (components of $\mathbf{X}(t)$) are given by

$$\langle X_i(t) \rangle_{\mathbf{X}(0)} = X_i(0) - t\sum_j M_{ij} X_j(0) + \mathcal{O}(t^2). \tag{4.6}$$

---

[6]Microscopic reversibility states that at thermodynamic equilibrium, the time correlation matrix satisfies $\langle X_j(0)X_i(\tau) \rangle_{eq} = \langle X_j(\tau)X_i(0) \rangle_{eq}$, this relationship follows from the invariance of the Hamiltonian dynamics under the transformación $t \to -t$ [5]. Note that we are here using the fact that the average value $\langle X_j(\tau) \rangle_{eq}$ is zero in thermodynamic equilibrium.

Multiplying this equation by $X_l(0)$ and taking mean values on the fluctuations, we get

$$\langle X_l(0) \langle X_i(t) \rangle_{\mathbf{X}(0)} \rangle = \langle X_l(0)X_i(0) \rangle - t \left\langle X_l(0) \sum_j M_{ij}X_j(0) \right\rangle + \mathcal{O}(t^2). \qquad (4.7)$$

Similarly, it follows that

$$\langle X_i(0) \langle X_l(t) \rangle_{\mathbf{X}(0)} \rangle = \langle X_l(0)X_i(0) \rangle - t \left\langle X_i(0) \sum_j M_{lj}X_j(0) \right\rangle + \mathcal{O}(t^2). \qquad (4.8)$$

Using microscopic reversibility we can set (4.7) equal to (4.8), and since

$$\langle X_l(0)X_i(0) \rangle = (\mathbf{g}^{-1})_{li}$$

we get

$$\mathbf{g}^{-1} \cdot \mathbf{M}^T = \mathbf{M} \cdot \mathbf{g}^{-1}. \qquad (4.9)$$

Here $\mathbf{M}^T$ indicates the transpose of $\mathbf{M}$, and thus we note that the matrix $\mathbf{L} \equiv \mathbf{M} \cdot \mathbf{g}^{-1}/k_B$ is symmetric. This is the desired relationship, since it makes possible to write the **mv** of currents $J_i = dX_i/dt$ in the form

$$\frac{d}{dt} \langle \mathbf{X}(t) \rangle_{\mathbf{X}(0)} = -\mathbf{M} \cdot \langle \mathbf{X}(t) \rangle_{\mathbf{X}(0)} = -k_B \, \mathbf{L} \cdot \mathbf{g} \cdot \langle \mathbf{X}(t) \rangle_{\mathbf{X}(0)}. \qquad (4.10)$$

Finally, using the definition of generalized forces (4.2) we find that, on average, the proportionality coefficients between the current and the generalized forces are characterized by a symmetric matrix $\mathbf{L}$, i.e.:

$$\frac{d}{dt} \langle \mathbf{X}(t) \rangle_{\mathbf{X}(0)} = -\mathbf{L} \cdot \langle \mathcal{F}(t) \rangle_{\mathbf{X}(0)}, \qquad (4.11)$$

where $\mathbf{L}$ is the matrix of the kinetic coefficients.

Exercise. In cases where there is no external magnetic field, it is easy to see that the equations of motion are even under time reversal, thus calculating stationary averages it makes no difference which of the variables X_k or X_j is taken at the earlier time. Show that the **mv** of the product $X_k(t+\tau)X_j(t)$ is equal to the **mv** of the product $X_j(t+\tau)X_k(t)$.

Excursus. In the case when some state variables $\{A_i\}$ are odd under time reversal $t \to -t$, one may introduce parity factors ϵ_l defined as $\epsilon_l = -1$ for odd variables, and $\epsilon_l = 1$ for even variables. Then, Onsager's Theorem is written in the generalized form:

$$\mathbf{L}_{ij} = \epsilon_i\epsilon_j\mathbf{L}_{ji}. \qquad (4.12)$$

Here the repeated indices should not be summed over.

Excursus. Consider two generalized forces with different physical origins: $\nabla n(r)$ a concentration gradient, and $\nabla T(r)$ a temperature gradient. By Onsager's Theorem $\nabla T(r)$ implies a heat flow, and vice versa $\nabla n(r)$ implies a stream of particles. Show that these physical processes, although different, have the same proportionality coefficients $\mathbf{L}_{ij}$. That is, using the symmetries of $\mathbf{L}_{ij}$ it is possible to obtain kinetic coefficients for situations in which it is very difficult to calculate them from first principles [1].

4.2 Entropy Production in the Linear Approximation

Using the quadratic form that characterizes the change of the total entropy, we can calculate the generalized forces $\mathcal{F}_i$. If at time t the degree of deviation from equilibrium of the system is given by the set $\{X_m(t)\}$, the generalized forces at that instant are

$$\mathcal{F}_i(t) = k_B \sum_l g_{il} X_l(t).$$

Furthermore, at equilibrium the entropy is maximum, then $\langle \mathcal{F}_i(t) \rangle_{eq} = 0$ for all i.

Using the definition of generalized currents $J_i = dX_i/dt$, the temporal variation of the total change of entropy ΔS_T, due to fluctuations, is given, using (4.4), by

$$\frac{dS}{dt} = -\sum_j \mathcal{F}_j J_j. \tag{4.13}$$

Assuming a linear response to be valid (i.e., that the currents $J_m = dX_m/dt$ are proportional to the generalized forces $J_m = -\sum_j \mathbf{L}_{mj} \mathcal{F}_j$), from (4.13) it follows that the derivative of the entropy with respect to time (entropy current) is

$$\frac{dS}{dt} = \sum_{ji} \mathbf{L}_{ij} \mathcal{F}_i \mathcal{F}_j.$$

This quantity characterizes the entropy production. Then $\sum_{ji} \mathbf{L}_{ij} \mathcal{F}_i \mathcal{F}_j$ has to be a positive definite quadratic form[7]; and therefore as the system approaches the equilibrium state, entropy must grow towards a maximum.

From thermodynamics it is possible to obtain expressions for the generalized forces $\mathcal{F}_i$, as well as for thermodynamic flows (and currents) according to the coefficients of proportionality $\mathbf{L}_{ij}$ (Onsager's kinetic coefficients).

---

[7]If $\mathbf{L}_{ij} \in \mathcal{R}_e$ and it is a *positive definite matrix*, then $\sum_{ji} \mathbf{L}_{ij} \mathcal{F}_i \mathcal{F}_j > 0$, $\forall \mathcal{F}_i, \mathcal{F}_j$ on the $\mathcal{R}_e$. About the concept of positive definite matrix, see Sect. 1.12 and note (7) of Chap. 6.

Exercise. Using that $\mathbf{L}_{ij}$ is a positive definite matrix, show that in general, for an $n \times n$ matrix, we have the relations

$$\mathbf{L}_{ii} > 0, \; \mathbf{L}_{ii}\mathbf{L}_{kk} > (\mathbf{L}_{ik})^2, \; \forall\{i, k\},$$

in addition to the Onsager relations, $\mathbf{L}_{ik} = \mathbf{L}_{ki}$.

Excursus. For a resistor maintained at constant temperature, J_e is the electric current and $\mathcal{F}$ is proportional to the applied field, and the quantity in expression (4.13) is proportional to the dissipated energy by Joule effect. The conductivity tensor of an anisotropic medium can be regarded as an example of Onsager's relation. An alternative derivation of this Onsager's symmetry has been derived by Stratonovich in *Nonlinear Nonequilibrium Thermodynamics I*, Springer-Verlag, Berlin, 1992. Further examples of applications of Onsager's symmetry will be given in the following sections.

4.2.1   Mechanocaloric Effect*

Consider a classic gas (with a single type of particle) enclosed in an insulated box divided into two compartments connected by a small hole. The mass and total energy of the system, M_T and U_T, are constant, and there may be fluctuations because of energy and mass transfer through the hole. If the system is close to equilibrium, we can calculate the variation of the total entropy as a function of variations ΔM_i and ΔU_i in each compartment ($\Delta M_1 = -\Delta M_2 = \Delta M$, $\Delta U_1 = -\Delta U_2 = \Delta U$). That is, ΔM and ΔU are random variables or macroscopic fluctuations; then the variation of the total entropy is written, to the second order, in the form

$$\Delta S_T = 2\left[\frac{1}{2}\left(\frac{\partial^2 S}{\partial U^2}\right)_M^o (\Delta U)^2 + \left(\frac{\partial}{\partial U}\left(\frac{\partial S}{\partial M}\right)_U^o\right)_M^o (\Delta U \Delta M) + \frac{1}{2}\left(\frac{\partial^2 S}{\partial M^2}\right)_U^o (\Delta M)^2\right].$$

The temporal change of the total variation of the entropy is given by

$$\frac{d\Delta S_T}{dt} = 2\left[\left(\frac{\partial^2 S}{\partial U^2}\right)_M^o \Delta U \frac{d\Delta U}{dt} + \left(\frac{\partial}{\partial U}\left(\frac{\partial S}{\partial M}\right)_U^o\right)_M^o \frac{d}{dt}(\Delta U \Delta M)\right.$$
$$\left. + \left(\frac{\partial^2 S}{\partial M^2}\right)_U^o \Delta M \frac{d\Delta M}{dt}\right]$$
$$= 2\left[\frac{d\Delta U}{dt}\left\{\left(\frac{\partial^2 S}{\partial U^2}\right)_M^o \Delta U + \left(\frac{\partial}{\partial M}\left(\frac{\partial S}{\partial U}\right)_M^o\right)_U^o \Delta M\right\}\right.$$
$$\left. + \frac{d\Delta M}{dt}\left\{\left(\frac{\partial^2 S}{\partial M^2}\right)_U^o \Delta M + \left(\frac{\partial}{\partial U}\left(\frac{\partial S}{\partial M}\right)_U^o\right)_M^o \Delta U\right\}\right]$$
$$\equiv 2\left[\frac{d\Delta U}{dt}\left\{\Delta\left(\frac{\partial S}{\partial U}\right)_M\right\} + \frac{d\Delta M}{dt}\left\{\Delta\left(\frac{\partial S}{\partial M}\right)_U\right\}\right].$$

So comparing with (4.4) we conclude that energy flow J_u and mass flow J_m are given by

$$J_u = \frac{d}{dt}\Delta U \quad \text{and} \quad J_m = \frac{d}{dt}\Delta M,$$

while the generalized forces are given by the expressions

$$\mathcal{F}_u = -\Delta\left(\frac{\partial S}{\partial U}\right)_M \quad \text{and} \quad \mathcal{F}_m = -\Delta\left(\frac{\partial S}{\partial M}\right)_U. \tag{4.14}$$

Moreover, from thermodynamics it is known that for a system with constant volume V, we have (see Appendix A)

$$dS = \frac{1}{T}dU - \frac{\mu}{T}dM \equiv \left(\frac{\partial S}{\partial U}\right)_M dU + \left(\frac{\partial S}{\partial M}\right)_U dM.$$

Then, comparing with (4.14) we show that the generalized forces are characterized by the temperature and chemical potential differences between the two compartments, i.e.:

$$\mathcal{F}_u = -\Delta\left(\frac{\partial S}{\partial U}\right)_M = -\Delta(\frac{1}{T}) \quad \text{and} \quad \mathcal{F}_m = -\Delta\left(\frac{\partial S}{\partial M}\right)_U = \Delta(\frac{\mu}{T}).$$

Onsager's relations $\mathbf{L} = \mathbf{L}^T$ enable us to check that the flows of energy and mass are closely linked even when the phenomena involved are physically different.

Excursus. In the linear regime, if we write that currents are proportional to generalized forces $J_j = -\sum_l L_{jl}\mathcal{F}_l$, it is possible to obtain expressions for the currents as a function of the variations of pressure and temperature. Using the Gibbs–Duhem equation $d\mu = -s\,dT + dP/\rho$, where s and ρ are the entropy and volume per unit mass, respectively, we can write the generalized forces in terms of the state variables T and P, i.e.:

$$\mathcal{F}_u = \frac{1}{T^2}\Delta T \quad \text{and} \quad \mathcal{F}_m = -\frac{h}{T^2}\Delta T + \frac{1}{\rho T}\Delta P,$$

where $h = sT + \mu$ is the enthalpy per unit mass. Then in the linear approximation the energy flow is

$$J_u = -L_{uu}\mathcal{F}_u - L_{um}\mathcal{F}_m = -L_{uu}\frac{\Delta T}{T^2} + L_{um}\left(\frac{h}{T^2}\Delta T - \frac{1}{\rho T}\Delta P\right),$$

and the mass flow is given by

$$J_m = -L_{mm}\mathcal{F}_m - L_{mu}\mathcal{F}_u = L_{mm}\left(\frac{h}{T^2}\Delta T - \frac{1}{\rho T}\Delta P\right) - L_{mu}\frac{\Delta T}{T^2}.$$

On the other hand, Onsager's relations state that $L_{mu} = L_{um}$. If $\Delta T = 0$ (i.e., the two chambers have the same temperature), but different pressure $\Delta P \neq 0$, we obtain:

$$J_u = -\frac{L_{um}}{\rho T}\Delta P \quad \text{and} \quad J_m = -\frac{L_{mm}}{\rho T}\Delta P,$$

from which we see that mass and energy flows are related through the condition $J_u = (L_{um}/L_{mm})\,J_m$. The constant L_{um}/L_{mm} determines the amount of energy per unit mass passing through the hole in the compartment. This proportionality constant can be calculated with different microscopic models [1]. However, it may happen that there are circumstances when it is not possible to obtain explicitly some of the kinetic coefficients. In such cases Onsager's relations are of invaluable utility.[8]

Guided Exercise (**Knudsen's Gas**). Consider a gas of free particles of mass m at temperature T, suppose that upon reaching the hole (communicating compartments as in the previous example) each of the particles can freely transport from one room to the other an amount of energy per unit mass $L_{um}/L_{mm} \equiv u^*$. In this case it is easy to calculate u^* using the Maxwell–Boltzmann statistics. That is, if we use the equilibrium distribution of velocities (in three dimensions) $P(\mathbf{v}) = (2\pi D_V)^{-3/2}\exp\left(-\mathbf{v}^2/2D_V\right)$, where $D_V \equiv k_B T/m$, the quantity u^* is given by

$$u^* = \frac{N_A}{m^*}\frac{\int_0^\infty dv_x \int_{-\infty}^{+\infty} dv_y \int_{-\infty}^{+\infty} dv_z \frac{m}{2}\mathbf{v}^2 n v_x P(\mathbf{v})}{\int_0^\infty dv_x \int_{-\infty}^{+\infty} dv_y \int_{-\infty}^{+\infty} dv_z\, n v_x P(\mathbf{v})},$$

where $n = N/V$ is the density of particles. Given that $\int_0^\infty dv_x\, v_x P(\mathbf{v}) = \sqrt{D_V/2\pi}$, $\int_0^\infty dv_x\, v_x^3 P(\mathbf{v}) = \sqrt{2D_V^3/\pi}$, $\int_{-\infty}^\infty dv_j\, v_j^2 P(\mathbf{v}) = D_V$, it follows immediately that $u^* = 2k_B T N_A/m^*$ is the energy transferred per unit mass. That is, we can write $L_{um}/L_{mm} = 2RT/m^*$, where $R = N_A k_B$ is the gas constant (N_A is Avogadro's number) and m^* is the molecular mass of the particles. We can also obtain a similar expression considering a hole size larger than the mean free path of "free" particles (Boyle's gas). In this case the particles have to push one other (do work) to pass through the hole, so u^* has a different expression from the previous case [1].

---

[8]Onsager symmetries are the starting point for analyzing entropy production in systems away from thermodynamic equilibrium. In the context of linear response theory it is possible to prove that the state of minimum entropy production is stable, but only within the linear regime. This was the basic idea that led to the concept of nonequilibrium phase transitions [7].

4.3 Onsager's Relations in an Electrical Circuit

Onsager's relations are macroscopic properties that make it possible to verify the principle of microscopic reversibility. Indeed, this was the first time that the principle of time invariance $t \rightarrow -t$ of Hamiltonian dynamics occupied a key role in the formulation of statistical mechanics. In order to construct the matrix of kinetic coefficients $\mathbf{L}_{ij}$ it is necessary to know the matrix $\mathbf{g}^{-1}$ that characterizes the fluctuations around thermodynamic equilibrium. A simple example that displays the symmetry of $\mathbf{L}_{ij}$ is an electric circuit RCL immersed in a thermal bath of ionic solution $\mathcal{B}$ (charge reservoir at temperature T).

We begin our analysis writing the phenomenological equation for the conservation of charge:

$$\frac{dq}{dt} = I - \gamma q + \Delta q(t). \tag{4.15}$$

In this equation $q(t)$ represents the instantaneous charge on the capacitor, I the electrical current in the RCL circuit, γq the loss of charge in the capacitor which is immersed in the thermal reservoir $\mathcal{B}$, and $\Delta q(t)$ the fluctuation term from the ion solution $\mathcal{B}$.

From Kirchhoff's laws we get the potential difference across the circuit:

$$L\frac{dI}{dt} + \frac{q}{C} + IR = \Delta V(t). \tag{4.16}$$

In this equation $L\,dI/dt$ represents the potential difference in the inductor and q/C and IR the voltage drop across the capacitor and resistor, respectively. Then $\Delta V(t)$ represents the voltage fluctuations due to the ion reservoir $\mathcal{B}$.

The fluctuation terms will be modeled by Gaussian white noises $\xi_l(t)$ of zero mean, so, in general, we have

$$\Delta q(t) = d_{11}\,\xi_1(t) + d_{12}\,\xi_2(t) \tag{4.17}$$

$$\Delta V(t) = d_{21}\,\xi_1(t) + d_{22}\,\xi_2(t),$$

where the correlation of the noises is denoted in the form: $\left\langle \xi_i(t)\xi_j(t')\right\rangle = \delta_{i,j}\,\delta(t-t')$.

If there are no fluctuations ($\Delta q(t) = \Delta V(t) = 0$), the dynamics of this RCL system (with $\gamma = 0$) is reduced to a deterministic equation for a resonant circuit. That is, since $I = dq/dt$ and $V_c = Lq$, we obtain [8]

$$\frac{d^2 V_c}{dt^2} + \frac{1}{LC}V_c + \frac{R}{L}\frac{dV_c}{dt} = 0.$$

But here we are interested in the study of **sde** (4.15) and (4.16) associated with the state variables $q(t)$ and $I(t)$ respectively, and their fluctuations. Since the circuit is linear we can write these equations in matrix form

$$\frac{d}{dt}\begin{pmatrix} q \\ I \end{pmatrix} = -\begin{pmatrix} \gamma & -1 \\ 1/LC & R/L \end{pmatrix}\begin{pmatrix} q \\ I \end{pmatrix} + \begin{pmatrix} d_{11} & d_{12} \\ d_{21} & d_{22} \end{pmatrix}\begin{pmatrix} \xi_1 \\ \xi_2 \end{pmatrix}. \tag{4.18}$$

Note that this equation defines the relaxation matrix **M**.

The probability distribution around thermodynamic equilibrium is

$$P_{eq}(q, I) \propto \exp(-W_{min}/k_B T),$$

where W_{min} is the minimum work (see Chap. 2), which, for the physical system we are studying, is given by

$$W_{min} = \frac{1}{2C}q^2 + \frac{L}{2}I^2 \equiv \frac{k_B T}{2}(\mathbf{X} \cdot \mathbf{g} \cdot \mathbf{X}) \quad \text{with} \quad \mathbf{X} \equiv (q, I). \tag{4.19}$$

Then the spontaneous forces (4.2), which restore equilibrium, are

$$\mathcal{F}_q = -\frac{\partial \Delta S_T}{\partial q} = \frac{1}{T}\frac{q}{C} \quad \text{and} \quad \mathcal{F}_I = -\frac{\partial \Delta S_T}{\partial I} = \frac{1}{T}LI. \tag{4.20}$$

The matrix $\mathbf{g}^{-1}$ characterizing the mean square fluctuations is given by

$$\mathbf{g}^{-1} = \begin{pmatrix} \langle qq \rangle & \langle qI \rangle \\ \langle Iq \rangle & \langle II \rangle \end{pmatrix} = \begin{pmatrix} k_B TC & 0 \\ 0 & k_B T/L \end{pmatrix}. \tag{4.21}$$

Onsager's relations (for even state variables) state that the combination $\mathbf{L} = \mathbf{M} \cdot \mathbf{g}^{-1}/k_B$ is symmetric. In our case, taking the **mv** of (4.18) to calculate **M**, and using (4.21) we obtain

$$\mathbf{L} = \begin{pmatrix} \gamma & -1 \\ 1/LC & R/L \end{pmatrix}\begin{pmatrix} TC & 0 \\ 0 & T/L \end{pmatrix} = \begin{pmatrix} \gamma TC & -T/L \\ T/L & TR/L^2 \end{pmatrix}, \tag{4.22}$$

a matrix that satisfies the generalized symmetries $L_{ij} = \epsilon_i \epsilon_j L_{ji}$. Note that under time reversal $t \to -t$ the state variables behave as $q \to q$ and $I \to -I$. Then, we note that $\epsilon \cdot \mathbf{M} \cdot \mathbf{g}^{-1} = (\mathbf{M} \cdot \mathbf{g}^{-1})^T \cdot \epsilon$. Here the matrix ϵ is defined by

$$\epsilon = \begin{pmatrix} 1 & 0 \\ 0 & -1 \end{pmatrix}.$$

From this example it is seen that if we express the **mv** flows as a function of generalized forces $\langle J_m \rangle = \langle dX_m / dt \rangle = -\sum_j \mathbf{L}_{mj} \langle \mathcal{F}_j \rangle$, the proportionality coefficients are precisely characterized by $\mathbf{M} \cdot \mathbf{g}^{-1} / k_B$. Using the definition of generalized forces $\mathcal{F}_i(t) = k_B \sum_l g_{il} X_l(t)$, we see from (4.18) and (4.20) that

$$\frac{d}{dt} \begin{pmatrix} \langle q \rangle \\ \langle I \rangle \end{pmatrix} = - \begin{pmatrix} \gamma TC & -T/L \\ T/L & TR/L^2 \end{pmatrix} \begin{pmatrix} \langle \mathcal{F}_q \rangle \\ \langle \mathcal{F}_I \rangle \end{pmatrix}. \tag{4.23}$$

Furthermore, it is also possible to check that both γ and R are dissipative coefficients, and therefore positive. The deterministic temporal variation of the energy of the system S is given by

$$\frac{dE}{dt} = LI \frac{dI}{dt} + \frac{q}{C} \frac{dq}{dt} \tag{4.24}$$

$$= LI \left(\frac{q/C + IR}{-L} \right) + \frac{q}{C} (I - \gamma q)$$

$$= - \left(RI^2 + \gamma q^2 / C \right),$$

an expression representing the energy loss in the resistive element and the capacitor. On the other hand, if we calculate the entropy current in the linear approximation, we obtain

$$\frac{dS}{dt} = \sum_{ji} \mathbf{L}_{ij} \mathcal{F}_i \mathcal{F}_j$$

$$= \left(\tfrac{1}{T} q/C \quad \tfrac{1}{T} LI \right) \begin{pmatrix} \gamma TC & -T/L \\ T/L & TR/L^2 \end{pmatrix} \begin{pmatrix} \tfrac{1}{T} q/C \\ \tfrac{1}{T} LI \end{pmatrix}$$

$$= \frac{1}{T} \left(RI^2 + \gamma q^2 / C \right) > 0.$$

This means that entropy production is balanced *exactly* by energy dissipation.

In the next section we will prove the (first) Fluctuation-Dissipation theorem. This result makes it possible to characterize the coefficients d_{ij} appearing in noise terms in the constituent **sde** (4.15), (4.16), and (4.17). In particular, we will show that $d_{12} = d_{21} = 0$, which is to say that the noise terms in charge and voltage are uncorrelated.

Often the inverse problem arises: that is, given the kinetic coefficient matrix $\mathbf{L}_{ij}$ and the equilibrium fluctuations matrix $\mathbf{g}^{-1}$, it may be necessary to calculate the linear relaxation matrix elements (macroscopic values) $\mathbf{M}$. In this case, Onsager's relations are also very useful.

4.4   Ornstein–Uhlenbeck Multidimensional Process

Now we study a linear system of n **sdes**, especially their fluctuations and relaxation to thermodynamic equilibrium. If we expect an exponential macroscopic relaxation behavior for the n variables $\{x_j\}$, then the deterministic dynamics will be governed by an ordinary differential equation of the form[9]

$$\frac{d}{dt}x_i = -\mathbf{M}_{ij}\, x_j, \quad \{i,j\} = 1, 2, \cdots, n. \tag{4.25}$$

Since in thermodynamic equilibrium the probability distribution is Gaussian and characterized by Einstein's distribution (see Chap. 2), it is natural to think that the whole set[10] of **sp** $\{x_j(t)\}$ will be Gaussian for all t. Then the associated **sde** will be a n-dimensional generalization of the Ornstein–Uhlenbeck process, as presented in the previous chapter when we analyzed the relaxation of the velocity of a Brownian particle. This time we present the inverse problem.

Let $\{x_m(t)\}$ be a stationary stochastic process (vector with n components) Gaussian and Markovian, then the propagator of the variables $\{x_m(t)\}$ satisfies

$$\frac{\partial}{\partial t}P(\{x_m\}, t) = \left[\frac{\partial}{\partial x_i}\mathbf{M}_{ij}\, x_j + \frac{1}{2}\mathbf{B}_{ij}\frac{\partial}{\partial x_i}\frac{\partial}{\partial x_j}\right]P(\{x_m\}, t), \tag{4.26}$$

with the initial condition[11] $P(\{x_m\}, t \to 0) \to \prod_{j=1}^{n} \delta\left(x_j - x_j(0)\right)$.

The solution at thermodynamic equilibrium is $P_{eq}(\{x_m\}) \propto \exp\left(-\frac{1}{2}\mathbf{x}\cdot\mathbf{g}\cdot\mathbf{x}\right)$, where the bilinear form $\mathbf{x}\cdot\mathbf{g}\cdot\mathbf{x}$ is given in terms of the change in total entropy. The problem is to find relations for the matrix $\mathbf{B}$ to ensure the relaxation to thermodynamic equilibrium (first Fluctuation-Dissipation theorem). Our goal is that the stationary solution of the **F-P** equation (4.26) should match $P_{eq}(\{x_m\})$.

Excursus (**Eigenfunctions for the Ornstein–Uhlenbeck Process.**) Solve, using eigenvalues analysis, a multidimensional **F-P** equation when the drift is linear and the diffusion matrix is constant and *positive definite*; that is, Eq. (4.26). This **sp** is extremely important and is the analog in mathematical difficulty, to the n-dimensional harmonic oscillator in quantum mechanics [4].

---

[9]Henceforth we use the repeated index notation to indicate a summation.

[10]In what follows we simplify the notation, when there is no ambiguity and will use the same letters to characterize both **sp** and **rv** (compare with Chap. 3).

[11]Note the simplifying notation used in the conditional probability.

4.4.1 The First Fluctuation-Dissipation Theorem

Given Eq. (4.26), we want to find the matrix $\mathbf{B}$ that ensures that the stationary probability of $\mathbf{F}$-$\mathbf{P}$ is Einstein's distribution. To solve this problem we introduce the distribution $P_{st}(\{x_m\}) \propto \exp\left(-\frac{1}{2}\mathbf{x} \cdot \mathbf{g} \cdot \mathbf{x}\right)$ in the stationary $\mathbf{F}$-$\mathbf{P}$ equation (4.26), that is:

$$0 = \left[2\frac{\partial}{\partial x_i}\mathbf{M}_{ij}\, x_j + \mathbf{B}_{ij}\frac{\partial}{\partial x_i}\frac{\partial}{\partial x_j}\right]P_{st}(\{x_m\}), \tag{4.27}$$

leading to

$$0 = 2\mathbf{M}_{ii} - \mathbf{B}_{ij}\mathbf{g}_{ji} + \left(\mathbf{B}_{ij}\mathbf{g}_{im}\mathbf{g}_{jn} - 2\mathbf{M}_{im}\mathbf{g}_{in}\right)x_m x_n. \tag{4.28}$$

Then for arbitrary (x_m, x_n), the first term of (4.28) and the quadratic form both cancel separately. Explicitly

$$2\sum_i \mathbf{M}_{ii} = \sum_{ij} \mathbf{B}_{ij}\mathbf{g}_{ji}. \tag{4.29}$$

Furthermore, using the symmetry of $\mathbf{g}$ and dummy indices we can write

$$0 = \left(\mathbf{B}_{ij}\mathbf{g}_{im}\mathbf{g}_{jn} - 2\mathbf{M}_{im}\mathbf{g}_{in}\right)x_m x_n$$
$$= \left(\mathbf{B}_{ij}\mathbf{g}_{im}\mathbf{g}_{jn} - \mathbf{M}_{im}\mathbf{g}_{in} - \mathbf{M}_{im}\mathbf{g}_{in}\right)x_m x_n,$$

where it is noted that

$$\mathbf{B}_{ij}\mathbf{g}_{im}\mathbf{g}_{jn} = \mathbf{g}_{ni}\mathbf{M}_{im} + \mathbf{g}_{mi}\mathbf{M}_{in}. \tag{4.30}$$

Multiplying (4.30) by $\mathbf{g}_{im}^{-1}$ and adding, we obtain

$$\mathbf{B}_{ij}\mathbf{g}_{jn} = \mathbf{g}_{im}^{-1}\mathbf{g}_{nl}\mathbf{M}_{lm} + \mathbf{g}_{im}^{-1}\mathbf{g}_{ml}\mathbf{M}_{ln} \tag{4.31}$$
$$= \mathbf{g}_{im}^{-1}\mathbf{g}_{nl}\mathbf{M}_{lm} + \mathbf{M}_{in}.$$

Note that setting $n = i$ and summing over i in (4.31) we get

$$\sum_{ij}\mathbf{B}_{ij}\mathbf{g}_{ji} = \sum_{ilm}\mathbf{g}_{im}^{-1}\mathbf{g}_{il}\mathbf{M}_{lm} + \sum_i \mathbf{M}_{ii} \tag{4.32}$$
$$= 2\sum_m \mathbf{M}_{mm},$$

which is precisely (4.29). All that is left now is satisfying (4.31). To solve the matrix
B, we can rewrite this condition in the form:

$$
\mathbf{B}_{ij} = \mathbf{g}_{jn}^{-1}\mathbf{g}_{im}^{-1}\mathbf{g}_{nl}\mathbf{M}_{lm} + \mathbf{g}_{jn}^{-1}\mathbf{M}_{in} \tag{4.33}
$$
$$
= \delta_{j,l}\,\mathbf{g}_{im}^{-1}\mathbf{M}_{lm} + \mathbf{M}_{in}\mathbf{g}_{nj}^{-1}
$$
$$
= \mathbf{M}_{jm}\mathbf{g}_{mi}^{-1} + \mathbf{M}_{in}\mathbf{g}_{nj}^{-1},
$$

that is to say,

$$
\mathbf{B} = \mathbf{M}\cdot\mathbf{g}^{-1} + \left(\mathbf{M}\cdot\mathbf{g}^{-1}\right)^{T}.
$$

Here the superscript T indicates transpose matrix.

Equation (4.33) is the expression we were looking for, representing the relationship between the fluctuation matrix **B** and dissipation matrix **M**, and ensures
that in the stationary state the probability distribution equals the distribution around
thermodynamic equilibrium.

Exercise (**Nyquist's Theorem.**) In Fig. 4.1 we represent an RCL circuit connected to a large neutral charge reservoir at temperature T, i.e., an ionic solution.
Using the equilibrium Boltzmann distribution of the state variables of the system
$\{I, q\}$:

$$
P_{\text{eq}}(q, I) \propto \exp\left(-\frac{LI^2}{2k_BT} - \frac{q^2}{2Ck_BT}\right).
$$

Show that the matrix **B** in the **F-P** equation associated with the **sde** (4.15)
and (4.16) is

$$
\mathbf{B} = 2k_BT\begin{pmatrix} \gamma C & 0 \\ 0 & R/L^2 \end{pmatrix},
$$

Fig. 4.1 Representation of
the electric RCL circuit
immersed in a thermal ionic
reservoir. The temperature
induces a stochastic voltage
$\xi(t)$ in the circuit. In addition,
in the capacitor there is a
fluctuating gain and loss of
charge into the reservoir

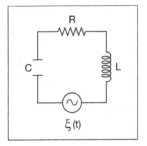

so that the coefficients d_{ij} in Eq. (4.17) are[1]

$$d_{12} = d_{21} = 0$$

$$d_{11} = \sqrt{2k_B T \gamma C}$$

$$d_{22} = \sqrt{2k_B T R / L}.$$

From these results we see that the fluctuating voltage is: $\Delta V(t) = \sqrt{2k_B TR}\, \xi_2(t)$, from which Nyquist's theorem follows: $\langle \Delta V(t)\,\Delta V(t')\rangle = 2k_B TR\,\delta(t-t')$). Note that the two sources of fluctuations are uncorrelated ($d_{12} = d_{21} = 0$), i.e., since they arise from different causes.

Example. Consider a Brownian particle of mass m in a harmonic potential, to be characterized by its speed $V(t) = \dot{X}(t)$ and position $X(t)$ in 1-dimension. The deterministic evolution[12] equation is

$$m\ddot{X} + \gamma \dot{X} + \kappa X = 0.$$

Then, in the variables X and $p = m\dot{X}(t)$, we have the pair of equations

$$\dot{X} = p/m \quad \text{and} \quad \dot{p} = -\gamma p/m - \kappa X,$$

and since the dynamics is linear, we can write

$$\frac{d}{dt}\begin{pmatrix} X \\ p \end{pmatrix} = \begin{pmatrix} 0 & 1/m \\ -\kappa & -\gamma/m \end{pmatrix}\begin{pmatrix} X \\ p \end{pmatrix}$$

$$\equiv -\mathbf{M}\begin{pmatrix} X \\ p \end{pmatrix}.$$

From the probability distribution around thermodynamic equilibrium we know that

$$-\frac{1}{2}\,\mathbf{x}\cdot\mathbf{g}\cdot\mathbf{x} = \frac{-1}{k_B T}\left(\frac{\kappa X^2}{2} + \frac{p^2}{2m}\right),$$

thus $\mathbf{g} = \frac{1}{k_B T}\begin{pmatrix} \kappa & 0 \\ 0 & 1/m \end{pmatrix}$. Then, using the definition of Onsager's matrix:

$$\mathbf{L} = \frac{1}{k_B}\mathbf{M}\cdot\mathbf{g}^{-1}$$

$$= \frac{1}{k_B}\begin{pmatrix} 0 & -1/m \\ \kappa & \gamma/m \end{pmatrix}\begin{pmatrix} k_B T/\kappa & 0 \\ 0 & k_B Tm \end{pmatrix}$$

$$= T\begin{pmatrix} 0 & -1 \\ +1 & \gamma \end{pmatrix}.$$

---

[12]Note that the dynamic system dissipates kinetic energy.

From the first Fluctuation-Dissipation theorem (4.33) it follows that

$$\mathbf{B} = \begin{pmatrix} 0 & -k_BT \\ +k_BT & \gamma k_BT \end{pmatrix} + \begin{pmatrix} 0 & k_BT \\ -k_BT & \gamma k_BT \end{pmatrix}$$

$$= \begin{pmatrix} 0 & 0 \\ 0 & 2\gamma k_BT \end{pmatrix},$$

then from (4.26) we conclude that the **F-P** equation for this system will be

$$\frac{\partial}{\partial t}P(X,p,t) = \left[-\frac{\partial}{\partial X}(p/m) + \frac{\partial}{\partial p}(\kappa X + \gamma p/m) + \gamma k_BT \frac{\partial^2}{\partial p^2} \right] P(X,p,t),$$

where $P(X, p, t \to t_0) \to \delta(X - X_0)\delta(p - p_0)$. Thanks to the Markovian nature of the process we are able to calculate any correlation function if we know the propagator $P(X, p, t)$, for example:

$$\langle X(t_1)X(t_2) \rangle, \quad \langle X(t_1)p(t_2) \rangle, \quad \langle p(t_1)p(t_2)p(t_3)p(t_4) \rangle, \quad \text{etc.}$$

Exercise. Show for a Brownian particle that the temporal variation of entropy is positive and is given by

$$\frac{1}{k_B}\frac{dS}{dt} = \frac{1}{k_B}\sum_{ij} L_{ij}\mathcal{F}_i\mathcal{F}_j = \frac{\gamma}{m}\frac{p^2/2m}{k_BT/2}.$$

That is, the entropy production is done at the expense of kinetic energy dissipation.

Guided Exercise. Consider the **sde** associated with the Ornstein–Uhlenbeck process, but suppose that the intensity of the stochastic fluctuations ϵ is unknown:

$$m\frac{dV}{dt} = -m\Gamma V(t) + \sqrt{\epsilon}\,\xi(t), \tag{4.34}$$

where $\xi(t)$ is a Gaussian white noise with mean-value zero.[13] Use Novikov's Theorem (see Chap. 1) to show that fluctuation in *thermal equilibrium* implies the constraint $\epsilon = 2m\Gamma k_BT$. From (4.34) it is possible to calculate the **mv** of the kinetic energy of the particle: simply multiply (4.34) by $V(t)$ and then take the **mv**, that is to say,

$$m\left\langle V(t)\frac{dV}{dt} \right\rangle = \frac{m}{2}\frac{d}{dt}\langle V(t)^2 \rangle \tag{4.35}$$

$$= -m\Gamma\langle V(t)^2 \rangle + \sqrt{\epsilon}\langle V(t)\xi(t) \rangle.$$

---

[13] In particular of intensity one; that is to say, $\langle \xi(t)\xi(t') \rangle = \delta(t - t')$.

Since the system is linear we know the formal solution of (4.34):

$$V(t) = \frac{\sqrt{\epsilon}}{m} \int_0^t e^{-(t-s)\Gamma} \xi(s) \, ds + V(0)e^{-\Gamma t}.$$

It is then possible to calculate the functional variation [see the advanced exercise 1.15.10, and Sect. 3.1.2]:

$$\frac{\delta V(t)}{\delta \xi(t')} = \frac{\sqrt{\epsilon}}{m} \int_0^t e^{-(t-s)\Gamma} \delta(s - t') \, ds = \frac{\sqrt{\epsilon}}{m} e^{-(t-t')\Gamma}, \quad t' < t.$$

Furthermore, Novikov's Theorem (in the continuous case) states that

$$\langle V(t)\xi(t)\rangle = \int_0^t \left\langle \xi(t)\xi(t')\right\rangle \left\langle \frac{\delta V(t)}{\delta \xi(t')}\right\rangle dt'$$

$$= \int_0^t \delta(t - t') \frac{\sqrt{\epsilon}}{m} e^{-(t-t')\Gamma} \, dt' = \frac{\sqrt{\epsilon}}{2m}.$$

Then, from (4.35) we have that

$$\frac{m}{2} \frac{d}{dt} \langle V(t)^2\rangle = -m\Gamma \langle V(t)^2\rangle + \frac{\epsilon}{2m}.$$

Integrating this equation we get for the mean square velocity

$$\langle V(t)^2\rangle = \frac{\epsilon}{m^2} \left(\frac{1 - e^{-2\Gamma t}}{2\Gamma}\right) + \langle V(0)^2\rangle \, e^{-2\Gamma t}.$$

In order that the steady state of this expression concurs with the thermodynamic result, it must hold (in 1-dimension)

$$\langle V^2\rangle = k_B T/m,$$

from which we conclude that $\epsilon = 2m\Gamma k_B T$. Take $m\Gamma = \gamma$ and compare with the previous example.

4.5 Canonical Distribution in Classical Statistics

Consider the movement of a Brownian particle (in 1-dimension) of mass m in an arbitrary potential $U(X)$ (its generalization to 3-dimension is immediate), and in the presence of a thermal bath $\mathcal{B}$, or reservoir at temperature T. The state variables are the position X and velocity V of the particle.

$$m\frac{dV}{dt} = -U'(X) - \gamma V + \sqrt{2\gamma k_B T}\xi(t) \quad \text{and} \quad \frac{dX}{dt} = V. \tag{4.36}$$

Here $\xi(t)$ is a δ-correlated Gaussian **sp** with mean value zero. This process, $\xi(t)$, represents the *hidden variable* of the reservoir which accounts for the random forces introduced by the thermal bath $\mathcal{B}$. Equation (4.36) is essentially a Langevin **sde** for the vector $(X(t), V(t))$. Here $U(X)$ is the classical potential, whose gradient results in a deterministic force, and the term $-\gamma V$ represents a frictional force on the particle (Stokes' law). While (4.36) is not a linear Langevin equation, its structure is not of the multiplicative type, as the **sp** $\xi(t)$ is not multiplied by any state variable. Therefore, it is indistinct to use either Stratonovich or Itô differential calculus [9, 10]. The factor $\sqrt{2\gamma k_B T}$ ensures that the stationary probability distribution matches the thermodynamic equilibrium distribution.[14] Note that the intensity of the fluctuation is proportional to the dissipation coefficient γ, this fact is just an elementary version of the Fluctuation-Dissipation theorem.

Exercise. Show that the system of **sde** (4.36) is associated with the **F-P** equation:

$$\frac{\partial}{\partial t}P(X, V, t) = \mathcal{L}(X, V, \partial_X, \partial_V)\, P(X, V, t) \tag{4.37}$$

$$\equiv \left[-\frac{\partial}{\partial X}V + \frac{1}{m}\frac{\partial}{\partial V}\left\{ U'(X) + \gamma V \right\} + \gamma k_B T/m^2 \frac{\partial^2}{\partial V^2} \right] P(X, V, t),$$

where $P(X, V, t \to t_0) \to \delta(X - X_0)\delta(V - V_0)$. This equation was introduced by Kramers to study temporal fluctuations around equilibrium [5].[15]

Exercise. For cases where the potential $U(X)$ implies natural boundary conditions,[16] prove by direct substitution that

$$P_{\text{st}}(X, V) = \mathcal{N} \exp\left(-\frac{U(X)}{k_B T} - \frac{mV^2}{2k_B T} \right), \tag{4.38}$$

is the stationary solution of (4.37). Note that $P_{\text{st}}(X, V)$ is just the canonical Boltzmann distribution. Knowledge of the propagator solution of (4.37) and the use of the stationary solution (4.38) makes possible the complete characterization of the process. In particular the study of temporal correlations around thermodynamic equilibrium of the mechanical system (4.36).

---

[14]That is, the principle of *detailed balance*; topic to be discussed in detail in the next section.

[15]In particular, the potential could be periodic: $U(X) = U(X+L)$. In general, the eigenvalues of the corresponding **F-P** operator fully characterize the temporal correlations: $\langle X(0)X(\tau)\rangle$, $\langle V(0)V(\tau)\rangle$, etc. [4].

[16]That is, $P(X = \pm\infty, V, t) = 0$ for all V, t.

4.6 Stationary Fokker–Planck Solution

The calculation of the stationary solution $P_{st}(\{x_m\})$ of the multidimensional equation **F-P** is a nontrivial problem. In principle, the analysis is reduced to finding the solution of the stationary equation:

$$0 = \left[-\frac{\partial}{\partial x_i} K_i(\{x_m\}) + \frac{\partial}{\partial x_i} \frac{\partial}{\partial x_j} D_{ij}(\{x_m\}) \right] P_{st}(\{x_m\}) \equiv \mathcal{L} P_{st}(\{x_m\}), \qquad (4.39)$$

where we have defined the operator $\mathcal{L}$ and used summation notation for repeated indices. The uniqueness of the stationary solution is ensured by the existence of the Lyapunov[17] function:

$$\mathcal{H}(t) = \int \cdots \int P_1(\{x_m\}, t) \ln \left(\frac{P_1(\{x_m\}, t)}{P_2(\{x_m\}, t)} \right) \prod_m dx_m. \qquad (4.40)$$

Here P_1 and P_2 are arbitrary solutions of the **F-P** equation with natural boundary conditions: $P_1(\{x_m\} \to \pm\infty, t) \to 0$ for all t.

If the drift $K_i(\{x_m\})$ is not singular[18] and the diffusion matrix $D_{ij}(\{x_m\})$ is *positive definite*[19] in the domain of $\{x_m\}$, then, it is possible to prove that $\mathcal{H}(t)$ is a Lyapunov function [4]. From this it follows that the stationary pdf is unique [11]. That is, the stationary state is independent of initial condition.[20]

Optional Exercise (**Lyapunov's Function**). Defining

$$R \equiv R(\{x_m\}, t) = P_1(\{x_m\}, t) / P_2(\{x_m\}, t),$$

it is possible to prove using (4.40) [see Appendix B for a similar demonstration] that

$$\mathcal{H}(t) = \int \cdots \int P_2(\{x_m\}, t) \, (R \ln R - R + 1) \prod_m dx_m \geq 0. \qquad (4.41)$$

---

[17] The concept of Lyapunov function $\mathcal{H}(t)$ was clearly introduced by Boltzmann to prove his famous H theorem. In that case, the function $\mathcal{H}(t) = \int P \ln P \, d\{x_m\}$ decreases over time if the solution of the Boltzmann equation P does not match the stationary solution. In general we can say that $\mathcal{H}(t)$ is a Lyapunov function if it is positive for all t and meets the condition $\frac{d}{dt}\mathcal{H}(t) < 0$; that is to say, $\mathcal{H}(t)$ is decreasing and is bounded from below.

[18] That is, the potential does not have "infinite walls" that inhibit spreading throughout the whole domain of $\{x_m\}$.

[19] That is to say, $\sum_{ij} D_{ij}(\{x_m\}) R_i R_j > 0$ throughout the domain of the variables $\{x_m\}$ and for all real numbers $R_i, R_j \neq 0$.

[20] In the case of nonstationary 2π-periodic Markov systems, it is also possible to prove under similar conditions, the uniqueness of the asymptotic solution which is periodic in time [12], see also advanced exercise 4.9.7.

On the other hand, if $D_{ij}(\{x_m\})$ is *positive definite* and $K_i(\{x_m\})$ has no singularities, using the structure of the differential operator $\mathcal{L}$ of the **F-P** it is possible to prove that [4]

$$\frac{d}{dt}\mathcal{H}(t) = -\int \cdots \int P_1(\{x_m\}, t) \sum_{ij} D_{ij}(\{x_m\}) \frac{\partial \ln R}{\partial x_i} \frac{\partial \ln R}{\partial x_j} \prod_m dx_m \leq 0,$$

where we have explicitly used summation notation. Hence $\mathcal{H}(t)$ decreases if $\partial \ln R / \partial x_i \neq 0$, in addition $d\mathcal{H}(t)/dt = 0$ only when $\ln R = \ln C$, where C is a constant. Thus we conclude that, asymptotically in time, R must be independent of $\{x_m\}$. Furthermore, since $\mathcal{H}(t)$ is bounded from below, the function $\mathcal{H}(t)$ does not decrease indefinitely. Then, when $\mathcal{H}(t)$ is close to its minimum value, $R \to C$, therefore

$$P_1(\{x_m\}, t \to \infty) \to C\, P_2(\{x_m\}, t \to \infty).$$

Finally, by normalization we find that the constant is 1; that is, $\mathcal{H}(t \to \infty) \to 0$. Thus completing the proof that the stationary solution is unique.

Exercise. Show that for the 1-dimensional Ornstein–Uhlenbeck process (see Sect. 3.12.1) we can also obtain the stationary solution $P_{st}(v)$ from the conditional probability $P(v, t \mid v_0, t_0)$ taking the initial condition at an infinitely remote time (i.e., in the limit $t_0 \to -\infty$).

Exercise. Show that a necessary condition for the existence of a stationary solution of **F-P** is that $K_i(\{x_m\})$ and $D_{ij}(\{x_m\})$ are coefficients independent of time. Hint: use the H-theorem and assume the existence of an eigenfunction expansion to write the propagator.

In this section we have explicitly used a different notation for the solution in thermodynamic equilibrium $P_{eq}(\{x_m\})$ and for the stationary solution of the **F-P**: $P_{st}(\{x_m\})$. Only if the system is closed and isolated both solutions coincide.

Guided Exercise. Consider a system S characterized by the **sde** (in the complex plane)

$$\dot{E} = (a - c)E - b\,|\,E\,|^2\, E + L(t), \qquad (4.42)$$

here the noise $L(t)$ is a complex white Gaussian **sp** that meets the conditions

$$\langle L(t) \rangle = 0, \quad \left\langle L(t)L(t') \right\rangle = 0 \quad \text{and} \quad \left\langle L(t)L^*(t') \right\rangle = \Gamma \delta(t - t'). \qquad (4.43)$$

In general, if the system S is away from equilibrium, the coefficient Γ, in principle, has no relation with the dissipation in the deterministic equation. Find the coefficients of drift $K_i(E, E^*)$ and diffusion $D_{ij}(E, E^*)$ that appear in the

F-P equation. Show that the stationary probability distribution is a solution of the equation

$$0 = \left[-\frac{\partial}{\partial E}(a - c - b \mid E \mid^2)E - \frac{\partial}{\partial E^*}(a - c - b \mid E \mid^2)E^* + \Gamma \frac{\partial}{\partial E} \frac{\partial}{\partial E^*} \right] P_{st}(E, E^*).$$
$$(4.44)$$

Hint: writing $L(t) = L_1(t) + iL_2(t)$ and using (4.43) it is simple to check that $\left\langle L_i(t) L_j(t') \right\rangle = \frac{\Gamma}{2} \delta_{i,j} \delta(t - t')$, $\{i, j\} = 1, 2$, then calculate $K_i(E, E^*)$ and $D_{i,j}(E, E^*)$ to write the **P-F** equation in the variables $\{E, E^*\}$. Note that $P(E, E^*, t)\, dEdE^* = P(E_1, E_2, t)\, dE_1 dE_2$ by calculating the Jacobian of the transformation. Furthermore, introducing polar coordinates, that is, using the change of variable $E = \sqrt{s} e^{i\phi}$, show that the stationary solution can also be obtained from the equation

$$0 = \left\{ -2\frac{\partial}{\partial s}(a - c - bs)s + \Gamma \left(\frac{\partial}{\partial s} s \frac{\partial}{\partial s} + \frac{1}{4s} \frac{\partial^2}{\partial \phi^2} \right) \right\} P_{st}(s, \phi). \qquad (4.45)$$

Hint: introduce the change of variable $E_1 + iE_2 = \sqrt{s}(\cos \phi + i \sin \phi)$ and use the definition of $K_i(s, \phi)$ and $D_{ij}(s, \phi)$ from the Stratonovich calculus, see Chap. 3.

Excursus. From the stationary solution $P_{st}(s, \phi)$ of (4.45) we can define a "potential function" but in general, this function does not coincide with the deterministic potential obtained eliminating the noise term in Langevin's equation, i.e., setting $\Gamma = 0$ in (4.42). This is an example of the difficulties arising from an indiscriminate use of Langevin equations, this issue can only be solved by introducing the Ω-expansion from the mesoscopic description of the laser in terms of the master equation, see van Kampen's book [5]. Hint: see guided exercise in Sect. 4.7.2.

4.6.1 The Inverse Problem

We know that given a Langevin equation once the differential stochastic calculus has been specified, both the drift coefficient $K_i(\{x_m\})$ and diffusion matrix $D_{ij}(\{x_m\})$ are univocally determined (see Chap. 3). However, the inverse problem is not, as there is freedom of choice. In general, in the inverse problem, if n is the number of stochastic variables, there is a degree of freedom in choosing the parameters of the Langevin **sde** $\dot{x}_j = h_j + g_{jl}\xi_l$, where generally h_j and g_{jl} are functions of $\{x_m\}$. In Chap. 3 we show that there is a relationship between drift and diffusion and the Langevin parameters. That is, drift and diffusion are given in the Stratonovich calculus by

$$K_i(\{x_m\}) = h_i + \frac{1}{2} g_{kj} \frac{\partial}{\partial x_k} g_{ij} \qquad (4.46)$$

$$D_{ij}(\{x_m\}) = \frac{1}{2}g_{ik}g_{jk},\tag{4.47}$$

where repeated indices are summed over. In addition, since $D_{ij}(\{x_m\})$ is symmetric, we note from (4.47) that we only have $\frac{1}{2}n(n+1)$ equations. That is, the degree of freedom to determine h_i and g_{kj} is characterized by

$$\{\text{Number of unknowns} = n+n^2\}-\{\text{number of equations} = n+\frac{n(n+1)}{2}\}=\frac{n(n-1)}{2}.$$

From which we conclude that it is only in the 1-dimensional case that the inverse problem is uniquely determined.

4.6.2   Detailed Balance

In Chap. 2 we saw the stationary description of fluctuations around thermodynamic equilibrium. But what about the time dependence of the fluctuations of the state variables? To describe the temporal correlations of these fluctuations it is necessary to use the concept of **sp**. In fact it is precisely within the framework of Markov processes that we are able to fully characterize the time dependence of these fluctuations.

We have emphasized that the stationary solution of **F-P** is a nontrivial problem for dimension larger than one. But, in particular, taking into account some restrictions, this stationary solution can be analyzed by quadratures. The crucial point is established by a symmetry condition called detailed balance. This condition is simply a reflection, at the macroscopic level, of invariance under time reversal $t \to -t$ in Hamiltonian systems.

In an isolated and closed system fluctuations around thermodynamic equilibrium are described by a **F-P** equation that satisfies the principle of detailed balance. In general, far from equilibrium this is not true. This principle states that at the stationary state, and for all elementary cells of the phase-space of the state variables $\{x_m\}$, the incoming and outgoing probability fluxes are equal (are the same). Mathematically detailed balance (for even variables) provides that

$$P(\{x_m\}, \tau \mid \{x'_m\}, 0)P_{eq}(\{x'_m\}) = P(\{x'_m\}, \tau \mid \{x_m\}, 0)P_{eq}(\{x_m\}).\tag{4.48}$$

If the state variables are odd under time reversal $t \to -t$, it is also possible to formulate[21] a generalization of (4.48). Here $P_{eq}(\{x_m\})$ represents the stationary probability distribution (around thermodynamic equilibrium).

---

[21]The proof of detailed balance in closed, isolated, classical systems can be seen in [13], for the quantum mechanical proof see [5] Chap. XVII.7.

If a one-dimensional **sp** takes the value x at time t and the value x' at time $t + \tau$, the probability of occurrence of this event is given by the two-times joint pdf $P_2(x', t + \tau; x, t)$. In particular, in thermodynamic equilibrium this joint probability distribution is given by[22]

$$P_2(x', t + \tau; x, t)_{\text{eq}} = P(x', t + \tau \mid x, t) P_{\text{eq}}(x, t),$$

where obviously $P_{\text{eq}}(x, t)$ must be independent of time. Then, the *principle of detailed balance* can be formulated as follows[23]

$$P_2(x', 0; x, \tau)_{\text{eq}} = P_2(x', \tau; x, 0)_{\text{eq}}. \tag{4.49}$$

Excursus. Let $Y(p, q)$ be a macroscopic observable (which in turn is a function of (p, q), all these being coordinates of the phase-space of the N bodies system). From the definition of the two-times joint pdf at thermodynamic equilibrium, it follows that

$$P_2(x', t; x, 0)_{\text{eq}} = \int \cdots \int \delta\left(x' - Y(p, q, t)\right) \delta\left(x - Y(p, q, 0)\right) P_{\text{eq}}(p, q) \, \mathcal{D}p \, \mathcal{D}q.$$

Note that here $P_{\text{eq}}(p, q) \propto \exp\left(-H(p, q)/k_B T\right)$. From the invariance of Hamilton's equations of the system under time reversal, and from the even parity of the observable $Y(p, q)$ under the transformation $t \to -t$, it can be shown that

$$P_2(x', 0; x, t)_{\text{eq}} = P_2(x', t; x, 0)_{\text{eq}}.$$

That is, the condition of detailed balance is proved by the principle of microscopic reversibility [5, 13].

Note Based on the principle of detailed balance (4.48) is possible to formulate necessary and sufficient conditions on the coefficients of drift and diffusion so that a continuous **sp**, characterized by a **F-P** equation, complies with the principle of detailed balance; see the first excursus in Sect. 4.7.2.

---

[22]Note that the principle of *Detailed-Balance* is not the same as the symmetric property that any joint pdf must fulfill, see Kolmogorov's hierarchy, Sect. 3.1.3. Consider for example, this symmetry for the two-time joint probability $P(A \cap B)$ (for the intersection of events $\{A, B\}$ here events refer to $\mathbf{X}(t + \tau)$ and $\mathbf{X}(t)$, then the two-time joint pdf of the **sp** is $P(\mathbf{X}(t + \tau); \mathbf{X}(t))$).

[23]In particular if the **sp** is discrete, Eq. (4.48) is immediately reduced to a simple relationship between the transition matrix elements W_{jm} (see Chap. 6) and the equilibrium pdf $P^{\text{eq}}(m)$, i.e., we can write

$$W_{jm} P^{\text{eq}}(m) = W_{mj} P^{\text{eq}}(j).$$

It is worth noting that here the convention of summing over repeated indices does not apply, see advanced exercise 4.9.3.

Exercise. Using the principle of detailed balance (microscopic reversibility) show that the correlation matrix (macroscopic) satisfies

$$\left\langle x_j(0)x_i(\tau)\right\rangle_{\text{eq}} = \left\langle x_j(\tau)x_i(0)\right\rangle_{\text{eq}}. \tag{4.50}$$

Note that here we are assuming that the **mv** $\left\langle x_j(\tau)\right\rangle_{\text{eq}}$ is zero, as expected, around thermodynamic equilibrium.

Excursus (**Magnetic Relaxation**). Consider a particle of volume v with a uniform magnetic moment **M** of constant magnitude M_s. If the instantaneous orientation of the vector **M** is described by stochastic angles $\theta \equiv \theta(t)$ and $\phi \equiv \phi(t)$, consider from Brown–Gilbert's **sde** the operator $\mathcal{L}$ associated with the **F-P** equation (see Sect. 3.19.1). Since the model describes a closed and isolated system, the stationary solution of **F-P** is the Boltzmann pdf

$$\mathcal{W}_{\text{eq}}(\theta, \phi) = \mathcal{N}e^{-\beta V(\theta,\phi)},$$

where $\beta = v/k_B T$ and $V(\theta, \phi)$ is the potential of the forces acting on the moment **M** (Fig. 4.2 shows a representative scheme). It is easy to show (by direct substitution) that this distribution is the stationary solution of the **F-P** equation if and only if $h' = \beta k'$. This is just the condition of detailed balance. Then, with this condition it is possible to relate the intensity of thermal fluctuations, ϵ, to the dissipation parameter, η, of Gilbert–Brown's model [14] as:

$$\epsilon = \frac{2\eta k_B T}{v}. \tag{4.51}$$

Fig. 4.2 Scheme representing the presence of the potential $V(\theta, \phi)$ acting on the magnetic moment **M**

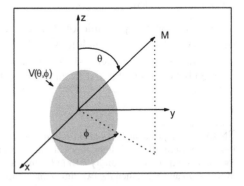

4.7 Probability Current

4.7.1 The One-Dimensional Case

In the one-dimensional case the stationary solution of the **F-P** equation is always solvable by quadratures. This follows immediately as, in the one-dimensional case, it is always possible to introduce a coordinate transformation in the Langevin equation so that the diffusion coefficient, in the associated **F-P** equation, is a constant [See last exercise in Sect. 3.18].

Let the one-dimensional **F-P** equation be[24]

$$\frac{\partial}{\partial t}P(x,t) = \left[-\frac{\partial}{\partial x}K(x) + \frac{1}{2}\frac{\partial^2}{\partial x^2}D(x)\right]P(x,t) = -\partial_x J(x,t). \tag{4.52}$$

In the stationary state the probability current $J_{st}(x)$ must be a constant. In particular, if the boundary condition imposes a zero current for some value of x', it follows that $J_{st}(x) = 0$ over the whole domain of x. From this we conclude that if $D(x)$ is not singular then

$$K(x)P_{st}(x) = K(x)\frac{D(x)}{D(x)}P_{st}(x) = \frac{1}{2}\frac{\partial}{\partial x}D(x)P_{st}(x).$$

This equation can be integrated for $D(x)P_{st}(x)$ as follows:

$$\int \frac{d[D(x)P_{st}(x)]}{D(x)P_{st}(x)} = 2\int \frac{K(x)}{D(x)}dx,$$

obtaining for the stationary solution the expression

$$P_{st}(x) = \frac{\mathcal{N}}{D(x)}\exp\left(2\int^x \frac{K(x')}{D(x')}dx'\right). \tag{4.53}$$

Excursus. Let the one-dimensional **F-P** equation be

$$\frac{\partial}{\partial t}P(x,t) = \left[-\frac{\partial}{\partial x}K(x) + \frac{1}{2}\frac{\partial^2}{\partial x^2}D(x)\right]P(x,t) \equiv \mathcal{L}P(x,t).$$

---

[24]Here the factor $\frac{1}{2}$ is just a notation, it is not associated with the derivation from the Langevin equation, see Sect. 3.18.

Since $\mathcal{L}$ is a differential operator, the principle of detailed balance cannot be analyzed in a simple way.[25] We can prove that the principle of detailed balance is satisfied iff [15]

$$K(x) = \frac{1}{2P_{st}(x)} \frac{d}{dx} P_{st}(x)D(x).$$

(4.54)

Then $P_{st}(x)$ is found by quadratures in the form (4.53). Note that this condition states that the probability current

$$J(x,t) = \left[K(x) - \frac{1}{2} \frac{\partial}{\partial x} D(x)\right] P(x,t)$$

is identically zero in the stationary state. This is because we have used a *natural* boundary condition for the pdf. The following *excursus* shows a nonnatural boundary condition (for a classification of the boundary conditions see advanced exercise 3.20.14).

In general, it can be shown that the canonical distribution of classical statistical mechanics is the stationary solution of **F-P** (4.39) iff $D_{ij}(\{x_m\})$ and $K_i(\{x_m\})$ are connected properly. In this case the **F-P** equation is multidimensional and will be discussed in the next section.

Excursus (**Josephson Junction**). A Josephson junction consists of two super-conductors separated by a thin layer of insulating material. This junction can be analyzed as a *parallel* electrical circuit with a certain resistance R and a capacitance C (RSJ model[26]). In the RSJ model the current in the junction $I_S = I_0 \sin \phi$, represents the supercurrent due to Cooper's pair, which by quantum tunneling *jumps* across the insulating material. Here ϕ is the phase difference of the quantum wave function on both sides of junction. Applying Kirchhoff's rules to the equivalent RSJ electrical circuit, we can decompose the current I across the junction in the form:

$$I = C\frac{dV}{dt} + \frac{V}{R} + I_S.$$

(4.55)

Consider now the voltage across the junction to be related to the temporal variation of the phase difference, i.e.: $d\phi/dt = 2eV/\hbar$. Then, in the overdamped limit, $C \to 0$, introducing the scale changes $t \to (2eRI_0/\hbar)t$ and $I(t) \to I/I_0$ we can rewrite the expression (4.55) as a differential equation for the phase ϕ :

$$\frac{d\phi}{dt} = -\sin \phi + I.$$

(4.56)

---

[25] As in the case of the master equation (this equation will be presented in Chap. 6) see note (23) in this chapter. Alternatively the symmetrization of the differential operator $\mathcal{L}$ is easier to observe, see optional exercise Sect. 4.8.4.

[26] The acronym RSJ comes from the English name: *Resistively Shunted Junction*.

Now if we consider the thermal noise generated in the resistance R, when the equivalent circuit RSJ is operating at a temperature different from zero, we can simply include a white additive Gaussian noise of zero mean[27] $\eta(t)$ in Eq. (4.55). Then, it turns out that instead of the deterministic equation (4.56), one obtains the following **sde** (with $\xi(t)$ a GWN):

$$\frac{d\phi}{dt} = -\sin\phi + I + \sqrt{\epsilon}\,\xi(t), \quad \text{where} \quad \langle\xi(t)\rangle = 0, \langle\xi(t)\xi(t+\tau)\rangle = \delta(\tau).$$
$$(4.57)$$

Optional Exercise. From the Langevin equation (4.57) obtain the corresponding **F-P** equation. By the principle of detailed balance it is possible to see that the intensity of the noise is given by $\epsilon = 4ek_BT/(\hbar I_0)$. Note that the physical meaning of the phase ϕ only makes sense in the interval $[0, 2\pi]$; hence the conditional probability $P(\phi, t \mid \phi_0, t_0)$ must be normalized in that interval and, of course, meets the periodicity condition

$$P(\phi, t \mid \phi_0, t_0) = P(\phi + 2\pi, t \mid \phi_0, t_0). \quad (4.58)$$

In the context of **F-P** a constraint like the one presented in (4.58) means typically a "nonnatural" boundary condition and needs a thorough study. This periodic boundary condition was originally analyzed by Stratonovich in 1958: Radiotekhnika, electronika 3, N° 4, 497 (see also p. 234, Vol. II of his book [9]); however, in this subject it is common to find only references to the work of V. Ambegaokar and B.I. Halpering [Phys. Rev. Lett. 22, 1364 (1969)]. A detailed analysis of the RSJ equivalent circuit can also be found in Risken's book [4].

4.7.2 Multidimensional Case

In general, away from thermodynamic equilibrium, and for the purpose of calculating the stationary solution of the multidimensional **F-P** equation, it is convenient to write Eq. (4.39) in terms of a probability current J_μ. Obviously, in the n-dimensional case the total stationary current does not necessarily cancel. For example, there could be a stationary state with a current flow having nonzero rotor.

In n-dimension the **F-P** equation can be written in the following compact way[28]:

$$\partial_t P + \partial_\mu J_\mu = 0, \quad (4.59)$$

---

[27] That is, $\langle\eta(t)\rangle = 0$ and $\langle\eta(t)\eta(t+\tau)\rangle = \dfrac{2k_BT}{R}\delta(\tau)$, see Nyquist's theorem in Sect. 4.4.1.

[28] Using the convention of summing over repeated indices.

where the probability current is given by

$$J_\mu = \left(K_\mu - \partial_\nu D_{\mu\nu}\right) P. \tag{4.60}$$

In general, $K_\mu \equiv K_\mu(\{q_\mu\})$ comes from the drift term and $D_{\mu\nu} \equiv D_{\mu\nu}(\{q_\mu\})$ from the diffusion term itself, both terms may depend on n-state variables $\{q_\mu\}$, but not explicitly on time because we are interested in a stationary **sp**. In order to obtain a stationary solution with potential structure, that is,

$$P_{st} \propto \exp\left(-F(\{q_\mu\})\right), \tag{4.61}$$

it is first necessary to split the drift term from Eq. (4.60) into two contributions: one dissipative, f_μ^d, and other non-dissipative, f_μ^o. This is,

$$K_\mu = f_\mu^d + f_\mu^o, \tag{4.62}$$

from which the dissipative current is defined (in principle this separation is unknown):

$$J_\mu^d = \left(f_\mu^d - \partial_\nu D_{\mu\nu}\right) P. \tag{4.63}$$

Then, the **F-P** equation shall be expressed in terms of two currents: one dissipative, the other not dissipative namely:

$$\partial_t P + \partial_\mu \left(f_\mu^o P\right) + \partial_\mu J_\mu^d = 0. \tag{4.64}$$

Proposition *For the stationary state* $\partial_t P_{st} = \left[-\partial_\mu J_\mu\right]_{st} = 0$, *and if the potential condition is met*

$$\left[J_\mu^d\right]_{st} = 0, \tag{4.65}$$

the divergence of the non-dissipative current $J_\mu^o = f_\mu^o P_{st}$ *is zero identically:*

$$\partial_\mu \left(f_\mu^o P_{st}\right) = 0. \tag{4.66}$$

By introducing the potential structure (4.61) for the stationary state P_{st} in (4.66) it follows that

$$\partial_\mu f_\mu^o = f_\mu^o \partial_\mu F, \tag{4.67}$$

which, in principle, can be solved by quadratures.

Note Imposing the nontrivial condition $\left[J_\mu^d\right]_{st} = 0$ implies that the functions f_μ^d, $D_{\mu\nu}$ and F must satisfy the consistency condition

$$f_\mu^d - \partial_\nu D_{\mu\nu} = -D_{\mu\nu}\partial_\nu F. \tag{4.68}$$

This relationship—called the potential condition— is difficult to solve, because the potential $F \equiv F(\{q_\mu\})$ is unknown. We can "guess" only occasionally the decomposition (4.62). On the other hand, it is possible to prove that if the **F-P** system satisfies the (more restrictive) principle of detailed balance [4, 16, 17], then, the above potential condition is met. Note that to calculate the dissipative component f_μ^d it is first necessary to know the nonequilibrium potential $F(\{q_\mu\})$. Often the reverse formulation is used: given the **F-P** equation, f_μ^d is tentatively proposed, from which, assuming that the potential condition (4.68) and (4.67) are true, the calculation of F is inferred by quadratures.

Exercise. Show that if the potential condition (4.68) is satisfied, $D_{\mu\nu}$ is invertible and f_μ^d is known. The potential F is given by integration in the form:

$$\partial_\alpha F(\{q_\gamma\}) = -D_{\alpha\mu}^{-1}\left(f_\mu^d - \partial_\nu D_{\mu\nu}\right).$$

Then, if we define the vector $A_\alpha \equiv -D_{\alpha\mu}^{-1}\left(f_\mu^d - \partial_\nu D_{\mu\nu}\right)$, it must satisfy the condition of zero rotor:

$$\partial_\mu A_\alpha = \partial_\alpha A_\mu.$$

Excursus (**Detailed Balance**.) Consider a closed, isolated system to be characterized by a **F-P** equation in which there are even and odd variables (under the temporal transformation $t \to -t$). For example, the position of a particle is an even variable, while its velocity is odd. If the quantities ϵ_μ are defined as prescribed

$$\epsilon_\mu = \begin{cases} 1, & \text{for even variables} \\ -1, & \text{for odd variables,} \end{cases}$$

It can be proved that the **F-P** equation complies with detailed balance iff [5]

$$D_{\mu\nu}(q) = \epsilon_\mu\epsilon_\nu D_{\nu\mu}(\epsilon q)$$

$$\epsilon_\mu K_\mu^{rev}(\epsilon q) = -K_\mu^{rev}(q)$$

$$\sum_\mu \partial_\mu\left(K_\mu^{rev}P_{eq}\right) = 0.$$

Note that when a product of the form $\epsilon_\mu K_\mu^{\text{rev}}(\epsilon q)$ or $\epsilon_\mu \epsilon_\nu D_{\nu\mu}(\epsilon q)$ appears, the convention of summing over repeated indices does not apply. Furthermore, the **F-P** equation has been redefined in the form

$$\partial_t P = -\partial_\mu K_\mu P + \partial_\mu \partial_\nu D_{\mu\nu} P \qquad (4.69)$$

$$= -\partial_\mu K_\mu^{\text{rev}} P + \partial_\mu P_{\text{eq}} D_{\mu\nu} \partial_\nu \frac{P}{P_{\text{eq}}}.$$

where

$$K_\mu^{\text{rev}} = K_\mu - \frac{1}{P_{\text{eq}}} \partial_\nu \left(D_{\mu\nu} P_{\text{eq}} \right). \qquad (4.70)$$

Since the system is closed and isolated, the stationary solution coincides with the equilibrium solution P_{eq}. Then, we can physically interpret each of the terms of the **F-P** equation (4.69). The first (deterministic), $-\partial_\mu K_\mu^{\text{rev}} P$, is precisely the Liouville operator, which is reversible. The second (purely dissipative) $\partial_\mu P_{\text{eq}} D_{\mu\nu} \partial_\nu \left[P/P_{\text{eq}} \right]$ accounts for the irreversible approach to the equilibrium state. Note that if the principle of detailed balance is satisfied there is a clear prescription for the separation of dissipative and non-dissipative currents. Then, in equilibrium, the dissipative current $P_{\text{eq}} D_{\mu\nu} \partial_\nu \left[P/P_{\text{eq}} \right]$ is zero.

Optional Exercise. If the principle of detailed balance is satisfied, show that for the stationary state there may be a nonzero non-dissipative current J_μ^o coming from the rotor of a potential vector: $J^o = \nabla \times A$.

Guided Exercise (**Nonequilibrium Potential.**) Consider the Langevin phenomenological equation (4.42) describing a laser within an electromagnetic cavity away from equilibrium. Using polar coordinates ones sees, from (4.45), that the deterministic equation ($\Gamma = 0$) for the modulus of the electric field $| E |^2 \equiv s$ is given by

$$\dot{s} = 2s(a - c - bs). \qquad (4.71)$$

Then, we can define a deterministic potential $V(s) = -(a - c)s^2 + 2bs^3/3$ so that (4.71) can be written in the potential form: $\dot{s} = -V'(s)$. On the other hand, we can solve the stationary state of the **F-P**, solution of (4.45), assuming that the "potential condition" (4.68) is met[29] proposing the following decomposition for the drift term K_μ:

$$f^d \equiv \left(f_s^d, f_\phi^d \right) = (2s(a - c - bs) + \Gamma, 0)$$

$$f^o \equiv \left(f_s^o, f_\phi^o \right) = (0, 0).$$

---

[29]By identifying the dissipative current, $J^d \equiv (J_s^d, J_\phi^d)$, and the non-dissipative one $J^o \equiv (J_s^o, J_\phi^o)$.

Hence it follows that the components of the dissipative current,

$$J_\mu^d = \left(f_\mu^d - \partial_v D_{\mu v}\right) P,$$

are

$$J_s^d(s, \phi, t) = [2s(a - c - bs) + \Gamma - \partial_s D_{ss}] P(s, \phi, t)$$
$$J_\phi^d(s, \phi, t) \equiv -\partial_\phi D_{\phi\phi} P(s, \phi, t).$$

Note, from (4.45) that $D_{s\phi} = D_{\phi s} = 0$, the matrix $D_{\mu v}$ depends on the state variable s and is given by

$$D_{\mu v} = \begin{pmatrix} \Gamma s & 0 \\ 0 & \frac{\Gamma}{4s} \end{pmatrix}. \tag{4.72}$$

Furthermore, the non-dissipative current is identically zero, because

$$J^o \equiv (f_s^o, f_\phi^o) P = (0, 0)).$$

Optional Exercise. Show that the stationary solution of the **F-P** equation of the previous exercise is

$$P_{st}(s) = \mathcal{N} \exp\left(\frac{2}{\Gamma}\left\{(a - c)s - \frac{1}{2}bs^2\right\}\right) \equiv \mathcal{N}e^{-F}.$$

Here $\mathcal{N}$ is the normalization constant. This solution shows that the nonequilibrium potential

$$F(s) \equiv -\frac{2}{\Gamma}\left\{(a - c)s - \frac{1}{2}bs^2\right\}$$

is different from the deterministic potential $V(s)$. This difference is due to the fact that the model did not originate as a systematic approximation to the actual equations of motion for a laser and therefore some of the details may be spurious.[30] In other words, if the laser system is described—at the mesoscopic level—by a master equation [M.O. Scully and W.E. Lamb, Phys. Rev. 159, 208 (1967)], the evolution equation with its fluctuations can only be properly solved invoking a systematic Ω-expansion, which in this case gives—to its lowest order—a nonlinear **F-P** equation. To get a short review of the Ω-expansion see advanced exercise 6.7.3.

---

[30]If fluctuations of a nonlinear system are taken into account by introducing *ad hoc* a noise term in the deterministic equation, the resulting **sde** can lead to wrong results. See the *Brillouin paradox* in van Kampen's book [5], p. 235.

Optional Exercise. From the laser stationary state of the previous exercise, calculate the most likely square module s. Interpret the meaning of the nonequilibrium potential $F(s)$ and the fact that it is independent of the state variable ϕ (angle).

Excursus (**Correlations Far from Equilibrium.**) If one considers the **F-P** equation for a laser within an electromagnetic cavity, see (4.42), we can use a "linear" approximation and verify that if $c - a \gg \sqrt{b\Gamma}$ the correlation function of the electric field is given by[31]

$$\left\langle\!\!\left\langle |\, E(t_1)\,|^2|\, E(t_2)\,|^2 \right\rangle\!\!\right\rangle_{\text{st}} = \frac{\Gamma^2}{4(c-a)^2}\exp\left(-2(c-a)\,|\,t_1 - t_2\,|\right).$$

While in the opposite case, that is, when the modulus s is around the most probable value $s_{\text{st}} = (a-c)/b$, the **F-P** equation can be linearized by setting: $s = s_{\text{st}} + \sigma$, then we can study fluctuations around this value. Show that now the correlation function is[32]

$$\left\langle\!\!\left\langle |\, E(t_1)\,|^2|\, E(t_2)\,|^2 \right\rangle\!\!\right\rangle_{\text{st}} = \frac{\Gamma}{2b}\exp\left(-2(a-c)\,|\,t_1 - t_2\,|\right).$$

4.7.3   Kramers' Equation

In order to exemplify the use of the separation of dissipative and non-dissipative probability currents, as a method for calculating the multidimensional stationary solution of the **F-P** equation, we solve here the stationary Kramers' equation introduced in Sect. 4.5.

Consider a Brownian particle of mass m in an arbitrary potential $U(X)$, the **F-P** equation will be (4.37). Then we immediately identify the gradient operator $\partial_\mu \equiv (\partial_X, \partial_V)$ and the total probability current[33] with

$$J = \left(V, -\frac{U'(X) + \gamma V}{m} - \varepsilon\partial_V\right)P, \tag{4.73}$$

where $\varepsilon \equiv \gamma k_B T/m^2$. Note that from the Kramers equation (4.37) it is possible "to guess" how to decompose K_μ into a dissipative part and a non-dissipative (purely mechanical part):

---

[31]When the intensity $|\, E(t)\,|^2$ is far below threshold, the system can be approximated by an Ornstein-Uhlenbeck process.

[32]When $(a - c) \gg \sqrt{b\Gamma}$, [5].

[33]That is the index μ can be used as X or V.

$$f^{\text{o}} \equiv \left(f_X^{\text{o}}, f_V^{\text{o}} \right) = \left(V, \frac{-U'(X)}{m} \right) \quad \text{and} \quad f^{\text{d}} \equiv \left(f_X^{\text{d}}, f_V^{\text{d}} \right) = \left(0, \frac{-\gamma V}{m} \right). \tag{4.74}$$

Obviously, this choice depends on the **F-P** system being studied (compare with the guided exercise in the previous section). Using (4.73) and (4.74) the following decomposition for the probability current is obtained:

$$J^{\text{o}} = \left(V, \frac{-U'(X)}{m} \right) P \quad \text{and} \quad J^{\text{d}} = \left(0, -\frac{\gamma V}{m} - \partial_V \varepsilon \right) P. \tag{4.75}$$

In this case the diffusion matrix

$$D_{\mu\nu} = \begin{pmatrix} 0 & 0 \\ 0 & \varepsilon \end{pmatrix} \tag{4.76}$$

is singular.[34]

Potential condition $f_\mu^{\text{d}} - \partial_\nu D_{\mu\nu} = -D_{\mu\nu}\partial_\nu F$ leads to the following equation for the potential $F(X, V)$ of the stationary solution:

$$\begin{pmatrix} 0 \\ \frac{-\gamma V}{m} \end{pmatrix} = -\begin{pmatrix} 0 & 0 \\ 0 & \varepsilon \end{pmatrix} \begin{pmatrix} \partial_X F \\ \partial_V F \end{pmatrix}. \tag{4.77}$$

Then, for the stationary state, from the condition $\left[J_\mu^{\text{d}} \right]_{\text{st}} = 0$ we conclude that $\gamma V / m = \varepsilon \partial_V F$. This equation can immediately be integrated, leading to

$$F(X, V) = \frac{\gamma}{2m\varepsilon} V^2 + C(X), \tag{4.78}$$

here $C(X)$ is a constant of integration that may depend on the variable X. On the other hand, in the stationary state, the vanishing of the divergence of the non-dissipative current, $\partial_\mu f_\mu^{\text{o}} = f_\mu^{\text{o}} \partial_\mu F$, leads to

$$\partial_X f_X^{\text{o}} + \partial_V f_V^{\text{o}} = \left(f_X^{\text{o}}, f_V^{\text{o}} \right) \cdot \left(\partial_X F(X, V), \partial_V F(X, V) \right). \tag{4.79}$$

Observing that $f^{\text{o}} \equiv \left(V, \frac{-U'(X)}{m} \right)$ this implies $\partial_\mu f_\mu^{\text{o}} = 0$, and from (4.79) we conclude that

$$0 = \left(V, \frac{-U'(X)}{m} \right) \cdot \left(\partial_X F(X, V), \partial_V F(X, V) \right).$$

---

[34] That is, this matrix has no inverse.

Using (4.78) we see that $C(X)$ is

$$C(X) = \frac{\gamma}{m^2 \varepsilon} U(X) + \mathcal{N},$$

thus, and noting that $\varepsilon \equiv \gamma k_B T / m^2$, we finally obtain for the potential F (apart from a constant $\mathcal{N}$ which is absorbed by normalization) the expression

$$F(X, V) = \frac{m}{2 k_B T} V^2 + \frac{1}{k_B T} U(X).$$

Then, it follows that the stationary solution of the Kramers equation is

$$P_{st} = \mathcal{N} \exp\left(-F(X, V)\right) \propto \exp\left(-\frac{m}{2 k_B T} V^2 - \frac{1}{k_B T} U(X)\right),$$

which, not surprisingly, coincides with the canonical solution of equilibrium statistical mechanics.

4.7.4   Generalized Onsager's Theorem*

In a previous section we argued that the principle of detailed balance implies that the potential condition[35] $f_\mu^d - \partial_v D_{\mu v} = -D_{\mu v} \partial_v F$ is met, and also that the divergence of the non-dissipative probability current is zero $\partial_\mu \left(f_\mu^o P_{st}\right) = 0$. Note that the reverse is not true.

In cases where the diffusion matrix is constant the potential condition is greatly simplified:

$$f_\mu^d = -D_{\mu v} \partial_v F. \tag{4.80}$$

Defining the generalized forces

$$\mathcal{F}_v = k_B \partial_v F$$

we can write (4.80) in the form

$$f_\mu^d = -\frac{D_{\mu v}}{k_B} \mathcal{F}_v, \tag{4.81}$$

---

[35]This condition is valid also for nonequilibrium **F-P** problems.

which is Onsager's relation (4.11). Moreover, in the particular (nonlinear) case where $\partial_\nu D_{\mu\nu} = 0$, the relation (4.81) is the nonlinear generalization of Onsager's theorem [17].

Excursus. Near thermodynamic equilibrium $F(\{q_\mu\})$ is the second variation of the internal energy. Expressed in terms of ΔT and ΔV we have $k_B T F = C_V/2T \, (\Delta T)^2 + \frac{C_V}{C_P} (\Delta V)^2 / 2V\gamma$, where $\gamma = -(\partial \log V/\partial P)_S$ is the adiabatic compressibility, C_V and C_P are the specific heats at constant volume and constant pressure, respectively. That is: $F = \frac{1}{2} g_{\mu\nu} q_\mu q_\nu$ is a *positive definite* quadratic form, thus $\mathcal{F}_\nu = \frac{k_B}{2} (g_{\mu\nu} q_\mu + g_{\nu\mu} q_\mu) = k_B g_{\nu\mu} q_\mu$. Therefore, in this approach: $f_\mu^d = -\frac{D_{\mu\nu}}{k_B} \mathcal{F}_\nu = -D_{\mu\nu} g_{\nu\alpha} q_\alpha = -M_{\mu\alpha} q_\alpha$, where $M_{\mu\alpha}$ is the damping matrix [17]. Note that some authors define generalized forces $\mathcal{F}_\nu$ with the opposite sign. See, for example, the work of M. Lax [Rev. of Mod. Phys. **32**, 25 (1960)].

4.7.5 Comments on the Calculation of the Nonequilibrium Potential*

In one dimension it is always possible to introduce a coordinate transformation so that the diffusion coefficient is constant. This greatly simplifies the calculation of the stationary pdf of the **F-P** equation. In n-dimensions the problem of finding a system of coordinates so that $D_{\mu\nu}$ is constant is a nontrivial problem, hence the usefulness of introducing a covariant representation for the **F-P** equation when calculating the stationary distribution [18].

In general, in situations far from equilibrium if the potential condition is not met, we cannot keep any of the conditions (4.67) or (4.68). Then we need to introduce perturbative methods for calculating the nonequilibrium potential $F(\{q_\mu\})$, in order to find the stationary solution of the **F-P** equation. In particular, the use of action and angle variables is important for calculating the nonequilibrium potential [19]. Alternatively another possibility is to detect the non-Hermitian part, $\mathcal{L}_N$, of the **F-P** operator, then it is feasible to study the full **F-P** problem in eigenfunctions of the Hermitian part $\mathcal{L}_H$ on such a basis $\mid q >$ that the transition elements $< q_1 \mid \mathcal{L}_N \mid q_2 >$ are as simple as possible. When these elements satisfy finite recurrence relations, particularly if they are tridiagonal, the method of continued fractions can be used to solve numerically the **F-P** problem in an efficient way [4]. Finally, it is worth noting that if the noise intensity is small (proportional to Langevin's forces), the path integral method makes it possible to obtain good approximations to the nonequilibrium potential [20].

Excursus. In the small noise approximation the propagator of the **F-P** equation can be calculated in terms of an "action" for the *most probable path*, this technique is similar to the calculation of the quantum mechanic propagator in terms of a classical action. The pioneer work on this topic was presented by R. Kubo, K. Matsuo and K. Kitahara, J. Stat. Phys. **9**, 51 (1973).

Symmetrization and Expansion in Eigenfunctions

As already discussed, if a stationary state exists, the distribution $P_{st}(\{q_\mu\})$ is the long-time asymptotic limit of the propagator of the system. We can get $P_{st}(\{q_\mu\})$ as the solution of the stationary **F-P** equation. Furthermore, if an analysis of eigenfunctions of the **F-P** operator $\mathcal{L}$ is performed, $P_{st}(\{q_\mu\})$ must be the eigenfunction with eigenvalue zero, of the boundary value problem[36] (where it is assumed that $\int_D \Phi_n(q_\mu)\,dq_\mu < \infty$)

$$\mathcal{L}\Phi_n = \lambda_n \Phi_n. \tag{4.82}$$

But it may happen that the set of eigenfunctions Φ_n is not complete; this is an extra difficulty usually encountered in nonequilibrium systems, since the operator $\mathcal{L}$ may not be normal[37]; see for example [21]. In what follows we will take the view that the set of eigenfunctions Φ_n is complete (this can always be justified by avoiding the degeneration of the eigenvalues via small perturbations[38] [4]. On the other hand, if the principle of detailed balance is satisfied, it is always possible to symmetrize the operator $\mathcal{L}$, which ensures the reduction to its diagonal form, see (4.105).

Optional Exercise. Show that in the one-dimensional case it is simple to transform the **F-P** operator into a Sturm–Liouville one, therefore different techniques to work out the propagator of the **F-P** can be used. Hint: a similar situation arises when calculating the mean first passage time, see Sect. 9.1.4.

4.8   Nonstationary Fokker–Planck Processes*

4.8.1   Eigenvalue Theory

We have shown in Chap. 3, from the Stratonovich stochastic calculus that every **sde** in the presence of white Gaussian noise results in a well-defined Markov process.[39] A continuous Markov process is completely characterized by its **F-P** operator $\mathcal{L}$, which can immediately be written down from its corresponding **sde**. If any parameter of the **sde** is time dependent, the process will not be stationary. In particular, if this dependence is 2π periodic, the **sp** is called nonstationary 2π-periodic Markov process. In this section we present a method for solving the

---

[36]If the domain of the distribution is D (with the corresponding boundary conditions), functions on which $\mathcal{L}$ operates should be integrable in D. That is, a space of integrable functions.

[37]That is, $\mathcal{L}$ is normal if it commutes with its hermitian adjoint $\mathcal{L}^*$, that is: $\mathcal{L}\mathcal{L}^* = \mathcal{L}^*\mathcal{L}$.

[38]In a similar context, see the example in Appendix B.1.

[39]This is true for any stochastic prescription for the differential calculus, for example, Stratonovich, Ito, etc.

Fokker–Planck dynamics when drift and /or diffusion coefficients are periodic functions in time. Consider the following **F-P** equation (in n-dimension)

$$\partial_t P(q,t) = \left[-\frac{\partial}{\partial q_\nu} K_\nu(q,t) + \frac{\epsilon}{2} \frac{\partial^2}{\partial q_\nu \partial q_\mu} D_{\nu\mu}(q,t) \right] P(q,t)$$

(4.83)

$$\equiv \mathcal{L}\left(q, \partial_q, t\right) P(q,t).$$

Here q represents the set of variables $(q_1, \ldots, q_n)$, and the sum over repeated indices is understood. It is assumed that the terms of drift and/or the diffusion matrix are periodic functions in time with discrete time translational invariance $t \to t + T$, i.e.:

$$K_\nu(q, t + T) = K_\nu(q, t)$$

(4.84)

$$D_{\nu\mu}(q, t + T) = D_{\nu\mu}(q, t).$$

We can use the ϵ parameter to measure the intensity of the noise. The propagator (conditional pdf) of the Fokker–Planck dynamics $P\left(q, t | q_0, t_0\right)$ is a solution of (4.83) and satisfies the initial condition

$$P\left(q, t_0 | q_0, t_0\right) = \delta\left(q - q_0\right).$$

(4.85)

If K_ν and $D_{\nu\mu}$ are independent of time, the dynamics of **F-P** is equivalent to an eigenvalue problem; then the propagator can be expanded in terms of a set of biorthonormal eigenfunctions of the operator $\mathcal{L}$. The need of the adjoint eigenfunctions is due to the fact that in general the operator $\mathcal{L}$ is neither normal nor Hermitian. In the particular case that the principle of detailed balance is satisfied, the **F-P** problem may be shown to be equivalent to a semi-negative defined *self-adjoint* eigenvalue problem [4, 21]; which shows the existence of a complete set of eigenfunctions with negative (or zero) eigenvalues, but in general it is not possible to prove the existence of a complete set of eigenfunctions of the operator $\mathcal{L}$. In this section we will show that the dynamics of a nonstationary T-periodic Markov process can also be studied as an eigenvalue problem, but such that the type of operator to be solved is an integral operator.[40] The following sections give some applications of this eigenvalue theory; i.e., first we deduce some relations between the eigenvalues and eigenfunctions, and then we show the relationship between these eigenvalues and certain quantities that characterize the dynamics of the system, such as: correlation functions, Lyapunov function, jumping times between attractors, etc.

---

[40]This method has more advantages when the system inevitably has to be solved numerically [12]. We show the mathematical difficulty is reduced to the analysis of eigenvalues of a Fredholm integral equation.

4.8.2   The Kolmogorov Operator

Any solution $f(q, t)$ the **F-P** operator (4.83) satisfies the compatibility condition[41]

$$f(q, t) = \int P\left(q, t | q', t'\right) f\left(q', t'\right) \, dq',$$

for all $t' \le t$.

Definition   The Kolmogorov operator is given, to $t_2 \ge t_1$, by

$$\mathcal{U}\left(t_2, t_1\right) : f(q) \rightarrow \int P\left(q, t_2 | q', t_1\right) f\left(q'\right) \, dq'; \tag{4.86}$$

that is, the evolution of any solution of the **F-P** equation (4.83) is obtained applying the Kolmogorov operator:

$$f\left(q, t_2\right) = \mathcal{U}\left(t_2, t_1\right) f\left(q, t_1\right). \tag{4.87}$$

This last equation is merely the Kolmogorov compatibility condition.

Proposition   *The Kolmogorov operator satisfies the semigroup law*

$$\mathcal{U}\left(t_1, t_1\right) = identity \tag{4.88}$$

$$\mathcal{U}\left(t_3, t_1\right) = \mathcal{U}\left(t_3, t_2\right) \mathcal{U}\left(t_2, t_1\right). \tag{4.89}$$

*If the **F-P** operator (4.83) is periodic in time, the Kolmogorov operator (4.86) has the periodicity*

$$\mathcal{U}\left(t_2 + T, t_1 + T\right) = \mathcal{U}\left(t_2, t_1\right). \tag{4.90}$$

The property (4.88) is inferred from the initial condition of the propagator (4.85), property (4.89) is proved from the Chapman–Kolmogorov equation (see Chap. 3), which is valid for any Markov process. On the other hand, from (4.83) and (4.84) it is easy to see that the propagator has the periodicity

$$P\left(q, t + T | q_0, t_0 + T\right) = P\left(q, t | q_0, t_0\right), \tag{4.91}$$

from which the property (4.90) follows immediately. Note that since in general the propagator is not symmetric under the transformation $q \leftrightarrow q_0$, Kolmogorov's operator is generally not self-adjoint; its adjoint operator is given by

---

[41]See Chap. 3.

$$\mathcal{U}(t_2, t_1)^+ : \phi(q) \rightarrow \int \phi(q') P(q', t_2|q, t_1) \, dq'. \tag{4.92}$$

Proposition *The Kolmogorov adjoint operator satisfies:*

$$\mathcal{U}(t_1, t_1)^+ = identity \tag{4.93}$$

$$\mathcal{U}(t_3, t_1)^+ = \mathcal{U}(t_2, t_1)^+ \, \mathcal{U}(t_3, t_2)^+ \,, \tag{4.94}$$

*and if the **F-P** operator (4.83) is periodic in time it follows that the adjoint meets*

$$\mathcal{U}(t_2 + T, t_1 + T)^+ = \mathcal{U}(t_2, t_1)^+ . \tag{4.95}$$

These three properties are inferred from the corresponding properties of the Kolmogorov operator.

If $\phi(q, t)$ is a solution of the Backward **F-P** *equation, that is*[42]*:*

$$\partial_t \phi(q, t) = \left[K_\nu(q, t) \frac{\partial}{\partial q_\nu} + \frac{\epsilon}{2} D_{\nu\mu}(q, t) \frac{\partial^2}{\partial q_\nu \partial q_\mu} \right] \phi(q, t) \tag{4.96}$$

$$\equiv \mathcal{L}\left(q, \partial_q, t\right)^+ \phi(q, t),$$

then its evolution backwards in time can be obtained by applying the Kolmogorov adjoint operator:

$$\phi(q, t_1) = \mathcal{U}(t_2, t_1)^+ \phi(q, t_2). \tag{4.97}$$

We can then call this equation the adjoint Kolmogorov compatibility condition.

4.8.3 Evolution Over a Period of Time

We are now interested in the eigenvalues problem of the operator $\mathcal{U}(t + T, t)$. Since the Kolmogorov operator is generally not self-adjoint, we need to find a biorthonormal set of eigenfunctions of $\mathcal{U}(t + T, t)$ and its adjoint $\mathcal{U}(t + T, t)^+$, that is to say:

$$\mathcal{U}(t + T, t) f_i(q, t) = k_i f_i(q, t) \tag{4.98}$$

---

[42]This *backward* equation is useful in the study of first passage "random" times across a given border see Chap. 9; also exercises in Sects. 3.18 and 6.6.

$$\mathcal{U}(t + T, t)^+ \phi_i(q, t + T) = k_i \phi_i(q, t + T) \tag{4.99}$$

$$\{\phi_i, f_j\} = \int \phi_i(q, t + T) f_j(q, t) \, dq = \delta_{i,j}. \tag{4.100}$$

Using the definitions (4.86) and (4.92), and properties (4.88)–(4.90) and (4.93)–(4.95), the following lemma is proved immediately:

Lemma

Insofar as $f(q, t)$ satisfies the Kolmogorov compatibility condition (4.87) and $\phi(q, t)$ satisfies the adjoint condition (4.97), it follows that

(a) If $f(q, t_0)$ is an eigenfunction of $\mathcal{U}(t_0 + T, t_0)$ with eigenvalue k, then $f(q, t)$ is an eigenfunction of $\mathcal{U}(t + T, t)$ with the same eigenvalue k for all t.
   If $\phi(q, t_0 + T)$ is an eigenfunction of $\mathcal{U}(t_0 + T, t_0)^+$ with eigenvalue k, then $\phi(q, t + T)$ is an eigenfunction of $\mathcal{U}(t + T, t)^+$ with same eigenvalue k for all t.
(b) The eigenfunctions $f_i(q, t)$ and $\phi_i(q, t)$ have the same Floquet structure

$$f_i(q, t) = e^{-\lambda_i t} g_i(q, t)$$

$$\phi_i(q, t) = e^{\lambda_i t} \gamma_i(q, t), \tag{4.101}$$

where the functions $g_i(q, t)$ and $\gamma_i(q, t)$ are periodic in t, that is,

$$g_i(q, t + T) = g_i(q, t)$$

$$\gamma_i(q, t + T) = \gamma_i(q, t), \tag{4.102}$$

and λ_i should be chosen so that the eigenvalues k_i have the form

$$k_i = e^{-\lambda_i T}.$$

(c) The integral $\int \phi(q, t + T) f(q, t) \, dq$ does not depend on time t, that is, the scalar product $\{\phi, f\}$ in (4.100) is well defined.

Proof (a) Since $\mathcal{U}(t + T, t)$ is periodic in t it is sufficient to show that for $t_0 + T > t > t_0$:

$$\mathcal{U}(t + T, t) f(q, t) = \mathcal{U}(t + T, t) \mathcal{U}(t, t_0) f(q, t_0)$$

$$= \mathcal{U}(t + T, t_0) f(q, t_0)$$

$$= \mathcal{U}(t + T, t_0 + T) \mathcal{U}(t_0 + T, t_0) f(q, t_0)$$

$$= \mathcal{U}(t + T, t_0 + T) k f(q, t_0)$$

$$= k \mathcal{U}(t, t_0) f(q, t_0)$$

$$= k f(q, t).$$

The proof for $\phi(q, t)$ is analogous. (b) Let be $k_i = e^{-\lambda_i T}$; then $g_i(q, t)$ is periodic in t:

$$
\begin{aligned}
g_i(q, t + T) &= e^{\lambda_i(t+T)} f_i(q, t + T) \\
&= e^{\lambda_i(t+T)} \mathcal{U}(t + T, t) f_i(q, t) \\
&= e^{\lambda_i(t+T)} k_i f_i(q, t) \\
&= e^{\lambda_i t} f_i(q, t) \\
&= g_i(q, t).
\end{aligned}
$$

The proof for $\phi_i(q, t)$ is completely analogous. (c) Let $t_2 > t_1$; then

$$
\begin{aligned}
\int \phi(q, t_1 + T) f(q, t_1) \, dq &= \int \left(\mathcal{U}(t_2 + T, t_1 + T)^+ \phi(q, t_2 + T) \right) f(q, t_1) \, dq \\
&= \int \left(\mathcal{U}(t_2, t_1)^+ \phi(q, t_2 + T) \right) f(q, t_1) \, dq \\
&= \int \phi(q, t_2 + T) \left(\mathcal{U}(t_2, t_1) f(q, t_1) \right) \, dq \\
&= \int \phi(q, t_2 + T) f(q, t_2) \, dq.
\end{aligned}
$$

Until now the lemma (parts a, b, and c) was only a conclusion on the basis of our periodic system (equivalent to the structure of the Floquet theorem [22]). If now we consider that our equations describe probability distributions of Markov processes, we arrive at the following conclusions [23]:

(d) There is always an eigenvalue $k_0 = 1$ ($\lambda_0 = 0$) with constant adjoint eigenfunction: $\phi_0(q, t) = \gamma_0(q, t) = 1$.
(e) The integrals for all the other eigenfunctions are all zero

$$
\int f_i(q, t) \, dq = \int g_i(q, t) \, dq = 0, \quad \text{for} \quad k_i \neq 1.
$$

(f) If the drift and diffusion matrix are not singular, the eigenvalue $k_0 = 1$ is not degenerate and its eigenfunction is the periodic asymptotic probability distribution $P_{as}(q, t) = f_0(q, t) = g_0(q, t)$.
(g) All other eigenvalues have moduli lower than 1:

$$
|k_i| < 1, \quad \text{that is,} \quad \mathcal{R}_e[\lambda_i] > 0, \quad \text{for} \quad i = 1, 2, \cdots.
$$

Proof (d) Since the propagator is a normalized probability density, we must have

$$\mathcal{U}(t+T,t)^+ 1 = \int P(q',t+T|q,t)\,dq' = 1.$$

(e) Because the **F-P** dynamics conserves the integral we get that

$$\int f_i(q,t+T)\,dq = \int f_i(q,t)\,dq = e^{-\lambda_i t}\int g_i(q,t)\,dq,$$

but the periodicity of $g_i(q,t)$ implies that

$$\int f_i(q,t+T)\,dq = e^{-\lambda_i(t+T)}\int g_i(q,t+T)\,dq = e^{-\lambda_i(t+T)}\int g_i(q,t)\,dq.$$

Both results are possible only if we have that $e^{-\lambda_i T}(=k_i) = 1$ or $\int g_i(q,t)\,dq = 0$, then we conclude that $\int f_i(q,t)\,dq = 0$. (f) Under these conditions the system for $t \to \infty$ approaches a single distribution $P_{as}(q,t)$. The eigenfunctions with eigenvalue 1 are precisely periodic functions satisfying (b); but $P_{as}(q,t)$ is the only such function (save a scalar factor). (g) Since under these conditions any solution of the **F-P** dynamics asymptotically approaches $P_{as}(q,t)$ for $t \to \infty$, any other eigenfunction must vanish for $t \to \infty$; so that, $|k_i|$ must be less than 1.[43]

From now on we order the eigenvalues in decreasing moduli as: $1 = k_0 > |k_1| > |k_2| > \ldots$, that is λ_i with increasing real part. So far nothing has been said about the completeness of the set of eigenfunctions. Indeed, this statement cannot be proved in general. But the parts (a) and (b) of the lemma show that from the existence of a complete set of eigenfunctions for some fixed value t_0, we infer its existence for all t. Henceforth, we will assume that such a complete set of eigenfunctions exists, so functions $f_i(q,t)$ and $\phi_i(q,t)$ satisfy:

$$\int \phi_i(q,t+T)f_j(q,t)\,dq = \delta_{i,j}$$

$$\sum_{i=0}^{\infty} \phi_i(q',t+T)f_i(q,t) = \delta(q'-q). \qquad (4.103)$$

We can now expand any function $h(q,t)$ that satisfies the Kolmogorov compatibility condition (4.87) in a series of eigenfunctions

$$h(q,t) = \sum_{i=0}^{\infty} A_i f_i(q,t) = \sum_{i=0}^{\infty} A_i e^{-\lambda_i t} g_i(q,t),$$

---

[43] An alternative proof is given in the advanced exercise 4.9.7. Assertion (g) can also be proved by defining a Lyapunov function [12], see the first optional exercise in Sect. 4.6, or in Appendix B.2 to understand a similar situation in the case of a stationary Markov **sp**.

where the coefficients A_i can be found from:

$$A_i = \{\phi_i, h\} = \int \phi_i(q, t + T)h(q, t)\, dq.$$

In particular, the propagator can be written as

$$P(q, t|q_0, t_0) = \sum_{i=0}^{\infty} A_i(q_0, t_0)e^{-\lambda_i(t-t_0)}g_i(q, t), \qquad (4.104)$$

where the coefficients $A_i(q_0, t_0)$ are periodic in t_0.

Exercise. Derive (4.104) from the periodicity of $P(q, t|q_0, t_0)$, (4.91), and the structure of $g_i(q, t)$ [part (b) of the lemma].

4.8.4 Periodic Detailed Balance

Proving the existence of a complete set of eigenfunctions of the Kolmogorov operator is an open question. The problem arises because on one hand the differential representation of the Kolmogorov operator involves the time ordering operator, which is difficult to handle (all operators involved must be chronologically ordered, see advanced exercise 4.9.9, and reference [5])

$$\mathcal{U}(t_2, t_1) = \vec{\mathcal{T}} \exp\left(\int_{t_1}^{t_2} \mathcal{L}(q, \partial_q, s)\, ds\right),$$

and on the other hand, in the integral representation of this operator (4.86), the kernel (i.e., the propagator) is generally not symmetrical. For stationary Markov processes it is possible to symmetrize the **F-P** operator assuming that detailed balance holds. This condition states that for variables q even under time reversal $t \to -t$, we have[44]

$$P(x, t|y, 0)P_{st}(y) = P(y, t|x, 0)P_{st}(x)$$

($P_{st}(x)$ is the stationary solution of the Markov process). This enables us to obtain self-adjoint **F-P** and Kolmogorov operators. In addition, under these conditions the symmetrized **F-P** operator is defined seminegative, so its eigenvalues are negative real numbers, from which it follows that the eigenvalues of the Kolmogorov operator are real between 0 and 1.

In the nonstationary T-periodic case we have no continuous temporal translation invariance, but indeed a discrete invariance $t \to t + T$. With a condition similar to

---

[44] The symbols x and y are equivalent to q and represent a set of even variables $\{q_1, q_2, \cdots\}$.

detailed balance we may symmetrize the Kolmogorov operator. This new condition is compatible with the discrete time translation invariance, and we call it the periodic detailed balance.

Definition The periodic detailed balance condition is met if

$$P(x, t + T|y, t)P_{\text{as}}(y, t) = P(y, t + T|x, t)P_{\text{as}}(x, t), \quad \forall x, y, t.$$

Proposition *If the condition of periodic detailed balance is satisfied, the Kolmogorov operator $\mathcal{U}(t + T, t)$ is self-adjoint under the scalar product*

$$\{\eta, \xi\} = \int \eta(x)\xi(x)/P_{\text{as}}(x, t) \, dx.$$

Proof

$$\{\eta, \mathcal{U}(t + T, t)\xi\} = \int \int \eta(x)P(x, t + T|y, t)\xi(y)/P_{\text{as}}(x, t) \, dx \, dy$$

$$= \int \int P(y, t + T|x, t)\eta(x)\xi(y)/P_{\text{as}}(y, t) \, dx \, dy$$

$$= \{\mathcal{U}(t + T, t)\eta, \xi\}.$$

Corollary *If periodic detailed balance is satisfied, there is a complete set of eigenfunctions for the Kolmogorov operator $\mathcal{U}(t + T, t)$.*

Optional Exercise (**Symmetrization of the Fokker–Planck Operator.**) Consider Kolmogorov's operator for the special case where the system is stationary (i.e., drift and diffusion matrix are independent of time). In general, using the Dyson representation for the Kolmogorov operator (see advanced exercise 4.9.9),

$$\mathcal{U}(t + \tau, t) = \vec{\mathcal{T}} \exp\left(\int_t^{t+\tau} \mathcal{L}(q, \partial_q, s) \, ds\right) = \vec{\mathcal{T}}\left\{1 + \sum_{n=1}^{\infty} \frac{1}{n!} \int_t^{t+\tau} dt_1 \int_t^{t+\tau} dt_2 \cdots \right.$$

$$\left. \cdots \int_t^{t+\tau} dt_n \, \mathcal{L}(q_\mu, \partial_\mu, t_1) \, \mathcal{L}(q_\mu, \partial_\mu, t_2) \cdots \mathcal{L}(q_\mu, \partial_\mu, t_n)\right\},$$

and the fact that for stationary processes $\mathcal{L}$ is independent of time, prove that if $\mathcal{U}(t + \tau, t)$ is symmetrizable, then so is $\mathcal{L}(q_\mu, \partial_\mu)$. This is equivalent to establishing that the condition of detailed balance supports a complete set of eigenfunctions for the **F-P** operator. Then detailed balance may also be formulated in terms of the symmetrization of the **F-P** operator:

$$\{\phi, \mathcal{L}\psi\} = \{\mathcal{L}\phi, \psi\}, \tag{4.105}$$

under the scalar product $\{\eta, \xi\} = \int \eta(x)\xi(x)/P_{st}(x) \, dx$. Then, if the system is stationary and the **F-P** equation satisfies the principle of detailed balance, show that $P_{st}(x) \equiv g_0(x)$ (eigenfunction with eigenvalue $k_0 = 1$).

Excursus (**Inverse Exponential Operator.**) Consider the following finite dimensional time-ordered operator: $\mathcal{U}(t + \tau, t) = \vec{\mathcal{T}} \exp\left(\int_t^{t+\tau} \mathbf{H}(s) \, ds\right)$, which is given in terms of a time-dependent matrix $\mathbf{H}(s)$. Write down explicitly the operator $\mathcal{U}(t + \tau, t)^{-1}$.

4.8.5 Strong Mixing

The concept of strong mixing was originally introduced as one of several conditions that an **sp** must meet to apply the central limit theorem. This problem does not arise here because, in general, our Markov process is not Gaussian. In this section we are more interested in the concept of "strong mixing" as a condition of asymptotic independence.

Let $q(t)$ be a realization of the nonstationary T-periodic Markov **sp**. A correlation function is defined, as usual, by

$$\langle\langle q(t)q(t')\rangle\rangle = \langle q(t)q(t')\rangle - \langle q(t)\rangle\langle q(t')\rangle, \tag{4.106}$$

where the symbol $\langle \cdots \rangle$ represents the ensemble average (see Chap. 3). In general, what usually interests us is the calculation of the correlation function in the limit of long times, that is, by using the asymptotic distribution, $P_{as}(q, t)$, to build the (asymptotic) two-time joint probability density [necessary for calculating (4.106)]. Then the (asymptotic) two-time second moment is given by[45]

$$\langle q(t)q(t')\rangle_{as} = \int \int qq' P(q', t'|q, t)P_{as}(q, t) \, dq \, dq',$$

where, without loss of generality, we have chosen $t' \geq t$. We can consider the correlation function as a function of the variables t and $\tau = t' - t$. Now, using the representation of the propagator in eigenfunctions of the Kolmogorov operator we get:

$$\langle\langle q(t + \tau)q(t)\rangle\rangle_{as} = \sum_{i=1}^{\infty} e^{-\lambda_i \tau} B_i(t, \tau), \tag{4.107}$$

---

[45] As in the previous sections, dq represents a n-dimensional set of differentials.

where the functions

$$B_i(t, \tau) = \int \int qq'A_i(q, t)g_i(q', t + \tau)g_0(q, t)\, dq\, dq' \qquad (4.108)$$

are periodic in t and τ.

Optional Exercise. Show (4.108) from the lemma.

The correlation function (4.107) is an oscillatory decreasing function of τ for any fixed value t, which tends to zero as $\tau \to \infty$. This fact proves the following corollary.

Corollary *All nonstationary T-periodic Markov processes characterized by a nonsingular drift and diffusion matrix are strong mixing.*

Correlation Function for Large τ

If we arrange the eigenvalues of the Kolmogorov operator in decreasing order, $1 = k_0 > |k_1| > |k_2| > \cdots$, and retain only the dominant terms in the series (4.107) we obtain for the asymptotic correlation

$$\langle\langle q(t + \tau)q(t)\rangle\rangle_{\text{as}} \approx e^{-\lambda_1 \tau}B_1(t, \tau). \qquad (4.109)$$

Then, after a period of time, the correlation function (as a function of τ) is damped by the factor $k_1 = e^{-\lambda_1 T}$. That is, the first eigenvalue of the Kolmogorov operator less than 1 characterizes the slope of decay $\langle\langle q(t + \tau)q(t)\rangle\rangle_{\text{as}}$ as a function of τ.

Excursus. We can see that the eigenvalue k_1 generalizes the concept of escape time, $\tau_K \approx \lambda_1^{-1}$, from an attractor for nonstationary periodic Markov systems [23]. In the case of a stationary process, the time τ_K is Kramer's activation time, see Chap. 9.

Lyapunov Function

Another interesting object that provides information on the dynamic evolution of a nonstationary T-periodic Markov process, is the Lyapunov function. Traditionally, this function was introduced to analyze the decay of the initial preparation of a stationary system.[46]

Consider the system prepared at point q_0 and time t_0. In this case the Lyapunov function is defined by

---

[46]This function was presented when we discussed irreversibility at the beginning of the chapter.

$$\mathcal{H}(t) = \int P(q, t | q_0, t_0) \ln \left(\frac{P(q, t | q_0, t_0)}{P_{as}(q, t)} \right) dq.$$

Note that this definition is independent of dimension, i.e., the number of elements of the set $\{q_v\}$. It is possible to prove that for all q_0 the function $\mathcal{H}(t)$ is not negative decreasing.[47] That is, the approach from the initial preparation of the system to the asymptotic state $P_{as}(q, t)$ is characterized by $\mathcal{H}(t)$ [12, 24]. Again, if we use an expansion in eigenfunctions of Kolmogorov's operator and take only the dominant decay behavior, we observe the following:

Corollary *In the long-time regime the Lyapunov function $\mathcal{H}(t)$ is decreasing oscillatory and decays in each time period by a factor $(k_1)^2$.*

The proof is left as an exercise for the reader. Note that in the asymptotic regime the decay of the correlation function is slower than the decay of the Lyapunov function.

Excursus (**Nuclear Magnetic Resonance.**) Consider a monodomain uniform particle with magnetic moment **M** of magnitude M_s (whose variation with temperature can be neglected), in the presence of a constant external field $\mathbf{H}_0$. If the instantaneous orientation of **M** is described by the angles $\theta \equiv \theta(t)$ and $\phi \equiv \phi(t)$ so that the magnetic moment components are

$$M_x = M_s \sin \theta \cos \phi$$
$$M_y = M_s \sin \theta \sin \phi$$
$$M_z = M_s \cos \theta,$$

consider from Brown–Gilbert's **sde** (see Sect. 3.19.1) the associated **F-P** operator [14]. In this case, by the principle of detailed balance, it is possible to relate the intensity of thermal fluctuations to the dissipation parameter in the model [see (4.51)]. On the other hand, if in addition to the magnetic field $\mathbf{H}_0$ there is a rotating radio frequency field $\mathbf{H}_1 = \mathbf{H}_1(t)$ (perpendicular to $\mathbf{H}_0$), it is of interest to study in these nonequilibrium conditions the behavior of the magnetic system; i.e., correlation functions, energy absorption, characteristic relaxation times (parallel and perpendicular to the direction of $\mathbf{H}_0$), etc. Then, for this model, a periodically modulated in time **F-P** operator is obtained. Consider, for example, the configuration of the total applied magnetic field:

$$\mathbf{H} = \mathbf{H}_0 + \mathbf{H}_1(t)$$
$$= \hat{\mathbf{x}} H_1 \cos(\omega t) + \hat{\mathbf{y}} H_1 \sin(\omega t) + \hat{\mathbf{z}} H_0,$$

---

[47] To prove that $\mathcal{H}(t)$ is a Lyapunov function we proceed in a way analogous to that in Sect. 4.6, and Appendix B.2.

Fig. 4.3 Configuration of the
total applied field,
$\mathbf{H} = \mathbf{H}_0 + \mathbf{H}_1(t)$, on the
magnetic moment $\mathbf{M}$

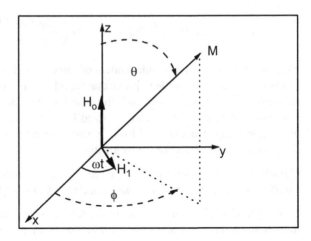

that is shown in Fig. 4.3. Using that $V(\theta, \phi) = -\mathbf{H} \cdot \mathbf{M}$, check that it is possible to
split the **F-P** operator in the form $\mathcal{L} = \mathcal{L}_0 + \mathcal{K}_1$, where $\mathcal{L}_0$ is the **F-P** operator for
the problem with the field $\mathbf{H}_0$; while $\mathcal{K}_1 = \mathcal{K}_1(t)$ is a drift term, periodic in time,
given by

$$\mathcal{K}_1 = \frac{-H_1 M_s}{\sin \theta} \frac{\partial}{\partial \theta} \left\{ \sin \theta \left[\left(h' \cos \theta \cos (\phi - \omega t) + g' \sin (\phi - \omega t) \right) \right] \right\}$$

$$- \frac{H_1 M_s}{\sin \theta} \frac{\partial}{\partial \phi} \left\{ \left(g' \cos \theta \cos (\phi - \omega t) - h' \sin (\phi - \omega t) \right) \right\}.$$

Then, from the theory of eigenvalues of the Kolmogorov operator we may characterize this nonstationary 2π-periodic Markov system.

Excursus (**Shapiro's Steps**). If in an electrical RSJ circuit [i.e., the Josephson
junction characterized by Eq. (4.56)] has besides the applied *dc* current I an *ac*
component, i.e., an alternating part in the form $I = I_{dc} + I_{ac} \cos (\omega t)$, the RSJ circuit
can be studied in terms of the frequency ω of the alternating applied current; thus
giving rise to the phenomenon of "phase locking" [25]. This fact leads to a step
structured graph for the I_{dc} current as a function of the potential difference V. That
is, these steps arise when the average, in one cycle, of the phase variation equals an
integer multiple of the external frequency $\langle \dot{\phi} \rangle = n\omega$. If now we consider thermal
fluctuations, the evolution of the phase is governed by the **sde** (4.57). Then we are
in the presence of a nonstationary 2π-periodic Markov process. In this case the
potential difference is given by

$$\frac{V}{RI_0} = \langle \dot{\phi} \rangle = \frac{1}{\Gamma} \int_0^\Gamma dt \int_0^{2\pi} \dot{\phi} \, P_{as}(\phi, t) \, d\phi, \qquad \Gamma = 2\pi/\omega.$$

From this it is possible to see that Shapiro's steps are rounded when the temperature
is nonzero [26].

Excursus (**Stochastic Resonance**.) The concept of "stochastic resonance" was introduced to describe the coincidence of two characteristic times in a multistable stochastic system modulated periodically in time. One of these scales is the time period established by the external modulation: $2\pi/\omega$; the other is the time "between jumps" (Kramers' activation time τ_K) induced by the stochastic nature of the system. Originally this concept was introduced for Markovian processes but it is also possible to generalize it to non-Markovian systems. Obviously in these cases a competing third time scale comes into play: the non-Markovian memory. It is interesting to note that the term "stochastic resonance" was originally introduced in an effort to uncover the mechanisms that explain the periodic occurrence of ice ages on Earth [27]. There are many mathematical techniques for studying such nonstationary 2π-periodic Markov processes. In particular, it is possible to efficiently analyze this phenomenon in the context of the path integral [28]. This problem can also be tackled using the eigenvalue theory of the Kolmogorov operator [24].

4.9 Additional Exercises with Their Solutions

4.9.1 Microscopic Reversibility

In cases that all the **rvs** $X_k(t)$ are even under transformation $t \rightarrow -t$, and using ergodicity of the correlation, it is simple to see that $\langle X_k(t+\tau)X_j(t)\rangle = \langle X_k(t)X_j(t+\tau)\rangle$. First note that using the change of variable $t' = -t$ we can write

$$\int_{-T}^{T} X_k(t+\tau)X_j(t)dt = -\int_{T}^{-T} X_k(-t'+\tau)X_j(-t')dt'$$

$$= \int_{-T}^{T} X_k(t'-\tau)X_j(t')dt', \qquad (4.110)$$

in the last line we have used that $X_l(t) = X_l(-t)$. Now, introducing the change of variable $t' - \tau = s$ we can write

$$\int_{-T}^{T} X_k(t'-\tau)X_j(t')dt' = \int_{-T-\tau}^{T-\tau} X_k(s)X_j(s+\tau)ds. \qquad (4.111)$$

Therefore taking the time average and comparing (4.110) and (4.111), we get

$$\overline{X_k(t+\tau)X_j(t)} = \lim_{T\to\infty} \frac{1}{2T} \int_{-T}^{T} X_k(t+\tau)X_j(t)dt$$

$$= \lim_{T\to\infty} \frac{1}{2T} \int_{-T-\tau}^{T-\tau} X_k(s)X_j(s+\tau)ds$$

$$= \overline{X_k(t)X_j(t+\tau)},$$

which is what we wanted to prove.

4.9.2   Regression Theorem

Let $\mathbf{X}(t)$ be an **sp** in n-dimension, it is simple to see that the following relation for the covariance matrix is true

$$\langle \mathbf{X}(t) \cdot \mathbf{X}^\top(t_0) \rangle = \int \int \mathbf{x} \cdot \mathbf{x}_0^\top P(\mathbf{x}, t; \mathbf{x}_0, t_0) \, d\mathbf{x} d\mathbf{x}_0$$

$$= \int \int \mathbf{x} \cdot \mathbf{x}_0^\top P(\mathbf{x}, t | \mathbf{x}, t_0) P(\mathbf{x}_0, t_0) \, d\mathbf{x} d\mathbf{x}_0$$

$$= \int \langle \mathbf{X}(t) \rangle_{\mathbf{X}_0} \cdot \mathbf{x}_0^\top P(\mathbf{x}_0, t_0) \, d\mathbf{x}_0.$$

Therefore if the **sp** has a stationary state, calling $(t - t_0) = \tau$, and assuming that the **mv** is zero (this does not restrict the proof), we can calculate the stationary correlation function as

$$\langle \mathbf{X}(\tau) \cdot \mathbf{X}^\top(0) \rangle_{\text{st}} = \int \langle \mathbf{X}(\tau) \rangle_{\mathbf{X}_0} \cdot \mathbf{x}_0^\top P_{\text{st}}(\mathbf{x}_0) \, d\mathbf{x}_0. \tag{4.112}$$

In the case of a Markov **sp** all the information is in the propagator $P(\mathbf{x}, t | \mathbf{x}, t_0)$ therefore we can compute $\langle \mathbf{X}(\tau) \rangle_{\mathbf{X}_0}$ at any time t_0. Convince yourself that this is not the case for a non-Markovian **sp**.

Now, from (4.112) we can write an evolution equation for the stationary correlation

$$\frac{d}{d\tau} \langle \mathbf{X}(\tau) \cdot \mathbf{X}^\top(0) \rangle_{\text{st}} = \int \frac{d}{d\tau} \langle \mathbf{X}(\tau) \rangle_{\mathbf{X}_0} \cdot \mathbf{x}_0^\top P_{\text{st}}(\mathbf{x}_0) \, d\mathbf{x}_0.$$

If the process is linear and noting that

$$\frac{d}{d\tau} \langle \mathbf{X}(\tau) \rangle_{\mathbf{X}_0} = -\mathbf{M} \cdot \langle \mathbf{X}(\tau) \rangle_{\mathbf{X}_0}, \quad \text{with} \quad \langle \mathbf{X}(0) \rangle_{\mathbf{X}_0} = \mathbf{X}_0,$$

we can write an evolution equation for the stationary correlation function

$$\frac{d}{d\tau} \langle \mathbf{X}(\tau) \cdot \mathbf{X}^\top(0) \rangle_{\text{st}} = -\mathbf{M} \cdot \langle \mathbf{X}(\tau) \cdot \mathbf{X}^\top(0) \rangle_{\text{st}},$$

which is the regression theorem in its simplest form. Then the solution is

$$\langle \mathbf{X}(\tau) \cdot \mathbf{X}^\top(0) \rangle_{\text{st}} = e^{-\mathbf{M}\tau} \cdot \langle \mathbf{X}(0) \cdot \mathbf{X}^\top(0) \rangle_{\text{st}}.$$

Exercise. If we apply this result to the Onsager problem, we conclude that time-correlations for macroscopic (thermodynamic) variables can we written as

$$\left\langle \mathbf{X}(\tau) \cdot \mathbf{X}^\top (0) \right\rangle_{\mathrm{st}} = e^{-\mathbf{M}\tau} \cdot \mathbf{g}^{-1},$$

where the matrix $\mathbf{g}^{-1}$ is characterized—around thermal equilibrium—by the Einstein pdf, see Chap. 2.

4.9.3 Detailed Balance in the Master Equation

The general concept of detailed balance was introduced in Sect. 4.6.2. This important symmetry adopts its simplest form in the context of the Master Equation (to be presented in detail in Chap. 6). Here we introduce it as a fundamental equation that describes the time evolution of the probability of a discrete stochastic process. Let this probability be denoted by $P_m(t)$ (conditioned to the initial state m_0 at time t_0). Then, if $W_{m,n}$ is the transition probability, per unit time, between states $n \to m$, we can represent the total *gain* and *loss* contributions to the time variation of $P_n(t)$ in the form:

$$\frac{dP_n(t)}{dt} = \sum_{m \in \mathcal{D}} W_{n,m} P_m(t) - P_n(t) \sum_{m \in \mathcal{D}} W_{m,n}, \tag{4.113}$$

here $\mathcal{D}$ represents the domain of all possible states of the discrete stochastic process. The *detailed balance* principle is a condition that is only applicable for closed and isolated systems, then equilibrium probability P_m^{eq} exists, and the principle implies

$$\frac{P_m^{\mathrm{eq}}}{P_n^{\mathrm{eq}}} = \frac{W_{m,n}}{W_{n,m}}, \quad \text{where } P_n^{\mathrm{eq}} \neq 0 \ \forall n. \tag{4.114}$$

From this symmetry it is easy to find a *potential* $\Phi(m)$ characterizing the stationary probability of the Master Equation (4.113). Writing $m = n_{j+1}$ and $n = n_j$ we can go from n_0 to n_N via a chain of intermediate states. If that stationary solution is unique (see Appendix B.2) we can write the solution in terms of a particular *path*

$$P_{n_N}(\infty) = P_{n_0}(t_0) \prod_{j=0}^{N-1} \frac{W_{n_{j+1},n_j}}{W_{n_j,n_{j+1}}}, \tag{4.115}$$

then, setting $P_m(\infty) \propto \exp(\Phi(m))$ in (4.115) we get the expression:

$$\Phi(n_N) = \sum_{j=0}^{N-1} \ln \left[\frac{W_{n_{j+1},n_j}}{W_{n_j,n_{j+1}}} \right] + \text{const.}$$

Fig. 4.4 Transition
probabilities in a three-level
atom without detailed
balance. The external "pump"
is represented by $W_{3,1} = \alpha$.
Transitions $W_{2,3}$ and $W_{1,2}$ are
from electron recombination

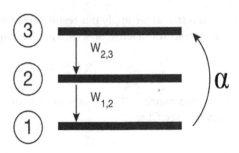

We note that by taking a suitable limit this potential could also be used for
continuous variables.

Let us now present a finite system that does not fulfill detailed balance, for
example the 3×3 system shown in Fig. 4.4. This model represents a *three level atom*
away from equilibrium and is characterized by the set of transition probabilities:

$$W_{3,1} = \alpha, \ W_{2,3} = 1, \ W_{1,2} = 1, \ W_{m,n} = 0 \text{ in any other case.}$$

From (4.113) it is easy to write down the evolution for each $P_n(t)$, $n = 1, 2, 3$. Then
at the stationary state, $dP_n(t)/dt = 0$, the normalized solution $P_n(\infty)$ is given by

$$\begin{pmatrix} P_3 \\ P_2 \\ P_1 \end{pmatrix}_{st} = \mathcal{N} \begin{pmatrix} 1 \\ 1 \\ 1/\alpha \end{pmatrix}, \ \mathcal{N} = (1/\alpha + 2)^{-1}.$$

From this solution show that the nonequilibrium probability current is: $J = \alpha/(2\alpha + 1)$.

4.9.4 Steady-State Solution of F-P (Case $J_\mu^{st} = 0, D_{\mu\nu} = \delta_{\mu\nu}D$)

We show here that when the stationary current is zero and if $D_{\mu\nu}$ is diagonal it is
always possible to find the stationary solution P_{st} by quadrature. In the stationary
state the divergence of the probability current $J_\mu^{st} = K_\mu P_{st} - \partial_\nu D_{\mu\nu} P_{st}$ vanishes.
However, in several dimensions adopting $J_\mu^{st} = 0$ is not the most general situation
(J_μ^{st} could be the rotor of some vector). Assuming $J_\mu^{st} = 0$ we can write in vector
notation

$$P_{st} \left[\vec{K} - \nabla D \right] = D \nabla P_{st}, \tag{4.116}$$

or

$$\frac{1}{P_{st}} \nabla P_{st} = \nabla (\ln P_{st}) = \left[\frac{\vec{K}}{D} - \frac{1}{D} \nabla D \right]. \tag{4.117}$$

Since the vector on the right-hand side is the gradient of the scalar field $\ln P_{st}$ it must be irrotational, and therefore we can show that

$$\frac{\partial}{\partial q_\mu}\left(\frac{K_\nu}{D} - \frac{(\partial q_\nu D)}{D}\right) = \frac{\partial}{\partial q_\nu}\left(\frac{K_\mu}{D} - \frac{(\partial q_\mu D)}{D}\right),$$

hence the vector $\left(\frac{\vec{K}}{D}\right)$ is also irrotational, and it is therefore expressible as the gradient of a "potential"

$$\frac{K\left(q_\mu\right)}{D\left(q_\mu\right)} = -\nabla U\left(q_\mu\right), \text{ with } U\left(\vec{q}\right) = -\int_{\vec{q}_0}^{\vec{q}} \frac{\vec{K}\left(\vec{q}\right)}{D\left(\vec{q}\right)} \cdot d\vec{q} + \text{const.}$$

Using (4.117) show that

$$\nabla U = -\nabla \ln\left(DP_{st}\right),$$

this equation can be immediately integrated: $U\left(\vec{q}\right) = -\ln\left(DP_{st}\right) + \text{const.}$ Then

$$P_{st}\left(\vec{q}\right) = \frac{\mathcal{N}}{D\left(\vec{q}\right)}e^{-U(\vec{q})}. \tag{4.118}$$

If the detailed balance condition is met $\nabla \cdot \vec{K}^{rev}P_{eq} = 0$, where $K_\mu^{rev} = K_\mu - \frac{1}{P_{eq}}\partial_\nu\left(D_{\mu\nu}P_{eq}\right)$ with $D_{\mu\nu} = \delta_{\mu\nu}D\left(q_\mu\right)$, recover (4.118).

4.9.5 Inhomogeneous Diffusion Around Equilibrium

Brownian motion diffusing in a three-dimensional, inhomogeneous anisotropic medium under an external force—with potential $U\left(\vec{r}\right)$—obeys an evolution equation that can be written in terms of the mobility tensor $\mu_{\mu\nu}\left(\vec{r}\right)$ and the diffusion matrix $D_{\mu\nu}\left(\vec{r}\right)$ in the form

$$\partial_t P\left(\vec{r}, t\right) = \nabla \cdot \{\mu \cdot (\nabla U) + D \cdot \nabla\} P\left(\vec{r}, t\right), \tag{4.119}$$

which of course has the stationary distribution $P_{eq}\left(\vec{r}\right) \propto \exp\left(-U\left(\vec{r}\right)/k_B T\right)$, because the Onsager generalized relations state that $\mu_{\mu\nu}\left(\vec{r}\right) = D_{\mu\nu}\left(\vec{r}\right)/k_B T$. Then using detailed balance write the generalized Fick's equation (4.119) in the usual **F-P** form.

4.9.6   Chain of N Rotators (The Stationary F-P distribution)

Consider a ring of N rotators interacting harmonically through the potential

$$U\left(\{\theta_n\}\right) = \frac{\varkappa}{2} \sum_{j=1}^{N} \left(\theta_{j+1} - \theta_j\right)^2, \quad \theta_{N+1} \equiv \theta_1. \tag{4.120}$$

If the rotators are strongly damped, subject to a Brownian motion and to an external torque τ, the **F-P** dynamics of the system of *variables* $\theta_j \in [0, 2\pi]$ is

$$\frac{\partial P\left(\{\theta_n\}, t\right)}{\partial t} = \frac{1}{M\gamma} \sum_{j=1}^{N} \left[\frac{\partial}{\partial \theta_j}\left(\frac{\partial U\left(\{\theta_n\}\right)}{\partial \theta_j} - \tau\right) + k_B T \frac{\partial^2}{\partial \theta_j^2}\right] P\left(\{\theta_n\}, t\right). \tag{4.121}$$

Convince yourself that (4.121) corresponds to the Brownian motion description of a strongly damped chain of N *harmonic* rotators.

Here we will find using the potential conditions of Sect. 4.7.2, the stationary solution $P_{\text{st}}\left(\{\theta_n\}\right)$ of the **F-P** dynamics (4.121), and we will calculate the average rotational speed for the whole rotator. The **F-P** (4.121) can be written in the form

$$\partial_t P\left(\{\theta_n\}, t\right) = -\sum_{\mu=1}^{N} \partial_\mu J_\mu\left(\{\theta_n\}, t\right), \tag{4.122}$$

$$J_\mu\left(\{\theta_n\}, t\right) = \left(K_\mu - \sum_{\nu=1}^{N} \partial_\nu D_{\mu\nu}\right) P\left(\{\theta_n\}, t\right),$$

where the current is given by

$$J_\mu\left(\{\theta_n\}, t\right) = -\left[\frac{1}{M\gamma}\left(\frac{\partial U}{\partial \theta_\mu} - \tau\right) + \sum_{\nu=1}^{N} \partial_\nu D_{\mu\nu}\right] P\left(\{\theta_n\}, t\right) \tag{4.123}$$

$$D_{\mu\nu} = \frac{K_B T}{M\gamma} \delta_{\mu\nu}. \tag{4.124}$$

The key step in the *potential condition* decomposition is guessing a clever splitting for the drift term $K_\mu = f_\mu^d + f_\mu^0$ into dissipative and non-dissipative parts. Choosing

$$f_\mu^d = \frac{-1}{M\gamma}\frac{\partial U}{\partial \theta_\mu}, \quad f_\mu^0 = \frac{+1}{M\gamma}\tau,$$

we get the final expression for the currents

$$J_\mu^d = \left(\frac{-1}{M\gamma} \frac{\partial U}{\partial \theta_\mu} - \sum_{\nu=1}^{N} \partial_\nu D_{\mu\nu} \right) P \qquad (4.125)$$

$$J_\mu^0 = \frac{\tau}{M\gamma} P \hat{\theta}_\mu \qquad (4.126)$$

Use a potential *form* $P_{st} \propto e^{-F}$ in the expressions $J_\mu^d\big|_{st} = 0$ and $\sum_{\mu=1}^{N} \partial_\mu J_\mu^0\big|_{st} = 0$, then

$$\left(J_\mu^d \big|_{st} = 0 \right) \Rightarrow \left(\frac{-1}{M\gamma} \frac{\partial U}{\partial \theta_\mu} - \sum_{\nu=1}^{N} \partial_\nu D_{\mu\nu} \right) e^{-F} = 0 \qquad (4.127)$$

$$\left(\sum_{\mu=1}^{N} \partial_\mu J_\mu^0 \big|_{st} = 0 \right) \Rightarrow \sum_{\mu=1}^{N} \partial_\mu \frac{\tau e^{-F}}{M\gamma} = 0. \qquad (4.128)$$

From (4.127) we get $\frac{\partial U}{\partial \theta_\mu} - k_B T \frac{\partial F}{\partial \theta_\mu} = 0$, which can be integrated for each θ_μ leading to

$$F(\{\theta_n\}) = \frac{U(\{\theta_n\})}{k_B T} + \text{constant.} \qquad (4.129)$$

From (4.128) we get

$$\sum_{\mu=1}^{N} \partial_\mu F = 0. \qquad (4.130)$$

Now we have to check (4.130) and also the consistency condition:

$$f_\mu^d - \sum_{\nu=1}^{N} \partial_\nu D_{\mu\nu} = -\sum_{\nu=1}^{N} D_{\mu\nu} \partial_\nu F. \qquad (4.131)$$

First, we prove (4.130) using (4.129)

$$\frac{1}{k_B T} \sum_{\mu=1}^{N} \frac{\partial U}{\partial \theta_\mu} = \frac{\varkappa}{2k_B T} \sum_{\mu=1}^{N} \frac{\partial}{\partial \theta_\mu} \sum_{j=1}^{N} (\theta_{j+1} - \theta_j)^2$$

$$= \frac{\varkappa}{2k_B T} \sum_{\mu=1}^{N} \left(4\theta_\mu - 2 \sum_{j \neq \mu}^{N} \theta_j \right) = 0, \qquad (4.132)$$

here the periodic condition $\theta_{N+1} \equiv \theta_1$ has been used. In addition (4.131) is easily checked to be true because $D_{\mu\nu}$ is diagonal, so we finally get the stationary pdf

$$P_{\text{st}}(\{\theta_n\}) = \mathcal{N} \exp\left(\frac{-U(\{\theta_n\})}{k_B T}\right), \tag{4.133}$$

where $\mathcal{N}$ is a normalization constant from:

$$\int_0^{2\pi} \cdots \int_0^{2\pi} d\theta_1 \cdots d\theta_N P_{\text{st}}(\{\theta_n\}) = 1.$$

We can calculate the total mean angular velocity $\omega = \left\langle \dot{\theta}_1 + \cdots + \dot{\theta}_N \right\rangle$ from the Langevin equation associated with the **F-P** dynamics (4.121), then we write[48]

$$\omega = \left\langle \sum_{\mu=1}^N \dot{\theta}_\mu \right\rangle = \left\langle \sum_{\mu=1}^N \left[\frac{-1}{M\gamma}\left(\frac{\partial U}{\partial \theta_\mu} - \tau\right) + \sqrt{\frac{2k_B T}{M\gamma}}\xi_\mu(t) \right] \right\rangle, \quad \mu = 1, \cdots, N,$$

here $\xi_\mu(t)$ is a Gaussian white noise: $\left\langle \xi_\mu(t)\xi_\nu(t') \right\rangle = \delta_{\mu,\nu}\,\delta(t-t')$. Using (4.132) we get

$$\omega = \sum_{\mu=1}^N \frac{\tau}{M\gamma} = \frac{N\tau}{M\gamma}. \tag{4.134}$$

This result can be interpreted from (4.126) in the stationary state. Let $p(\theta_j)$ be the marginal stationary pdf of the variable θ_j, by symmetry this pdf must be a constant, thus

$$p(\theta_j) = \int_0^{2\pi} \cdots \int_0^{2\pi} P_{\text{st}}(\{\theta_n\}) \prod_{\mu=1}^N d\theta_{\mu\neq j} = \frac{1}{2\pi}.$$

From (4.126), the j-component of the (marginal) stationary current J^0

$$\left. J^0 \cdot \hat{\theta}_j \right|_{\text{st}} = \left. J^0_j \right|_{\text{st}}$$

is

$$\left. J^0_j \right|_{\text{st}} = \frac{\tau}{M\gamma}p(\theta_j) = \frac{\tau}{M\gamma 2\pi}.$$

---

[48]Note that in the strongly damped limit we pass from the joint pdf $P(\{\theta_j\}, \{\omega_j\}, t)$ to the joint pdf $P(\{\theta_j\}, t)$ assuming a fast thermalization of variables $\omega_j \equiv \dot{\theta}_j$ to a nonzero value.

This quantity is the average time required for θ_j to increase by 2π, hence

$$2\pi J_j^0\big|_{st} = \langle \dot{\theta}_j \rangle = \frac{\tau}{M\gamma},$$

and so adding all N variables we get (4.134) again.

4.9.7 Asymptotic Solution of the F-P Dynamics for Long Times

A necessary condition for a solution of the F-P equation to be stationary is that the drift and the diffusion terms must be independent of time. Furthermore a Lyapunov function of the **F-P** equation exists if the drift is not singular and the diffusion is positive definite over the whole domain. This last consideration, on the existence of a unique stationary solution, can also be extended for 2π-periodic nonstationary Markov processes [12]. Alternatively, proving that Kolmogorov's operator has eigenvalues satisfying $|k_i| \equiv |e^{-\lambda_i T}| < 1$, and assuming that its eigenfunctions are complete this fact implies the existence of a unique asymptotic (2π-periodic) solution of the **F-P** equation (i.e., the eigenfunction ϕ_0). To prove this (i.e., $\mathcal{R}_e\lambda_i > 0$, $i = 1, 2, \cdots$) we start from the definition of the adjoint Kolmogorov operator (see Sect. 4.8.2):

$$k_i\phi_i(q, t+T) = \int \phi_i(q', t+T) P(q', t+T \,|\, q, t)\, dq' \tag{4.135}$$

Now we use that the propagator is nonnegative for all q and q' and satisfies normalization to 1, and denote q by q_m if q is such that $|\phi_i(q, t+T)| = $ max. Then from (4.135)

$$|k_i|\, |\phi_i(q_m, t+T)| = \left| \int \phi_i(q', t+T) P(q', t+T \,|\, q_m, t)\, dq' \right|$$

$$\leq \int |\phi_i(q', t+T)| P(q', t+T \,|\, q_m, t)\, dq'$$

$$\leq \int |\phi_i(q_m, t+T)| P(q', t+T \,|\, q_m, t)\, dq'$$

$$\leq |\phi_i(q_m, t+T)|,$$

so therefore $|k_i| \leq 1$. This means that the eigenvalues can be ordered as $1 = k_0 > |k_1| > |k_2| > \cdots$, i.e., λ_i with increasing real part [24].

Show that a similar argument can be used for the analysis of the time-independent **F-P** operator. If there are pure imaginary eigenvalues the equilibrium (stationary state) is not uniquely determined. If there are no such eigenvalues the solution is unique and the Markovian **sp** is said to be *ergodic*.

4.9.8 2π-Periodic Nonstationary Markov Processes

Consider a linear stochastic dynamics under the action of a time-periodic driving force term

$$\frac{dX}{dt} = -(a - \cos \omega t) X + \xi(t), \quad a > 1,$$

here $\xi(t)$ is a zero mean GWN, then **sp** $X(t)$ is Markovian because $\xi(t)$ is white. Under the constraint $a > 1$ we can interpret this periodic modulation as a particle in a time-dependent harmonic potential. The first moment follows in a straightforward form

$$\frac{d \langle X \rangle}{dt} = -(a - \cos \omega t) \langle X \rangle,$$

In general the moments can be calculated from the realizations of the process:

$$X(t) = X(0) \exp\left[-\int_0^t (a - \cos \omega s)\, ds\right] + \int_0^t \exp\left[-\int_{-t'}^t (a - \cos \omega s)\, ds\right] \xi(t')\, dt'.$$

In particular we can calculate the cumulant, using that $\langle \xi(t)\, \xi(t') \rangle = \delta(t - t')$, and $t_2 \geq t_1$ we get

$$\langle\langle X(t_1) X(t_2) \rangle\rangle = \langle X(t_1) X(t_2) \rangle - \langle X(t_1) \rangle \langle X(t_2) \rangle$$

$$= \int_0^{t_1} \exp\left[-a\left(t_1 - t_2'\right) - \left(\sin \omega t_2' - \sin \omega t_1\right)/\omega\right]$$

$$\times \exp\left[-a\left(t_2 - t_2'\right) - \left(\sin \omega t_2' - \sin \omega t_2\right)/\omega\right] dt_2'$$

In the asymptotic regime $\{t_2 \gg 0, t_1 \gg 0\}$, such that $t_2 - t_1 = \tau \geq 0$ we should expect a modulated decreasing periodic behavior for this correlation function (see Sect. 4.8.5). In order to see this behavior it is convenient to introduce the change of variable: $t_2 \rightarrow t_1 + \tau$, therefore the correlation function adopts the simpler expression

$$\langle\langle X(t_1) X(t_1 + \tau) \rangle\rangle = \exp\left[\frac{2}{\omega} \sin \omega\, (t_1 + \tau/2) \cos\left(\frac{\omega \tau}{2}\right)\right]$$

$$\times \int_\tau^{2t_1 + \tau} \frac{dy}{2} \exp\left[-ay - \frac{2}{\omega} \sin \omega\left(t_1 + \frac{\tau}{2} - \frac{y}{2}\right)\right],$$

unfortunately this last integral cannot be solved analytically. Nevertheless taking the limit $t_1 \gg \tau$ it is simple to realize that this correlation is a decreasing function of τ and periodic in the argument t_1, but it is hard to verify if the **sp** $X(t)$ fulfills Doob's Theorem.

Exercise. Calculate the first $\langle X(t) \rangle$ and second moment $\left\langle X(t)^2 \right\rangle$ of the 2π-periodic Gaussian **sp** $X(t)$.

Exercise. Argue that if we could write the correlator of a 2π-periodic Markov process in the form: $\rho(t_2, t_1) = f(t_2 - t_1) g(t_2)/g(t_1)$ where $f(z) \leq 1$ and $g(z + 2\pi/\omega) = g(z)$, the property $\rho(t_3, t_1) = \rho(t_3, t_2)\rho(t_2, t_1)$ can be fulfilled when $f(z) = e^{-\beta z}$, but the condition $|\rho(t_2, t_1)| < 1$ cannot be met because $g(z)$ is periodic.

4.9.9 Time-Ordered Exponential Operator

Consider the following time-ordered exponential operator

$$\mathcal{U}(t, 0) = \overrightarrow{\mathcal{T}} \exp\left(\int_0^t \mathbf{H}(s)\, ds \right), \tag{4.136}$$

where $\mathbf{H}(s)$ is any general time-dependent operator (for a time-dependent matrix, see Chap. 6). The interesting point is to expand watching carefully to keep chronological order in (4.136). This can be done noting:

$$\overrightarrow{\mathcal{T}}\left(1 + \int_0^t \mathbf{H}(s)\,ds + \frac{1}{2!}\left(\int_0^t \mathbf{H}(s)\,ds \right)^2 + \frac{1}{3!}\left(\int_0^t \mathbf{H}(s)\,ds \right)^3 + \cdots \right)$$

$$= 1 + \int_0^t \mathbf{H}(s_1)\,ds_1 + \int_0^t ds_1 \int_0^{s_1} \mathbf{H}(s_1)\mathbf{H}(s_2)\,ds_2$$

$$+ \int_0^t ds_1 \int_0^{s_1} ds_2 \int_0^{s_2} \mathbf{H}(s_1)\mathbf{H}(s_2)\mathbf{H}(s_3)\,ds_3 + \cdots. \tag{4.137}$$

Show that the *inverse* operator can be represented in the form

$$\overleftarrow{\mathcal{T}} \exp\left(-\int_0^t \mathbf{H}(s)\,ds \right) = 1 - \int_0^t \mathbf{H}(s_1)\,ds_1 + \int_0^t ds_1 \int_{s_1}^t \mathbf{H}(s_1)\mathbf{H}(s_2)\,ds_2$$

$$- \int_0^t ds_1 \int_{s_1}^t ds_2 \int_{s_2}^t \mathbf{H}(s_1)\mathbf{H}(s_2)\mathbf{H}(s_3)\,ds_3 + \cdots. \tag{4.138}$$

Therefore, prove that

$$\left(\overrightarrow{\mathcal{T}} \exp\left(\int_0^t \mathbf{H}(s)\,ds \right) \right)^{-1} = \overleftarrow{\mathcal{T}} \exp\left(-\int_0^t \mathbf{H}(s)\,ds \right). \tag{4.139}$$

Hint: to prove (4.137) apply straightforward perturbation theory to the evolution equation: $d\mathcal{U}(t,0)/dt = \mathbf{H}(t) \cdot \mathcal{U}(t,0)$. To prove (4.139) multiply both expansions (4.137) and (4.138) to check

$$\left(\overrightarrow{\mathcal{T}} \exp\left(\int_0^t \mathbf{H}(s)\,ds\right)\right)^{-1} \cdot \overrightarrow{\mathcal{T}} \exp\left(\int_0^t \mathbf{H}(s)\,ds\right) = \mathbf{I},$$

here $\mathbf{I}$ is the identity operator. These results are very useful in solving the problem when $\mathbf{H}(t)$ is a stochastic matrix.

References

1. L.E. Reichl, *A Modern Course in Statistical Physics*, 2nd edn. (Edward Arnold, Austin, 1992)
2. I. Prigogine, *Las Leyes del Caos* (Crítica, Grijalbo Mondadori S.A., Barcelona, 1997); T. Rothman, E.C.G Sudarshan, *Doubt and Certainty* (Perseus Books, Massachusetts, 1998)
3. N.G. van Kampen, *Views of a Physicist, Selected Papers of N.G. van Kampen*, ed. by P.H.E. Meijer (World Scientific, Singapore, 2000)
4. H. Risken, *The Fokker-Planck Equation* (Springer, Berlin, 1984)
5. N.G. van Kampen, *Stochastic Process in Physics and Chemistry*, 2nd edn. (North-Holland, Amsterdam, 1992)
6. L. Onsager, Reciprocal relations in irreversible processes I. Phys. Rev. **37**, 405 (1931)
7. G. Nicolis, I. Prigogine, *Self-Organization in Nonequilibrium Systems* (Wiley, New York, 1977)
8. R.P. Feynman, R.B. Leighton, M. Sands, *The Feynmann Lectures on Physics*, vol. 2 (Fondo Educativo Interamericano, Addison-Wesley, Bogotá, New York, 1972)
9. R.L. Stratonovich, *Topics in the Theory of Random Noise*, vols. 1 and 2 (Gordon and Breach, New York, 1963)
10. C.W. Gardiner, *Handbook of Stochastic Methods* (Springer, Berlin, 1983)
11. J.L. Lebowitz, P.G. Bergmann, Irreversible gibbsian ensembles. Ann. Phys. **1**(1), 1–23 (1957)
12. M.O. Cáceres, A. Becker, L. Kramer, On the asymptotic probability distribution for a supercritical bifurcation swept periodically in time. Phys. Rev. A **43**, 6581–6591 (1991)
13. E.P. Wigner, Derivations of Onsager's reciprocal relations. J. Chem. Phys. **22**, 1912–1915 (1954)
14. W.F. Brown, Thermal fluctuations of a single-domain particle. Phys. Rev. **130**, 1677 (1963)
15. S.R. de Groot, N.G. van Kampen, On the derivation of reciprocal relations between irreversible processes. Physica **21**, 39 (1954)
16. R. Graham, *Springer Tracts in Modern Physics* (Springer, Berlin, 1973)
17. C.P. Enz, Fokker-Planck description of classical systems with application to critical dynamics. Phys. A **89**, 1 (1977)
18. R. Graham, Covariant formulation of non-equilibrium statistical thermodynamics. Z. Phys. **B26**, 397 (1977)
19. Z. Schuss, *Theory and Applications of Stochastic Differential Equations* (Wiley, New York, 1980)
20. R. Graham, T. Tél, Weak-noise limit of Fokker-Planck models and non-differentiable potentials for dissipative dynamical systems. Phys. Rev. **31**, 1109 (1985)
21. W. Horsthemke, R. Lefever, *Noise-Induced Transitions* (Springer, Berlin, 1984)
22. J.H. Shirley, Solution of the Schrödinger equation with a Hamiltonian periodic in time. Phys. Rev. B **138**, 979 (1965)

23. A. Becker, M.O. Cáceres, L. Kramer, Correlation function in stochastic periodically driven instabilities. Phys. Rev. A **46**, R4463 (1992); Correlation function in the order-disorder transition on Raleigh-Bénard convection. Chaos Solitons Fractals **6**, 27 (1995)
24. M.O. Cáceres, A.M. Lobos, Theory of eigenvalues for periodic non-stationary Markov processes: the Kolmogorov operator and its applications. J. Phys. A Math. Gen. **39**, 1547 (2006)
25. B.Ya. Shapiro, I. Dayan, M. Gitterman, G. Weiss, Exact calculation of Shapiro step sizes for pulse-driven Josephson junctions. Phys. Rev. B **46**, 8349 (1992)
26. A.K. Chattah, C.B. Briozzo, O. Osenda, M.O. Cáceres, Signal-to-noise ratio in stochastic resonance. Mod. Phys. Lett. B **10**, 1085 (1996)
27. R. Benzi, A. Sutera, A. Vulpiani, The mechanism of stochastic resonance. J. Phys. A Math. Gen. **14**, 453 (1981)
28. A.K. Chattah, C.B. Briozzo, O. Osenda, M.O. Cáceres, Signal-to-noise ratio in stochastic resonance. Mod. Phys. Lett. B **10**, 1085 (1996)

Chapter 5
Irreversibility and Linear Response

When a system S interacts with an external environment,[1] we may wish to know how the forces associated with such interaction may change the fluctuations in our system of interest. When a system is in a stationary state (or thermodynamic equilibrium) and it is moved away from this equilibrium by applying external forces, it is usually of interest to study the relaxation (return to steady state) when these forces cease to act. Also, if they are weak, it is possible to introduce a perturbation theory in some appropriate smallness parameter. This approach is the origin of linear response theory. In general, this theory can be formulated for systems in equilibrium, as well as in nonequilibrium.[2] In the remainder of this chapter, we will focus more on the first case; however, the Fokker-Planck dynamics also makes it possible to formulate a linear response theory for systems that are far from thermodynamic equilibrium [1, 2].

5.1 Wiener-Khinchin's Theorem

This theorem states the relationship between the time-dependent correlation of the fluctuations and the spectral density of the stationary **sp** [1, 3, 4]. If the vector $\mathbf{X}(t)$ is a **sp** (defined in the time interval $[0, T]$) with zero mean and mean square fluctuation $\langle \mathbf{X}(t)^2 \rangle$, it is reasonable to expand a realization of the **sp** $\mathbf{X}(t)$ in a Fourier series in the same time interval. For example, for the i-th component of vector $\mathbf{X}(t)$, we can

---

[1]The fact that the system is physical, chemical, etc. is not crucial for the development of linear response theory.

[2]That is, the system S may originally be away from thermodynamic equilibrium; still we generally always assume that, at the time of application of external forces, the system was in a steady state.

© Springer International Publishing AG 2017

M.O. Cáceres, *Non-equilibrium Statistical Physics with Application to Disordered Systems*, DOI 10.1007/978-3-319-51553-3_5

write[3]

$$\mathbf{X}_i(t) = \sum_{n=1}^{\infty} \mathbf{X}_i(n) \sin\left(\frac{n\pi}{T}t\right) \tag{5.1}$$

with

$$\mathbf{X}_i(n) = \frac{2}{T} \int_0^T dt\, \mathbf{X}_i(t) \sin\left(\frac{n\pi}{T}t\right), \tag{5.2}$$

where now $\mathbf{X}_i(n)$ are random values depending on the i-th component of the function to be integrated in (5.2) (realization of the **sp** $\mathbf{X}(t)$). Then, a natural question arises: how is the intensity of fluctuations distributed over the $\mathbf{X}_i(n)$ amplitudes? If the process is stationary and ergodic, the answer is given by the Wiener-Khinchin theorem. In order to prove this theorem, first, we introduce a new notation.

We present the problem in terms of the Fourier integral. Let $\mathbf{X}(t, T)$ be a new **sp** related to $\mathbf{X}(t)$ as modified by

$$\mathbf{X}(t, T) = \begin{cases} \mathbf{X}(t) \text{ if } \mid t \mid < T \\ 0 \quad \text{if } \mid t \mid > T, \end{cases} \tag{5.3}$$

in such a way that $\lim_{T\to\infty} \mathbf{X}(t, T) = \mathbf{X}(t)$ for all $t \in (-\infty, +\infty)$. Here $\mathbf{X}(t)$ is, in general, a vector whose components represent fluctuations of state variables present in the mesoscopic description of a system $\mathcal{S}$. The Fourier transform of the **sp** $\mathbf{X}(t, T)$ is well defined and it is given by[4]

$$\mathbf{X}(\omega, T) = \int_{-\infty}^{+\infty} \mathbf{X}(t, T) \exp(i\omega t)\, dt = \int_{-T}^{+T} \mathbf{X}(t, T) \exp(i\omega t)\, dt. \tag{5.4}$$

Since fluctuations are real,[5] we deduce that $\mathbf{X}^*(\omega, T) = \mathbf{X}(-\omega, T)$. The spectral matrix is defined as the intensity of the Fourier mode of $\mathbf{X}(t)$; i.e., in matrix notation (element i, j)

$$S_{\mathbf{X}_i\mathbf{X}_j}(\omega) = \lim_{T\to\infty} \frac{1}{2T} \mathbf{X}_i^*(\omega, T)\mathbf{X}_j(\omega, T). \tag{5.5}$$

---

[3]Here, we use the Fourier sine series since we can always extend and antisymmetrize the function $\mathbf{X}(t)$ to the interval $[-T, T]$.

[4]But not so the integral $\int_{-\infty}^{\infty} e^{i\omega t} \mathbf{X}(t)\, dt$, because the function $\mathbf{X}(t)$ may not necessarily go to zero for$\mid t \mid \to \infty$. This difficulty was overcome by Wiener (1930) in a classical work which was the origin of a whole branch of mathematics called: *Generalized Harmonic Analysis*.

[5]The quantity $\mathbf{X}^*(\omega, T)$ indicates the conjugate of the vector $\mathbf{X}(\omega, T) \in C$.

Using the fact that fluctuations are real and making the change of variable $s = t + \tau$, we can write (5.5) in the form

$$S_{XX}(\omega) = \lim_{T \to \infty} \frac{1}{2T} \int_{-T}^{+T} dt\, \mathbf{X}(t, T) \exp(-i\omega t) \int_{-T}^{+T} ds\, \mathbf{X}(s, T) \exp(i\omega s) \quad (5.6)$$

$$= \lim_{T \to \infty} \int_{-T}^{T} d\tau \int_{-T}^{T} \mathbf{X}(t, T)\mathbf{X}(t + \tau, T) \exp(i\omega \tau)\, \frac{dt}{2T}$$

$$= \int_{-\infty}^{\infty} \exp(i\omega \tau) \left\{ \lim_{T \to \infty} \frac{1}{2T} \int_{-T}^{T} \mathbf{X}(t, T)\mathbf{X}(t + \tau, T)\, dt \right\}\, d\tau.$$

Here $\mathbf{X}(\tau)\mathbf{X}$ represents the dyadic matrix.[6] If the process is stationary and ergodic, we can replace the time average, in brackets, by the average in distribution, that is:

$$\lim_{T \to \infty} \frac{1}{2T} \int_{-T}^{T} \mathbf{X}(t, T)\mathbf{X}(t + \tau, T)\, dt = \langle \mathbf{X}(\tau)\mathbf{X} \rangle_{\mathrm{st}}. \quad (5.7)$$

Then, from (5.6) and (5.7) the Wiener-Khinchin theorem follows:

$$S_{XX}(\omega) = \int_{-\infty}^{\infty} \langle \mathbf{X}(\tau)\mathbf{X} \rangle_{\mathrm{st}} \exp(i\omega \tau)\, d\tau = 2 \int_{0}^{\infty} \langle \mathbf{X}(\tau)\mathbf{X} \rangle_{\mathrm{st}} \cos(\omega \tau)\, d\tau. \quad (5.8)$$

That is, the spectral density of the **sp** $\mathbf{X}(t)$ is the Fourier transform of the correlation function. This theorem has a variety of applications in physics, chemistry, electronics, irreversible thermodynamics, and in the analysis of fluctuations far from equilibrium.[7] One of the most notable applications will be seen in the context of linear response for systems slightly away from thermodynamic equilibrium.

Guided Exercise. Based on the property of a stationary **sp**, for each element of the correlation matrix $\langle \mathbf{X}(\tau)\mathbf{X} \rangle$, we have

$$\langle \mathbf{X}_j(\tau)\mathbf{X}_l(0) \rangle = \langle \mathbf{X}_j(t + \tau)\mathbf{X}_l(t) \rangle = \langle \mathbf{X}_l(-\tau)\mathbf{X}_j(0) \rangle.$$

That is, $\langle \mathbf{X}(\tau)\mathbf{X} \rangle = \langle \mathbf{X}(-\tau)\mathbf{X} \rangle^T$. Then, using $S_{XX}(\omega) = \int_{-\infty}^{\infty} e^{i\omega \tau} \langle \mathbf{X}(\tau)\mathbf{X} \rangle_{\mathrm{st}}\, d\tau$, we deduce that $S_{XX}(\omega) = S_{XX}^*(\omega)^T$, that is, Hermitian. On the other hand, because $\langle \mathbf{X}(\tau)\mathbf{X}(0) \rangle \in \mathcal{R}_e$, we can see that $S_{XX}^*(\omega) = S_{XX}(-\omega)$. Then, we can conclude that the spectral density is a real matrix, symmetric and even in the argument ω.

---

[6] The expression $\mathbf{X}(t)\mathbf{X}(0)$ indicates the matrix formed by the product of all its components, i.e., $\{\mathbf{X}_j(t)\mathbf{X}_l(0)\}$. For example: let $\mathbf{A}$ and $\mathbf{B}$ be two vectors in 2-dimension; then, the dyadic product $\mathbf{AB}$ represents the (2×2) matrix:

$$\mathbf{AB} = \begin{pmatrix} a_1 b_1 & a_1 b_2 \\ a_2 b_1 & a_2 b_2 \end{pmatrix},$$

therefore, obviously $\mathbf{AB} \neq \mathbf{BA}$.

[7] In general, in a nonequilibrium situation, the stationary solution is not the canonical equilibrium distribution.

Exercise. Show from the definition (assuming that the process $\mathbf{X}(t)$ is stationary and ergodic)

$$\langle \mathbf{X}(t,T)\mathbf{X}(t+\tau,T)\rangle_T = \frac{1}{2T}\int_{-T}^{T} \mathbf{X}(t',T)\mathbf{X}(t'+\tau,T)\,dt', \quad \forall\{\tau,t\},$$

and using the inverse transform of (5.4), i.e.,

$$\mathbf{X}(t,T) = \frac{1}{2\pi}\int_{-\infty}^{+\infty} \mathbf{X}(\omega,T)\exp(-i\omega t)\,d\omega,$$

that:

$$\langle \mathbf{X}(0)\mathbf{X}(\tau)\rangle_{\text{st}} = \lim_{T\to\infty} \langle \mathbf{X}(t,T)\mathbf{X}(t+\tau,T)\rangle_T = \frac{1}{2\pi}\int_{-\infty}^{\infty} S_{\mathbf{X}\mathbf{X}}(\omega)\exp(-i\omega\tau)\,d\omega,$$

where

$$S_{\mathbf{X}\mathbf{X}}(\omega) \equiv \lim_{T\to\infty} \mathbf{X}^{*}(\omega,T)\mathbf{X}(\omega,T)\big/2T.$$

Note (Ergodicity) We say that a stationary **sp** is ergodic in mean:

$$\text{ms} - \lim_{T\to\infty} \frac{1}{2T}\int_{-T}^{T} \mathbf{X}(t)\,dt = \langle \mathbf{X}\rangle_{\text{st}},$$

if the mean-value of the **sp** converges in mean square.[8] This convergence is met if the correlation function of the **sp** decays to zero fast enough, i.e.:

$$\int_{0}^{\infty} |\langle\langle \mathbf{X}(t)\mathbf{X}(t+\tau)\rangle\rangle|\,d\tau < \infty. \tag{5.9}$$

A sufficient condition for this inequality to be fulfilled is that the correlation function decays exponentially, see advanced exercises Sect. 5.5.

5.2   Linear Response, Susceptibility

Suppose we apply external forces $\mathbf{F} = (F_1, F_2, \cdots, F_n)$ to a system S at a given time. These forces will be coupled to state variables $(A_1, A_2, \cdots, A_n)$ and will cause variations around their equilibrium values. If these variations are small and proportional to the applied forces (linear response), we can write (in general $\mathbf{Y}(t)$ is a vector of n components)

---

[8]On the different definitions of convergence, see Sect. 3.8, and reference [5].

$$\langle \mathbf{Y}(t) \rangle_{\mathbf{F}} = \int_{-\infty}^{+\infty} \mathbf{K}(t - t') \cdot \mathbf{F}(t') \, dt' = \int_{-\infty}^{+\infty} \mathbf{K}(\tau) \cdot \mathbf{F}(t - \tau) \, d\tau. \tag{5.10}$$

Note that the variation at time t, that is $\langle \mathbf{Y}(t) \rangle_{\mathbf{F}}$, can depend only on previous forces $\mathbf{F}(t - \tau)$ (principle of causality). Actually, for the n components of $\langle \mathbf{Y}(t) \rangle_{\mathbf{F}}$, the matrix $\mathbf{K}(t)$ is real ($n \times n$) and commonly referred to as the response matrix of the system $\mathcal{S}$. Here, $\langle \mathbf{Y}(t) \rangle_{\mathbf{F}}$ represents the variations at time t in the presence of forces $\mathbf{F}$ acting since times before t. By the principle of causality, it follows that $\mathbf{K}(\tau) = 0$ if $\tau < 0$. On the other hand, the Fourier transform of (5.10) will be:

$$\langle \mathbf{Y}(\omega) \rangle_{\mathbf{F}} = \int_{-\infty}^{+\infty} e^{i\omega t} \langle \mathbf{Y}(t) \rangle_{\mathbf{F}} \, dt \tag{5.11}$$

$$= \mathcal{F}_{\omega}[\mathbf{K}(t)] \cdot \mathcal{F}_{\omega}[\mathbf{F}(t)]$$

$$\equiv \chi(\omega) \cdot \mathbf{F}(\omega),$$

note that $\chi(\omega)$ generally is a matrix; then, $\chi(\omega) \cdot \mathbf{F}(\omega)$ indicates a matrix product with the vector $\mathbf{F}(\omega)$.

Equation (5.11) indicates that Fourier's mode $\mathbf{F}(\omega)$ (the external force) will cause a response only at the same frequency ω. This is only true under the assumption of linear response, the quantity $\chi(\omega)$ is called susceptibility and only depends on fluctuations near equilibrium.[9] The Fluctuation-Dissipation theorem (microscopic), proved by Kubo in 1957 [6], and independently by Lax in 1958 [7], establishes the important relationship between the susceptibility and temporal correlations in equilibrium [8]. This remarkable theorem shows the excellence of the simple, fundamental, and original ideas established by Onsager [9]:

> If a system is taken away from equilibrium by the action of small external forces, temporal fluctuations caused by such an action should not be very different from the fluctuations around equilibrium.

Examples:

1. "Dirac-δ" pulse. Assuming that the time dependence of the external force is an infinite pulse at time $t = 0$, that is, $\mathbf{F}(t) = \mathbf{F}\delta(t)$, and $\mathbf{K}(\tau)$ is the response function of $\mathcal{S}$ (for example, a paramagnetic response); then, the variation of the magnetization in the presence of a Dirac-δ force at $t = 0$ will be:

$$\langle \mathbf{Y}(t) \rangle_{\delta} = \int_{-\infty}^{+\infty} \delta(t - t') \, \mathbf{K}(t') \cdot \mathbf{F} \, dt' = \mathbf{K}(t) \cdot \mathbf{F}. \tag{5.12}$$

---

[9]That is, the response function of the system $\mathbf{K}(t)$ depends only on fluctuations around equilibrium.

2. "Step" pulse. Assuming that the time dependence of the external force is a step function of intensity $| \mathbf{F} |$ starting at time $t = 0$ [i.e., $\mathbf{F}(t) = \mathbf{F}\,\Theta(t)$] and the response function of the system is as in the previous example, $\mathbf{K}(\tau)$; the variation of the magnetization in the presence of a step force will be[10]

$$\langle \mathbf{Y}(t) \rangle_{\Theta} = \int_{-\infty}^{+\infty} \Theta(t - t')\, \mathbf{K}(t') \cdot \mathbf{F}\, dt' = \int_0^t \mathbf{K}(t') \cdot \mathbf{F}\, dt'. \qquad (5.13)$$

Then, from (5.12) and (5.13) we see the following relationship between the different temporal variations:

$$\frac{d}{dt} \langle \mathbf{Y}(t) \rangle_{\Theta} = \mathbf{K}(t) \cdot \mathbf{F} = \langle \mathbf{Y}(t) \rangle_{\delta}. \qquad (5.14)$$

Exercise. Suppose that an external field $\mathbf{H}$ (constant) is applied from $t = 0$ to a paramagnetic substance, i.e., $\mathbf{F}(t) = \mathbf{H}\,\Theta(t)$. If the variation of the magnetization is $\langle \mathbf{Y}(t) \rangle_{\Theta} = \mathbf{m}\left(1 - e^{-t/\tau}\right)$, use (5.14) and calculate what the change in the magnetization would be, for the same $\mathcal{S}$ system, if the external force were a Dirac-δ force.

5.2.1 The Kramers-Kronig Relations*

Experimentally, only the imaginary part of the susceptibility $\chi(\omega)$ is determined since it is directly related to energy dissipation. For this reason, it is important to note that Kramers-Kronig's relation establishes a connection between the imaginary part and the real part of the susceptibility $\chi(\omega) = \chi_1(\omega) + i\chi_2(\omega)$, namely:

$$\chi_1(u) = \frac{1}{\pi} \mathcal{P} \int_{-\infty}^{\infty} \frac{\chi_2(\omega)}{\omega - u}\, d\omega \qquad (5.15)$$

$$\chi_2(u) = \frac{-1}{\pi} \mathcal{P} \int_{-\infty}^{\infty} \frac{\chi_1(\omega)}{\omega - u}\, d\omega,$$

with $u \in \mathcal{R}_e$ (symbol $\mathcal{P} \int_{-\infty}^{\infty} \cdots$ means Cauchy's principal value).

To prove these relationships, it is only necessary to use the principle of causality and the concept of analytic continuation in the complex plane (ε, ω). Then, it follows that using $z = \omega + i\varepsilon$ with $\varepsilon > 0$ in the Fourier transform of the response function $\mathbf{K}(\tau)$, and the fact that $\mathbf{K}(-| \tau |) = 0$, we obtain:

$$\chi(z) = \int_0^{\infty} \mathbf{K}(t) \exp(izt)\, dt = \int_0^{\infty} \mathbf{K}(t)\, e^{-\varepsilon t} \exp(i\omega t)\, dt. \qquad (5.16)$$

---

[10]If working with frequencies, it is convenient to use the spectral representation of $\Theta(t)$. See note (12).

By separating the real from the imaginary part it follows that

$$\chi_1(\varepsilon, \omega) + i\chi_2(\varepsilon, \omega) = \int_0^\infty \mathbf{K}(t)\, e^{-\varepsilon t} \cos(\omega t)\, dt + i \int_0^\infty \mathbf{K}(t)\, e^{-\varepsilon t} \sin(\omega t)\, dt. \tag{5.17}$$

From this expression it is easy to see that the function $\chi(z)$ is analytic because it meets Cauchy's condition. That is, if $\varepsilon \geq 0$, which is the condition for the integral (5.16) to be well defined

$$\frac{\partial \chi_1}{\partial \omega} = \frac{\partial \chi_2}{\partial \varepsilon} \quad \text{and} \quad \frac{\partial \chi_1}{\partial \varepsilon} = -\frac{\partial \chi_2}{\partial \omega}. \tag{5.18}$$

That is, $\chi(z)$ is analytic in the upper half plane (ε, ω). Defining the auxiliary function

$$f(z) = \frac{\chi(z)}{z - u}, \tag{5.19}$$

with $u \in \mathcal{R}_e$, we note that

$$\oint_C f(z)\, dz = 0, \tag{5.20}$$

if the contour C is in the upper half-plane (ε, ω) and if, in addition, we "avoid" the simple pole at u on the real axis; see Fig. 5.1. Choosing this contour for (5.20), it follows that we can write the following expression:

$$0 = \oint_C \frac{\chi(z)}{z - u}\, dz = \left[\int_{-\infty}^{u-r} \frac{\chi(\omega)}{\omega - u}\, d\omega + \int_{u+r}^\infty \frac{\chi(\omega)}{\omega - u}\, d\omega \right] + \tag{5.21}$$

$$+ \int_\pi^0 \frac{\chi(u + re^{i\phi})}{u + re^{i\phi} - u} i r e^{i\phi}\, d\phi + \int_0^\pi \frac{\chi(Re^{i\phi})}{Re^{i\phi} - u} i R e^{i\phi}\, d\phi.$$

Fig. 5.1 Diagrammatic representation of the contour of the line integral C. The radius of the large semicircle is R while the radius of the semicircle centered in u is r

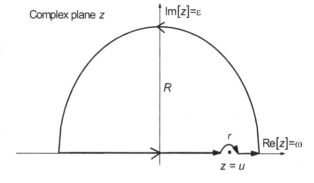

In the limit $R \to \infty$ the last term vanishes because $\lim_{R\to\infty} \chi(Re^{i\phi}) \to 0$, on the semicircular contour of the upper plane.[11] On the other hand, in the limit $r \to 0$ (small semicircle), we just get Cauchy's principal value of the integral.

$$\mathcal{P} \int_{-\infty}^{\infty} d\omega \, \frac{\chi(\omega)}{\omega - u} \equiv \lim_{r\to 0} \left[\int_{-\infty}^{u-r} \frac{\chi(\omega)}{\omega - u} \, d\omega + \int_{u+r}^{\infty} \frac{\chi(\omega)}{\omega - u} \, d\omega \right]$$

$$= -\lim_{r\to 0} \left[\int_{\pi}^{0} \frac{\chi(u + re^{i\phi})}{u + re^{i\phi} - u} ire^{i\phi} \, d\phi \right] = \lim_{r\to 0} i \int_{0}^{\pi} \chi(u + re^{i\phi}) \, d\phi$$

$$= i\pi\chi(u),$$

and noting that $\chi(u) = \chi_1(u) + i\chi_2(u)$, it follows that

$$\chi_1(u) + i\chi_2(u) = \frac{1}{i\pi} \mathcal{P} \int_{-\infty}^{\infty} \frac{\chi_1(\omega) + i\chi_2(\omega)}{\omega - u} \, d\omega.$$

Taking the real and the imaginary part of this complex equation, Kramers-Kronig's relations (5.15) are obtained.

Note From the fact that $\mathbf{K}(t)$ is real, it follows that $\chi(\omega) = \chi(-\omega)^*$. Then,

$$\chi_1(\omega) + i\chi_2(\omega) = \chi_1(-\omega) - i\chi_2(-\omega),$$

from which it follows that the imaginary part is odd and the real part is even.

Example (**"Dirac-δ" absorption**). Suppose that the imaginary part of the susceptibility has a simple structure with two peaks (Dirac delta functions) at frequencies $\omega = \pm\Omega$. That is, the imaginary part of the susceptibility could have a structure of the form:

$$\chi_2(\omega) = \text{constant} \, (\delta(\omega - \Omega) - \delta(-\omega - \Omega)).$$

Using one of the Kramers-Kronig relations, we can get the real part of the susceptibility as:

$$\chi_1(u) = \frac{\text{constant}}{\pi} \mathcal{P} \int_{-\infty}^{\infty} \frac{(\delta(\omega - \Omega) - \delta(-\omega - \Omega))}{\omega - u} \, d\omega$$

$$= \text{constant} \left(\frac{1}{\Omega - u} + \frac{1}{\Omega + u} \right).$$

Thus as expected, χ_1 is an even function. In Fig. 5.2 both functions are shown.

---

[11] To see this fact use: $z = Re^{i\phi}$ in (5.16).

Fig. 5.2 The real (*solid line*) and imaginary parts (*arrows*) in the example of Dirac-δ absorption. The arrow at frequency $\omega = \Omega$ represents the absorption spectrum of the system; the second arrow anti-symmetrizes the imaginary part of the susceptibility $\chi(\omega)$

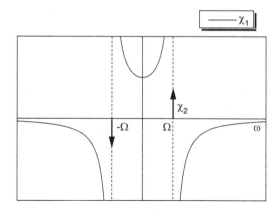

5.2.2 Relaxation Against a Discontinuity at $t = 0$

We now will apply the linear response theory to study explicitly the time dependence of fluctuations in the presence of a constant force **F** which is *turned off* at time $t = 0$; that is to say

$$\mathbf{F}(t) = \begin{cases} \mathbf{F} & \text{if} \quad t < 0 \\ 0 & \text{if} \quad t > 0. \end{cases} \tag{5.22}$$

Noting that the Fourier transform of $\mathbf{F}(t)$ is

$$\mathbf{F}(\omega) = \int_{-\infty}^{\infty} \mathbf{F}(t) \exp{(i\omega t)} \; dt = \mathbf{F} \int_{-\infty}^{0} \exp{(i\omega t)} \; dt, \tag{5.23}$$

it follows that $\mathbf{F}(\omega)$ can also be written in terms of the Laplace transform $\mathcal{L}_u [\bullet]$, namely:

$$\mathbf{F}(\omega) = \mathbf{F} \lim_{u \to 0^+} \int_0^{\infty} \exp{(-i\omega t)} \; e^{-ut} \; dt = \mathbf{F} \lim_{u \to 0^+} \mathcal{L}_u [\exp{(-i\omega t)}] \tag{5.24}$$

$$= \mathbf{F} \lim_{u \to 0^+} \left(\frac{1}{i\omega + u} \right).$$

This is equivalent to[12]

$$\mathbf{F}(\omega) = \mathbf{F} \lim_{u \to 0} \left(\frac{-i}{\omega - iu} \right) = \left[\mathcal{P} \left(\frac{1}{i\omega} \right) + \pi \delta(\omega) \right] \mathbf{F}, \tag{5.25}$$

---

[12]This expression corresponds to the spectral representation of the step function:

$$\theta(t - t') = - \lim_{u \to 0^+} \int_{-\infty}^{+\infty} \frac{e^{-i\omega(t-t')}}{\omega + iu} \frac{d\omega}{2\pi i}.$$

where we have used the *operational* identity [6]

$$\lim_{u \to 0} \frac{1}{\omega' - \omega \mp iu} = \mathcal{P}\left(\frac{1}{\omega' - \omega}\right) \pm i\pi\delta(\omega' - \omega). \tag{5.26}$$

Using (5.25) and the inverse Fourier transform of (5.11), the following time dependent behaviors are obtained (see guided exercise):

$$\langle \mathbf{Y}(t) \rangle_{\mathbf{F}} = \chi(0) \cdot \mathbf{F} \qquad\qquad\qquad \text{if} \qquad t < 0$$

$$\tag{5.27}$$

$$\langle \mathbf{Y}(t) \rangle_{\mathbf{F}} = \frac{1}{i\pi} \mathcal{P} \int_{-\infty}^{\infty} \frac{\chi(\omega) \cdot \mathbf{F}}{\omega} \cos(\omega t) \, d\omega \qquad \text{if} \qquad t > 0.$$

That is, when the constant force $\mathbf{F}$ is *acting*, the system behavior is constant over time; while for $t > 0$ (when the force is turned off at $t = 0$), the system behavior is time dependent: *the state variables relax to their equilibrium values.* For increasing times, the $\cos(\omega t)$ factor in the integrand of (5.27) makes the function $\langle \mathbf{Y}(t) \rangle_{\mathbf{F}}$ monotonically decreasing.

Exercise. Using $\mathbf{F}(t) = \mathbf{F}\,\Theta(-t)$ in (5.10), show that the temporal variation $\langle \mathbf{Y}(t) \rangle_{\mathbf{F}}$ in the presence of the step force (5.22) can be expressed in the form:

$$\langle \mathbf{Y}(t) \rangle_{\mathbf{F}} = \left(\int_0^\infty \mathbf{K}(t') \, dt'\right) \cdot \mathbf{F} \qquad \text{if} \qquad t < 0$$

$$\langle \mathbf{Y}(t) \rangle_{\mathbf{F}} = \left(\int_t^\infty \mathbf{K}(t') \, dt'\right) \cdot \mathbf{F} \qquad \text{if} \qquad t > 0.$$

Compare with (5.27) and physically interpret this result.

Guided Exercise (**Proof of 5.27**). To prove this result we need to define two auxiliary functions: $f_+(z, t)$ and $f_-(z, t)$. Then, for $z = \omega + i\varepsilon$, it follows that

$$f_+(z, t) = \frac{e^{izt}\chi(z)}{z - u} \implies \lim_{\varepsilon \to \infty} f_+(z, t) = \frac{e^{(i\omega - \varepsilon)t}\chi(\omega + i\varepsilon)}{(\omega + i\varepsilon - u)} \to 0 \qquad \text{if} \qquad t > 0$$

$$\tag{5.28}$$

$$f_-(z, t) = \frac{e^{-izt}\chi(z)}{z - u} \implies \lim_{\varepsilon \to \infty} f_-(z, t) = \frac{e^{-(i\omega - \varepsilon)t}\chi(\omega + i\varepsilon)}{(\omega + i\varepsilon - u)} \to 0 \qquad \text{if} \qquad t < 0.$$

Then, integrating over a closed path the functions $f_\pm(z, t)$ (using their analytical conditions depending on the sign of t), in the same way as in Sect. 5.2.1, the following relations are obtained:

$$e^{iut}\chi(u) = \frac{1}{i\pi}\mathcal{P}\int_{-\infty}^{\infty}\frac{e^{i\omega't}\chi(\omega')}{(\omega'-u)}\,d\omega' \qquad \text{if} \qquad t>0$$

$$\tag{5.29}$$

$$e^{-iut}\chi(u) = \frac{1}{i\pi}\mathcal{P}\int_{-\infty}^{\infty}\frac{e^{-i\omega't}\chi(\omega')}{(\omega'-u)}\,d\omega' \qquad \text{if} \qquad t<0.$$

The time dependence of the variation $\langle\mathbf{Y}(t)\rangle_{\mathbf{F}}$ is given by the inverse Fourier transform; that is to say, using (5.11) it must be that

$$\langle\mathbf{Y}(t)\rangle_{\mathbf{F}} = \int_{-\infty}^{\infty}\langle\mathbf{Y}(\omega)\rangle_{\mathbf{F}}\,e^{-i\omega t}\,\frac{d\omega}{2\pi} = \int_{-\infty}^{\infty}\chi(\omega)\cdot\mathbf{F}(\omega)\,e^{-i\omega t}\,\frac{d\omega}{2\pi}. \tag{5.30}$$

Moreover, replacing $\mathbf{F}(\omega) = \left[\mathcal{P}\left(\frac{1}{i\omega}\right) + \pi\delta(\omega)\right]\mathbf{F}$ in (5.30), we obtain

$$\langle\mathbf{Y}(t)\rangle_{\mathbf{F}} = \frac{1}{2\pi}\int_{-\infty}^{\infty}\left[\mathcal{P}\left(\frac{1}{i\omega}\right) + \pi\delta(\omega)\right]\chi(\omega)\cdot\mathbf{F}\,e^{-i\omega t}\,d\omega$$

$$= \left\{\mathcal{P}\int_{-\infty}^{\infty}\chi(\omega)\left(\frac{1}{i\omega}\right)e^{-i\omega t}\,\frac{d\omega}{2\pi} + \int_{-\infty}^{\infty}\chi(\omega)\delta(\omega)\,e^{-i\omega t}\,\frac{d\omega}{2}\right\}\cdot\mathbf{F} \tag{5.31}$$

$$= \frac{1}{2\pi}\mathcal{P}\int_{-\infty}^{\infty}\frac{\chi(\omega)\cdot\mathbf{F}}{i\omega}\,e^{-i\omega t}\,d\omega + \frac{\chi(\omega=0)\cdot\mathbf{F}}{2}.$$

From (5.29) we observe that

$$\chi(0) = \frac{1}{i\pi}\mathcal{P}\int_{-\infty}^{\infty}\frac{e^{i\omega't}\chi(\omega')}{\omega'}\,d\omega' \qquad \text{if} \qquad t>0$$

$$\tag{5.32}$$

$$\chi(0) = \frac{1}{i\pi}\mathcal{P}\int_{-\infty}^{\infty}\frac{e^{-i\omega't}\chi(\omega')}{\omega'}\,d\omega' \qquad \text{if} \qquad t<0;$$

then, (5.31) can be written in two ways according to t being positive or negative. Using $\chi(0)$ from (5.32), we have for positive times:

$$\langle\mathbf{Y}(t)\rangle_{\mathbf{F}} = \frac{1}{2\pi}\mathcal{P}\int_{-\infty}^{\infty}\frac{\chi(\omega)\cdot\mathbf{F}}{i\omega}\,e^{-i\omega t}\,d\omega + \frac{1}{2\pi i}\mathcal{P}\int_{-\infty}^{\infty}\frac{e^{i\omega't}\chi(\omega')\cdot\mathbf{F}}{\omega'}\,d\omega' \tag{5.33}$$

$$= \frac{1}{i\pi}\mathcal{P}\int_{-\infty}^{\infty}\frac{\chi(\omega)\cdot\mathbf{F}}{\omega}\cos(\omega t)\,d\omega \qquad \text{if} \qquad t>0.$$

While for negative times, we obtain

$$\langle \mathbf{Y}(t)\rangle_{\mathbf{F}} = \frac{1}{2\pi}\mathcal{P}\int_{-\infty}^{\infty}\frac{\chi(\omega)\cdot\mathbf{F}}{i\omega}\,e^{-i\omega t}\,d\omega + \frac{1}{2\pi i}\mathcal{P}\int_{-\infty}^{\infty}\frac{e^{-i\omega' t}\chi(\omega')\cdot\mathbf{F}}{\omega'}\,d\omega'$$

(5.34)

$$= \frac{1}{i\pi}\mathcal{P}\int_{-\infty}^{\infty}\frac{\chi(\omega)\cdot\mathbf{F}}{\omega}\,e^{-i\omega t}\,d\omega = \chi(0)\cdot\mathbf{F} \qquad \text{if} \qquad t < 0,$$

which are the expressions we were looking for.

Each system $\mathcal{S}$ is characterized by a susceptibility $\chi(\omega)$. As discussed below, this function is calculated studying fluctuations (correlations) at equilibrium; a fact that greatly reduces the effort required to obtain the time dependence of a system out of equilibrium by the action of external forces. In the next section we will discuss a simple example: a classical particle in a harmonic potential immersed in a fluid (thermal bath) at temperature T. Then, using linear response theory, we can study fluctuations when the particle deviates from equilibrium by the action of an arbitrary external force $\mathbf{F}(t)$.

Exercise. Show that for $t > 0$, that the variation $\langle \mathbf{Y}(t)\rangle_{\mathbf{F}}$ in the presence of a step force (5.22) can be expressed in terms of the imaginary part of $\chi(\omega)$ in the following way

$$\langle \mathbf{Y}(t)\rangle_{\mathbf{F}} = \frac{1}{\pi}\mathcal{P}\int_{-\infty}^{\infty}\frac{\chi_2(\omega)\cdot\mathbf{F}}{\omega}\cos(\omega t)\,d\omega.$$

(5.35)

5.2.3 Power Dissipation

Part of the work done by external forces on the $\mathcal{S}$ system is transformed by it into thermal energy, which is then dissipated to the environment. The work done by $\mathbf{F}(t)$ to change $\mathbf{Y}(t)$ by an amount $d\mathbf{Y}$ is given by[13]

$$dW = -\mathbf{F}(t)\cdot d\mathbf{Y}.$$

(5.36)

The average power absorbed by the environment is just the average of dW/dt, i.e.:

$$\mathcal{P}(t) \equiv \left\langle\frac{dW}{dt}\right\rangle_{\mathbf{F}} = -\mathbf{F}(t)\cdot\left\langle\frac{d\mathbf{Y}}{dt}\right\rangle_{\mathbf{F}}$$

(5.37)

$$= -\mathbf{F}(t)\cdot\frac{d}{dt}\int_{-\infty}^{\infty}\mathbf{K}(t-t')\cdot\mathbf{F}(t')\,dt'.$$

---

[13]The minus sign comes from considering the work done on environment. Also, note that the dimension of the thermodynamic variables $\mathbf{Y}$ and the generalized forces $\mathbf{F}(t)$ must be such that $[\mathbf{F}(t)\cdot d\mathbf{Y}]$ has the dimension of energy.

This equation can also be written in terms of Fourier transforms of $\mathbf{K}(\tau)$ and $\mathbf{F}(\tau)$, i.e.:

$$\mathcal{P}(t) = i \left(\frac{1}{2\pi}\right)^2 \int_{-\infty}^{+\infty} d\omega \int_{-\infty}^{+\infty} d\omega'\, \omega'\, \mathbf{F}(\omega) \cdot \chi(\omega') \cdot \mathbf{F}(\omega')\, \exp\left(-i(\omega + \omega')t\right).$$
(5.38)

From expression (5.38), it is easy to check that the power dissipated by the system $\mathcal{S}$ is proportional to the imaginary part of the susceptibility $\chi(\omega)$. To understand this fact, we propose the following exercises:

Exercise (**Singular External Excitation**). Let $\mathbf{F}(t) = \mathbf{F}\,\delta(t)$ be an external Dirac-δ force in time (hence $\mathbf{F}(\omega) = \mathbf{F} = $ constant). Using (5.38) and making use of the symmetries of $\chi(\omega)$, show that the average of the total absorbed energy is

$$W_{\text{abs}} = \int_{-\infty}^{+\infty} \mathcal{P}(t)\, dt = \frac{-1}{2\pi} \int_{-\infty}^{+\infty} \omega\, \chi_2(\omega) \circ \mathbf{FF}\, d\omega.$$
(5.39)

Here $\chi_2(\omega) \circ \mathbf{FF}$ denotes the matrix product $\mathbf{F} \cdot \mathcal{I}_m[\chi(\omega)] \cdot \mathbf{F}$. The negative sign comes from considering the power absorbed by the medium.

Exercise (**Monochrome Excitation**). Consider a monochromatic external force $\mathbf{F}(t) = \mathbf{F}\cos(\Omega_0 t)$. Then, $\mathbf{F}(\omega) = \pi\,[\delta(\omega + \Omega_0) + \delta(\omega - \Omega_0)]\mathbf{F}$. Show that the absorbed power varies in time according to

$$\mathcal{P}(t) = \frac{i}{4} \left[\Omega_0 \left(e^{-i2\Omega_0 t} + 1\right) \chi(\Omega_0) - \Omega_0 \left(e^{i2\Omega_0 t} + 1\right) \chi(-\Omega_0)\right] \circ \mathbf{FF}.$$

Show that the average absorbed power in a cycle is proportional to $\chi_2(\Omega_0)$ and is given by

$$\langle \mathcal{P}(t) \rangle_{\text{cycle}} \equiv \frac{\Omega_0}{2\pi} \int_0^{2\pi/\Omega_0} \mathcal{P}(t)\, dt = -\frac{\Omega_0}{2}\, \chi_2(\Omega_0) \circ \mathbf{FF}.$$
(5.40)

Since this power consumption can be measured experimentally, it is possible to obtain $\chi_2(\Omega_0)$ as a function of Ω_0 and, by using the Kramers-Kronig relations, we can rebuild the real part of the susceptibility, i.e., $\chi_1(\Omega_0)$.

5.3 Dissipation and Correlations

5.3.1 Brownian Particle in a Harmonic Potential

Consider, as system $\mathcal{S}$, a classical particle immersed in a fluid at temperature T and linked to the container wall by a spring with natural frequency ω_o.

If the particle is massive, the Brownian particle approximation is applicable; then, we denote the degrees of freedom of the fluid through a white Gaussian **sp** $\xi(t)$ with zero mean and intensity one. In this way, the medium (or fluid) is taken into account by the action of two forces on the particle: *the friction* proportional to the velocity of the massive particle $-m\Gamma\dot{X}(t)$, and the other term corresponding to a stochastic force $\sqrt{\epsilon}\,\xi(t)$, where the coefficient $\epsilon = 2m\Gamma k_B T$ ensures the detailed balance principle.

The dynamics of the particle (in 1-dimension) will be represented by a **sde** which, in the presence of an external scalar force $F(t)$, is

$$m\ddot{X}(t) + m\omega_o^2 X(t) = -m\Gamma\dot{X}(t) + \sqrt{\epsilon}\,\xi(t) + F(t). \tag{5.41}$$

The external force may generally be an arbitrary function of time. Applying the **mv** over the ensemble of all realizations of **sp** $\xi(t)$ on (5.41) and in the presence of external force $F(t)$, we get:

$$m\left\langle\ddot{X}(t)\right\rangle_F + m\omega_o^2\left\langle X(t)\right\rangle_F = -m\Gamma\left\langle\dot{X}(t)\right\rangle_F + F(t). \tag{5.42}$$

Using linear response theory, from (5.10), (5.11) and particularizing for a time dependent external force of the form[14] $F(t) = F\delta(t)$, we get

$$\langle X(t)\rangle_F = K(t)F, \tag{5.43}$$

or in Fourier representation

$$\langle X(\omega)\rangle_F = \chi(\omega)F. \tag{5.44}$$

On the other hand, taking the Fourier transform of (5.42) and comparing with (5.44), we immediately obtain that:

$$\chi(\omega) = \frac{1}{m\left(\omega_o^2 - \omega^2 - i\Gamma\omega\right)}. \tag{5.45}$$

This susceptibility, $\chi(\omega)$, can be separated into real and imaginary parts, and thus check Kramers-Kronig's relations. In the context of linear response theory, formula (5.45) enables us to study the fluctuations of the system $\mathcal{S}$ against any external excitation $F(t)$.

Exercise (**Dissipation**). Show that the imaginary part of the susceptibility (5.45) tends to zero if the friction coefficient decreases sufficiently. That is, the medium does not absorb energy when $\Gamma = 0$.

---

[14]As simple as possible, remember that $\chi(\omega)$ does not depend on the choice of $F(t)$.

Exercise. Consider the external force $\mathbf{F}(t) = \mathbf{F}\,\delta(t)$ applied to the same system. Using (5.45), show that the average of the total energy absorbed by the medium is

$$
W_{\text{abs}} = \int_{-\infty}^{+\infty} \mathcal{P}(t)\, dt =
$$

$$
= \frac{-1}{2\pi} \int_{-\infty}^{+\infty} \frac{\Gamma F^2 \omega^2 / m}{\left(\omega_o^2 - \omega^2\right)^2 + \Gamma^2 \omega^2}\, d\omega.
$$

Optional Exercise (**Response Function for Harmonic Brownian Motion**). Calculate the response function $\mathbf{K}(t)$ of the physical system characterized by **sde** (5.41). Show that for $t < 0$ we have that $\mathbf{K}(t) = 0$ (the line integral has to be closed in the upper half plane and there are no poles!). Show that for $t > 0$

$$
\mathbf{K}(t) = \frac{e^{-\Gamma t/2} \sin\left(t\sqrt{\omega_o^2 - \frac{1}{4}\Gamma^2}\right)}{m\sqrt{\omega_o^2 - \frac{1}{4}\Gamma^2}},
$$

is obtained. Hint: the poles are $2\omega_\pm = -i\Gamma \pm (4\omega_o^2 - \Gamma^2)^{1/2}$ and the line integral must be closed in the lower half plane for $t > 0$.

Exercise. Study fluctuations of a harmonic Brownian motion when the external excitation is monochromatic: $\mathbf{F}(t) = \mathbf{F}\cos(\Omega_0 t)$.

5.3.2 Brownian Particle in the Presence of a Magnetic Field

Consider a plasma of particles at temperature T and in the presence of a constant magnetic field $\mathbf{B}$. As system S we take a *test* particle of charge q_e and mass m in the presence of the field $\mathbf{B}$ (in the z direction, i.e., $\mathbf{B} = B\hat{\mathbf{z}}$), and under the influence of random particle collisions from the medium (the plasma) [10, 11]. Newton's equation for this test particle will have a contribution of the Lorentz force $\propto \mathbf{V} \times \mathbf{B}$ in addition to the stochastic collisions term $\mathbf{a}(t)$ which produces random accelerations, and a frictional force $-\nu\mathbf{V}$. That is, we can write an **sde** for the velocity vector $\mathbf{V}$ of the test particle in the form

$$
\frac{d\mathbf{V}}{dt} = \Omega\, \mathbf{V} \times \hat{\mathbf{z}} - \nu\, \mathbf{V} + \mathbf{a}(t), \tag{5.46}
$$

where $\Omega = q_e B / mc$ (c = speed of light) is the Larmor frequency[15]; and the stochastic vector $\mathbf{a}(t)$ is a GWN of zero mean $[\langle a_j(\tau_1)a_l(\tau_2)\rangle = \epsilon\,\delta_{j,l}\,\delta(\tau_1 - \tau_2)]$.

---

[15]See related models in the exercises of Sects. 3.20.10 and 3.20.14.

Without the stochastic term $\mathbf{a}(t)$, the particle dynamics describes a damped helical motion about the axis of symmetry $\hat{\mathbf{z}}$.

Equation (5.46) can be written in matrix form:

$$\frac{d\mathbf{V}}{dt} = \mathbf{A} \cdot \mathbf{V} + \mathbf{a}(t), \tag{5.47}$$

where the matrix $\mathbf{A}$ is defined by

$$\mathbf{A} = \begin{pmatrix} -v & \Omega & 0 \\ -\Omega & -v & 0 \\ 0 & 0 & -v \end{pmatrix},$$

then, we note that equation (5.47) is of a multidimensional Ornstein-Uhlenbeck type [see Sect. 4.4], i.e., this **sp** is fully characterized from the solution of the corresponding **F-P** equation.

Exercise. Find the **F-P** operator corresponding to the **sde** (5.47).

To solve this system, in this case, we will use an alternative method[16] to the **F-P** scheme. This method introduces naturally the concept of Green's function which will be frequently used in the next chapters. Instead of solving equation (5.47), we propose the following matrix problem

$$\frac{d\mathbf{G}}{dt} = \mathbf{A} \cdot \mathbf{G}, \quad \text{with} \quad \mathbf{G}(t = 0) = \mathbf{1}, \tag{5.48}$$

where $\mathbf{1}$ is the identity matrix. The solution of (5.48) is[17]

$$\mathbf{G}(t) = e^{-vt} \begin{pmatrix} \cos \Omega t & \sin \Omega t & 0 \\ -\sin \Omega t & \cos \Omega t & 0 \\ 0 & 0 & 1 \end{pmatrix}. \tag{5.49}$$

Then, the solution of the inhomogeneous problem (5.47) is given by

$$\mathbf{V}(t) = \mathbf{G}(t) \cdot \mathbf{V}(0) + \int_0^t \mathbf{G}(t - \tau) \cdot \mathbf{a}(\tau) \, d\tau. \tag{5.50}$$

Exercise. Show by direct substitution and using (5.48), that (5.50) is a solution of (5.47).

Due to the symmetry of the problem, the correlation perpendicular to the field **B** can be analyzed with any V_x or V_y, component, whereas the correlation in the

---

[16]This method is similar to that used in Chap. 3 when the Ornstein-Uhlenbeck **sp** in 1-dimension was solved. Note that the Green function method is only applicable to linear problems.

[17]This fact can be easily checked by direct substitution.

direction parallel to the applied field will be different. Consider, then, one of the perpendicular correlations $\langle V_x(t_1)V_x(t_2)\rangle$. If we assume that initially the plasma was in equilibrium, it means that we also have to average over the initial condition $\mathbf{V}(0)$. For this purpose, we use the Maxwell-Boltzmann distribution:

$$P(\mathbf{V}(0)) = \frac{1}{(2D_V\pi)^{3/2}} \exp\left(\frac{-\mathbf{V}(0)^2}{2D_V}\right).$$

To calculate the correlation $\langle V_x(t_1)V_x(t_2)\rangle$ we have to take the average over the initial conditions and over all realizations of $\mathbf{sp\,a}(t)$; that is, using (5.50), we have

$$\langle V_x(t_1)V_x(t_2)\rangle = \sum_{j,l} \mathbf{G}(t_1)_{x,j}\, \mathbf{G}(t_2)_{x,l}\, \langle V_j(0)V_l(0)\rangle \tag{5.51}$$

$$+ \int_0^{t_1} d\tau_1 \int_0^{t_2} d\tau_2 \sum_{j,l} \mathbf{G}(t_1-\tau_1)_{x,j}\, \mathbf{G}(t_2-\tau_2)_{x,l}\, \langle a_j(\tau_1)a_l(\tau_2)\rangle.$$

Here, we have used that $\langle V_j(0)\rangle = 0$; then, noting that $\langle a_j(\tau_1)a_l(\tau_2)\rangle = \epsilon\, \delta_{j,l}\, \delta(\tau_1-\tau_2)$ and $\langle V_j(0)V_l(0)\rangle = \frac{1}{2}V_T^2\,\delta_{j,l}$, where $V_T^2 = 2D_V = 2k_BT/m$, we finally get:

$$\langle V_x(t_1)V_x(t_2)\rangle = \frac{1}{2}\left\{\frac{\epsilon}{\nu}\,e^{-\nu|t_1-t_2|} + \left(V_T^2 - \frac{\epsilon}{\nu}\right)e^{-\nu(t_1+t_2)}\right\}\cos\left(\Omega(t_1-t_2)\right). \tag{5.52}$$

The parallel correlation $\langle V_z(t_1)V_z(t_2)\rangle$ can be obtained immediately from (5.52) taking $\Omega = 0$, that is:

$$\langle V_z(t_1)V_z(t_2)\rangle = \frac{1}{2}\left\{\frac{\epsilon}{\nu}\,e^{-\nu|t_1-t_2|} + \left(V_T^2 - \frac{\epsilon}{\nu}\right)e^{-\nu(t_1+t_2)}\right\}. \tag{5.53}$$

Exercise. Verify that $\langle V_y(t_1)V_y(t_2)\rangle = \langle V_x(t_1)V_x(t_2)\rangle$, and the expression (5.53).

Noting that the expressions for the correlations show a temporal structure that is not stationary, which is not justifiable physically, it follows that we must adjust the fluctuation parameter ϵ so that this term is canceled. Imposing equality

$$V_T^2 = \frac{\epsilon}{\nu}$$

involves an important relationship between the fluctuation and dissipation parameters. Therefore, the velocity correlations are written finally in the form:

$$\langle V_x(0)V_x(\tau)\rangle = \langle V_y(0)V_y(\tau)\rangle = \frac{V_T^2}{2}\exp(-\nu\,|\,\tau\,|)\cos\left(\Omega\tau\right) \tag{5.54}$$

$$\langle V_z(0)V_z(\tau)\rangle = \frac{V_T^2}{2}\exp(-\nu\,|\,\tau\,|), \tag{5.55}$$

Fig. 5.3 The *dotted line* represents the envelope $e^{-\nu\tau}$. The *solid curve* corresponds to a value $\Omega/\nu = 4\pi$ while the *dashed line* to a value $\Omega/\nu = \pi$

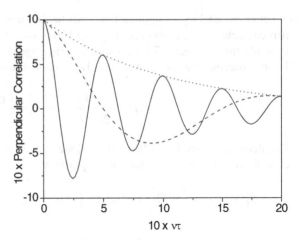

From these relations, one can see that the number of oscillations of the perpendicular correlation before decaying by a factor $1/e$ depends on the magnitude Ω/ν. In Fig. 5.3 two graphs of $2\langle V_\perp(0)V_\perp(\tau)\rangle/V_T^2$ corresponding to the values $\Omega/\nu = \{4\pi, \pi\}$ are shown.

5.4 On the Fluctuation-Dissipation Theorem

The Fluctuation-Dissipation theorem links the susceptibility $\chi(\omega)$ with the stationary correlation function at equilibrium. On the other hand, the Wiener-Khinchin theorem connects this correlation $\langle\langle \mathbf{X}(t)\mathbf{X}(0)\rangle\rangle_0$ with the spectral density $S_{\mathbf{XX}}(\omega)$; then, it is possible to study the temporal fluctuations of thermodynamic equilibrium perturbing slightly the system with external forces $\mathbf{F}(t)$.

There are several examples of application of this theorem, in particular, we will focus on the study of the electrical conductivity in amorphous materials (Chap. 8), where the Fluctuation-Dissipation theorem, which is crucial for understanding transport in disordered materials is still valid, but the diffusion approximation is not. In the context of linear response theory, the susceptibility $\chi(\omega)$ is independent of the external force applied to the system $\mathcal{S}$. On the other hand, using the theory of stochastic processes it is possible to calculate the time dependent variation $\langle \mathbf{X}(t)\rangle_{\mathbf{F}} - \langle \mathbf{X}(t)\rangle_0$ when the system is out of equilibrium[18]; particularly when the external force is a step function at $t = 0$. From these two facts, it is possible to relate the susceptibility to the correlation function at equilibrium. That is, it is possible to analyze nonequilibrium fluctuations in terms of correlations at equilibrium. This is the so-called Green-Callen's Theorem [12].

---

[18]Caused by the application of a small external force.

5.4.1   Theorem II: The Green-Callen's Formula

Suppose a (classical) physical system is in thermodynamic equilibrium from ancient times to the instant $t_0 = 0$. Then, for times $t \in [-\infty, t_0]$, the probability distribution that characterizes S is the canonical distribution, and it is given by equilibrium statistical mechanics.[19]

If we are interested in measuring a macroscopic quantity $\mathbf{Y} \equiv Y(q, p)$, its probability distribution in equilibrium will be[20]

$$P_0(y) = \frac{1}{Z_0} \int \cdots \int \mathcal{D}q\, \mathcal{D}p\, \delta(y - \mathbf{Y}) \exp(-H_0/k_B T). \qquad (5.56)$$

Here $\mathcal{D}q\, \mathcal{D}p$ is the elemental volume of phase-space of system S, and $Y(q, p)$ is the thermodynamic variable under study. Z_0 is the normalization constant that is calculated from the Hamiltonian of the N-body system S, $H_0 \equiv H_0(q, p)$, i.e.:

$$Z_0 = \int \cdots \int \mathcal{D}q\, \mathcal{D}p\, \exp(-H_0/k_B T). \qquad (5.57)$$

Consider as an example a paramagnetic substance, i.e., $H_0(q, p) = H(q, p) - B\mu \sum_i \cos\theta_i$, where θ_i is the angle between the external magnetic field $\mathbf{B}$ and the i-th moment. We can take $\mathbf{Y} = \mu \sum_i \cos\theta_i$ as the random variable representing the total magnetization of the sample and $H(q, p)$ as the part of the Hamiltonian without external magnetic field $\mathbf{B}$ [4]. Then, if at time $t = t_0$ the interaction of S is slightly altered with the external medium by an amount ΔB, and if the total Hamiltonian is modified in the form $H_0(q, p) \to H_0(q, p) - \Delta B\mathbf{Y}$, the probability distribution (5.56) will change according to

$$P_{\Delta B}(y) = \frac{1}{Z_{\Delta B}} \int \cdots \int \mathcal{D}q\, \mathcal{D}p\, \delta(y - \mathbf{Y}) \exp(-(H_0 - \Delta B\mathbf{Y})/k_B T). \qquad (5.58)$$

In this case, the normalization constant is

$$Z_{\Delta B} = \int \cdots \int \mathcal{D}q\, \mathcal{D}p\, \exp(-(H_0 - \Delta B\mathbf{Y})/k_B T). \qquad (5.59)$$

For small ΔB, we can approximate $P_{\Delta B}(y)$ expanding $\exp(\Delta B\mathbf{Y}/k_B T)$ in power series. Then, from (5.58), and to first order in ΔB, we get

---

[19]In the canonical ensemble, the probability distribution (at thermodynamic equilibrium) at temperature T is given by the Gibbs distribution. That is, the probability that the system is in a macroscopic energy state E is given by $\rho(E) = Z_0^{-1} e^{-E/k_B T}$, the constant Z_0 is the normalization factor [see Appendix F].

[20]Here, the set $\{p, q\}$ represents the canonical coordinates of the N bodies of the system S in study.

$$P_{\Delta B}(y) = \frac{\int \cdots \int \mathcal{D}q \, \mathcal{D}p \, \delta(y - \mathbf{Y}) \exp\left(-H_0/k_B T\right)\left(1 + \Delta B \mathbf{Y}/k_B T + \mathcal{O}(\Delta B^2)\right)}{\int \cdots \int \mathcal{D}q \, \mathcal{D}p \, \exp\left(-H_0/k_B T\right)\left(1 + \Delta B \mathbf{Y}/k_B T + \mathcal{O}(\Delta B^2)\right)}.$$
(5.60)

Using the density $\delta(y - \mathbf{Y})$ which appears in the integrand of (5.60), it is easy to see that to first order in ΔB we obtain

$$P_{\Delta B}(y) = \frac{(1 + \Delta By/k_B T) \int \cdots \int \mathcal{D}q \, \mathcal{D}p \, \delta(y - \mathbf{Y}) \exp\left(-H_0/k_B T\right)}{(1 + \Delta B \langle \mathbf{Y} \rangle_0 / k_B T) \, Z_0} + \mathcal{O}(\Delta B^2),$$
(5.61)

an expression which can be simplified further if we use the definition of $P_0(y)$. That is, using (5.56) we see that to first order in ΔB, we can write:

$$P_{\Delta B}(y) = P_0(y)(1 + \Delta By/k_B T)(1 - \Delta B \langle \mathbf{Y} \rangle_0 / k_B T) + \mathcal{O}(\Delta B^2) \qquad (5.62)$$
$$\simeq P_0(y)(1 + \Delta By/k_B T - \Delta B \langle \mathbf{Y} \rangle_0 / k_B T).$$

Note that up to $\mathcal{O}(\Delta B)$ the probability (5.62) is properly normalized. Let us go back to the initial problem of calculating the time dependence of the macroscopic quantity $\langle \mathbf{Y}(t) \rangle_F$ when the external force $\mathbf{F}$ is constant over time until the instant $t = t_0$ when it is modified.

We can calculate $\langle \mathbf{Y}(t) \rangle_F$ using the theory of stochastic processes as follows: if we denote $\langle \mathbf{Y}(t) \rangle_{F|t_0}$ by the first moment conditioned to having the value y_0 at t_0, it is known that

$$\langle \mathbf{Y}(t) \rangle_{F|t_0} = \int y P(y, t \mid y_0, t_0) \, dy;$$
(5.63)

then, the quantity of interest shall be the average $\langle \mathbf{Y}(t) \rangle_{F|t_0}$ over the initial condition, i.e.,

$$\langle \mathbf{Y}(t) \rangle_F = \int dy_0 \, P_1(y_0, t_0) \int y P(y, t \mid y_0, t_0) \, dy.$$
(5.64)

If we identify $P_1(y_0, t_0)$ with $P_{\Delta B}(y_0)$, using (5.62) we can immediately approximate (5.64) (for $\Delta B \ll 1$) in the form[21]:

$$\langle \mathbf{Y}(t) \rangle_F \simeq \int \int y P(y, t \mid y_0, t_0) P_0(y_0)(1 + \Delta By_0/k_B T - \Delta B \langle \mathbf{Y} \rangle_0 / k_B T) \, dy \, dy_0,$$
(5.65)

whence we get

$$\langle \mathbf{Y}(t) \rangle_F = \langle \mathbf{Y}(t) \rangle_0 + \frac{\Delta B}{k_B T}\{\langle \mathbf{Y}(t)\mathbf{Y}(t_0) \rangle_0 - \langle \mathbf{Y}(t) \rangle_0 \langle \mathbf{Y}(t_0) \rangle_0\} + \mathcal{O}(\Delta B^2). \qquad (5.66)$$

---

[21]The subscript 0 indicates "at equilibrium."

Then, using the definition of correlation function, we have shown that to first order in ΔB

$$\langle \mathbf{Y}(t) \rangle_{\mathbf{F}} - \langle \mathbf{Y}(t) \rangle_0 \simeq \frac{\Delta B}{k_B T} \langle\langle \mathbf{Y}(t)\mathbf{Y}(t_0) \rangle\rangle_0 . \tag{5.67}$$

The quantity $\langle \mathbf{Y}(t) \rangle_{\mathbf{F}} - \langle \mathbf{Y}(t) \rangle_0$ measures the change "away from thermodynamic equilibrium" (in the presence of an external step force at time t_0). According to (5.67), this quantity is given in terms of the correlation function in equilibrium $\langle\langle \mathbf{Y}(t)\mathbf{Y}(t_0) \rangle\rangle_0$ and it is proportional to the external variation ΔB. That is, the correlation function—determined in equilibrium—governs the relaxation from slightly perturbed equilibrium. If we use the result of linear response theory (5.27), when $\mathbf{F}(t) = \Delta B \, \Theta(-t)$, we can write, by considering $t_0 = 0$ and $t > 0$, the following equality to order $\mathcal{O}(\Delta B)$

$$\langle \mathbf{Y}(t) \rangle_{\mathbf{F}} - \langle \mathbf{Y}(t) \rangle_0 = \Delta B \frac{1}{i\pi} \mathcal{P} \int_{-\infty}^{\infty} \frac{\chi(\omega)}{\omega} \cos(\omega t) \, d\omega. \tag{5.68}$$

Then, by comparing this expression with (5.67), we obtain the desired result[22]

$$\langle\langle \mathbf{Y}(t)\mathbf{Y} \rangle\rangle_0 = \frac{k_B T}{i\pi} \mathcal{P} \int_{-\infty}^{\infty} \frac{\chi(\omega)}{\omega} \cos(\omega t) \, d\omega. \tag{5.69}$$

That is, a relationship between the susceptibility for a system out of equilibrium $\chi(\omega)$, and the correlation function in thermodynamic equilibrium $\langle\langle \mathbf{Y}(t)\mathbf{Y} \rangle\rangle_0$.

Exercise. For the matrix case and using that the **sp** $\mathbf{Y}(t)$ is stationary: $\langle\langle \mathbf{Y}(t)\mathbf{Y} \rangle\rangle_0 = \langle\langle \mathbf{Y}(-t)\mathbf{Y} \rangle\rangle_0^T$; then, what symmetry properties for the matrix $\chi(\omega)$ can be inferred?

5.4.2 Nyquist's Formula*

The thermal motion of electrical charges in the resistance of an arbitrary circuit induces in it a stochastic electromotive force[23] (Johnson's noise [Phys. Rev. **32**, 97, (1928)]). Consider a current $I(t)$ flowing through the resistance R and inductance L of a given circuit.[24] In this case, the equation governing the temporal variation of the electric current is:

$$L\frac{dI}{dt} = -RI + \mathcal{E}(t), \quad \text{with} \quad \langle \mathcal{E}(t) \rangle = 0, \quad \langle \mathcal{E}(t)\mathcal{E}(t+\tau) \rangle = G_{\mathcal{E}}(\tau), \tag{5.70}$$

---

[22]Here, we have used that around equilibrium the **sp** is homogeneous in time and Markovian [4]; then, $\langle \mathbf{Y}(t)\mathbf{Y}(t_0) \rangle_0 = \int \int y y_0 \, P_{eq}(y, t; y_0, t_0) \, dy \, dy_0 = \int \int y y_0 \, P(y, t - t_0 \mid y_0) P_0(y_0) \, dy \, dy_0$. The response of "mature" materials could also be characterized using a non-Markovian approach.

[23]Electromotive force is also called *emf* (denoted by $\mathcal{E}$ and measured in volt).

[24]See Sects. 4.3 and 4.4.1 for other similar electric circuits.

where $G_\mathcal{E}(\tau)$ is the correlation function that characterizes the stochastic *emf* $\mathcal{E}(t)$. From this equation, it is easy to see that for long times the **mv** of the current vanishes. On the other hand, we see that asymptotically the **mv** of the square of the current is not canceled. Moreover, for sufficiently long times, the **mv** $\langle I(t)^2 \rangle$ can be approximated by[25]

$$\langle I(t)^2 \rangle \simeq \langle I(0)^2 \rangle e^{-2Rt/L} + \frac{1}{2RL} \left(1 - e^{-2Rt/L} \right) G_\mathcal{E}(\omega = 0),$$

where $G_\mathcal{E}(\omega)$ is the Fourier transform of the correlation function of the *emf*. Then, in the limit $t \to \infty$, we get

$$\langle I^2 \rangle_{st} = \frac{1}{2RL} G_\mathcal{E}(\omega = 0).$$

However, since the quantity $\frac{1}{2}L \langle I^2 \rangle_{st}$ is the magnetic energy stored in the inductance L, if we apply the equipartition law to this system, we get that $\frac{1}{2}L \langle I^2 \rangle_{st} = \frac{1}{2}k_B T$; from which it follows that

$$G_\mathcal{E}(\omega = 0) = 2k_B T R. \tag{5.71}$$

Again as in Theorem I (Sect. 4.4.1) and Theorem II (Sect. 5.4.1), we are now faced with a relationship between fluctuations (in the voltage) and dissipation (in resistance). This formula was derived by Nyquist [13] to explain Johnson's experiment. Also Nyquist showed, using physical arguments, that the power spectrum of the *emf* was approximately constant (up to very high frequencies). From this fact follows the more general formula

$$G_\mathcal{E}(\omega) = 2k_B T R. \tag{5.72}$$

Excursus. Consider the relationship $I(\omega) = \mathcal{E}(\omega)/Z(\omega)$ where $Z(\omega)$ is the total impedance of a circuit,[26] together with the definition of the power spectrum (5.5) and the generalized Nyquist's formula for the *emf* spectrum

$$G_\mathcal{E}(\omega) = 2k_B T R(\omega).$$

Show, applying Rice's method, that the current spectrum is[27]

$$S_{II}(\omega) = 2k_B T \, |Z(\omega)|^{-2} \, \mathcal{R}_e[Z(\omega)] \tag{5.73}$$

$$= 2k_B T \, \mathcal{R}_e[Y(\omega)],$$

---

[25]To solve this problem for the case of white noise (the Markovian limit) see the Ornstein-Uhlenbeck model of Sect. 4.4.1.

[26]That is, $Z(\omega) = R(\omega) + iX(\omega)$, where $R(\omega)$ is the generalized resistance and $X(\omega)$ the generalized reactance.

[27]See Rice's method in advanced exercise 5.5.2.

where

$$Y(\omega) = \frac{1}{Z(\omega)}.$$

Subsequently, Callen and Welton [8] generalized Nyquist's formula for quantum systems. In that case, formula (5.73) takes the form

$$S_{II}(\omega) = \hbar\omega \coth\left(\frac{\hbar\omega}{2k_BT}\right) \mathcal{R}_e[Y(\omega)]. \tag{5.74}$$

Interestingly, we can also use Nyquist's formula for systems other than electrical circuits if we identify the admittance $Y(\omega)$ with the susceptibility of a mechanical system [14].

Exercise. Study the expression (5.74) in the limits $\hbar\omega \gg k_BT$ and $\hbar\omega \ll k_BT$ and compare with Nyquist's formula. Consider the RCL electrical circuit presented in Sect. 4.3, using that the diffusion matrix of the **F-P** equation is[28] $\mathbf{B} = k_B(\mathbf{L} + \mathbf{L}^T)$. Show that in the limit of a white stochastic voltage $\Delta V(t)$, the intensity of the correlation is consistent with Nyquist's formula:

$$\langle \Delta V(t+\tau)\Delta V(t)\rangle = 2k_BTR\,\delta(\tau).$$

In 1957, Kubo proved, based on Liouville-Neumann's dynamics for the density matrix, that the susceptibility $\chi(\omega)$ may be associated microscopically with a correlation function in thermodynamic equilibrium. This (third) theorem will be discussed in Chap. 8. Furthermore, it can be considered that Kubo's formula is a consequence of the Callen-Welton theorem [14].

5.5 Additional Exercises with Their Solutions

5.5.1 Spectrum of the Dichotomic Process

Here, we are going to calculate the spectrum of the (symmetric) dichotomic **sp** $\mathbf{X}(t)$ using the Wiener-Khinchin theorem (5.8)

$$S_{\mathbf{XX}}(\omega) = \int_{-\infty}^{\infty} \langle \mathbf{X}(t)\mathbf{X}(t+\tau)\rangle_{\mathrm{st}} \exp(i\omega\tau)\,d\tau \tag{5.75}$$

The stationary correlation function $\langle \mathbf{X}(t)\mathbf{X}(t+\tau)\rangle_{\mathrm{st}}$ can be calculated knowing the propagator of the system which was presented in Chap. 3. Here, we present an

---

[28]See Sect. 4.4.1.

alternative procedure based on the fact that for a given realization of the dichotomic **sp** the quantity $X(t)X(t + \tau)$ can only have two values $\pm\Delta^2$. Noting that in the isotropic case the transition rate is characterized by the constant α (see Chap. 6); then, in a time interval $[t, t + \tau]$ there could be n jumps (with n odd or even) characterized by the Poisson probability $P(n, \tau) = (\alpha\tau)^n e^{-\alpha\tau}/n!$ Thus, the average of the quantity $X(t)X(t + \tau)$ can be written in the form

$$\langle \mathbf{X}(t)\mathbf{X}(t + \tau) \rangle = \Delta^2 \sum_{n=0,2,4,\cdots}^{\infty} P(n, \tau) - \Delta^2 \sum_{n=1,3,5,\cdots}^{\infty} P(n, \tau)$$

$$= \Delta^2 \sum_{n=0,2,4,\cdots}^{\infty} (-\alpha\tau)^n \frac{\exp(-\alpha\tau)}{n!}$$

$$= \Delta^2 \exp(-2\alpha\tau), \quad \tau \geq 0. \tag{5.76}$$

Using (5.76) in (5.75) and noting that

$$\langle \mathbf{X}(t)\mathbf{X}(t + \tau) \rangle = \langle \mathbf{X}(t)\mathbf{X}(t - \tau) \rangle = \Delta^2 e^{-2\alpha|\tau|},$$

we get

$$S_{\mathbf{XX}}(\omega) = \int_{-\infty}^{\infty} \Delta^2 \exp(-2\alpha|\tau|) \exp(i\omega\tau)\, d\tau$$

$$= \frac{4\alpha\Delta^2}{4\alpha^2 + \omega^2},$$

which is a Lorentzian spectral density. Sometimes the Dichotomic **sp** is called the random telegraph signal. Compare this process with the random rectified **sp** presented in Sect. 3.20.13.

5.5.2 On the Rice Method

Here, we are going to use the Wiener-Khinchin formula in reverse; that is, to calculate the correlation function of a stationary ergodic **sp** given its spectral density. Consider the Ornstein-Uhlenbeck **sp** $\mathbf{V}(t)$; then, by using Rice's method,[29] we can calculate the spectrum of the process from its **sde**

$$m\frac{dV}{dt} = -m\Gamma V + \sqrt{\epsilon}\,\xi(t), \quad \text{with} \quad \langle \xi(t)\xi(t + \tau) \rangle = \delta(\tau), \tag{5.77}$$

---

[29]This method was presented by Rice S.O. in 1944, and it is very useful when dealing with linear systems. See Selected Papers on Noise and Stochastic Process, ed. N. Wax (Dover, N.Y. 1954).

where m is the mass of the particle, Γ the dissipation parameter, ϵ the intensity of the fluctuations and $\xi(t)$ is a GWN with mean value zero. Taking the Fourier transform in (5.77), we get

$$m(-i\omega + \Gamma) V(\omega) = \sqrt{\epsilon}\, \xi(\omega). \tag{5.78}$$

The spectrum of **sp** $V(t)$ can be obtained first taking the square modulus of $V(\omega)$; then, from (5.78), we get

$$|V(\omega)|^2 = \frac{\epsilon}{m^2} \left| \frac{\xi(\omega)}{(-i\omega + \Gamma)} \right|^2$$

$$= \frac{\epsilon}{m^2(\omega^2 + \Gamma^2)} |\xi(\omega)|^2 ; \tag{5.79}$$

therefore, using (5.5) $S_{VV}(\omega) = \left[\epsilon/m^2(\omega^2 + \Gamma^2)\right] S_\xi(\omega)$, so we get

$$S_{VV}(\omega) = \frac{\epsilon}{m^2(\omega^2 + \Gamma^2)}, \tag{5.80}$$

inserting (5.80) in the inverse of (5.8) we finally obtain

$$\langle \mathbf{V}(t)\mathbf{V}(t+\tau)\rangle_{\text{st}} = \frac{1}{2\pi} \int_{-\infty}^{\infty} S_{VV}(\omega) \exp(-i\omega\tau)\, d\omega$$

$$= \frac{1}{2\pi} \int_{-\infty}^{\infty} \frac{\epsilon \exp(-i\omega\tau)}{m^2(\omega^2 + \Gamma^2)}\, d\omega$$

$$= \frac{\epsilon}{\pi m^2} \int_{0}^{\infty} \frac{\cos(\omega\tau)}{(\omega^2 + \Gamma^2)}\, d\omega$$

$$= \frac{\epsilon}{2m^2\Gamma} e^{-\Gamma\tau}.$$

Then using the detailed balance condition $\epsilon = 2m\Gamma k_B T$ we get the expected stationary correlation function of the Ornstein-Uhlenbeck **sp** $V(t)$, see (5.96).

5.5.3 Ergodicity in Mean of the Mean-Value

Consider a stationary **sp** $X(t)$, $\forall t \in [-\infty, \infty]$ and define the random number

$$X_T = \frac{1}{2T} \int_{-T}^{T} \mathbf{X}(t)\, dt. \tag{5.81}$$

Clearly the random character of X_T depends on each realization of the **sp** $\mathbf{X}(t)$. To see if

$$\text{ms} - \lim_{T \to \infty} \frac{1}{2T} \int_{-T}^{T} \mathbf{X}(t)\, dt = \langle \mathbf{X} \rangle_{\text{st}}, \tag{5.82}$$

it is necessary to prove that

$$\lim_{T \to \infty} \left\langle (X_T - \langle \mathbf{X} \rangle_{\text{st}})^2 \right\rangle_{\text{st}} \to 0. \tag{5.83}$$

Proof: First note that

$$\langle X_T \rangle_{\text{st}} = \left\langle \frac{1}{2T} \int_{-T}^{T} \mathbf{X}(t)\, dt \right\rangle_{\text{st}} = \langle \mathbf{X} \rangle_{\text{st}}.$$

Then,

$$\begin{aligned} \langle X_T^2 \rangle_{\text{st}} &= \frac{1}{4T^2} \left\langle \left(\int_{-T}^{T} dt_1 \int_{-T}^{T} dt_2 \mathbf{X}(t_1)\mathbf{X}(t_2) \right) \right\rangle_{\text{st}} \\ &= \frac{1}{4T^2} \int_{-T}^{T} dt_1 \int_{-T}^{T} dt_2 \left[C(t_1 - t_2) + \langle \mathbf{X} \rangle_{\text{st}}^2 \right]. \end{aligned}$$

Considering that the correlation $C(t_1 - t_2)$ is symmetric and introducing the change of variables: $t = t_1, \tau = t_1 - t_2$, we write

$$\begin{aligned} 4T^2 \left\langle (X_T - \langle \mathbf{X} \rangle_{\text{st}})^2 \right\rangle_{\text{st}} &= \int_{-T}^{T} dt_1 \int_{-T}^{T} dt_2 C(t_1 - t_2) \tag{5.84} \\ &= 2 \int_{-T}^{T} dt_1 \int_{-T}^{t_1} dt_2 C(|t_1 - t_2|) \\ &= 2 \int_{-T}^{T} dt \int_{0}^{t+T} d\tau C(\tau). \end{aligned}$$

Integrating by parts and introducing a new change of variable, we finally get

$$\begin{aligned} 2 \int_{-T}^{T} dt \int_{0}^{t+T} d\tau C(\tau) &= 2 \left\{ \left[t \int_{0}^{t+T} d\tau C(\tau) \right]_{t=-T}^{t=T} - \int_{-T}^{T} tC(t+T)\, dt \right\} \\ &= 2 \int_{-T}^{T} (2T - \tau) C(\tau)\, d\tau = 2T \int_{-2T}^{2T} \left(1 - \frac{|\tau|}{2T} \right) C(\tau)\, d\tau. \end{aligned}$$

Thus, from (5.84) we conclude that

$$\left\langle (X_T - \langle \mathbf{X} \rangle_{st})^2 \right\rangle_{st} = \frac{1}{2T} \int_{-2T}^{2T} \left(1 - \frac{|\tau|}{2T} \right) C(\tau) \, d\tau. \tag{5.85}$$

Therefore, (5.82) follows if

$$\lim_{T \to \infty} \frac{1}{2T} \int_{-2T}^{2T} \left(1 - \frac{|\tau|}{2T} \right) C(\tau) \, d\tau \to 0. \tag{5.86}$$

Then, we say that the **sp** $\mathbf{X}(t)$ is mean-ergodic iff (5.86) is met. In this case, the time average of $\mathbf{X}(t)$ computed from a single realization is close to $\langle \mathbf{X} \rangle_{st}$. Then, the mean-ergodicity of a process depends on the behavior of the correlation function $C(\tau)$ for large τ.[30]

Counter-Examples:

1) Suppose that we define the **sp** $\mathbf{X}(t) = \Omega$, where Ω is a **rv** characterized by the pdf $P(\Omega)$ on $\mathcal{D}_\Omega$. Then the ensemble of realizations of the **sp** is a family of straight lines in the course of time. Show that in this case the random quantity

$$X_T = \frac{1}{2T} \int_{-T}^{T} \mathbf{X}(t) dt,$$

is not mean-ergodic.

2) Consider two mean-ergodic **sps** $\mathbf{X}_1(t)$ and $\mathbf{X}_2(t)$ with $\langle \mathbf{X}_1(t) \rangle = \eta_1$ and $\langle \mathbf{X}_2(t) \rangle = \eta_2$, and form the new **sp** $\mathbf{Y}(t) = \mathbf{X}_1(t) + \Omega \, \mathbf{X}_2(t)$, where Ω is an equiprobable binary **rv** independent of $\mathbf{X}_2(t)$. Show that the **sp** $\mathbf{Y}(t)$ is not mean-ergodic.

5.5.4 Ergodicity in Mean of Other "Statistical Quantities"

Introducing similar reasoning, it is possible to prove that the function

$$C_T(\tau) = \frac{1}{T} \int_0^T \mathbf{X}(t)\mathbf{X}(t + \tau) \, dt$$

converges in mean square to the stationary correlation $\langle\langle \mathbf{X}(t)\mathbf{X}(t + \tau) \rangle\rangle_{st} = C(|\tau|)$, iff the following condition is met

$$\lim_{T \to \infty} \frac{1}{2T} \int_{-2T}^{2T} \left(1 - \frac{|\lambda|}{2T} \right) \rho(\tau, \lambda) \, d\lambda \to 0.$$

---

[30]The sufficient condition (5.9) follows immediately.

Where $\rho(\tau, \lambda)$ is a four-time correlation defined as:

$$\langle \mathbf{X}(t+\lambda+\tau)\mathbf{X}(t+\tau)\mathbf{X}(t+\lambda)\mathbf{X}(t)\rangle_{st} - \langle \mathbf{X}(t)\mathbf{X}(t+\tau)\rangle_{st}^2 = \rho(\tau, \lambda).$$

Also other quantities can be studied in a similar way; see [5].

5.5.5 More on the Fluctuation-Dissipation Theorem

In (5.69), the Fluctuation-Dissipation theorem was written for thermodynamics variables. If it is not the case, we can use an alternative formula.

In general, the mean value of the vector $\mathbf{X}(t)$ conditioned to the initial value $\mathbf{X}(0) = \mathbf{X}_0$ is

$$\langle \mathbf{X}(t)\rangle_{\mathbf{X}_0} = \int \mathbf{X} P(\mathbf{X}, t | \mathbf{X}_0, 0) \, d\mathbf{X}. \tag{5.87}$$

If there is a time-dependent force $\mathbf{F}(t)$, this expression changes to

$$\langle \mathbf{X}(t)\rangle_{\mathbf{X}_0;\mathbf{F}(t)} = \int \mathbf{X} P(\mathbf{X}, t | \mathbf{X}_0, \mathbf{F}(t)) \, d\mathbf{X}.$$

Here we have explicitly included the dependence of the external force in the conditional pdf. If we take the average over all possible initial conditions $\mathbf{X}_0$, we arrive at

$$\langle \mathbf{X}(t)\rangle_{\mathbf{F}(t)} = \int \int \mathbf{X} P(\mathbf{X}, t | \mathbf{X}_0, \mathbf{F}(t)) \; P(\mathbf{X}_0, \mathbf{F}(t)) \, d\mathbf{X} d\mathbf{X}_0$$

$$= \int \langle \mathbf{X}(t)\rangle_{\mathbf{X}_0;\mathbf{F}(t)} \; P(\mathbf{X}_0, \mathbf{F}(t)) \, d\mathbf{X}_0. \tag{5.88}$$

Now, consider the case when the time-dependent force is

$$\mathbf{F}(t) = \mathbf{F}_0 \Theta(-t).$$

As we have studied in Sect. 5.2.2, in this case, the force is turned off at $t = 0$; then, if a linear regression applies (as in Sect. 4.1 Onsager's analysis) we know that

$$\langle \mathbf{X}(t)\rangle_{\mathbf{X}_0} = e^{-\mathbf{M}t} \cdot \mathbf{X}_0, \; t \geq 0.$$

Then, we can write (5.88) in the form

$$\langle \mathbf{X}(t)\rangle_{\mathbf{F}_0} = \int e^{-\mathbf{M}t} \cdot \mathbf{X}_0 \, P(\mathbf{X}_0, \mathbf{F}_0) \, d\mathbf{X}_0$$

$$= e^{-\mathbf{M}t} \cdot \langle \mathbf{X}(0)\rangle_{\mathbf{F}_0}, \; t \geq 0. \tag{5.89}$$

Using the results from linear response theory (5.27), we can calculate $\langle \mathbf{X}(t) \rangle_{\mathbf{F}_0}$ for $t \lessgtr 0$. In particular $\langle \mathbf{X}(0) \rangle_{\mathbf{F}_0} = \chi(0) \cdot \mathbf{F}_0$; therefore, we get the formula

$$\langle \mathbf{X}(t) \rangle_{\mathbf{F}_0} = e^{-\mathbf{M}t} \cdot \chi(0) \cdot \mathbf{F}_0, \ t \geq 0 \tag{5.90}$$

and

$$\langle \mathbf{X}(t) \rangle_{\mathbf{F}_0} = \frac{1}{i\pi} \mathcal{P} \int_{-\infty}^{\infty} \frac{\chi(\omega) \cdot \mathbf{F}_0}{\omega} \cos(\omega t) \ d\omega, \ t \geq 0. \tag{5.91}$$

Comparing (5.90) with (5.91), we conclude that

$$e^{-\mathbf{M}t} \cdot \chi(0) = \frac{1}{i\pi} \mathcal{P} \int_{-\infty}^{\infty} \frac{\chi(\omega)}{\omega} \cos(\omega t) \ d\omega. \tag{5.92}$$

On the other hand, the regression theorem (see Sect. 4.9.2) states that

$$C_{\mathbf{xx}}(t) \equiv \langle \mathbf{X}(t) \mathbf{X}(0) \rangle = e^{-\mathbf{M}t} \cdot \langle \mathbf{X}(0) \mathbf{X}(0) \rangle .$$

Here, we have used the dyadic notation for $\langle \mathbf{X}(0) \mathbf{X}(0) \rangle$. Then, applying $\chi(0)^{-1}$ from the right to equation (5.92), we arrive at

$$e^{-\mathbf{M}t} = \frac{1}{i\pi} \mathcal{P} \int_{-\infty}^{\infty} \frac{\chi(\omega) \cdot \chi(0)^{-1}}{\omega} \cos(\omega t) \ d\omega. \tag{5.93}$$

Finally, applying $\langle \mathbf{X}(0) \mathbf{X}(0) \rangle$ to (5.93), we get a relationship between the susceptibility and the equilibrium correlation, i.e., the Fluctuation-Dissipation theorem

$$C_{\mathbf{xx}}(t) = \frac{1}{i\pi} \mathcal{P} \int_{-\infty}^{\infty} \frac{\chi(\omega) \cdot \chi(0)^{-1} \cdot \langle \mathbf{X}(0) \mathbf{X}(0) \rangle}{\omega} \cos(\omega t) \ d\omega, \ t > 0. \tag{5.94}$$

Example (**The temporally forced Ornstein-Uhlenbeck process**). Consider a free particle (in 1-dimension) to be characterized by the Langevin equation. If in addition the particle is subject to the action of an external force $\mathbf{F}(t)$, the description of the system $\mathcal{S}$ will be given by the **sde**

$$m\frac{dV}{dt} = -m\Gamma V + \mathbf{F}(t) + \sqrt{\epsilon} \, \xi(t), \quad \text{with} \quad \langle \xi(t)\xi(t+\tau) \rangle = \delta(\tau), \tag{5.95}$$

where m is the mass of the particle, Γ the dissipation parameter, ϵ the intensity of the fluctuations of a GWN $\xi(t)$ with mean zero. In order to use (5.94), we first take the Fourier transform and the **mv** over the GWN of (5.95); then, we get

$$m(-i\omega + \Gamma) \langle V(\omega) \rangle_{\mathbf{F}} = \mathbf{F}(\omega).$$

From the linear response relation: $\langle V(\omega)\rangle_F = \chi(\omega) F(\omega)$ we arrive at an expression for the susceptibility

$$\chi(\omega) = \frac{1}{m}(\Gamma - i\omega)^{-1}.$$

Then, from the Fluctuation-Dissipation theorem (5.94), we can write

$$\langle V(t)V(0)\rangle_{eq} = \frac{\chi(0)^{-1}\langle V(0)V(0)\rangle_{eq}}{i\pi}P\int_{-\infty}^{\infty}\frac{\cos(\omega t)}{\omega m(\Gamma - i\omega)}d\omega.$$

Noting that $\chi(0)^{-1} = \Gamma m$ and $\langle V(0)V(0)\rangle_{eq} = k_B T/m$ we get

$$\langle V(t)V(0)\rangle_{eq} = \frac{k_B T\Gamma}{i\pi m}P\int_{-\infty}^{\infty}\frac{(\Gamma + i\omega)\cos(\omega t)}{\omega(\Gamma^2 + \omega^2)}d\omega$$

$$= \frac{2k_B T\Gamma}{\pi m}\int_{0}^{\infty}\frac{\cos(\omega t)}{(\Gamma^2 + \omega^2)}d\omega.$$

Here we have used that the real part of $\chi(\omega)$ is an even function, therefore the imaginary part of the integral vanishes. On the other hand, in the real part of the integral, we do not need to take the Cauchy principal value because there is no singularity at $\omega = 0$. Using that

$$\mathcal{F}(u) = \int_{0}^{\infty}(b^2 + \omega^2)^{-1}\cos(u\omega)\,d\omega = \frac{\pi}{2b}e^{-bu}, \quad \text{for} \quad u \geq 0,$$

we arrive at the final expression

$$\langle V(t)V(0)\rangle_{eq} = \frac{k_B T}{m}e^{-\Gamma t}, \quad \text{for} \quad t \geq 0, \tag{5.96}$$

in accordance with the expected result of Sect. 3.12.1.

Exercise. From the **sde**

$$m\,dV/dt = -m\Gamma V + \sqrt{\epsilon}\,\xi(t),$$

show that the equilibrium correlation $\langle\langle V(t)V(0)\rangle\rangle_{eq}$ of the Ornstein-Uhlenbeck process is effectively (5.96) if the noise intensity satisfies the principle of detailed balance ($\epsilon = 2m\Gamma k_B T$), see guided exercise (using Novikov's Theorem) in Sect. 4.4.1.

5.5.6 The Half-Fourier Transform of Stationary Correlations

In linear response theory, as well as in the study of stationary quantum fluctuations (see Appendix I), it is sometimes necessary to know the half-Fourier transform of a correlation function $C(-\tau)$ in terms of its Fourier's transform:

$$h(\omega) = \int_{-\infty}^{\infty} C(-\tau) e^{-i\omega\tau} d\tau \tag{5.97}$$

$$C(-\tau) = \frac{1}{2\pi} \int_{-\infty}^{\infty} h(\omega) e^{+i\omega\tau} d\omega. \tag{5.98}$$

Assuming the symmetry $C(-\tau) = C^*(\tau)$ it is possible to prove that

$$\int_{0}^{\infty} C(-\tau) e^{-i\omega\tau} d\tau = \frac{1}{2} h(\omega) + iS(\omega), \tag{5.99}$$

where $h(\omega) \in \mathcal{R}_e$ and function $S(\omega)$ is given in terms of Hilbert's transform:

$$S(\omega) = \frac{1}{2\pi} P \int_{-\infty}^{\infty} \frac{h(u)}{(u - \omega)} du. \tag{5.100}$$

To prove (5.99) we proceed in the following way. First introduce (5.98) in (5.99) and interchange the order of integrals, then we get

$$\int_{0}^{\infty} \left\{ \frac{1}{2\pi} \int_{-\infty}^{\infty} h(\omega') e^{+i\omega'\tau} d\omega' \right\} e^{-i\omega\tau} d\tau = \frac{1}{2\pi} \int_{-\infty}^{\infty} h(\omega') d\omega' \int_{0}^{\infty} e^{+i\omega'\tau - i\omega\tau} d\tau, \tag{5.101}$$

now using the identity:

$$\int_{0}^{\infty} e^{+i(\omega'-\omega)\tau} d\tau = \pi\delta(\omega' - \omega) + iP\left(\frac{1}{\omega' - \omega}\right), \tag{5.102}$$

where the operator $P\left(\frac{1}{\omega'-\omega}\right)$ represents Cauchy's principal value, from (5.101) we get

$$\int_{0}^{\infty} C(-\tau) e^{-i\omega\tau} d\tau = \frac{1}{2} h(\omega) + \frac{i}{2\pi} P \int_{-\infty}^{\infty} \frac{h(\omega')}{(\omega' - \omega)} d\omega',$$

from which (5.99) follows.

Exercise. Assume that $C(-\tau) \in \mathcal{R}_e$ and prove that $S(\omega = 0) = 0$.

Exercise. Also a relation for the quantity:

$$\int_{0}^{\infty} \tau C(-\tau) \exp(-i\omega\tau) d\tau$$

can be found in a similar way. Using the operational formula

$$\int_0^\infty \tau e^{+i(\omega'-\omega)\tau} d\tau = -i\pi\delta'\left(\omega'-\omega\right) + \mathcal{P}'\left(\frac{1}{\omega'-\omega}\right),$$

with definitions:

$$\int_{-\infty}^\infty f\left(\omega'\right)\delta'\left(\omega'-\omega\right) d\omega' = -f'\left(\omega\right)$$

and

$$\int_{-\infty}^\infty f\left(\omega'\right)\mathcal{P}'\left(\frac{1}{\omega'-\omega}\right) d\omega' = -\mathcal{P}\int_{-\infty}^\infty \frac{f'\left(\omega'\right)}{\left(\omega'-\omega\right)} d\omega'.$$

Show that

$$\int_0^\infty \tau C\left(-\tau\right) e^{-i\omega\tau} d\tau = r^R\left(\omega\right) + i r^I\left(\omega\right),$$

where

$$r^R\left(\omega\right) = -\frac{1}{2\pi}\mathcal{P}\int_{-\infty}^\infty \frac{h'\left(\omega'\right)}{\left(\omega'-\omega\right)} d\omega'; \quad r^I\left(\omega\right) = \frac{h'\left(\omega'\right)}{2}.$$

References

1. L.E. Reichl, *A Modern Course in Statistical Physics,* 2nd edn. (Edward Arnold Publ. Ltd, Austin, 1992)
2. H. Risken, *The Fokker-Planck Equation* (Springer, Berlin, 1984)
3. N. Wiener, Acta Math. **55**, 117 (1930); A. Khinchin, Math. Ann. **109**, 604 (1934)
4. N.G. van Kampen, *Stochastic Process in Physics and Chemistry,* 2dn edn. (North-Holland, Amsterdam, 1992)
5. C.W. Gardiner, *Handbook of Stochastic Methods* (Springer, Berlin, 1983)
6. R. Kubo, M. Toda, N. Hashitsume, *Statistical Physics II: Nonequilibrium Statistical Mechanics* (Springer, Berlin, 1985)
7. M. Lax, Rev. Mod. Phys. **32**, 25 (1960)
8. H.B. Callen, T.A. Welton, Phys. Rev. **83**, 34 (1951)
9. L. Onsager, Phys. Rev. **37**, 405 (1931)
10. R. Balescu, *Statistical Dynamics* (Imperial College Press, London, 1997)
11. B. Kurşunoğlu, Ann. Phys. **17**, 259 (1962)
12. H.B. Callen, R.F. Greene, Phys. Rev. **86**, 702 (1952)
13. H. Nyquist, Phys. Rev. **32**, 110 (1928)
14. R.L. Stratonovich, *Nonlinear Nonequilibrium Thermodynamics I* (Springer, Berlin, 1992)

Chapter 6
Introduction to Diffusive Transport

The study of the transport of different entities has been vital in different areas of exact sciences, as well as in Biology, Sociology, Economics, etc. Robert Brown, an eminent botanist who first named the cell nucleus, in 1827 was studying a suspension of pollen particles in water under the microscope and noticed that from within the pollen were released amyloplasts (starch organelles) and spherosomes (lipid organelles), tiny particles that executed an incessant erratic motion, which he attributed, correctly, to collisions with the water molecules. He confirmed also that this phenomenon existed in both organic and inorganic matter.[1] Then came Einstein, Smoluchowski, Wiener, Fokker, Planck, Kramers, etc., who with their physical and mathematical contributions laid the foundations of the theory of diffusive transport phenomena.

In previous chapters we have studied the process of Fokker-Planck transport, which is a more general Markovian process with continuous diffusive character.[2] In this chapter we introduce and discuss Markovian stochastic processes[3] which can be defined by a discrete random variable.[4]

6.1 Markov Chains

We can define a Markov chain as follows [1, 2]. Consider a Markov **sp** where time is discrete, $t_i \in \mathcal{N}$, and elements of the sample space or domain, $\{x_i\}$, form a discrete

---

[1]Hence this erratic movement or transport of particles with rapid oscillations is called Brownian motion.

[2]The **F-P** equation naturally includes dragging (or drift) and the presence of a potential in the diffusive transport of a continuous spatio temporal model.

[3]Processes that evolve in a "continuous" or "discrete" time.

[4]That is, the sample space is a finite or infinite set of discrete states.

© Springer International Publishing AG 2017
M.O. Cáceres, *Non-equilibrium Statistical Physics with Application to Disordered Systems*, DOI 10.1007/978-3-319-51553-3_6

and ordered set of values; then, the Chapman-Kolmogorov equation adopts a matrix character. Assuming that the conditional probability is a homogeneous function in time we get

$$P(x_i, t_i \mid x_j, t_j) = P_{t_i - t_j}(x_i \mid x_j). \tag{6.1}$$

We can simplify the notation if we identify $x_i \rightarrow s$ and $t_i \rightarrow n$, where s and n are both discrete. Using this new notation we have:

$$\mathbf{T}_n(s \mid s') \equiv P_{t_i - t_j}(x_i \mid x_j), \quad n = t_i - t_j, \; x_i = s, \; x_j = s'. \tag{6.2}$$

Then we can write the Chapman-Kolmogorov equation in the form[5]

$$\mathbf{T}_{n+m} = \mathbf{T}_n \cdot \mathbf{T}_m, \tag{6.3}$$

where $\mathbf{T}_n$ is the matrix of elements $\mathbf{T}_n(s \mid s')$. By direct substitution we see that the solution of the matrix Eq. (6.3) is

$$\mathbf{T}_n = (\mathbf{T}_1)^n \equiv \mathbf{T}_1^n. \tag{6.4}$$

Note that we have solved a particular case of the Chapman-Kolmogorov equation. Here $\mathbf{T}_1$ is the transition probability matrix in the discrete time unit; then, the whole Markov chain (which is temporally homogeneous) will be characterized by specifying a model for $\mathbf{T}_1$. It is important to see that the solution (6.4) is useful to the extent that we can diagonalize the matrix $\mathbf{T}_1$.[6]

6.1.1 Properties of $\mathbf{T}_1$

The transition probability matrix is characterized by the following properties:

1. All its elements are positive.[7] This follows from the very definition of transition probability (6.2), that is,

$$\mathbf{T}_1(s \mid s') = P(x_i, t_i + 1 \mid x_j, t_i) \geq 0, \; \forall x_i, x_j.$$

---

[5] See optional exercise in Sect. 3.3.1.

[6] That is, one wishes to solve the following problem: $\mathbf{T}_1 \mathbf{X}_\lambda = \lambda \mathbf{X}_\lambda$, where X_λ represents a possible eigenvector corresponding to the eigenvalue λ.

[7] Note the difference between this condition and the concept of *positive definite* matrix. In the first case we have $\mathbf{T}_1(j \mid l) \geq 0$ for all j, l; while a $(n \times n)$ matrix is *positive definite* over the real numbers if the following condition is met: $\sum_{jl}^n \mathbf{T}_1(j \mid l) \, k_j k_l > 0$ for all the set of real numbers $\{k_l\}$. When this condition is ≥ 0 we say that the matrix is *positive semi-definite*.

2. The sum of elements of any column is equal to unity. This follows from (6.2) by normalizing the conditional probability:

$$\sum_s \mathbf{T}_1(s \mid s') = \sum_{x_i} P(x_i, t_i + 1 \mid x_j, t_i) = 1, \quad \forall s'.$$

Guided Exercise (**Invariant State**). Let $\mathbf{T}_1$ be an arbitrary Markov matrix of dimension $(n \times n)$. The property (2) ensures that there is a vector $\psi = (1, 1, \cdots, 1)$ of dimension n such that under left-multiplication of matrix $\mathbf{T}_1$, the result is again the vector ψ; that is: $\psi \cdot \mathbf{T}_1 = \psi$. This result has important implications as it indicates an eigenvector ϕ_1 (to the right) having eigenvalue 1 and representing an invariant state (probability); that is to say: $\mathbf{T}_1 \cdot \phi_1 = \phi_1$. Hence the usefulness of solving the eigenvectors problem associated with a Markov matrix. The analysis of the degeneration of the eigenvalue 1 is related to the study of the relaxation to the stationary state, this is presented in Appendix B.

Optional Exercise (**Markov Chain** 2×2). Let the transition matrix $\mathbf{T}_1$ be

$$\mathbf{T}_1 = \begin{pmatrix} 1 - a & b \\ a & 1 - b \end{pmatrix}, \tag{6.5}$$

where $0 < a \le 1$ and $0 < b \le 1$. Show that the solution of a Markov chain characterized by the matrix (6.5) is

$$\mathbf{T}_n = \frac{1}{b/a + 1} \begin{pmatrix} (b/a) + (1 - a - b)^n & (b/a)(1 - (1 - a - b)^n) \\ (1 - (1 - a - b)^n) & 1 + (b/a)(1 - a - b)^n \end{pmatrix}.$$

Show that in the asymptotic limit $n \to \infty$ the conditional probability matrix is

$$\lim_{n \to \infty} \mathbf{T}_n = \begin{pmatrix} \frac{1}{a/b+1} & \frac{1}{a/b+1} \\ \frac{1}{b/a+1} & \frac{1}{b/a+1} \end{pmatrix}, \tag{6.6}$$

i.e., (6.6) is the stationary probabilistic solution of the *discrete-time* dichotomic noise. Calculate the correlation function $\langle\langle s(n) s(0) \rangle\rangle_{\text{st}}$ and study the limit when $n \to \infty$. Hint: the matrix $\mathbf{T}_1$ has eigenvalues $\lambda_l = \{1, 1 - a - b\}$ and (unnormalized) eigenvectors $\{(b/a, 1), (-1, 1)\}$ respectively. The value $a + b = 1$ shows a critical situation, in this case interpret the behaviors of elements of matrix $\mathbf{T}_n$.

Guided Exercise (**Sylvester's Criterion**). Let the transition matrix be

$$\mathbf{T}_1 = \begin{pmatrix} 1 & 0,1 & 0,2 \\ 0 & 0,1 & 0,8 \\ 0 & 0,8 & 0 \end{pmatrix}. \tag{6.7}$$

Obviously this matrix characterizes a 3×3 Markov chain. However, this is not a positive definite matrix. To study whether a matrix is *positive definite* or not, Sylvester's criterion is often very useful, namely a matrix is positive definite if all its principal minor determinants are positive. Let us use this criterion for matrix (6.7). In this case it is easy to see that by eliminating row 1 and column 1 we get $\det_{11} = -0,64$; deleting row 2 and column 2 the result is zero; deleting row 3 and column 3 we get $\det_{33} = 0,1$; the main determinant gives $\det[\mathbf{T}_1] = -0,64$; and $\mathbf{T}_1(s \mid s) \geq 0, \forall s$. From all these results it follows that the Sylvester criterion is not met, therefore the matrix (6.7) is not positive definite. On the other hand, in this case we can easily see that its eigenvalues are $\{1; 0,8515; -0,7515\}$.

Guided Exercise (**Markov Chain** 3×3). A problem of interest in biology is to determine the genetic evolution of a given genotype g_j, i.e., whether there will be or not descendants of g_j after n generations. Suppose a Markov matrix $\mathbf{T}_1$ models the occurrence of possible genotypes after one generation. If $\mathbf{T}_1$ is characterized by a 3×3 matrix of the form:

$$\mathbf{T}_1 = \begin{pmatrix} 1 & 1/4 & 1/18 \\ 0 & 1/2 & 4/9 \\ 0 & 1/4 & 1/2 \end{pmatrix}, \tag{6.8}$$

this matrix indicates that in one generation there may not be descendants of type g_2 or g_3 from type g_1. Determine the conditional probability $\mathbf{T}_n(j \mid l)$, i.e., the probability that a particular genotype g_l will have descendants of type g_j after n generations. In this case it is advisable to study the eigenvalues and eigenvectors of the matrix (6.8), which are

$$\begin{aligned} \lambda_1 &= 1 \quad , \mathbf{X}_1 = (1,0,0) \\ \lambda_2 &= 5/6 \,, \mathbf{X}_2 = (-7/3, 4/3, 1) \\ \lambda_3 &= 1/6 \,, \mathbf{X}_3 = (1/3, -4/3, 1). \end{aligned} \tag{6.9}$$

From this expression we observe that only one eigenvector (corresponding to the eigenvalue 1) has all its components greater than or equal to zero. This result is just a particular case of Perron-Frobenius' Theorem establishing certain properties of the steady state of a Markov matrix[8]; in particular the positivity of the eigenvector associated with the eigenvalue $\lambda_1 = 1$. From (6.9) we can find the transformation matrix[9] $\mathbf{U}$ and its inverse $\mathbf{U}^{-1}$, that is,

---

[8]A simple presentation of this theorem can be seen in van Kampen's book [1]. In Appendix B an alternative approach related to this fact is presented.

[9]Note the particular way in which the matrix $\mathbf{U}$ is constructed, that is, "hanging" eigenvectors in an orderly manner.

$$\mathbf{U} = \begin{pmatrix} 1 & -7/3 & 1/3 \\ 0 & 4/3 & -4/3 \\ 0 & 1 & 1 \end{pmatrix} \quad \text{and} \quad \mathbf{U}^{-1} = \begin{pmatrix} 1 & 1 & 1 \\ 0 & 3/8 & 1/2 \\ 0 & -3/8 & 1/2 \end{pmatrix}.$$

Calculate the matrix $\mathbf{T}_n(j \mid l)$ using that $\mathbf{T}_D^n = \left(\mathbf{U}^{-1} \cdot \mathbf{T}_1 \cdot \mathbf{U}\right)^n$, then:

$$\mathbf{T}_n = \mathbf{U} \cdot \begin{pmatrix} 1 & 0 & 0 \\ 0 & 5/6 & 0 \\ 0 & 0 & 1/6 \end{pmatrix}^n \cdot \mathbf{U}^{-1}.$$

Calculate the elements of the matrix $\mathbf{T}_n(j \mid l)$ (the conditional probability) as $n \to \infty$ (i.e., after infinite generations). Show that this probability is characterized by the matrix:

$$\lim_{n \to \infty} \mathbf{T}_n \to \begin{pmatrix} 1 & 1 & 1 \\ 0 & 0 & 0 \\ 0 & 0 & 0 \end{pmatrix}.$$

This means that only genotype g_1 is stable, in the sense that is independent of the initial condition; that is, if $\mathcal{P} = (\mathcal{A}, \mathcal{B}, \mathcal{C})$ is an arbitrary initial condition, it follows that

$$\lim_{n \to \infty} \mathbf{T}_n \cdot \mathcal{P} = (1, 0, 0).$$

6.2 Random Walk

A Random Walk (RW) is an **sp** $X_\Omega(t)$ in which time $t \in \mathcal{N}$ and the domain of the process is in a lattice (finite or infinite). Then an RW is a special case of Markov chain. Usually an RW is defined by a recurrence relation between the probability $P(s, n)$ to be on site[10] s in step (time) n, and the probability $P(s', n-1)$ to be somewhere else s' in the previous step. This recurrence relation can be obtained from the marginal probability of the joint probability of 2-times $P_2(s, n; s', n')$ and the use of the (formal) solution of a Markov chain (6.4).

From the definition of a temporarily homogeneous RW and using the concept of marginal probability it follows that

---

[10] Here we use the letter s to characterize a "site" on a regular $1D$ lattice, but its generalization to an arbitrary number of dimensions is obvious. On the other hand, it is worth noting that when we speak of a large number of dimensions we always refer to hyper cubic lattices; analysis to other regular lattices is also feasible [2]. RWs on a hexagonal two-dimensional lattice have been used to describe superionic conductors; see advanced exercise 7.5.3, or C.B. Briozzo, C.E. Budde, M.O. Cáceres, Phys. Rev A, 39, 6010 (1989).

$$P(s,n) = \sum_{s'} P_2(s,n;s',n')$$

$$= \sum_{s'} P(s,n \mid s',n') P(s',n')$$

$$= \sum_{s'} \mathbf{T}_1^{n-n'}(s \mid s') P(s',n').$$

Then, using the Chapman-Kolmogorov equation we get:

$$P(s,n) = \sum_{s'} \sum_{s''} \mathbf{T}_1(s \mid s'') \mathbf{T}_1^{n-1-n'}(s'' \mid s') P(s',n') \qquad (6.10)$$

$$= \sum_{s''} \mathbf{T}_1(s \mid s'') P(s'',n-1).$$

This is the recurrence relation we sought, connecting the probability $P(s,n)$ with $P(s'',n-1)$ through the transition probability $\mathbf{T}_1(s \mid s'')$, the so-called matrix of jumps (or hopping matrix). Different RWs can be put in correspondence with different structured models of *Hopping*.

If $\mathbf{T}_1(s \mid s') = \mathbf{T}_1(s-s')$, then there is translational invariance in the sample space, and in this particular case, expression (6.10) is further simplified. That is, in this case the function $\mathbf{T}_1(s-s')$ characterizes the length of the jump, so now the difference $s-s'$ is a **rv** and (6.10) is reduced to

$$P(s,n) = \sum_{s'} \mathbf{T}_1(s-s')P(s',n-1). \qquad (6.11)$$

This is a typical recurrence relation that can be easily solved by Fourier representation ($s \to k$). This is an important advantage relative to solution (6.4), when the interest is in solving an RW on a lattice of arbitrary dimension.

Example (**The Usual Random Walk**). If the lattice involves an infinite 1D sample space (i.e., $s \in [0,\pm 1,\pm 2,\cdots] \equiv \mathcal{Z}$), the transition probabilities are to first neighbors, and the hopping structure is modeled by $\mathbf{T}_1(s-s') = p\delta_{s,s'-1} + q\delta_{s,s'+1}$ (where $p+q=1$). Then, thanks to the spatial convolution in (6.11), it is immediate to take the discrete Fourier transform[11]; that is:

$$P(k,n) \equiv \sum_s \exp(iks) P(s,n) = \sum_s \exp(iks) \sum_{s'} \mathbf{T}_1(s-s')P(s',n-1) \qquad (6.12)$$

$$= \sum_s \sum_{s'} \exp(ik(s-s')) \exp(iks') \mathbf{T}_1(s-s')P(s',n-1).$$

---

[11]It is the same Fourier transform presented in Chap. 1 when we introduced the characteristic function. The difference now is that we *add* rather than *integrate* due to the discrete nature of the sample space of the **rv** s.

Now, introducing the change of variable $s'' = s - s'$ we get

$$P(k, n) = \sum_{s''} \exp(iks'') \mathbf{T}_1(s'') \sum_{s'} \exp(iks') P(s', n - 1) \tag{6.13}$$

$$= \mathbf{T}_1(k) P(k, n - 1),$$

where $\mathbf{T}_1(k) = p\exp(-ik) + q\exp(+ik)$ is the characteristic function of the hopping probability $\mathbf{T}_1(s - s')$. In the symmetric case ($p = q = 1/2$) we get $\mathbf{T}_1(k) = \cos(k)$. Note that even in this particular case, it is still necessary to solve the recurrence relation

$$P(k, n) = \cos(k) P(k, n - 1).$$

This equation can be solved using the technique of generating functions.

6.2.1 Generating Functions

The general problem involves solving recurrence relations of the form

$$P(k, n + 1) = \sum_{m=0}^{n} \mathbf{L}_m(k) P(k, m), \tag{6.14}$$

where $(n, m) \in [0, 1, 2, 3, \cdots]$.

Equation (6.14) can be solved by defining the generating function

$$G(k, Z) = \sum_{n=0}^{\infty} Z^n P(k, n), \tag{6.15}$$

from which it follows that

$$P(k, n) = \frac{1}{n!} \left[\frac{\partial^n}{\partial Z^n} G(k, Z) \right]_{Z=0}. \tag{6.16}$$

The case we are now interested in is defined by $\mathbf{L}_m(k) = \delta_{m,n} \mathbf{T}_1(k)$. More general cases where $\mathbf{L}_m(k)$ has contributions with $m \neq n$ are non-Markovian chains [3]; see, for example, Sect. 8.5.1.

Taking $\mathbf{L}_m(k) = \delta_{m,n} \mathbf{T}_1(k)$, multiplying (6.14) by Z^{n+1}, and summing over all n we get:

$$\sum_{n=0}^{\infty} Z^{n+1} P(k, n + 1) = G(k, Z) - P(k, n = 0)$$

$$= \sum_{n=0}^{\infty} Z^{n+1} \mathbf{T}_1(k) P(k, n)$$

$$= Z\mathbf{T}_1(k) G(k, Z),$$

from which it follows that

$$G(k, Z) = \frac{P(k, n = 0)}{1 - Z\mathbf{T}_1(k)};$$

(6.17)

then Eq. (6.16) is reduced to

$$P(k, n) = \frac{1}{n!} \left[\frac{\partial^n}{\partial Z^n} \frac{P(k, n = 0)}{1 - Z\mathbf{T}_1(k)} \right]_{Z=0}.$$

(6.18)

Alternatively, Eq. (6.17) can also be written as

$$G(k, Z) = P(k, n = 0) \sum_{n=0}^{\infty} Z^n \left(\mathbf{T}_1(k)\right)^n,$$

from where we immediately see that

$$P(k, n) = \left(\mathbf{T}_1(k)\right)^n P(k, n = 0),$$

(6.19)

which is just a particular case of solution (6.4). Finally, introducing the inverse discrete Fourier transform, the probability in real space is found to be[12]

$$P(s, n) = \frac{1}{2\pi} \int_{-\pi}^{\pi} \left(\mathbf{T}_1(k)\right)^n P(k, n = 0) \exp(-iks) \, dk.$$

(6.20)

Depending on the analyticity properties of the characteristic function $\mathbf{T}_1(k)$ around $k = 0$, it is possible to model different behaviors in an RW. This will be discussed in detail in the following sections.

6.2.2   Moments of a Random Walk

Once the function $\mathbf{T}_1(k)$ is given (which characterizes the lattice structure and the jumps model on it) the problem of a Markov chain is completely solved from the solution (6.20). In particular, it may happen that we are only interested in moments of displacement of the RW, that is,

$$\langle s(n)^p \rangle = \sum_s s^p P(s, n); \quad \forall p = 1, 2, \cdots.$$

---

[12]In general, since $s \in \mathcal{Z}$, we can show that if we use $\frac{1}{2\pi} \int_0^{2\pi} e^{ik(s-s')} \, dk = \delta_{s,s'}$, we can also write $P(s, n) = \frac{1}{2\pi} \int_0^{2\pi} e^{iks} P(k, n) \, dk$. That is to say, eigenfunctions e^{iks} are complete for $k \in [0, 2\pi]$.

In this case moments can be studied directly from $P(k, n)$, without resorting to the inverse Fourier transform, as follows

$$\langle s(n)^p \rangle = \sum_s \frac{\partial^p}{\partial (ik)^p} \exp(iks) P(s, n) \Big|_{k=0} = \frac{\partial^p}{\partial (ik)^p} P(k, n) \Big|_{k=0}$$

$$= \frac{\partial^p}{\partial (ik)^p} \left(\mathbf{T}_1(k) \right)^n P(k, n = 0) \Big|_{k=0}.$$

Example (**Symmetrical One Step Random Walk in 1D**). For a 1D lattice and in the symmetrical case $\mathbf{T}_1(k) = \cos(k)$; then, if we take as an initial condition $P(s, n = 0) = \delta_{s,0}$, it follows that odd moments are zero, but not even moments. For example, for the first moment we have

$$\langle s(n) \rangle = \frac{\partial}{\partial (ik)} (\cos(k))^n \Big|_{k=0} = \frac{-1}{i} n \sin(k) (\cos(k))^{n-1} \Big|_{k=0} = 0,$$

while for the second we obtain

$$\langle s(n)^2 \rangle = \frac{\partial^2}{\partial (ik)^2} (\cos(k))^n \Big|_{k=0} = n.$$

This is precisely the typical behavior of a diffusive transport; i.e., linear in the discrete time n. Furthermore, the proportionality coefficient is the diffusion coefficient of the medium, which is usually defined as

$$d = \frac{\langle s(n)^2 \rangle}{2n}.$$

In arbitrary dimension, and for the isotropic case, $d = \langle | s(n) |^2 \rangle / (2Dn)$, where D is the dimension of the space.

Exercise (**Poisson's Random Walk**). Consider an RW in an infinite 1D lattice

$$P(s, n) = \sum_{s' \in \mathbb{Z}} \mathbf{T}_1(s - s') P(s', n - 1), \quad P(s, n = 0) = \delta_{s,0}, \tag{6.21}$$

where the hopping structure is

$$\mathbf{T}_1(s - s') = \mathcal{K} e^{-a} \left(\frac{a^{|s-s'|}}{|s-s'|!} - \delta_{s,s'} \right), \quad a > 0.$$

Note that for this RW model the likelihood of making a jump of zero length is zero, because $\mathbf{T}_1(0) = 0$. Show that the structure function is

$$\mathbf{T}_1(k) = \mathcal{K} e^{-a} \left(\exp\left(ae^{ik}\right) + \exp\left(ae^{-ik}\right) - 2 \right),$$

where the normalization constant is

$$\mathcal{K} = \frac{1}{2(1 - e^{-a})}.$$

Show that the first moment is zero, while the second is

$$\langle s(n)^2 \rangle = \frac{a(1 + a)n}{(1 - e^{-a})}.$$

Note that even when very long jumps are allowed, the behavior of the second moment remains linear in discrete time n. Why is it so? Study the diffusion coefficient in the asymptotic limit $a \ll 1$.

Exercise (**Random Walk with Geometric Jumps**). Consider an RW in an infinite 1D lattice when the structure of jumps is given by a geometric probability distribution (sometimes called Pascal's distribution)

$$\mathbf{T}_1(s - s') = \mathcal{K}\left(\gamma^{|s - s'|} - \delta_{s,s'}\right), \quad \text{with} \quad 0 < \gamma < 1.$$

In this case the very long jumps are weighted by a nonzero probability of occurrence $\propto \gamma^{|s - s'|}$. Show that the normalization constant is $\mathcal{K} = (1 - \gamma)/2\gamma$ and the RW probability is given by

$$P(s, n) = \frac{(1 - \gamma)^n}{2\pi} \int_{-\pi}^{\pi} \left(\frac{\cos(k) - \gamma}{1 - 2\gamma \cos(k) + \gamma^2}\right)^n P(k, n = 0) \exp(-iks) \, dk.$$

Calculate the first and second moments of the displacement of the RW versus the discrete time (number of steps n). Is it a diffusive behavior? Why is it not possible to formulate a similar model in which $\gamma \geq 1$?

Exercise (**Random Walk in a 2D Lattice with Drift**). Consider an RW in a 2D lattice characterized by the bidimensional jump structure:

$$\mathbf{T}_1(s - s') = \frac{1}{4}(1 + B)(\delta_{s,s'+e_1} + \delta_{s,s'+e_2}) + \frac{1}{4}(1 - B)(\delta_{s,s'-e_1} + \delta_{s,s'-e_2}),$$

where lattice sites can be written in terms of vectors in the plane, in the form $s = s_1 e_1 + s_2 e_2$. So,

$$\mathbf{T}_1(\mathbf{k}) = \frac{1}{4}(1 + B)\left(e^{ik_1} + e^{ik_2}\right) + \frac{1}{4}(1 - B)\left(e^{-ik_1} + e^{-ik_2}\right).$$

Here $\mathbf{k}$ is a vector[13] in two dimensions. The meaning of the quantity B comes from the structure function; that is, there is a drift to $45°$ from axis e_1. From the definition of p-th moment [and using the initial condition $P(\mathbf{s}, n = 0) = \delta_{0,\mathbf{s}}$] we conclude that

$$\langle s_1(n)^p \rangle = \frac{\partial^p}{\partial(ik_1)^p} P(\mathbf{k}, n)\bigg|_{\mathbf{k}=0} = \frac{\partial^p}{\partial(ik_1)^p} (\mathbf{T}_1(\mathbf{k}))^n \bigg|_{\mathbf{k}=0}.$$

Show that the second moment in the direction e_1 or e_2 is given by

$$\langle s_1(n)^2 \rangle = \langle s_2(n)^2 \rangle = \frac{n(n-1)B^2 + 2n}{4}.$$

Then the total second moment is given by $\langle |\,\mathbf{s}(n)\,|^2 \rangle = \frac{n(n-1)B^2+2n}{2}$. Note that if $B \neq 0$, the second moment, for $n \gg 1$, has a quadratic behavior in time. Why is it so? From this result and for the case $B = 0$ (with zero drift), we observe that the diffusion coefficient is

$$d = \frac{\langle |\,\mathbf{s}(n)\,|^2 \rangle}{2Dn} = \frac{1}{4},$$

where, of course, we have used that $D = 2$.

Excursus (**Lévy's RW in 1D**). Consider that the jump structure in Fourier space is given by Weierstrass' function:

$$\mathbf{T}_1(k) = \sum_s \mathbf{T}_1(s)\, e^{iks} = \frac{a-1}{a} \sum_{l=0}^{\infty} \frac{1}{a^l} \cos(b^l k), \quad a > 1, \ b > 0.$$

The RW associated with this model has a statistically self-similar jump structure for the visited points in the lattice, and this set of points conforms a cluster shaped fractal domain if the condition $b^2/a > 1$ is met. In [4] it is proved, defining $\mu = \ln a / \ln b$ and for $0 < \mu < 1$, that the fraction μ can be associated, for all time, with the fractal dimension of the "clusters" of points that the 1D Lévy's RW has visited.[14] That is, this RW has associated a probability of Lévy's type for the **rv** of the jump events in the real space [4, 5]. Remember that this characteristic function $\mathbf{T}_1(k)$ is not differentiable at $k = 0$ if $b^2/a > 1$; see Sect. 1.4.2. Therefore, if $b^2/a > 1$, the dispersion of this RW is not defined. On the concept of fractal dimension see Appendix H.

---

[13]Here, for convenience, we have used letters in "bold" to denote a $2D$ vector in the plane.

[14]In Chap. 1 we have mentioned that the RW generated by the Weierstrass function, for the jump structure, is called the Lévy RW.

Optional Exercise (**Cauchy's Random Walk**). Compare the Lévy RW (defined on a discrete sample space) with the Cauchy RW (defined on the continuum); see exercises in Sects. 3.3.1 and 1.4.2.

6.2.3   Realizations of a Fractal Random Walk*

Consider a composite model of two Lévy's RW, independent of each other, one for each coordinate axis $\{\hat{\mathbf{x}}, \hat{\mathbf{y}}\}$ of the 2D lattice. From the expression for a jump probability[15] of length x in the axis $\hat{\mathbf{x}}$

$$\mathbf{T}_1(x) = \frac{a-1}{2a} \sum_{m=0}^{\infty} 1/a^m \, (\delta_{x,b^m} + \delta_{x,-b^m}), \quad a > 1, \ b > 0,$$

and similarly for a jump of length y in the axis $\hat{\mathbf{y}}$, it is possible to generate numerically a 2-dimensional Lévy's RW. In a completely analogous way, a conventional 2D RW is generated by a probability of a jump of unit length, using, for example, $\mathbf{T}_1(s) = \frac{1}{2}(\delta_{s,1} + \delta_{s,-1})$ for each coordinate axis $\{\hat{\mathbf{x}}, \hat{\mathbf{y}}\}$ in a regular lattice.

Note that in the case of a Lévy RW there is an additional difficulty because of the commensurate/incommensurate nature of the jump lengths and the location of sites in the lattice. To achieve a numerical realization of a Lévy RW it is therefore necessary to take an integer value of b to ensure that a jump of length $x = b^m$, $\forall m$ shall always fall somewhere on the $\hat{\mathbf{x}}$ axis sites of the net, etc. (Though of course not all sites are generally accessible topologically.) From all these comments concerning the construction of an RW realization, we see that if the condition $b^2/a > 1$ is met, the realization of a Lévy RW shows a very different structure from any realization of a diffusive process (usual RW in the lattice, or the Wiener process in the continuous case). In particular, in the following three figures we show realizations of a walk starting at the origin of coordinates $(0, 0)$ and over a recorded time of $n = 10^4$ steps. In the three figures (i.e., in each realization) the lattice parameter was taken as unity for both coordinate axes $\{\hat{\mathbf{x}}, \hat{\mathbf{y}}\}$. In Fig. 6.1 case $(b = 2, a = 3)$ is shown, ensuring the divergence of the second moment of $\mathbf{T}_1(x)$ (or $\mathbf{T}_1(y)$). In this figure it is also possible to see that the realization of the particle leaves the "diffusive domain," making long trips outside the "scale" of the typical diffusion length $\mathcal{O}(\sqrt{n})$ (where n indicates the number of steps in the RW). On the other hand, the occurrence of a structure of clusters for visited sites is very noticeable, as noted previously. For the particular case $(b = 2, a = 4)$ a transition to a different regime is observed (but still long unusual excursions in a truly diffusive realization persist), this is shown in Fig. 6.2. Finally, in Fig. 6.3 the case $(b = 2, a = 5)$ is shown, i.e., a situation in which the second moment of the one-step jump probability $\mathbf{T}_1(x)$ (or $\mathbf{T}_1(y)$) does not diverge.

---

[15]That is to say, the Weierstrass probability.

Fig. 6.1 Realization of a
Lévy random walk on the 2D
lattice (x, y). The number of
steps taken by the walker is
$n = 10^4$. The parameters of
the characteristic function
(Weierstrass function for each
coordinate axis) involve:
$b^2/a = 4/3$. The *solid lines*
are for easy viewing of long
RW excursions

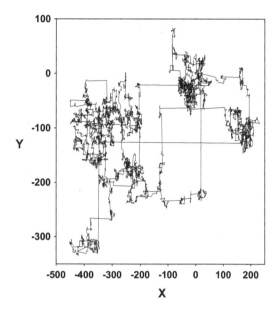

Fig. 6.2 Here the parameters
of the characteristic functions
involve: $b^2/a = 1$

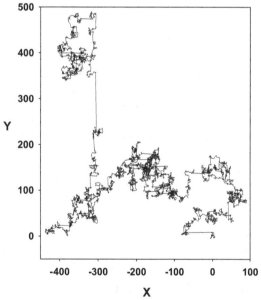

Realizations of a Lévy RW always display this type of behavior, with steep
trips if the parameters (b, a) are both greater than unity. Furthermore, the condition
$b^2/a > 1$ is necessary (but not sufficient) to obtain a structure of clusters (at all
times) for the visited sites in the lattice [4, 5]. The next exercise presents a simple
discussion related to the realizations of a usual RW.

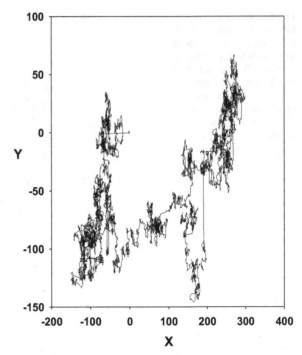

Fig. 6.3 Here the parameters of the characteristic functions involve: $b^2/a = 4/5$

Guided Exercise (**Limit of a Usual Random Walk**). Consider a 1D RW to be characterized by a hopping structure (in Fourier space) $\mathbf{T}_1(k)$. Show that if there is a limit in which the structure of jumps takes a form that approximates $\cos(k)$, in that limit the Markov chain tends to the usual RW (symmetrical). For example, if the jump structure is the Weierstrass one $\mathbf{T}_1(k) = \frac{a-1}{a} \sum_{l=0}^{\infty} \frac{1}{a^l} \cos(b^l k)$, we see that for $b \to 1$ with $a > 1$ we get $\mathbf{T}_1(k) \to \cos(k)$; that is, in this limit we recover the diffusive behavior of a conventional RW. Study the relevant limit for other models of RW, for example with geometric or Poisson jumps.

A different analysis is to consider the following problem [4, 5]: Given an arbitrary structure function $\mathbf{T}_1(k)$ under what conditions does the RW behave diffusively at long times? In a different context and in the continuous-space case, it is interesting also to check the existence of a diffusive regime for non-Markovian processes; see *excursus* in Sect. 3.14.

Excursus (**Continuous Limit**). Consider a 1D RW to be characterized by the Weierstrass function. Let Δ and τ be spatial and temporal length scales, respectively. Then, if in Lévy's RW we take $a \sim 1 + \alpha\Delta$, $b \sim 1 + \beta\Delta$ and $n = t/\tau$ in such a way that Δ and τ approach zero with the proviso that:

$$\mu = \frac{\ln a}{\ln b} \quad \text{and} \quad \frac{\Delta^\mu}{\tau}$$

are constant quantities [5], the probability density $P(X, t)$ of finding the particle (or the RW in the continuous case) at position X and at instant t converges to the probability distribution of Lévy's flight. That is to say, the characteristic function is $P(k, t) \sim \exp(-Ct \mid k \mid^{\mu})$, where C is a constant. In the second *excursus* of Sect. 3.14 we present an introduction to the analysis of Lévy's flights; see also Appendix H.2.

6.3 Master Equation (Diffusion in the Lattice)

Solving the Chapman-Kolmogorov integral equation is a difficult task; see Sect. 3.3.1. However, sometimes it is possible to make a differential analysis. In particular, when the process is of continuous-time and the domain of the random variables is discrete (finite or infinite) it is easy to find a differential equation for the evolution of the conditional probability of the system. Using the shorthand notation

$$P(i, t \mid j, t') \equiv P(x_i, t \mid x_j, t'),$$

consider a situation in which the propagator (conditional probability) for short times has the following infinitesimal behavior [1]:

$$P(i, t + \Delta t \mid j, t) = \left(1 - \sum_m W_{m,j}(t) \, \Delta t\right) \delta_{i,j} + W_{i,j}(t) \, \Delta t + \mathcal{O}(\Delta t^{\theta}), \quad (6.22)$$

with $\theta > 1$. Obviously the probability (6.22) satisfies the normalization condition, and for $\Delta t = 0$ the adequate initial condition is obtained. The physical meaning of $W_{i,j}(t)$ is a transition probability (per unit time[16]) from the j-th to the i-th state at time t. Introducing (6.22) in the Chapman-Kolmogorov equation, the evolution equation for the propagator $P(i, t + \Delta t \mid j, t)$ is obtained. To prove this fact, we first write the Chapman-Kolmogorov equation in matrix form:

$$P(i, t + \Delta t \mid j, t_0) = \sum_l P(i, t + \Delta t \mid l, t) P(l, t \mid j, t_0), \quad (6.23)$$

where we have used the temporal ordering $t + \Delta t \geq t > t_0$. If we now apply the short times approximation (6.22) in (6.23), we get to order $\mathcal{O}(\Delta t)$

$$P(i, t + \Delta t \mid j, t_0) = \sum_l \delta_{i,l} \left(1 - \sum_m W_{m,l}(t) \, \Delta t\right) P(l, t \mid j, t_0) \quad (6.24)$$

---

[16]Note that the diagonal elements $W_{j,j}(t)$ are not involved in the definition (6.22), and that the only restriction on $W_{j,l}(t)$ is to be positive.

$$+ \sum_l W_{i,l}(t) \, \Delta t \, P\left(l, t \mid j, t_0\right)$$

$$= \left(1 - \sum_m W_{m,i}(t) \, \Delta t\right) P(i, t \mid j, t_0) + \Delta t \sum_l W_{i,l}(t) \, P(l, t \mid j, t_0).$$

Then we can build an incremental differential for the conditional probability

$$\frac{1}{\Delta t}\left[P\left(i, t + \Delta t \mid j, t_0\right) - P\left(i, t \mid j, t_0\right)\right] = -P\left(i, t \mid j, t_0\right) \sum_m W_{m,i}(t) \qquad (6.25)$$

$$+ \sum_l W_{i,l}(t) \, P\left(l, t \mid j, t_0\right),$$

so that in the $\Delta t \to 0$ limit it leads to a differential equation of first order called the Master Equation (EM).

$$\frac{d}{dt}P(i, t \mid j, t_0) = \sum_l W_{i,l}(t) \, P(l, t \mid j, t_0) - P(i, t \mid j, t_0) \sum_m W_{m,i}(t). \qquad (6.26)$$

Exercise. Note that the ME was obtained for the propagator $P(i, t \mid j, t_0)$, not for the 1-time probability $P(i, t)$ even though it satisfies the same evolution equation. Show from Eq. (6.26) that terms that come from the transition probabilities $W_{i,i}(t)$ vanish in the sum.

In a compact form the ME is written as

$$\frac{d}{dt}P(i, t \mid j, t_0) = \sum_l \mathbf{H}_{i,l} P(l, t \mid j, t_0), \qquad (6.27)$$

where we have defined the elements of the matrix $\mathbf{H}$ in the form

$$\mathbf{H}_{i,l} \equiv W_{i,l} - \delta_{i,l} \sum_m W_{m,i}. \qquad (6.28)$$

The ME is a balance equation (for the i-th state) between gain and loss for each state. In general we will always be interested in situations where $W_{s,s} = 0$, this choice is related to the problem of permanence in a given state s, which is usually the case when we work with a lattice.[17] To differentiate this situation we will use the notation s instead of the notation i-th for the state, then the transition element $W_{s,s}$ will not be considered.

---

[17]That is, we do not consider those physical processes in which the particle "jumps" from a site s to the same site s in a finite time interval.

In the matrix representation also it is often useful to introduce the notation $<bra|$ and $|ket>$, where $|s>$ represents a state in site s of the D-dimensional lattice, i.e.[18]

$$\mathbf{H} = \sum_{s,s'(\neq)} |s> W_{s,s'} <s'| - \sum_s |s> \Gamma_s <s|. \qquad (6.29)$$

Here $\Gamma_s = \sum_{s'} W_{s',s}$ and $W_{s',s}$ indicates the probability per unit time to make a transition from site s to site s'.

Exercise. Show from (6.28) that the diagonal matrix elements of $\mathbf{H}$ are

$$\mathbf{H}_{l,l} = W_{l,l} - \sum_m W_{m,l} = -\sum_{m(\neq l)} W_{m,l}.$$

On the other hand, from (6.29) we note (using Kronecker's delta $<s|s'> = \delta_{s,s'}$) that the diagonal elements are

$$\mathbf{H}_{s,s} = -\sum_{s'} W_{s',s}.$$

In general, $\mathbf{H}$ may depend on time through the transition elements $W_{s,s'}(t)$, but the ME can always be written in a compact form[19] in terms of the conditional probability matrix $\mathbf{P}$ as:

$$\dot{\mathbf{P}} = \mathbf{H} \cdot \mathbf{P}, \qquad (6.30)$$

where it is understood that both $\mathbf{P}$ and $\mathbf{H}$ are matrices of the same dimension.

Exercise. In the general case when $\mathbf{H} = \mathbf{H}(t)$, the formal solution of (6.30) is

$$\mathbf{P}(t) = \hat{\mathcal{T}} \left(\exp \int_0^t \mathbf{H}(s) \, ds \right) \cdot \mathbf{P}(0),$$

here the expansion of the exponential operator has to be ordered chronologically; see advanced exercises 4.9.9.

It is to be noted that $\mathbf{H}$ satisfies two important properties that characterize all Markov matrices $\mathbf{H}$ for discrete stochastic processes continuous in time [even in the case when $\mathbf{H} = \mathbf{H}(t)$ is explicitly a function of time], namely:

$$\sum_s \mathbf{H}_{s,s'} = 0, \qquad (6.31)$$

---

[18] The notation $<bra|$ and $|ket>$ comes from the word "braket" (parentheses) and was introduced by Dirac; see advanced exercise 6.7.1.

[19] In general, we use the notation $\dot{\mathbf{P}}$ to indicate time derivative, i.e., $\dot{\mathbf{P}} \equiv d\mathbf{P}/dt$. Furthermore $\mathbf{H} \cdot \mathbf{P}$ indicates a matrix product.

and

$$\mathbf{H}_{s,s'} \geq 0, \quad \text{if} \quad s \neq s'. \tag{6.32}$$

Property (6.31) is inferred by normalizing the conditional probability[20]

$$\frac{d}{dt} \sum_s \mathbf{P}(t)_{s,s'} = 0 = \sum_s \frac{d\mathbf{P}(t)_{s,s'}}{dt} = \sum_s \left(\sum_{s_2} \mathbf{H}_{s,s_2} \cdot \mathbf{P}(t)_{s_2,s'} \right),$$

that is to say:

$$0 = \sum_{s_2} \left(\sum_s \mathbf{H}_{s,s_2} \right) \cdot \mathbf{P}(t)_{s_2,s'}, \quad \forall \mathbf{P}(t)_{s_2,s'} \quad \Rightarrow \quad \sum_s \mathbf{H}_{s,s'} = 0.$$

Condition (6.32) follows from the definition of probability (6.28).

In the stationary case ($W_{s',s}$ are independent of time), and from the condition of temporal homogeneity $P(s, t + t_0 \mid s', t_0) = P(s, t \mid s', 0)$, it is convenient to introduce the following notation

$$P(s, t + t_0 \mid s', t_0) = \, <s \mid \mathbf{P}(t) \mid s'> \, \equiv \mathbf{P}(t)_{s,s'}. \tag{6.33}$$

Excursus. Note that the properties (6.31) and (6.32) are not sufficient to ensure that there exists a matrix $\mathbf{U}$ such that $\mathbf{U}^{-1} \cdot \mathbf{H} \cdot \mathbf{U}$ is diagonal [something similar happens with the Markov matrix $\mathbf{T}_1$, see Sect. 6.1.1]. Only the principle of detailed balance[21] allows to symmetrize $\mathbf{H}$ under a certain scalar product.[22] This means that in general we can only ensure a similarity transformation to the Jordan form. We propose exercises to clarify this further.

Exercise. Interpret the differences between the Markov matrix $\mathbf{T}_1$ (Markov chains with discrete time) and the matrix $\mathbf{H}$ (for continuous-time Markov processes); in particular the normalizations: $\sum_s \mathbf{T}_1(s \mid s') = 1$ and $\sum_s \mathbf{H}_{s,s'} = 0$ respectively. Hint: assuming a characteristic time scale Δt write $\mathbf{T}_1 = e^{\mathbf{H}\Delta t}$ and expand the exponential operator.

---

[20] As for Markov chains, this property ensures that the matrix $\mathbf{H}$ has a *left* eigenvector $\psi = (1, 1, \cdots, 1)$ with eigenvalue 0. Concerning the problem of degeneration of this eigenvalue see Appendix B.

[21] In advanced exercise 4.9.3 we have already stressed that in the discrete case the principle of detailed balance is very easy to write: $W_{j,l}P_{eq}(l) = W_{l,j}P_{eq}(j)$. The repeated index convention does not apply here.

[22] In Chap. 4 we analogously proved that the F-P operator is always diagonalizable if the principle of detailed balance is met.

Excursus. If the matrix $\mathbf{H}$ is not reducible (see Appendix B), properties (6.31) and (6.32) are sufficient to ensure that there is a single stationary state, and thus the relaxation to it from any initial condition $\mathcal{P}$. That is: $\lim_{t \to \infty} e^{\mathbf{H}t} \cdot \mathcal{P} \to \mathbf{P}_{st}$.

6.3.1 Formal Solution (Green's Function)

The solution of (6.30) for the stationary case, i.e., when $\mathbf{H}$ does not explicitly depend on time, can formally be written using the Laplace representation

$$\mathbf{P}(u) = \int_0^\infty e^{-ut} \mathbf{P}(t) \, dt. \tag{6.34}$$

Introducing this transformation into (6.30) we obtain

$$u\mathbf{P}(u)_{s,s'} - \mathbf{P}(t=0)_{s,s'} = \sum_{s_2} \mathbf{H}_{s,s_2} \mathbf{P}(u)_{s_2,s'}. \tag{6.35}$$

From this expression it follows, using $\mathbf{P}(t=0) = \mathbf{1}$, that

$$\mathbf{P}(u) = [u\mathbf{1} - \mathbf{H}]^{-1}, \tag{6.36}$$

where $\mathbf{1}$ indicates the identity matrix of the same dimension as the matrix $\mathbf{H}$. Given the similarity between this solution and the Green function in quantum mechanics, $G(u) = [u\mathbf{1} - \mathbf{H}]^{-1}$, the matrix $\mathbf{H}$ is often called the master Hamiltonian of the RW, but no extrapolation should be made from this similarity of names. The matrix solution in the temporal representation has to satisfy the initial condition $\mathbf{P}(t=0) = \mathbf{1}$, which in the Laplace variable implies

$$\mathbf{P}(u \to \infty) = \frac{1}{u}\mathbf{1},$$

in agreement with (6.36).

Optional Exercise (**Stationary Case**). Find the matrix element $\mathbf{P}_{s,s'}(t)$ of the formal solution when the matrix $\mathbf{H}$ is an explicit function of time. Hint: use the time-ordered exponential operator.

Exercise (**Dichotomous Process**). Let a two-level system $(\Delta, -\Delta)$ be characterized by the following master Hamiltonian:

$$\mathbf{H} = \begin{pmatrix} -\gamma & \alpha \\ \gamma & -\alpha \end{pmatrix}.$$

That is, α is the transition probability per unit time to make the "jump" $(-\Delta) \longmapsto$ (Δ) and γ for the transition $(\Delta) \longmapsto (-\Delta)$. Calculate the conditional probability $P(t)$, compare this solution with the expression given in the optional exercise of Sect. 3.3.1. Calculate the stationary correlation function $\langle\langle s(t_1)s(t_2)\rangle\rangle$, and interpret the *correlation time* in terms of the parameters α and γ. Hint: as in Sect. 6.1.1 diagonalize the matrix $\mathbf{H}$.

Example (**Non-diagonalizable H**). Let a three level system $(\downarrow, \circ, \uparrow)$ be characterized by the master Hamiltonian:

$$\mathbf{H} = \begin{pmatrix} -1 & 1 & 3 \\ 1/2 & -2 & 0 \\ 1/2 & 1 & -3 \end{pmatrix}. \tag{6.37}$$

It is easy to see that indeed (6.37) satisfies the requirements (6.31) and (6.32). However, this matrix cannot be diagonalized, so it is not possible to ensure that we could find a basis of eigenvectors. Furthermore, it is also easily verified that $\mathbf{H}$ does not satisfy the principle of detailed balance; i.e., this matrix characterizes indeed a situation away from equilibrium. Note that in the state $(\circ)$ there is only gain induced through the transition probability: $(\downarrow) \longmapsto (\circ)$. That is, in the stationary state there will be a current different from zero. However, even in this nonequilibrium situation, the solution of ME, for all t, can be obtained by studying the Jordan form of (6.37), and from it we can also get the corresponding stationary probability.

Excursus (**Jordan's Form**). Consider the equation $\dot{\mathbf{P}} = \mathbf{H} \cdot \mathbf{P}$ with the master Hamiltonian characterized by (6.37). Show that the eigenvalues of $\mathbf{H}$ are only two $\lambda_i (= 0, -3)$, here the degenerate eigenvalue is (-3) and does not support two linearly independent eigenvectors. It is easy to see that matrices

$$C = \begin{pmatrix} 4 & -1 & -1 \\ 1 & 1/2 & 1 \\ 1 & 1/2 & 0 \end{pmatrix} \quad \text{and} \quad C^{-1} = \begin{pmatrix} 1/6 & 1/6 & 1/6 \\ -1/3 & -1/3 & 5/3 \\ 0 & 1 & -1 \end{pmatrix}$$

lead $\mathbf{H}$ to its Jordan form $C^{-1} \cdot \mathbf{H} \cdot C = \begin{pmatrix} 0 & 0 & 0 \\ 0 & -3 & 1 \\ 0 & 0 & -3 \end{pmatrix}$. Then using the spectral expansion:

$$\mathbf{H} = \sum_i \lambda_i \mathcal{P}_i + \mathcal{N}_i,$$

where $\mathcal{P}_i$ and $\mathcal{N}$ are their respective projectors and nilpotents, and using the expansion of a matrix function:

$$f(A) = \sum_i f(\lambda_i)\mathcal{P}_i + f'(\lambda_i)\mathcal{N}_i + \frac{1}{2!}f''(\lambda_i)\mathcal{N}_i^2 + \frac{1}{3!}f'''(\lambda_i)\mathcal{N}_i^3 + \cdots$$

show that the conditional probability is given by

$$\mathbf{P}(t) \equiv e^{\mathbf{H}t} = \begin{pmatrix} 2/3 & 2/3 & 2/3 \\ 1/6 & 1/6 & 1/6 \\ 1/6 & 1/6 & 1/6 \end{pmatrix} + e^{-3t} \begin{pmatrix} 1/3 & -2/3 & -2/3 \\ -1/6 & 5/6 & -1/6 \\ -1/6 & -1/6 & 5/6 \end{pmatrix}$$

$$+ te^{-3t} \begin{pmatrix} 0 & -1 & 1 \\ 0 & 1/2 & -1/2 \\ 0 & 1/2 & -1/2 \end{pmatrix}. \tag{6.38}$$

We can also make an alternative perturbative analysis, without using the Jordan form, from a set of biorthonormal eigenvectors (see Appendix B). Using (6.38) calculate the stationary solution $\mathbf{P}(t \to \infty)$. At the stationary state, what is the level that has the highest probability of being occupied? Do we expect this result intuitively?

6.3.2 Transition to First Neighbors

In the general case of an arbitrary transition matrix $\mathbf{H}$, the ME cannot be solved in closed form, so it is necessary to use different approximations [1]. A very simple and extremely applicable model is the so-called transitions to first neighbors model. Consider an infinite $1D$ lattice, in this case ME can be written in the form:

$$\frac{dP(s,t \mid s_0, t_0)}{dt} = v_{s+1}P(s+1, t \mid s_0, t_0) + \mu_{s-1}P(s-1, t \mid s_0, t_0) - (v_s + \mu_s)P(s, t \mid s_0, t_0),$$
$$\tag{6.39}$$

where $s \in \mathcal{Z}$. Depending on the model, functions v_s and μ_s have different symmetries; this is discussed in detail in the following sections. In Fig. 6.4 a schematic representation of Eq. (6.39) is displayed. Furthermore, it is noteworthy that if v_s and μ_s depend on s, only the linear case ($v_s \propto s$, $\mu_s \propto s$) can be solved in a general form [6].

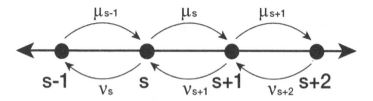

Fig. 6.4 Schematic representation of a $1D$ RW with asymmetric next neighbor transitions dependent of the sites

6.3.3   Solution of the Homogeneous Problem in One Dimension

Consider a homogeneous and symmetrical system, with natural boundary conditions and transitions to first neighbors in the $1D$ lattice. Since space is homogeneous it follows that $(\nu_s = \mu_s = \mu)$, then (6.39) is reduced to

$$\frac{dP(s, t \mid s_0, t_0)}{dt} = \mu P(s + 1, t \mid s_0, t_0) + \mu P(s - 1, t \mid s_0, t_0) - 2\mu P(s, t \mid s_0, t_0). \tag{6.40}$$

Obviously, because of the homogeneous nature of space and time, the conditional probability has translational invariance in the initial condition (s_0, t_0), then from now on we will take $(s_0 = 0, t_0 = 0)$. The solution of (6.40) can be obtained using the Z-transform. This method is very useful in solving other similar cases of finite difference equations,[23] which is why we introduce this technique here. First the generating function $F(Z, t)$ is defined in the form

$$F(Z, t) = \sum_{s=-\infty}^{\infty} Z^s P(s, t), \quad |Z| < 1; \tag{6.41}$$

then, using (6.41) in (6.40), the latter is reduced to

$$\frac{1}{\mu} \partial_t F(Z, t) = \left(Z + Z^{-1} - 2 \right) F(Z, t), \tag{6.42}$$

whose solution is obvious:

$$F(Z, t) = \Omega(Z) \exp \left(\mu t \left[Z + Z^{-1} - 2 \right] \right). \tag{6.43}$$

Using the initial condition $P(s, t = 0) = \delta_{s,0}$, the arbitrary function Ω takes the form $\Omega(Z) = 1$. Then $P(s, t)$ is given by the inverse Z-transform, that is:

$$P(s, t) = \frac{1}{2\pi i} \oint_{|Z|=1} Z^{-s-1} F(Z, t) \, dZ. \tag{6.44}$$

Alternatively, we can find $P(s, t)$ expanding $F(Z, t)$ in powers of Z. Then, from (6.43) we obtain

$$F(Z, t) = \exp(-2\mu t) \sum_{(k,m)=0}^{\infty} \frac{(t\mu)^{k+m}}{k! m!} Z^{k-m}. \tag{6.45}$$

---

[23] For example, when μ_s, ν_s are linear polynomials in s, and the system has boundary conditions [1].

Introducing the change of variable $k - m = s$ in (6.45) we can replace $\sum_{k=0}^{\infty} \rightarrow \sum_{s=-\infty}^{\infty}$; then, using (6.41) we note that

$$P(s,t) \equiv P(s,t \mid 0,0) = \exp(-2t\mu) \sum_{m=0}^{\infty} \frac{(t\mu)^{2m+s}}{(m+s)!\, m!} = \exp(-2t\mu)\, I_{|s|}(2t\mu).$$
(6.46)

Here $I_s(z)$ is the n-th modified Bessel's function. For the purpose of completeness we now introduce the Laplace transform of (6.46). We will use this formula several times when we solve by perturbation the problem of diffusion in disordered media.

$$P(s,u) = (2\mu)^{|s|} \left(u^2 + 4\mu u\right)^{-1/2} \left(u + 2\mu + \left(u^2 + 4\mu u\right)^{1/2}\right)^{-|s|}; \quad s \in \mathbb{Z}.$$
(6.47)

We now present different notations, all equivalent to one another, employed throughout the text:

$$P(s,u) \equiv\, <s \mid \mathbf{P}(u) \mid 0> =\, <s \mid [u\mathbf{1} - \mathbf{H}]^{-1} \mid 0> \equiv\, <s \mid G(u) \mid 0>,$$

note that according to (6.40) here we have used that $W_{s,s} = 0$ and $W_{s',s} = \mu$ if $s' = s \pm 1$; that is, a homogeneous model where transitions are allowed only to first neighbors.[24]

Exercise. Show that if $(s_0 \neq 0, t_0 \neq 0)$ the general solution is

$$P(s,t \mid s_0,t_0) = \exp\left(-2(t-t_0)\mu\right) I_{|s-s_0|}\left(2(t-t_0)\mu\right).$$

Exercise (**Diffusion in Continuous Space**). From the solution in the infinite lattice (6.46), obtain through an appropriate limit process the solution of the diffusion equation in continuous space (Wiener process, see Chap. 3):

$$P(x,t) = \frac{1}{\sqrt{4\pi dt}} \exp(-x^2/4dt),$$

where d is the diffusion coefficient, i.e., $\langle x(t)^2 \rangle = 2dt$. Hint: use

$$I_{|s|}(\beta) = \frac{e^\beta}{\sqrt{2\pi\beta}} \left(1 + \frac{\frac{1}{4} - s^2}{2\beta} + \cdots\right); \quad \beta \to \infty,$$

and replace $s = x/a$, where a is the lattice parameter.

---

[24]In matrix notation we can write $\mathbf{H}_{s,s'} \equiv W_{s,s'} - \delta_{s,s'} \sum_m W_{m,s} = \mu(1 - \delta_{s,s'}) - 2\mu\delta_{s,s'}$ with $s = \{s' \pm 1, s'\}$.

Optional Exercise (**Asymmetric Transitions**). Consider an RW in a 1D lattice in the particular situation in which $\mu_s = A$ and $\nu_s = B$ for all sites s in the lattice [see Fig. 6.4]. Show that the Green function of the asymmetric problem, $G(u)$ is given by

$$< n \mid G(u) \mid m >= \left(\frac{A}{B}\right)^{\frac{n-m}{2}} < n \mid G^0(u + \beta) \mid m >,$$

where $G^0(u)$ is the homogeneous Green function of the 1D symmetric problem (6.47) with $\mu = \sqrt{AB}$ and $\beta = A + B - 2\sqrt{AB}$. Hint: use the Z-transform and the change of variable $X = \sqrt{B/AZ}$ to obtain a similar expansion to that in (6.45). The n-dimensional anisotropic and asymmetric case is solved in [7].

6.3.4 Density of States, Localization

In order to study the possible localization of *classical* particles diffusing in a lattice, it is convenient to introduce the concept of density of states of Fourier modes . Therefore, we solve here again the case of a homogeneous 1D RW ($\nu_s = \mu_s = \mu$), but using the concept of density of state (DOS). If we introduce the Fourier transform of the conditional probability[25] (with $s_0 = 0, t_0 = 0$)

$$\lambda(k, t) = \sum_{s=-\infty}^{\infty} e^{iks} P(s, t) \tag{6.48}$$

in (6.40), an equation is obtained for Fourier mode $\lambda(k, t)$

$$\partial_t \lambda(k, t) = 2\mu \left(\cos(k) - 1\right) \lambda(k, t). \tag{6.49}$$

Using the initial condition $P(s, t) = \delta_{s,0}$, which implies $\lambda(k, 0) = 1$, the solution for (6.49) will be

$$\lambda(k, t) = \exp\left(2\mu t \left[\cos(k) - 1\right]\right). \tag{6.50}$$

Then, the probability of a homogeneous RW in real space is given by

$$P(s, t) = \frac{1}{2\pi} \int_{-\pi}^{\pi} e^{-iks} \lambda(k, t) \, dk. \tag{6.51}$$

In particular, relaxation of the initial condition is described by the quantity

$$P(0, t) = \frac{1}{2\pi} \int_{-\pi}^{\pi} e^{-\gamma(k)t} \, dk, \tag{6.52}$$

---

[25]This is equivalent of having "diagonalized" the ME (6.40).

where $\gamma(k) \equiv 2\mu[1 - \cos(k)]$ can be interpreted as a relaxation density of mode k. Note that defining a probability density (uniform) for the k mode; that is, $\Pi(k) = 1/2\pi$, we have[26]

$$P(0,t) = \int_{-\pi}^{\pi} e^{-\gamma(k)t} \Pi(k) \, dk = \int_{0}^{4\mu} e^{-\gamma t} \rho(\gamma) \, d\gamma; \quad \gamma \geq 0. \tag{6.53}$$

Then the relaxation density $\rho(\gamma)$ is given by the expression

$$\rho(\gamma) = \langle \delta(\gamma - 2\mu[1 - \cos(k)]) \rangle_{\Pi(k)} = \frac{1}{\pi} \left(4\mu\gamma - \gamma^2\right)^{-1/2}. \tag{6.54}$$

This relaxation density of the initial position of the RW is analogous to the DOS of a classical system of phonons, as well as to the DOS of a *Tight-Binding 1D* Hamiltonian in quantum mechanics [8].

Finally it is worth noting that we can write the relaxation of the initial condition (6.52) in the form

$$P(0,t) = \frac{1}{\pi} \int_{0}^{4\mu} e^{-\gamma t} \left[\gamma(4\mu - \gamma)\right]^{-1/2} \, d\gamma. \tag{6.55}$$

When $t \to \infty$ the most important contribution to this integral is given by the behavior of the integrand near $\gamma \sim 0$. Then, we can see that $\lim_{\gamma \to 0} \rho(\gamma) \to \mathcal{O}(\gamma^{-1/2})$, which is not enough to offset the exponential term, and avoid thus, in the limit of long times, the vanishing of (6.55).

In disordered systems, depending on the degree of disorder, it may happen that a localization of the particle that is spreading appears. In this case the time relaxation shows an asymptotic behavior different from the diffusive case[27] $P(0, t \to \infty) \sim t^{-1/2}$. The analysis of localization is one of the central issues in the study of transport in amorphous materials. This topic will be discussed in the next chapter.

6.4 Models of Disorder

In general, models of disorder are studied considering an RW in a nD regular lattice, but assuming that transitions between different lattice sites are random variables. That is, each transition element $W_{s,s'}$ is a **rv** and has associated a certain probability distribution. Those cases where this probability distribution is narrow and satisfies that Prob.$[W_{s,s'} = 0] \to 0$, are considered models of weak disorder [i.e., the inverse

---

[26]Using the theorem of transformation of random variables; see Chap. 1.

[27]This asymptotic behavior can be estimated approximating (6.55) with a Laplace transform $\sim \int_{0}^{\infty} e^{-xp} x^{\nu} \, dx = \Gamma(\nu + 1)/p^{\nu+1}$, $\nu > -1$, $\mathcal{R}e[p] > 0$. A rigorous analysis of this asymptotic result is presented, using the Tauberian-Abelian theorems in Sect. 7.3.2.

moments $\left\langle W_{s,s'}^{-p} \right\rangle$ are defined for all $p = 1, 2, 3 \cdots$]. Another working hypothesis for solving the problem of transport in disordered media is to consider that $W_{s,s'}$ [for each pair of sites s, s'] are **sirv**, except for adjacent sites which they could be correlated depending on the symmetry of the model.

From now on we define two models of disorder with transition to next neighbors, these models frequently appear in the bibliography.

- Bond disorder model,[28] this model satisfies the symmetry

$$W_{s+1,s} \equiv \mu_s = \nu_{s+1} \equiv W_{s,s+1}$$

(6.56)

$$W_{s,s-1} \equiv \mu_{s-1} = \nu_s \equiv W_{s-1,s}.$$

That is, we assume that between sites s and s' (adjacent) there is a barrier of random height; see Fig. 6.5.

- Site disorder model[29] this model satisfies the symmetry

$$W_{s+1,s} \equiv \mu_s = \nu_s \equiv W_{s-1,s}.$$

(6.57)

That is, we assume that at the site s there is a random well depth; see Fig. 6.5.

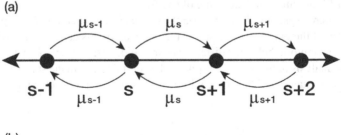

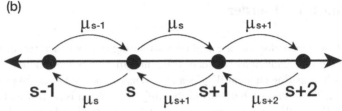

Fig. 6.5 (a) Schematic representation of a $1D$ RW with bond disorder. (b) With site disorder

---

[28] Sometimes called *Random-barrier*.

[29] Sometimes called *Random-trap*.

Exercise (**Translation Operators**). The notation of models (6.56) and (6.57) can be simplified, further, introducing translation operators (*Step operator*.[30])

$$E^{\pm} P(s, t) = P(s \pm 1, t). \tag{6.58}$$

Show that (6.39) can be written in the form

$$\frac{d}{dt} P(s, t) = \left(E^+ - 1 \right) v_s P(s, t) + \left(E^- - 1 \right) \mu_s P(s, t). \tag{6.59}$$

That is to say, in infinite systems we can think of this operator as the expansion

$$E_s^+ = 1 + \partial_s + \frac{1}{2!} \partial_s^2 + \frac{1}{3!} \partial_s^3 + \cdots$$

This notation (creation and destruction) is very useful in the perturbative analysis of transport in disordered media.

Exercise (**Disorder in** $1D$). Show explicitly that models of bond-disorder and site-disorder correspond to the following master Hamiltonians

$$\mathbf{H}_{BD} = \mathbf{H}_0 + (1 - E^-) \, \xi_s \left(E^+ - 1 \right) \qquad \text{bond-disorder}$$
$$\tag{6.60}$$
$$\mathbf{H}_{SD} = \mathbf{H}_0 + \left(E^- + E^+ - 2 \right) \xi_s \qquad \text{site-disorder}.$$

Here ξ_s are **rv** with zero mean and $\mathbf{H}_0 = \mu \left(E^- + E^+ - 2 \right)$ is the homogeneous master Hamiltonian (normal diffusion). Then the transition element (*Hopping*) $W_{s+1,s} = \mu + \xi_s$ can take different symmetries depending on the model. Note that the only condition on the statistics of the **rv** ξ_s is that the quantity $\mu + \xi_s$ is always positive. A configuration of the disorder will be characterized by the statistical weight associated with the random vector $\{\cdots \xi_{-2}, \xi_{-1}, \xi_0, \xi_1, \xi_2 \cdots\}$. In particular, it is assumed that all ξ_i (for each lattice site) are **sirv**, this simplifies greatly the study of disordered systems.

Exercise (**Diffusion Coefficient**). From the master Hamiltonian (ordered case) $\mathbf{H}_0 = \mu \left(E^- + E^+ - 2 \right)$, calculate the second moment of displacement of the RW and show that in the absence of disorder in $1D$ the diffusion coefficient is $d = \mu$.

Exercise (**Transpose Matrix**). From the master Hamiltonian

$$\mathbf{H}_0 = \mu \left(1 - E^- \right) \left(E^+ - 1 \right),$$

show that $\mathbf{H}_{BD}$; that is, the case with bond-disorder may alternatively be written as

$$\mathbf{H}_{BD} = - \left(E^- - 1 \right) \mathbf{B} \left(E^+ - 1 \right),$$

---

[30] In more than 1 dimension there will be a set of operators $E_j^{\pm}$ for each coordinate axis j.

where **B** is a diagonal random matrix. Interpret this result in terms of the operation of transposing the matrix $\mathbf{H}_{BD}$.

Exercise. From the master Hamiltonian for site-disorder, show that $\mathbf{H}_{SD}$ can also be written in the form $\mathbf{H}_{SD} = \left(1 - E^+\right)\left(E^- - 1\right)\mathbf{B}$.

Optional Exercise. Show that the study of the transpose of the master Hamiltonian $\mathbf{H}_{SD}$, i.e., the ME [9]

$$\dot{\mathbf{P}} = \left[\left(1 - E^+\right)\left(E^- - 1\right)\mathbf{B}\right]^T \cdot \mathbf{P},$$

corresponds to solving an electrical *ladder* circuit of nonrandom conductances and random capacitances $C_s \equiv 1/\xi_s$, where $< s \mid \mathbf{P}(u) \mid 0 > \equiv P(s, u)$ represents the potential on a node.

6.4.1   Stationary Solution

In an infinite domain with natural boundary conditions, the stationary solution of the RW $P(s, t \rightarrow \infty \mid s_0, t_0) \equiv P_{st}(s)$ is zero, because the particle is diffusing without restraint and $P_{st}(s)$ must be normalized throughout the infinite lattice. Let us see now the stationary solution of an ME in a finite domain to show that this mainly depends on the symmetries involved. The presentation of an RW on a finite domain will be introduced in detail in Sect. 6.5.

For example, for a fixed configuration $\{W_{s',s}\}$ in a site-disorder model, see Fig. 6.6. The stationary solution is

$$P(s, t \rightarrow \infty) \equiv P_{st}(s) \propto \frac{1}{W_{s',s}}.$$

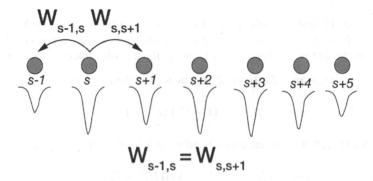

Fig. 6.6 Schematic representation of a 1D RW with site disorder or random trap

To prove this result simply write the stationary ME $0 = \mathbf{H}_{SD} \cdot \mathbf{P}$, in components (here we make abuse of notation in using operators $E^{\pm}$ which are defined for infinite domains), that is:

$$
\begin{aligned}
0 &= \left[\mathbf{H}_0 + (E^+ + E^- - 2)\xi_s\right] P_{st}(s) \\
&= (\mu + \xi_{s-1})P_{st}(s-1) + (\mu + \xi_{s+1})P_{st}(s+1) - 2(\mu + \xi_s)P_{st}(s) \\
&= W_{s,s-1}P_{st}(s-1) + W_{s,s+1}P_{st}(s+1) - (W_{s+1,s} + W_{s-1,s})\, P_{st}(s),
\end{aligned}
$$

from this, the solution

$$
P_{st}(s) = \frac{\text{const.}}{W_{s',s}} = \frac{\text{const.}}{\mu + \xi_s}
$$

follows by direct substitution.

While in a model of bond-disorder, for a fixed configuration $\{W_{s',s}\}$ the stationary solution is homogeneous in the finite lattice

$$
P(s, t \to \infty) \equiv P_{st}(s) = \text{const.}
$$

Once again, to prove this result we write the stationary ME $0 = \mathbf{H}_{BD} \cdot \mathbf{P}$, in components, that is:

$$
\begin{aligned}
0 &= \left[\mathbf{H}_0 + (1 - E^-)\xi_s(E^+ - 1)\right] P_{st}(s) \\
&= (\mu + \xi_{s-1})P_{st}(s-1) + (\mu + \xi_s)P_{st}(s+1) - (2\mu + \xi_{s-1} + \xi_s)P_{st}(s) \\
&= W_{s,s-1}P_{st}(s-1) + W_{s,s+1}P_{st}(s+1) - (W_{s+1,s} + W_{s-1,s})\, P_{st}(s).
\end{aligned}
$$

Now, from the symmetries of bond-disorder $W_{s+1,s} = W_{s,s+1}$ and $W_{s,s-1} = W_{s-1,s}$, it follows that

$$
0 = W_{s,s-1}\left(P_{st}(s-1) - P_{st}(s)\right) + W_{s,s+1}\left(P_{st}(s+1) - P_{st}(s)\right),
$$

whence we observe that $P_{st}(s) = \text{const.}$ is the solution.

Note that in both models (bond or site disorder) the constant is obtained by normalization in the finite domain.

Exercise (**Unique Stationary Solution**). Show that in general for a system with first neighbor transitions in a finite domain $s \in \mathcal{D} \equiv [-L, L]$ without absorbing sites within $\mathcal{D}$ [i.e., $W_{s\pm 1,s} \neq 0$, $\forall s$], there will always be a unique stationary solution; see Appendix B. On the other hand, note that if in Fig. 6.4 we take $v_s = \mu_{s-1} = 0$, the domain is separated into "disconnected" parts, and so in this case the stationary solution is not unique.

6.4.2   Short Times

It is of particular interest to know the behavior at short and long times of the **mv** of the propagator of the RW. In general, as seen in Sect. 6.3.1, the Green function of the problem (for any of the above models of disorder) takes the form

$$G(u) = [u\mathbf{1} - \mathbf{H}]^{-1} . \tag{6.61}$$

To study the behavior at short times it is convenient to write (6.61) in the temporal representation; i.e., taking the inverse of the Laplace transform we get

$$G(t) = \exp [\mathbf{H}t] . \tag{6.62}$$

The average over all configurations of disorder of $\{W_{s',s}\}$ is given by the **mv**

$$\langle G(t) \rangle = \langle \exp [\mathbf{H}t] \rangle = \exp \sum_{n=1}^{\infty} \frac{t^n}{n!} \langle\langle \mathbf{H}^n \rangle\rangle , \tag{6.63}$$

where in the last equality we have invoked an expansion in cumulants[31] of the random matrix $\mathbf{H}$. In the short time limit this expansion is a good perturbation expansion in the smallness parameter t. Then for sufficiently short times we can approximate

$$\langle G(t \sim 0) \rangle \simeq \exp [t \langle \mathbf{H} \rangle] . \tag{6.64}$$

Using (6.60) we can write, for a $1D$ bond-disorder model

$$\langle \mathbf{H}_{BD} \rangle = \langle (1 - E^-) \mathbf{B} (E^+ - 1) \rangle , \tag{6.65}$$

and given that $\mathbf{B}$ is a diagonal random matrix,

$$\mathbf{B} = \begin{pmatrix} \cdot & 0 & \cdot & \cdot & \cdot \\ 0 & B_{-1} & 0 & \cdot & \cdot \\ \cdot & 0 & B_0 & 0 & \cdot \\ \cdot & \cdot & 0 & B_1 & 0 \\ \cdot & \cdot & \cdot & 0 & \cdot \end{pmatrix} ,$$

we get

$$\langle \mathbf{H}_{BD} \rangle = (1 - E^-) \langle \mathbf{B} \rangle (E^+ - 1) . \tag{6.66}$$

---

[31] In Chap. 1 the definition of cumulant is introduced.

Furthermore, since B_s are equally distributed **rv**, it follows that $\langle \mathbf{B} \rangle = \langle B \rangle \mathbf{1}$, then

$$\langle \mathbf{H}_{BD} \rangle = (1 - E^-) \langle \mathbf{B} \rangle (E^+ - 1) = \langle B \rangle (E^+ + E^- - 2). \tag{6.67}$$

Note that in the case of site-disorder the analysis is completely equivalent, because

$$\langle \mathbf{H}_{SD} \rangle = \langle (1 - E^+) (E^- - 1) \mathbf{B} \rangle = \langle B \rangle (E^+ + E^- - 2). \tag{6.68}$$

Then, for both models of disorder we get that in the limit $t \to 0$

$$\langle G(t \sim 0) \rangle \simeq \exp\left[t \langle \mathbf{H} \rangle\right] = \exp\left[t \langle B \rangle (E^+ + E^- - 2)\right]. \tag{6.69}$$

From (6.69) we see that for short times an effective master Hamiltonian may be defined, $\mathbf{H}_{\mathrm{eff}} = \langle \mathbf{H} \rangle = \langle B \rangle (E^+ + E^- - 2)$, therefore comparing with the third exercise of Sect. 6.4, it follows that for short times the system behaves as if it were in the presence of an effective diffusion coefficient

$$\mu = \langle B \rangle.$$

6.4.3 Long Times

The study of $\langle G(t) \rangle$ for long times requires a more careful analysis. For this purpose let us start with the bond-disorder model $1D$. In this case the master Hamiltonian can be written alternatively in the form

$$\mathbf{H}_{BD} = -\left(E^+ - 1\right) \mathbf{B} \left(E^- - 1\right).$$

For long times it is more convenient to study the behavior of the **mv** of the Green function in the Laplace representation, that is[32]:

$$G(u) = \left[u\mathbf{1} + (E^+ - 1) \mathbf{B} (E^- - 1)\right]^{-1} \tag{6.70}$$

$$= (E^- - 1)^{-1} \mathbf{B}^{-1} \left[u\mathbf{B}^{-1} + (E^- - 1) (E^+ - 1)\right]^{-1} (E^- - 1).$$

If we assume that all inverse moments of random variables B_s exist, we can define the quantity $T = \langle 1/B_s \rangle$ which is independent of the location of the s-th site. Using this fact we can write the random matrix $\mathbf{B}^{-1}$ in the form:

$$\mathbf{B}^{-1} = (\mathbf{T} - \mathbf{Q}). \tag{6.71}$$

---

[32]To get this result shift expression $(E^+ - 1)$ to the left, $\mathbf{B}(E^- - 1)$ to the right and use that $(\mathbf{AC})^{-1} = \mathbf{C}^{-1}\mathbf{A}^{-1}$, etc.

Here the (diagonal) random matrix $\mathbf{Q}$ has zero mean value $\langle \mathbf{Q} \rangle = 0$ and $\mathbf{T}$ is a nonrandom matrix (also diagonal). Then it holds that $\langle \mathbf{B}^{-1} \rangle = \langle \mathbf{T} - \mathbf{Q} \rangle = \mathbf{T} = T\mathbf{1}$. Thus we can write $G(u)$ in the form

$$G(u) = (E^- - 1)^{-1} (\mathbf{T} - \mathbf{Q}) [u (\mathbf{T} - \mathbf{Q}) - \Delta]^{-1} (E^- - 1), \qquad (6.72)$$

where we have introduced the notation $\Delta = -(E^- - 1)(E^+ - 1)$. Defining the (nonrandom) Green function $G_0(u) = [u\mathbf{T} - \Delta]^{-1}$, we can write (6.72) in the form

$$G(u) = (E^- - 1)^{-1} (\mathbf{T} - \mathbf{Q}) (1 - G_0 u \mathbf{Q})^{-1} G_0 (E^- - 1) \qquad (6.73)$$

$$= (E^- - 1)^{-1} [\mathbf{T} (G_0 + G_0 u \mathbf{Q} G_0 + G_0 u \mathbf{Q} G_0 u \mathbf{Q} G_0 + \cdots)$$

$$- \mathbf{Q} (G_0 + G_0 u \mathbf{Q} G_0 + G_0 u \mathbf{Q} G_0 u \mathbf{Q} G_0 + \cdots)] (E^- - 1).$$

If we now take the average over disorder and consider only the first contribution in the Laplace variable, $u \sim 0$, we get

$$\langle G(u) \rangle = (E^- - 1)^{-1} \langle \mathbf{T} (G_0 + G_0 u \mathbf{Q} G_0 + G_0 u \mathbf{Q} G_0 u \mathbf{Q} G_0 + \cdots) \rangle \qquad (6.74)$$

$$- \langle \mathbf{Q} (G_0 + G_0 u \mathbf{Q} G_0 + G_0 u \mathbf{Q} G_0 u \mathbf{Q} G_0 + \cdots) \rangle (E^- - 1)$$

$$= (E^- - 1)^{-1} [\langle \mathbf{T} G_0 \rangle - \langle \mathbf{Q} G_0 \rangle] (E^- - 1) + \mathcal{O}(u)$$

$$= (E^- - 1)^{-1} \mathbf{T} G_0 (E^- - 1) + \mathcal{O}(u).$$

Here we have used that $\langle \mathbf{T} G_0 \rangle = \mathbf{T} G_0$, because $\mathbf{T}$ is nonrandom, and that $\langle \mathbf{Q} G_0 \rangle = \langle \mathbf{Q} \rangle G_0 = 0$. Rewriting G_0 in terms of E^- and E^+ operators, we obtain to the order $\mathcal{O}(u)$

$$\langle G(u \sim 0) \rangle = (E^- - 1)^{-1} \mathbf{T} [u\mathbf{T} - \Delta]^{-1} (E^- - 1) \qquad (6.75)$$

$$= (E^- - 1)^{-1} [u\mathbf{1} - T^{-1}\Delta]^{-1} (E^- - 1)$$

$$= [u\mathbf{1} + T^{-1} (E^+ - 1) (E^- - 1)]^{-1} = [u\mathbf{1} - T^{-1}\Delta]^{-1}.$$

This means that for long times the system behaves as if there were an effective diffusion coefficient

$$\mu = T^{-1} \equiv \langle B_s^{-1} \rangle^{-1}.$$

Note that this behavior for long times only makes sense if the quantity $\langle B_s^{-1} \rangle$ is finite. In the event that this does not happen we are in the presence of a type of disorder, commonly called "strong" disorder that results in an anomalous behavior for the diffusion. That is, in that case there will not be a long time regime in which the diffusion coefficient is a constant.

Optional Exercise (**Higher Order Corrections**). Using the model of bond-disorder, study the "asymptotic" behavior of the diffusion process in the next higher order of the small parameter u. Note that the following contribution in the perturbation expansion (6.74) is the term $\langle QG_0 uQG_0 \rangle$. Then, only if the quantity $\langle Q^2 \rangle = \langle \mathbf{B}^{-2} \rangle - \langle \mathbf{B}^{-1} \rangle^2$ is finite will this perturbative order make sense. Show that the following correction to the diffusion coefficient is of order $\mathcal{O}(\sqrt{u})$, and it is related to the variance of the inverse moment of the **sirv** B_s. In general for the perturbation expansion (6.74) to make sense the condition $\langle B_s^{-p} \rangle \neq \infty$, is required for all $p = 1, 2, 3, \cdots$. The same result is also true for a site disorder model if the initial condition of the RW is δ_{s,s_0}.

Exercise (**Site-Disorder**). Calculate for long times the effective diffusion coefficient in $1D$ in a site-disorder model. Hint: Calculate $\langle G(u \sim 0) \rangle$ considering that

$$\mathbf{H}_{SD} = - \left(E^+ - 1 \right) \left(E^- - 1 \right) \mathbf{B};$$

see also Sect. 7.2.5.

6.5 Boundary Conditions in the Master Equation

6.5.1 Introduction

Often it is necessary to know the solution of an ME with absorbing or reflecting boundary conditions at any of the edges of the domain. This problem should not be confused with the situation where a site is absorbent or reflective.[33]

To fix this idea, consider a $1D$ system with M sites ($s = 1, 2, \cdots, M$) and assume that site 1 is absorbent. Then, if $M = 3$, the matrix $\mathbf{H}$ may have the structure

$$\mathbf{H} = \begin{pmatrix} 0 & c & d-f \\ 0 & -(c+d) & f \\ 0 & d & -d \end{pmatrix}. \tag{6.76}$$

Analyzing the properties of $\mathbf{H}$ we noted that the conditions imposed on (6.76) are to meet $d - f \geq 0, c \geq 0, f \geq 0, d \geq 0$. The solution of the Markov system $\dot{\mathbf{P}} = \mathbf{H} \cdot \mathbf{P}$ is the matrix $\mathbf{P}(t) = \exp(t\mathbf{H})$, which does not represent the solution of a boundary value problem.

---

[33]That is, a site is absorbing if the transition probability from this site to any other place is zero; $\mu_{s*} = v_{s*} = 0$, in Fig. 6.4 we represent the $1D$ case. A site is reflecting if the transition probability from that site is zero in either of the two directions. For example, if the site is reflective on the left edge of the domain, $\mu_{s*} > 0$, $v_{s*} = 0$; see Fig. 6.4.

Example (**Absorbent Site**). Consider the matrix (6.76) associated with the dynamics of the states $\{\downarrow, \circ, \uparrow\}$. From the analysis of eigenvectors $\mathbf{H} \cdot \phi_i = \lambda_i \phi_i$ we see that there are three eigenvalues $\lambda_i = \{0, \frac{-1}{2}(c + 2d \pm \sqrt{c^2 + 4df})\}$, which always meet the general condition $\mathcal{R}_e[\lambda_i] \leq 0$; see Appendix B. The dynamic analysis of the ME can be studied diagonalizing (if possible) $\mathbf{H}$. For example, for the particular case where $c = f = 1, d = 2$, the eigenvalues are $\{-4, -1, 0\}$, then the solution is easy to obtain:

$$\mathbf{P}(t) = e^{-4t}\begin{pmatrix} 0 & 0 & 0 \\ 0 & 2/3 & -1/3 \\ 0 & -2/3 & 1/3 \end{pmatrix} + e^{-t}\begin{pmatrix} 0 & -1 & -1 \\ 0 & 1/3 & 1/3 \\ 0 & 2/3 & 2/3 \end{pmatrix} + \begin{pmatrix} 1 & 1 & 1 \\ 0 & 0 & 0 \\ 0 & 0 & 0 \end{pmatrix}. \tag{6.77}$$

Considering this expression it is immediate to see that states $\{\circ, \uparrow\}$ are depopulated at the expense of the absorption in the state $\{\downarrow\}$. For an arbitrary initial condition $\mathcal{P}(0) = (\mathcal{A}, \mathcal{B}, \mathcal{C})$, the solution versus time is obtained by multiplying (6.77) by the vector $\mathcal{P}(0)$; i.e., the components of the vector solution $\mathcal{P}(t) = \mathbf{P}(t) \cdot \mathcal{P}(0)$ are

$$P_\downarrow(t) = \mathcal{A} + \mathcal{B}(1 - e^{-t}) + \mathcal{C}(1 - e^{-t})$$

$$P_\circ(t) = \frac{\mathcal{B}}{3}(2e^{-4t} + e^{-t}) + \frac{\mathcal{C}}{3}(e^{-t} - e^{-4t})$$

$$P_\uparrow(t) = \frac{2\mathcal{B}}{3}(e^{-t} - e^{-4t}) + \frac{\mathcal{C}}{3}(2e^{-t} + e^{-4t}).$$

This solution obviously does not satisfy $P_\downarrow(t) = 0$ for all t.

6.5.2 The Equivalent Problem

Consider an infinite lattice with an absorbent site, for example, $s = l$. Suppose further that the initial condition of RW is $P(s, t_0 \mid s_0, t_0) = \delta_{s,s_0}$ with $s_0 \geq l + 1$. Then, from the ME (6.39) or (6.59) we see that the occupation probability $P(l+1, t \mid s_0, t_0)$ satisfies the equation[34]

$$\frac{dP(l+1, t \mid s_0, t_0)}{dt} = \nu_{l+2}P(l+2, t \mid s_0, t_0) - (\nu_{l+1} + \mu_{l+1})P(l+1, t \mid s_0, t_0). \tag{6.78}$$

If we are interested in solving a one-step ME for $s \geq l$, we see that Eq. (6.78) is equivalent to imposing the boundary condition $P(l, t \mid s_0, t_0) = 0, \forall t$.

On the other hand, considering an infinite lattice with a reflective site, for example $s = l$, and assuming that $P(s, t_0 \mid s_0, t_0) = \delta_{s,s_0}$, with $s_0 \geq l$; from the

---

[34]See note (33).

ME (6.39) we see that the occupation probability $P(l, t \mid s_0, t_0)$ satisfies the equation

$$\frac{dP(l, t \mid s_0, t_0)}{dt} = v_{l+1}P(l+1, t \mid s_0, t_0) - \mu_l P(l, t \mid s_0, t_0). \tag{6.79}$$

So if we are interested in the solution of ME for $s \geq l$, we see that Eq. (6.79) is equivalent to imposing the boundary condition

$$v_l P(l, t \mid s_0, t_0) = \mu_{l-1}P(l-1, t \mid s_0, t_0), \quad \forall t.$$

Note the difference between the concepts of *special sites* and *special boundary conditions*. In general, if we are interested in solving the ME restricted to a finite domain with boundary conditions, it is always possible to introduce special sites where the transition probabilities change in order to emulate the required boundary conditions.[35]

A method for solving the ME in a restricted domain $s \in \mathcal{D} \equiv [l_1, l_2]$ with boundary conditions is to introduce dummy variables d_s for the initial probabilities of a special process being in $s \notin \mathcal{D}$, i.e.:

$$P(s, t_0 \mid s_0, t_0) = d_s, \quad \forall s \notin \mathcal{D}. \tag{6.80}$$

Then the problem in the restricted domain $s \in \mathcal{D} \equiv [l_1, l_2]$ is reduced to calculating the propagator[36] $P(s, t \mid s_0, t_0)$ for the unrestricted problem, $s \in [-\infty, +\infty]$ together with the initial conditions (6.80) and $P(s, t_0 \mid s_0, t_0) = \delta_{s, s_0}, \forall s_0 \in \mathcal{D}$. The values d_s are imposed so that the required boundary conditions are met [6].

Another technique is to use the method of images so as to meet—at all times—absorbing or reflecting boundary conditions, somewhere on the lattice [2, 10]. This method will be explained in detail in the following sections.

Exercise (**Two Absorbing Sites**). Write a generic matrix **H** (5×5), with transitions to first next neighbors and two absorbent sites at both ends of the domain; i.e., $W_{2,1} = W_{4,5} = 0$. Consider now the dynamics in the "subdomain" of sites $\{2, 3, 4\}$.

Exercise (**Two Reflecting Sites**). Write a generic matrix **H** (5×5) with transitions to first next neighbors, where sites $\{2, 4\}$ [next to the right and left edges domain] satisfy $W_{1,2} = 0, W_{3,2} > 0$ and $W_{5,4} = 0, W_{3,4} > 0$. Consider now the dynamics in the "subdomain" of sites $\{2, 3, 4\}$.

---

[35]There is a huge variety of possible boundary conditions and special sites, but here we are not going into this analysis; see for example [1, 6].

[36]Using the technique of generating function.

6.5.3   Limbo Absorbent State

When considering models with transitions to first neighbors there is an artifice, introducing a *limbo* state, which enables us to find easily the solution of the ME with absorbing boundary condition [1].

To fix ideas, let us study a system of $2L+1$ sites $s \in \mathcal{D} \equiv [-L, L]$. Now consider two absorbing states at sites $\pm(L+1) \notin \mathcal{D}$, this fact dictates that there can be no transition from outside the domain of interest $\mathcal{D}$. In this case the matrix $\mathbf{H}$ is given by[37]

$$\mathbf{H} = \begin{pmatrix} -(\mu_{-L} + \nu_{-L}) & \nu_{-L+1} & 0 & 0 & 0 \\ \mu_{-L} & -(\mu_{-L+1} + \nu_{-L+1}) & \cdots & 0 & 0 \\ 0 & \mu_{-L+1} & \cdots & \nu_{L-1} & 0 \\ 0 & 0 & \cdots -(\mu_{L-1} + \nu_{L-1}) & \nu_L \\ 0 & 0 & 0 & \mu_{L-1} & -(\mu_L + \nu_L) \end{pmatrix}.$$

$$(6.81)$$

Where we have used that the transition probability from the site s to the right is denoted by μ_s, and the transition probability from that site s to the left by ν_s. Note that, by construction, this matrix $\mathbf{H}$ cannot meet the condition $\sum_{s' \in \mathcal{D}} \mathbf{H}_{s',s} = 0$ at the edges of $\mathcal{D}$. This happens because at the edges of the domain, $-L$ and L, there is a "loss of probability" since sites $s^* = \pm(L+1)$ are absorbent. Only if we considered the extended domain $\mathcal{D} + \{\pm s^*\}$ would the total probability be preserved. In particular, we can see that the probability of being in the *limbo* sites $\pm s^*$ increases over time at the expense of decreased likelihood of staying within $\mathcal{D}$, or $\sum_{s' \in \mathcal{D}} \mathbf{P}_{s',s}(t)$. Thus, working with a matrix of type (6.81) we can emulate absorbing boundary conditions immediately outside $\mathcal{D}$.

Example. For a finite and homogeneous system with 3 sites ($L = 1$) we can immediately see that $\mathbf{H}$ may be diagonalized; the three eigenvalues are $\lambda_i / \mu = \{-2, -2 \pm \sqrt{2}\}$. Then, in the domain of interest $\mathcal{D}$, the conditional probability matrix is given by $e^{t\mathbf{H}}$, i.e.:

$$\mathbf{P}(t) = e^{-2\mu t} \begin{pmatrix} 1/2 & 0 & -1/2 \\ 0 & 0 & 0 \\ -1/2 & 0 & 1/2 \end{pmatrix} + e^{-\mu t(2+\sqrt{2})} \begin{pmatrix} 1/4 & -1/\sqrt{2^3} & 1/4 \\ -1/\sqrt{2^3} & 1/2 & -1/\sqrt{2^3} \\ 1/4 & -1/\sqrt{2^3} & 1/4 \end{pmatrix}$$

$$+ e^{-\mu t(2-\sqrt{2})} \begin{pmatrix} 1/4 & 1/\sqrt{2^3} & 1/4 \\ 1/\sqrt{2^3} & 1/2 & 1/\sqrt{2^3} \\ 1/4 & 1/\sqrt{2^3} & 1/4 \end{pmatrix}.$$

---

[37] See note (33).

This is the desired solution and satisfies absorbing boundary conditions immediately outside $\mathcal{D}$. Therefore, we have solved a special boundary condition introducing limbo sites $s^* = \pm(L+1)$.

6.5.4 Reflecting State

Now consider reflecting boundary conditions at the edges of $\mathcal{D}$. In this case we place reflecting sites states at $s = \pm L$, then between these and the next sites: $s^* = \pm(L+1)$, there is no net contribution from outside the domain $\mathcal{D}$ or out of it. Then the matrix $\mathbf{H}$ is given by[38]

$$\mathbf{H} = \begin{pmatrix} -\mu_{-L} & \nu_{-L+1} & 0 & 0 & 0 \\ \mu_{-L} & -(\mu_{-L+1}+\nu_{-L+1}) & \cdots & 0 & 0 \\ 0 & \mu_{-L+1} & \cdots & \nu_{L-1} & 0 \\ 0 & 0 & \cdots & -(\mu_{L-1}+\nu_{L-1}) & \nu_L \\ 0 & 0 & 0 & \mu_{L-1} & -\nu_L \end{pmatrix}. \qquad (6.82)$$

Note that, unlike the matrix (6.81), in this case the sum of any of the columns (6.82) gives zero. This ensures normalization of probability within $\mathcal{D}$ for all t. The solution of the ME with a matrix $\mathbf{H}$ given by (6.82) can be considered to emulate reflecting boundary conditions at the edges of the domain $\mathcal{D}$. Then we can use the master Hamiltonian (6.82) to study a finite problem with reflective boundaries.

Example (**Homogeneous 3 × 3 Model**). For a homogeneous system with 3 sites $(L = 1)$ from (6.82) we can immediately see that $\mathbf{H}$ may be diagonalized. Its eigenvalues are $\lambda_i/\mu = \{0, -1, -3\}$ and the conditional probability is given by

$$\mathbf{P}(t) = e^{t\mathbf{H}} = e^{-3\mu t}\begin{pmatrix} 1/6 & -1/3 & 1/6 \\ -1/3 & 2/3 & -1/3 \\ 1/6 & -1/3 & 1/6 \end{pmatrix} + e^{-\mu t}\begin{pmatrix} 1/2 & 0 & -1/2 \\ 0 & 0 & 0 \\ -1/2 & 0 & 1/2 \end{pmatrix}$$

$$+ \frac{1}{3}\begin{pmatrix} 1 & 1 & 1 \\ 1 & 1 & 1 \\ 1 & 1 & 1 \end{pmatrix}.$$

This solution shows, as expected, that the stationary state is homogeneous.

---

[38] As in the previous case, to each site $s \in \mathcal{D}$ corresponds a transition probability μ_s to the right and ν_s to the left. See note (33).

Optional Exercise (**Model with Asymmetric Disorder**). A particularly interesting situation arises when the system is not homogeneous. Using the model (6.82)
in the case of a (3×3) matrix, the eigenvalues different from zero are

$$2\lambda_i = -(\mu_1 + \mu_2 + v_3 + v_2) \pm \sqrt{(\mu_1 + \mu_2 + v_2 + v_3)^2 - 4(\mu_1\mu_2 + v_3\mu_1 + v_2v_3)}.$$

Once again we see that $\mathcal{R}_e[\lambda_i] \leq 0$ for all $\{\mu_i, v_i\} \geq 0$. In particular, if the transition
probabilities $\{\mu_i, v_i\} \geq 0$ are **rv** with positive support, the system represents a finite
model with asymmetric disorder (local drift). Note that even this simple model
can present difficulties to be solved. Consider some random feature for variables
$\{\mu_i, v_i\}$ and calculate the **mv** of the conditional probability over all configurations
of disorder.

$$\langle \mathbf{P}(t) \rangle_{\text{Disorder}} = \langle e^{t\mathbf{H}} \rangle_{\Pi(\{\mu_i, v_i\})} \equiv \langle e^{t\mathbf{H}} \rangle. \tag{6.83}$$

Optional Exercise (**Binary Disorder**). Calculate, in the Laplace representation,
the **mv** of the conditional probability (6.83) over the disorder, in the case where
transition probabilities $\{\mu_i, v_i\}$ are **sirv** that can only take the values $\{a, b\}$ with
equal probability $(\frac{1}{2})$. To calculate the **mv** $\langle \mathbf{P}(u) \rangle$ compute each configuration of the
disorder. Note that the statistical weight for any of the configurations of disorder will
be $\frac{1}{16}$, since $\{\mu_i, v_i\}$ is a set of **sirv**. Show that for some configurations of disorder
the master Hamiltonian takes the form:

$$\mathbf{H} = \begin{pmatrix} -a & a & 0 \\ a & -a-b & a \\ 0 & b & -a \end{pmatrix} \quad \text{configuration:} \quad \mu_1 = v_2 = v_3 = a \,, \ \mu_2 = b$$

$$\mathbf{H} = \begin{pmatrix} -a & b & 0 \\ a & -2b & a \\ 0 & b & -a \end{pmatrix} \quad \text{configuration:} \quad \mu_1 = v_3 = a \,, \ \mu_2 = v_2 = b$$

$$\mathbf{H} = \begin{pmatrix} -a & b & 0 \\ a & -b-a & a \\ 0 & a & -a \end{pmatrix} \quad \text{configuration:} \quad \mu_1 = \mu_2 = v_3 = a \,, \ v_2 = b,$$

$$\tag{6.84}$$

then we conclude that it is easy to build any other configuration. Show that if some
of the transition probabilities $\{a, b\}$ are zero an anomalous effect occurs in the time
behavior of the RW propagator.

Exercise (**Site-Disorder**). Show that the master Hamiltonian

$$\mathbf{H} = \begin{pmatrix} -\mu_1 & \mu_2 & 0 \\ \mu_1 & -2\mu_2 & \mu_3 \\ 0 & \mu_2 & -\mu_3 \end{pmatrix} \tag{6.85}$$

corresponds to a (3×3) model of site-disorder without drift and reflecting boundary conditions. Use the behavior of the Laplace transform when $u \to 0$ to study the stationary solution of the problem. Calculate the propagator (for $u \sim 0$) when $\{\mu_i\}$ are **sirv** that can take values $\{a, b\}$ with equal probability.

Exercise (**Bond-Disorder**). Show, for a (3×3) system with reflecting boundary conditions, that a model with bond-disorder has a master Hamiltonian of the form

$$\mathbf{H} = \begin{pmatrix} -\mu_1 & \mu_1 & 0 \\ \mu_1 & -(\mu_1 + \mu_2) & \mu_2 \\ 0 & \mu_2 & -\mu_2 \end{pmatrix}. \tag{6.86}$$

Furthermore, for each configuration of disorder the eigenvalues of $\mathbf{H}$ are

$$\lambda_i = \{0, -(\mu_1 + \mu_2) \pm \sqrt{\mu_1^2 + \mu_2^2 - \mu_2 \mu_1}\},$$

we see that $\mathcal{R}_e[\lambda_i] \le 0$ for all values of $\{\mu_i\} \ge 0$. If $\{\mu_i\}$ are **sirv** that can take values $\{a, b\}$ with equal probability, there will be only four configurations of disorder. Once again, for each configuration of disorder, the Green function $G(u) = (u\mathbf{1} - \mathbf{H})^{-1}$ can be calculated explicitly. As we have previously said it is easy to recognize that the four configurations have the same statistical weight $(\frac{1}{4})$. Calculate for long times the **mv** of the Green function, that is, in the Laplace representation $\langle G(u \sim 0) \rangle \sim (u\mathbf{1} - \mathbf{H}_{\text{eff}})^{-1}$, then evaluate $\mathbf{H}_{\text{eff}}$.

6.5.5 Boundary Conditions (Method of Images)

Often it is necessary to know the conditional probability $\mathcal{P}(s, t \mid s_0, t_0)$ for systems of arbitrary length (finite domain) and/or absorbing or reflecting boundary conditions. If the solution of the problem of infinite length with natural boundary conditions (infinite domain) is known $P^0(s, t \mid s_0, t_0)$, then by the technique of summing images (negatives and/or positives) it is possible to calculate the solution to the problem for special contours.

To fix ideas, let us start with the simplest case: A semi-infinite straight $s \in (-\infty, +L]$ with an absorbing boundary condition at the site L. In this case the image to be used is "negative." That is, consider the "plane" $s = L$, and place an image of the RW with respect to that plane and consider the "specular evolution" of the RW [in this way when the RW reaches the plane $s = L$, it will be immediately annihilated –for all times– by the "negative" image[39]]. For example, if at instant t the RW is at n sites from the edge of the domain [i.e., in site $s = L - n$] the location of the image will be in $-[L - n - 2L] = L + n$. In this case it is clear that

---

[39] See the next *excursus* "Reflection principle."

if $P^0(s, t \mid s_0, t_0)$ is the conditional probability for the infinite system $s \in (-\infty, \infty)$ and we add a "negative" image, the desired solution is given by

$$\mathcal{P}_{abs}(s, t \mid s_0, t_0) = P^0(s, t \mid s_0, t_0) - P^0(-s + 2L, t \mid s_0, t_0), \quad \forall s, s_0 \in (-\infty, L].$$
(6.87)

Note that this solution satisfies the fundamental property of a propagator[40]

$$\lim_{t \to t_0} \mathcal{P}_{abs}(s, t \mid s_0, t_0) \to \delta(s, s_0),$$

as can be seen from the fact that

$$\lim_{t \to t_0} \mathcal{P}_{abs}(s, t \mid s_0, t_0) = \delta(s, s_0) - \delta(-s + 2L, s_0) = \delta(s, s_0), \quad \text{if } \{s, s_0\} \in (-\infty, L].$$

On the other hand, (6.87) meets the absorbing condition in $s = L$ for all times t.

The semi-infinite case with reflecting boundary condition is analogous to (6.87), but with a "positive" image to emulate any trajectory "bouncing-back" on the edge of the domain; that is,

$$\mathcal{P}_{ref}(s, t \mid s_0, t_0) = P^0(s, t \mid s_0, t_0) + P^0(-s + 2L, t \mid s_0, t_0), \quad \forall s, s_0 \in (-\infty, L].$$
(6.88)

Thus, in the event that there is a reflective condition at $+L$, a similar construction follows for the "positive" image. On the other hand, the probability current is zero in L, as can be shown using (6.88):

$$[J_{s,s_0}(t)]_{s=L} \propto \mathcal{P}_{ref}(L - 1, t \mid s_0, t_0) - \mathcal{P}_{ref}(L + 1, t \mid s_0, t_0) = 0.$$

Note that solution (6.88) is properly normalized in the semi infinite straight $s \in (-\infty, L]$.

Exercise. Using $\sum_{s=-\infty}^{\infty} P^0(s, t \mid s_0, t_0) = 1$, show that the normalization of the probability $\mathcal{P}_{ref}(s, t \mid s_0, t_0)$ is compensated in $(-\infty, L]$ by the contribution of the "area" which is provided by the term $\sum_{s=-\infty}^{L} P^0(-s + 2L, t \mid s_0, t_0)$.

6.5.6   Method of Images in Finite Systems*

Consider a 1D finite system characterized by a discrete domain $\mathcal{D} = [-L, L]$, with $2L + 1$ sites. Here we build a solution that meets special boundary conditions at the edges of $\mathcal{D}$.

---

[40]We denote the Kronecker delta indistinctly by δ_{s,s_0} or $\delta(s, s_0)$; see exercises in Sect. 1.4.

In Sect. 1.4.3 we discussed a similar situation when studying a toroidal lattice. There the concept of discrete Fourier series appeared in a natural way. In that chapter we also noted that this probability could be constructed in terms of the probability— on the infinite lattice—using the method of images, which we now will present in detail.

Periodic Boundary Conditions

As for the toroidal lattice[41], consider here the case of a $1D$ lattice with periodic conditions where $s \in \mathcal{D} = [-L, L]$. Then, properly placing positive images throughout the infinite lattice, the probability with periodic boundary conditions on $\mathcal{D}$ can be written in the form

$$\mathcal{P}^L_{\text{per}}(s, t \mid s_0, t_0) = \sum_{q=-\infty}^{\infty} P^0(s + q(2L+1), t \mid s_0, t_0), \quad \forall s, s_0 \in [-L, L]. \quad (6.89)$$

Since the domain $\mathcal{D}$ is discrete, there will be $2L + 1$ available sites; hence the need to place mirror images with respect to $[-L, L]$.

Exercise. Interpret graphically the location of positive images in (6.89).

On the other hand, we can use discrete Fourier series to get an alternative expression for $\mathcal{P}^L_{\text{per}}(s, t \mid s_0, t_0)$. Notice first that

$$\mathcal{P}^L_{\text{per}}(s, t \mid s_0, t_0) = \sum_{q=-\infty}^{\infty} \frac{1}{2\pi} \int_0^{2\pi} dk \, \exp\left(-ik\left[s - s_0 + q(2L + 1)\right]\right) \tilde{P}^0(k, t - t_0).$$

$$(6.90)$$

In this expression we have used the notation

$$\tilde{P}^0(k, t) \equiv \sum_s e^{iks} P^0(s, t \mid 0, 0), \quad (6.91)$$

which is the Fourier transform of the homogeneous solution in the infinite $1D$ lattice. Then

$$\mathcal{P}^L_{\text{per}}(s, t \mid s_0, t_0) = \frac{1}{2\pi} \int_0^{2\pi} dk \, \tilde{P}^0(k, t - t_0) \exp\left(-ik(s - s_0)\right) \sum_{q=-\infty}^{\infty} \exp\left(-ikq(2L + 1)\right).$$

$$(6.92)$$

---

[41] A toroidal lattice in nD is defined by the symmetry: $P(s_1, s_2, \cdots, s_n) = P(s_1 + N_1, s_2 + N_2, \cdots, s_n + N_n)$, where N_i are arbitrary integer numbers and $s_i \in [1, 2, \cdots, N_i]$.

Now, if we use Poisson's sum $\sum_{q=-\infty}^{\infty} \exp(2\pi imq) = \sum_{q=-\infty}^{\infty} \delta(m+q)$, we obtain

$$
\begin{aligned}
\mathcal{P}_{\text{per}}^{L}(s,t \mid s_0,t_0) &= \frac{1}{2\pi} \int_0^{2\pi} dk\, \tilde{P}^0(k, t-t_0)\, e^{-ik(s-s_0)} \sum_{q=-\infty}^{\infty} \delta\left(\frac{k(2L+1)}{2\pi} - q\right) \\
&= \frac{1}{2\pi} \int_0^{2\pi} dk\, \tilde{P}^0(k, t-t_0)\, e^{-ik(s-s_0)} \sum_{q=-\infty}^{\infty} \frac{2\pi}{2L+1} \delta\left(k - \frac{2\pi q}{2L+1}\right) \\
&= \frac{1}{2L+1} \sum_{q=1}^{2L+1} \tilde{P}^0\left(\frac{2\pi q}{2L+1}, t-t_0\right) \exp\left(-i\frac{2\pi q}{2L+1}(s-s_0)\right). \quad (6.93)
\end{aligned}
$$

The last equality follows from the fact that the values of $q = 0$ and $q = 2L+1$ give the same Fourier's weight, then we can adopt $q \in [1, 2L+1]$. Moreover, with this choice, the transformation (6.91) can be considered as a simple linear transformation of $P(s,t) \to \tilde{P}(k,t)$ through the matrix $\mathcal{R}_{ks} = e^{iks}$, with $k = \frac{2\pi q}{2L+1}$ and $q \in [1, (2L+1)]$; then, the inverse transformation is associated with the matrix $\mathcal{R}_{ks}^{-1}$.

Guided Exercise (**Diffusion with Periodic Boundary Conditions in** $1D$). Here we want to build a simple example using the formula (6.93). To this end we take

$$
\tilde{P}^0(k, t-t_0) = \exp(2\mu(t-t_0)[\cos(k) - 1]),
$$

that is to say, the solution of the homogeneous ME without drift, in an infinite $1D$ lattice. From (6.93) we get for a finite lattice with periodic boundary conditions on $\mathcal{D}$

$$
\begin{aligned}
\mathcal{P}_{\text{per}}^{L}(s,t \mid s_0,t_0) &= \frac{1}{2L+1} \sum_{q=1}^{2L+1} \exp\left(2\mu(t-t_0)\left[\cos\left(\frac{2\pi q}{2L+1}\right) - 1\right]\right) \\
&\times \exp\left(-i\frac{2\pi q}{2L+1}(s-s_0)\right). \quad (6.94)
\end{aligned}
$$

From this expression is easy to see that the stationary state is

$$
\mathcal{P}_{\text{per}}^{L}(s, t \to \infty \mid s_0, t_0) = \frac{1}{2L+1}.
$$

Compare with the result of Sect. 1.4.3.

Absorbing Boundary Conditions

This is a more complicated situation than in the previous section, because each edge domain reflects an image which in turn is reflected back in the opposite edge. In conclusion we must add an infinite set of positive and/or negative images, depending on the type of contour to consider. If the $1D$ finite system has absorbing boundary conditions, the solution in $\mathcal{D}$ is given by [10, 11]

$$
\mathcal{P}_{abs}^{L}(s, t \mid s_0, t_0) = \sum_{q=-\infty}^{\infty} \left[P^0(s + 4q(L+1), t \mid s_0, t_0) \right. \tag{6.95}
$$

$$
\left. - P^0(-s - (4q+2)(L+1), t \mid s_0, t_0) \right], \quad \forall s, s_0 \in [-L, L].
$$

Guided Exercise (**Fourier Representation, Absorbing Case**). Here we show that the conditional probability (6.95) can be written in the alternative form:

$$
\mathcal{P}_{abs}^{L}(s, t \mid s_0, t_0) = \frac{1}{4(L+1)} \sum_{q=-2(L+1)}^{2(L+1)} \tilde{P}^0 \left(\frac{q\pi}{2(L+1)}, t - t_0 \right) \tag{6.96}
$$

$$
\times \left[\exp\left(i\frac{q\pi (-s + s_0)}{2(L+1)} \right) - \exp\left(i\frac{q\pi (s + s_0 + 2(L+1))}{2(L+1)} \right) \right].
$$

This expression is useful for obtaining numerical results, because the sum is finite. Furthermore, from this expression for the propagator, one can immediately verify that

$$
\mathcal{P}_{abs}^{L}(L + 1, t \mid s_0, t_0) = \mathcal{P}_{abs}^{L}(-L - 1, t \mid s_0, t_0) = 0.
$$

To prove (6.96) from (6.95) we use the fact that $P^0(s, t \mid s_0, t_0)$ is homogeneous in space and time, then it follows that we can write

$$
\mathcal{P}_{abs}^{L}(s, t \mid s_0, t_0) = \sum_{q=-\infty}^{\infty} \frac{1}{2\pi} \int_{-\pi}^{\pi} dk \, e^{-ik(s+4q(L+1)-s_0)} \, \tilde{P}^0(k, t - t_0) \tag{6.97}
$$

$$
- \sum_{q=-\infty}^{\infty} \frac{1}{2\pi} \int_{-\pi}^{\pi} dk \, e^{-ik(-s-(4q+2)(L+1)-s_0)} \, \tilde{P}^0(k, t - t_0),
$$

that is,

$$
\mathcal{P}_{abs}^{L}(s, t \mid s_0, t_0) = \frac{1}{2\pi} \int_{-\pi}^{\pi} dk \, \tilde{P}^0(k, t - t_0) \left[e^{ik(-s+s_0)} \sum_{q=-\infty}^{\infty} e^{-ik4q(L+1)} \right.
$$

$$
\left. - e^{ik(s+s_0)} \sum_{q=-\infty}^{\infty} e^{ik(4q+2)(L+1)} \right]. \tag{6.98}
$$

If we now use the Poisson sum, it follows that

$$
\mathcal{P}_{abs}^{L}(s, t \mid s_0, t_0) = \frac{1}{2\pi} \int_{-\pi}^{\pi} dk \, \tilde{P}^0(k, t - t_0) \left[e^{ik(-s+s_0)} \sum_{q=-\infty}^{\infty} \delta\left(\frac{4k(L+1)}{2\pi} - q\right) \right.
$$
$$
\left. - e^{ik(s+s_0+2(L+1))} \sum_{q=-\infty}^{\infty} \delta\left(\frac{4k(L+1)}{2\pi} + q\right) \right].
$$

Since the integral is defined within the domain $k \in [-\pi, \pi]$, we find that the only terms of the infinite sum $\sum_{q=-\infty}^{\infty}$ that contribute to the summation are those that fulfill $q \in [-2(L+1), 2(L+1)]$; from which the expression (6.96) is inferred.

Exercise. Using (6.96) show that $\mathcal{P}_{abs}^{L}(\pm(L+1), t \mid s_0, t_0) = 0$. Interpret graphically the location of positive and negative images.

Optional Exercise (**Sum of Images**). Using the explicit construction of the images appearing in (6.95), show that $\mathcal{P}_{abs}^{L}(s, t \mid s_0, t_0)$ satisfies the absorbing boundary conditions required in $s = \pm(L+1)$. We can also consider, similarly, mixed boundary conditions (absorbing and reflecting) on $\mathcal{D}$ [12]. An even more complex situation is to get the solution $\mathcal{P}_{abs}^{L}(s, t \mid s_0, t_0)$ in a finite domain with absorbing boundary conditions and in the presence of *drift* [13].

Excursus (**Reflection Principle**). It is to be noted that the method of images [10] is based on the applicability of the principle of reflection for Markov processes homogeneous in t [14]. This principle states that for every trajectory of a Markov **sp** crossing some arbitrary level $s = s^*$, at a given time t^*, there is a reflected path, originating at $s = s^*$, from that moment t^*, which is the specular reflection and identical to the one considered. Chandrasekhar's method is based on the fact that adding realizations corresponds to calculating the probability distribution.[42]

We can also use an alternative method to that employed by Chandrasekhar, but without applying the reflection principle, to build *stochastic realizations* for non-Markovian processes with periodic and/or reflecting boundary conditions [15]. Unfortunately, this (*window*) method cannot be used for absorbing boundary conditions! On the other hand, it is easy to imagine why the reflection principle[43] does not apply to non-Markovian processes. For example, consider the limiting case of a process with infinite memory, obviously there is no such a reflected path. The *window* stochastic picture is explained in the next section.

---

[42]From this it follows that in the absorbing case, the negative image "annihilates" $\forall t \geq t^*$ all realizations going out of the domain.

[43]A detailed presentation of the *reflection principle* can be seen in: W. Feller, *An Introduction to Probability Theory and its Applications*, Second Edition, New York, John Wiley and Sons (1971).

Reflecting Boundary Conditions

Consider a 1D system with reflecting boundary conditions at the edges of $\mathcal{D}$. Then, if the sites $\pm(L + 1)$ are the places where the paths are reflected-back, the solution with reflecting boundary conditions is given by an expression analogous to the absorbing case, but with positive images.

Optional Exercise. Using a similar construction to that shown in (6.95), show graphically that reflecting boundary conditions at the edges of $\mathcal{D}$, can be obtained by adding appropriate images.

In the reflecting case the sum of images is always positive, and also the location of images in the lattice is different from the absorbing case, the result in this case is:

$$\mathcal{P}_{\text{ref}}^{L}(s, t \mid s_0, t_0) = \sum_{q=-\infty}^{\infty} \left[P^0(s + 2q(2L + 1), t \mid s_0, t_0) \right. \tag{6.99}$$

$$\left. + P^0(-s + (2q + 1)(2L + 1), t \mid s_0, t_0) \right], \quad \forall s, s_0 \in [-L, L].$$

Exercise (**Fourier Representation, Reflecting Case**). Show that the conditional probability (6.99) can also be written in terms of the discrete Fourier transform in the form[44]:

$$\mathcal{P}_{\text{ref}}^{L}(s, t \mid s_0, t_0) = \frac{1}{2(2L + 1)} \sum_{q=1}^{2(2L+1)} \tilde{P}^0\left(\frac{q\pi}{(2L + 1)}, t - t_0 \right) \exp\left(-i \frac{q\pi}{(2L + 1)} (s - s_0) \right)$$

$$\times \left[1 + \exp(iq\pi) \right],$$

Exercise. Consider the domain $\mathcal{D} = [-L, L]$ and show that the conditional probability (6.99) satisfies the fundamental property of any propagator

$$\lim_{t \to t_0} \mathcal{P}_{\text{ref}}^{L}(s, t \mid s_0, t_0) \to \delta(s, s_0).$$

Excursus (**Mixed Boundary Conditions** in 1D). In order to solve the conditional probability with reflecting boundary condition at $s = -(L + 1)$ and absorbing at $s = (L + 1)$ we introduce a shorthand notation: $\tilde{P}^0(s, t \mid s_0, t_0) \equiv G_{s,s_0}^0$. Then the conditional probability is given by the following sum of images (positive and negative) [12]:

---

[44]The proof of this result is entirely analogous to (6.96).

$$\mathcal{P}^L_{\text{ref-abs}}(s, t \mid s_0, t_0) = G^0_{00} + \sum_{q=0}^{\infty} (-1)^q G^0_{s+(2L+1)+q(4L+3),-s_0} + \sum_{q=1}^{\infty} (-1)^q G^0_{-s+q(4L+3),-s_0}$$

$$+ \sum_{q=0}^{\infty} (-1)^{q+1} G^0_{2L+2+q(4L+3)-s,s_0} + \sum_{q=1}^{\infty} (-1)^q G^0_{s+q(4L+3),s_0}, \quad \forall s, s_0 \in [-L, L].$$

From this expression show that $\mathcal{P}^L_{\text{ref-abs}}(L+1, t \mid s_0, t_0) = 0$. Furthermore, the reflecting boundary condition is satisfied between sites $-(L+1)$ and $-L$.

Optional Exercise (2D **Mixed Boundary Conditions**). Find the conditional probability in a 2D square lattice when the sides parallel to the x direction are reflecting contours, while those parallel to the y direction are absorbing. Write the solution in real space, in terms of the two-dimensional Fourier transform of the homogeneous problem with natural boundary conditions. Hint: as in Sect. 6.3.4, solve the 2D ME by using the transformation

$$\lambda(k_1, k_2, t) = \sum_{x=-\infty}^{\infty} \sum_{y=-\infty}^{\infty} e^{ik_1 x} e^{ik_2 y} P^0(x, y, t)$$

Continuous Systems

Reflecting boundary conditions at every edge of a continuous domain $\mathcal{D}$ are treated by a construction of images similar to that used in (6.99). The following result applies when the sample space is continuous; in this case the solution is written in the form

$$\mathcal{P}^L_{\text{ref}}(X, t \mid X_0, t_0) = \sum_{q=-\infty}^{\infty} \left[P^0(X + 4qL, t \mid X_0, t_0) \right. \tag{6.100}$$

$$\left. + P^0(-X + (4q+2)L, t \mid X_0, t_0) \right], \quad \forall X, X_0 \in [-L, L].$$

From this expression we can consider the following method for representing graphically all realizations involved in the sum (6.100). We start defining as "positive images"

$$X_q^+(t) = X_0(t) + 4qL, \tag{6.101}$$

all identical realizations which originate from the true initial condition in the range of interest $\mathcal{D} = [-L, L]$. From (6.101) we can see, clearly, that for each $q \in \{\pm 1, \pm 2, \cdots\}$ we obtain an identical realization to the original one $X_0(t)$ but with an initial condition[45]: $X_q^+(0) = X_0(0) + 4qL \notin \mathcal{D}$. We define as "negative images"

---

[45]The subscript 0 indicates that the initial condition of the realization $X_0(t)$ is within $\mathcal{D}$. In the next section we prove that this result applies to both diffusive or non-diffusive realizations.

Fig. 6.7 Portion of a 1D lattice that represents diagrammatically the location of images $X_q^{\pm}(t)$ to obtain reflecting boundary conditions in $\mathcal{D}$. The *double arrows* indicate the domain of interest, $\mathcal{D}$, where the original realization $X_0(t)$ begins

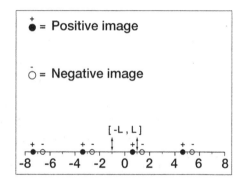

$$X_q^-(t) = -X_0(t) + 4qL + 2L, \tag{6.102}$$

all *specular* realizations originating in the *true realization* in the range of interest $\mathcal{D}$. Hence we observe that for each value of $q \in \{\pm 1, \pm 2, \cdots\}$, we get a specular realization from $X_0(t)$ with initial condition $X_q(0) = -X_0(0) + 4qL + 2L \notin \mathcal{D}$. In Fig. 6.7 a—static—representation for the location of some of the images depicted by $X_q^{\pm}(t)$ for the case $L = 1$ is shown.

Exercise. In the case when there are periodic boundary conditions in the continuous interval $\mathcal{D} = [-L, L]$, interpret graphically the location of the positive images $X_q^+(t) = X_0(t) + q2L$. Compare with the discrete case, when there are $2L+1$ sites in $\mathcal{D}$, see (6.89).

6.5.7 Method of Images for Non-diffusive Processes*

The Window Stochastic Picture

Now, consider a continuous domain characterized by some interval $\mathcal{D} = [-L, L]$ with reflecting boundary conditions at the edges. For each stochastic realization $X_0(t)$, of an arbitrary unbounded **sp**, we can see from (6.101) and (6.102), that during the time evolution of all images $X_q^{\pm}(t \geq 0)$ there will be a *single* realization within $\mathcal{D}$. Thus, from the sum of the stochastic realizations (6.101) and (6.102): $X_q^+(t) + X_q^-(t)$ when they meet $X_q^{\pm}(t) \in \mathcal{D}$, we will have built the following *window* **sp**:

$$\chi(t) = \sum_{q=-\infty}^{\infty} \mathcal{W}\left[X_q^+(t)\right] X_q^+(t) + \mathcal{W}\left[X_q^-(t)\right] X_q^-(t), \tag{6.103}$$

where $\mathcal{W}[z]$ is the "window" function:

$$\mathcal{W}[z] = \begin{cases} 1 & \text{if } -L < z < L \\ \frac{1}{2} & \text{if } z = \pm L \\ 0 & \text{if } |z| > L. \end{cases} \tag{6.104}$$

That is, the *window* **sp** $\chi(t)$ defined in (6.103) simulates reflecting boundary conditions at the edges of the domain $\mathcal{D}$. Note that this construction is not restricted to Markov processes and can be used to calculate an arbitrary n-times characteristic function [15].

Analogously, from the "positive images"

$$X_q^+(t) = X_0(t) + 2qL, \tag{6.105}$$

the *window* **sp** $\chi(t)$ defined by

$$\chi(t) = \sum_{q=-\infty}^{\infty} \mathcal{W}[X_q^+(t)] X_q^+(t), \tag{6.106}$$

satisfies periodic boundary conditions on $\mathcal{D}$.

Remark Note that the *window* stochastic picture cannot be used to emulate absorbing boundary conditions.

Periodic Boundary Conditions

Definition 1 Let the **sde** be

$$\frac{dX_0(t)}{dt} = U'(X_0) + \eta(t), \quad X_0(t) \in [-\infty, +\infty] \tag{6.107}$$

with $U'(X_0) \equiv dU/dX_0$, $U(X_0) = U(X_0 + 2L)$ a periodic potential, and $\eta(t)$ an arbitrary noise. The conditional probability distribution associated with (6.107) and satisfying *periodic* boundary conditions on $\mathcal{D}$ can be built in terms of the method of images as

$$P(x, t) = \mathcal{W}(x) \sum_{q=-\infty}^{+\infty} P_0(x + 2qL, t) \tag{6.108}$$

where $\mathcal{W}(x)$ is the window function (6.104), and $P_0(x, t)$ is the corresponding conditional pdf of the unbounded **sde** (6.107) with natural boundary conditions.

Theorem 1 *Let $X_0(t)$ be any stochastic realization which is a solution of the unbounded sde (6.107). The realizations of the window sp (6.106) satisfy periodic boundary conditions on $\mathcal{D}$. Furthermore the conditional pdf of sp $X(t)$ is given by the distribution (6.108).*

Reflecting Boundary Conditions

Definition 2 Let the **sde** be

$$\frac{dX_0(t)}{dt} = U'(X_0) + \eta(t), \quad X_0(t) \in [-\infty, +\infty] \tag{6.109}$$

with $U'(X_0) \equiv dU/dX_0$, $U(X_0) = U(X_0 + 2L)$ an *even* periodic potential, and $\eta(t)$ an arbitrary noise. The conditional probability distribution associated with (6.109) and satisfying *reflecting* boundary conditions on $\mathcal{D}$ can be built in terms of the method of images as

$$P(x, t) = W(x) \sum_{q=-\infty}^{+\infty} P_0(x + 4qL, t) + P_0(-x + 4qL + 2L, t), \tag{6.110}$$

as before, $W(x)$ is the window function (6.104), and $P_0(x, t)$ is the corresponding conditional pdf of the unbounded **sde** (6.109) with natural boundary conditions.

Theorem 2 *Let $X_0(t)$ be any stochastic realization which is a solution of the unbounded sde (6.109). The realizations of the window sp (6.103) satisfy reflecting boundary conditions on $\mathcal{D}$. Furthermore the conditional pdf of sp $X(t)$ is given by the distribution (6.110).*[46]

6.6 Random First Passage Times*

The study of the "escape time" from a given domain is an issue of interest in many areas of research in physics, chemistry, biology, etc.; for example, we may want to know the average time to leave the domain of attraction of a metastable state; or to know the average of the different random moments t_j, in which the **sp** $\xi(t)$ takes for the first time a certain value $\xi(t_j) = L$. In Chaps. 3 and 4 we already mentioned that this problem was related to the analysis of the adjoint **F-P** operator. In general, the formalism for studying the random first passage times through a border of a domain is a nontrivial problem, both for continuous systems (in the formalism of **F-P**) and for discrete systems (in the formalism of ME). However, if the transport process is characterized by a Markov chain that has only transitions to first neighbors, the study of the first passage time problem is considerably simplified. Here we present the formalism for calculating the probability density of the first passage time for the case of a Markov process[47] in the $1D$ lattice. Consider a Markov process to be

---

[46]The proof of these theorems, and other interesting results concerning applications to non-Markov finite systems can be seen in [15].

[47]This problem can also be formulated for some non-Markovian processes [1].

characterized by an ME and also assume that we are interested in knowing the first passage times distribution through the boundary of the domain $\mathcal{D} = [-L, L]$. To study the probability distribution of the first passage time, $f_{s_0}(t \mid t_0)$, we should first define the probability $F_{s_0}(t \mid t_0)$ of staying in the domain $\mathcal{D}$ until time t (without having left it previously), if the walk started at site $s_0 \in \mathcal{D}$ at the instant t_0. To compute this probability, commonly called chance of survival, we need to note that the realizations do not return to the domain $\mathcal{D}$ if they had left it at earlier time, then we can write

$$F_{s_0}(t \mid t_0) = \sum_s \mathcal{P}_{abs}^L(s, t \mid s_0, t_0), \quad \forall s_0 \in \mathcal{D} \qquad (6.111)$$

where $\mathcal{P}_{abs}^L(s, t \mid s_0, t_0)$ is the solution of the ME with absorbing boundary conditions at the sites $\pm(L+1)$ [see Sect. 6.5.6]; that is, if the particle reached the border of $\mathcal{D}$ at the random instant t_j, it is immediately removed by the boundary conditions imposed. Thus, by (6.111) we can ensure a "good counting" of the realization of interest to us. Differentiating with respect to time $F_{s_0}(t \mid t_0)$ and since the Green function of ME is also a solution of the adjoint ME: $\partial_t \mathbf{P} = \mathbf{P} \cdot \mathbf{H}$, we obtain the evolution equation for the survival probability:

$$\begin{aligned}
\partial_t F_{s_0}(t \mid t_0) &= \sum_s \partial_t \mathcal{P}_{abs}^L(s, t \mid s_0, t_0) \\
&= \sum_s \sum_{s'} \mathcal{P}_{abs}^L\left(s, t \mid s', t_0\right) \mathbf{H}_{s', s_0} \\
&= \sum_{s'} \mathbf{H}_{s_0, s'}^T \sum_s \mathcal{P}_{abs}^L\left(s, t \mid s', t_0\right) \\
&= \sum_{s'} \mathbf{H}_{s_0, s'}^T F_{s'}(t \mid t_0),
\end{aligned}$$

where $\mathbf{H}^T$ indicates the transposed matrix; i.e., in compact notation $\partial_t F = \mathbf{H}^T \cdot F$, together with the initial condition $F_s(t_0 \mid t_0) = 1, \forall s \in \mathcal{D}$, and the "dummy" boundary conditions $F_{-(L+1)}(t) = F_{(L+1)}(t) = 0, \forall t$. Hence we see that the survival probability satisfies the adjoint ME; without loss of generality we can take $t_0 = 0$.

The survival probability is a decreasing function of time, then it follows that the first-passage time probability distribution is given by $f_{s_0}(t \mid 0) = -\partial_t F_{s_0}(t \mid 0)$; then the **mv** of the random first passage time is

$$T_{s_0} = \int_0^\infty t f_{s_0}(t \mid 0)\, dt. \qquad (6.112)$$

If it is met that $t f_{s_0}(t \mid 0) \to 0$ for $t \to \infty$; then, integrating by parts (6.112) we get

$$T_{s_0} = \int_0^\infty F_{s_0}(t \mid 0)\, dt = \mathcal{L}_u\left[F_{s_0}(t \mid 0)\right]_{u=0}.$$

That is, the average time, or mean "escape time," is given in terms of the Laplace transform of the survival probability F. We can see that this is an important relationship that lets us find easily the characterization of T_{s_0} for different models of disorder [11]. In general, it is possible to obtain coupled equations for all moments of the first-passage time distribution:

$$T_{s_0}^{(n)} = \int_0^\infty t^n f_{s_0}(t \mid 0) \, dt = -[t^n F_{s_0}(t \mid 0)]_0^\infty + n \int_0^\infty t^{n-1} F_{s_0}(t \mid 0) \, dt,$$

here we have integrated by parts. If the surface term vanishes[48] we get

$$T_{s_0}^{(n)} = n \int_0^\infty t^{n-1} F_{s_0}(t \mid 0) \, dt. \tag{6.113}$$

Exercise. Multiplying the adjoint equation $\partial_t F = \mathbf{H}^T \cdot F$ by t^{n-1} and integrating by parts, show from (6.113) that moments satisfy the recurrence relation:

$$\sum_{s'} \mathbf{H}_{s,s'}^T \, T_{s'}^{(n)} = -n T_s^{(n-1)}, \quad n \ge 1 \quad \text{with} \quad T_s^{(0)} = 1, \tag{6.114}$$

together with the boundary conditions: $T_{\pm(L+1)}^{(n)} = 0$.

6.6.1 Survival Probability

The mean first passage time is an important quantity in the study of diffusive systems controlled by chemical reactions, since its inverse measures the constant (or rate) of reaction [16]. It is also possible to prove that T_{s_0} is exactly the inverse of the Kramers' activation time for time-homogeneous **sp**, see Chap. 9.

If the transport is characterized by an asymmetric homogeneous ME[49] the master Hamiltonian is given by $\mathbf{H} = B\left(E^+ - 1\right) + A\left(E^- - 1\right)$. Then the evolution equation for the survival probability is $\partial_t F = \left[A\left(E^+ - 1\right) + B\left(E^- - 1\right)\right] \cdot F$. In the Laplace representation this equation can be written in the form

$$\left[AE^+ + BE^- - (u + A + B)\,\mathbf{1}\right] F_s(u) = -F_s(t = 0). \tag{6.115}$$

This equation must be solved with the initial condition:

$$F_s(t = 0) = \begin{cases} 1 \ \forall s \in \mathcal{D} \\ 0 \ \forall s \notin \mathcal{D}, \end{cases}$$

---

[48] Affirmation that is true if $T_{s_0}^{(n)}$ is a finite quantity.

[49] See Eq. (6.39) and the optional exercise in Sect. 6.3.3. In [17] the problem of the survival probability is solved in a disordered asymmetric medium.

together with the boundary condition

$$F_{-(L+1)}(u) = F_{(L+1)}(u) = 0, \quad \forall u > 0. \tag{6.116}$$

To solve the recurrence relation (6.115), it is convenient to introduce the following notation: $\psi_{s+1} \equiv F_s(u)$; then we can write

$$\left[A(E^+)^2 - (u + A + B) E^+ + B1 \right] \psi_s = -1. \tag{6.117}$$

This is a second order inhomogeneous linear equation. A particular solution of the inhomogeneous equation is $\psi_s^p = 1/u$, and proposing $\psi_s^h = x^s$ as a solution of the homogeneous equation, a quadratic equation is obtained for the magnitude x:

$$x^2 - (r + 1 + \gamma)x + \gamma = 0, \quad \text{where} \quad r = u/A, \gamma = B/A < 1.$$

The roots of this equation are

$$x_{1,2} = \frac{1}{2} \left(r + 1 + \gamma \pm \sqrt{(r + 1 + \gamma)^2 - 4\gamma} \right)$$

and meet $x_1 x_2 = \gamma$ with $x_1 \geq 1$. Then, the general solution of (6.117) is

$$\psi_s(u) = \frac{1}{u} + c_1 x_1^s + c_2 x_2^s.$$

The constants c_1 and c_2 are set using boundary conditions (6.116). Then, finally, using $\psi_{s+1} \equiv F_s(u)$ the following expression is obtained:

$$F_s(u) = \frac{1}{u} \left(1 - \frac{\left(x_1^{2L+2} - \gamma^{2L+2} \right) x_1^{s-L-1} + \left(\gamma^{2L+2} - x_2^{2L+2} \right) x_2^{s-L-1}}{x_1^{2L+2} - x_2^{2L+2}} \right), \quad s \in \mathcal{D}.$$

Exercise (**Mean Time First Passage**). From the Laplace transform of the survival probability, show that T_{s_0} is given by (with the drift to the right)

$$T_{s_0} = \frac{L + 1 - s_0}{A(1 - \gamma)} - \frac{2L + 2}{A(1 - \gamma)} \frac{\gamma^{s_0} - \gamma^{L+1}}{\gamma^{-L-1} - \gamma^{L+1}}, \quad \text{with} \quad s_0 \in [-L, L], \ \gamma \in [0, 1].$$

On the other hand, in the "pure" diffusive case, $A = B = \mu$ (limit $\gamma \to 1$), show that the expression

$$T_{s_0} = \left[(L + 1)^2 - s_0^2 \right] \Big/ 2\mu.$$

is obtained. This quadratic behavior in the system length-scale is typical for diffusive processes. In the presence of a small drift $\epsilon = (1 - \gamma) \sim 0$, the mean escape time has a linear contribution of the form

$$T_{s_0} \simeq \frac{1}{2\mu} \left[(L+1)^2 - s_0^2 \right] \left(1 - \frac{2s_0 - 3}{6} \epsilon \right).$$

Hint: use that $\gamma^q \simeq 1 - q\epsilon + \frac{1}{2}q(q-1)\epsilon^2 + \cdots$, $\forall q \in \mathcal{N}$. Show that in the unidirectional hopping case $B = 0$, the mean escape time is given by $T_{s_0} = [L + 1 - s_0]/\mu$.

Optional Exercise (**Adjoint Master Equation**). Consider the following accumulative function

$$\mathcal{F}_{s_0}(t \mid 0) = \int_0^t f_{s_0}(t' \mid 0) \, dt',$$

that is, $\mathcal{F}_{s_0}(t \mid 0)$ is the probability that the first passage time across the border ∂D [with initial condition s_0 at t_0] is less than t. From this definition it follows that

$$\mathcal{F}_{s_0}(t + \Delta t \mid 0) = \sum_{s \in \mathcal{D}} \mathcal{F}_s(t \mid 0) \, P^0(s, \Delta t \mid s_0, 0) \quad \text{with} \quad \mathcal{F}_{s_0}(t = 0 \mid 0) = 0, \ \forall s_0 \in \mathcal{D}.$$

$$(6.118)$$

Thus, $\mathcal{F}_{s_0}(t + \Delta t \mid 0)$ is given in terms of the probability of spreading from site s_0 to site s in the time interval Δt, multiplied by probability that the first passage time—from the site s—does not exceed t, summed over all possibilities with the constraint of always staying within $\mathcal{D}$. Use $P^0(s, \Delta t \mid s_0, 0) = [\exp \Delta t \, \mathbf{H}]_{s, s_0}$ and show that in the limit $\Delta t \to 0$ the evolution equation for the probability $\mathcal{F}_{s_0}(t \mid 0)$ is

$$\frac{\partial \mathcal{F}_{s_0}(t \mid 0)}{\partial t} = \sum_{s \in \mathcal{D}} \mathbf{H}_{s_0, s}^T \cdot \mathcal{F}_s(t \mid 0), \quad \text{where} \quad \mathcal{F}_{s_0}(0 \mid 0) = 0.$$

Excursus (**Adjoint F-P Equation**). If the Markov process were continuous, show that the evolution equation for $\mathcal{F}_{s_0}(t \mid 0)$ would be the backward **F-P** equation,[50] subject to appropriate boundary conditions. Hint: Replace in (6.118) $\sum_{s \in \mathcal{D}}$ by an integral, and shift variable $s = s_0 + \Delta s$, with $\mathcal{F}_s(t \mid 0) \to \mathcal{F}(s; t \mid 0)$ continuous in s; that is to say:

$$\mathcal{F}(s_0; t + \Delta t \mid 0) = \int_{\mathcal{D}'} \mathcal{F}(s_0 + \Delta s; t \mid 0) \, P^0(s_0 + \Delta s, \Delta t \mid s_0, 0) \, d(\Delta s),$$

$$(6.119)$$

where $\mathcal{D}'$ is a proper domain of integration. Expand $\mathcal{F}(s_0 + \Delta s; t \mid 0)$ in Taylor's series for $\Delta s \sim 0$, and take $\Delta t \to 0$. Then, using $\partial_t \mathcal{F}(s; t \mid 0) = f(s; t \mid 0)$ obtain the evolution equation for the first passage time probability density:

---

[50]The *Backward* **F-P** equation was presented in Chaps. 3 and 4 (the differential operator is the adjoint of the **F-P** operator). Its connection with the first passage time theory will be presented in Chap. 9.

$$\frac{\partial f(s;t\mid 0)}{\partial t} = K(s)\frac{\partial f(s;t\mid 0)}{\partial s} + D(s)\frac{\partial^2 f(s;t\mid 0)}{\partial s^2}. \tag{6.120}$$

From this equation it is possible to obtain a set of equations for moments $T_s^{(n)}$ [analogously to what was done in (6.114)]. In general, the boundary conditions on ∂D will be derived from the integral Eq. (6.119); there is a very wide variety of processes that lead to multidimensional Markov process with the usual boundary conditions [18]:

$$T_s^{(n)} = 0, \ n \geq 1, \ \forall s \in \partial D.$$

In Chap. 9 using the backward **F-P** equation the mean first passage time for a Brownian particle in a two-dimensional potential (force) is solved analytically.

6.7 Additional Exercises with Their Solutions

6.7.1 On the Markovian Chain Solution

The general solution of a Markovian chain is given in terms of the one-step transition matrix $\mathbf{T}_1(s, s')$. In particular when there is translational invariance $\mathbf{T}_1(s, s') \rightarrow \mathbf{T}_1(s - s')$, the solution of an RW in 1D can be written in the form (6.20). This result represents the fact that the matrix $\mathbf{T}_1(s, s')$ has been diagonalized in the Fourier basis, that is, from (6.4) we get

$$P(s, n) = \frac{1}{2\pi} \int_{-\pi}^{\pi} (\mathbf{T}_1(k))^n \exp(-iks) \, dk, \tag{6.121}$$

here we have used the initial condition $P(s, 0) = \delta_{s,0}$. If the 1D RW has a symmetric next neighbor hopping structure: $\mathbf{T}_1(k) = \cos(k)$; then, from (6.121) and using Newton's binomial formula for $(\mathbf{T}_1(k))^n = \frac{1}{2^n}\left(e^{ik} + e^{-ik}\right)^n$ we get

$$P(s, n) = \frac{1}{2\pi 2^n} \int_{-\pi}^{\pi} \sum_{j=0}^{n} \frac{n!}{j!\,(n-j)!} \exp\left[-2ik(n-j) + ik(n-s)\right] dk. \tag{6.122}$$

Noting that $\frac{1}{2\pi}\int_{-\pi}^{\pi} e^{-ik(s-s')}dk = \delta_{s,s'}$ for $s, s' \in \mathbb{Z}$, we can work out (6.122). Assume, for example, that $s > 0$, then we see that for $s \in [0, n]$ we get

$$P(s, n) = \frac{1}{2\pi 2^n} \sum_{j=0}^{n} \frac{n!\,2\pi}{j!\,(n-j)!}\delta_{n+s,2j}$$

$$= \frac{n!}{2^n \left(\frac{n+s}{2}\right)!\left(\frac{n-s}{2}\right)!},$$

which is the expected result; compare with Chap. 1. In general note that when n is even the only possible values of s are with $|s| \leq n$ and when n is odd s is odd with $|s| \leq n$.

Exercise. Show explicitly that the element k, k' of the matrix $\mathbf{T}_1$; that is, $\mathbf{T}_1(k, k')$ can be written in the form

$$\mathbf{T}_1(k, k') = \langle k| \mathbf{T}_1 |k'\rangle$$

$$= \sum_{s, s' \in \mathbb{Z}} \langle k |s\rangle \langle s| \mathbf{T}_1 |s'\rangle \langle s'| k'\rangle$$

$$= \cos(k)\delta \left(k - k'\right),$$

where $\mathbf{T}_1(k) = \cos(k)$, and we have used the notation

$$|k\rangle = \sum_{s \in \mathbb{Z}} e^{iks} |s\rangle$$

$$|s\rangle = \frac{1}{2\pi} \int_{-\pi}^{+\pi} e^{-iks} |k\rangle \, dk$$

$$1 = \sum_{s \in \mathbb{Z}} |s\rangle \langle s|$$

$$\delta_{s, s'} = \langle s'| s\rangle .$$

Exercise. Expanding $\mathbf{T}_1(k) = \cos(k)$ for small k in (6.121), setting $q = k\sqrt{n}$ into the integral, and taking $n \to \infty$ obtain a Gaussian form for $P(s, n)$; that is:

$$P(s, n) \simeq (2\pi n)^{-1/2} \exp\left(-s^2/2n\right) .$$

Now, suppose that a and τ are the parameters characterizing dimensions of space and time respectively: $x = as$ and $t = \tau n$ then identifying $P(s, n)ds = P(x, t)dx$ with $ds = 1$ we can recover the Gaussian distribution $P(x, t) = (4\pi dt)^{-1/2} \exp\left(-x^2/4dt\right)$, with $d = a^2/2\tau$.

6.7.2 Dichotomic Markovian Chain

The general solution of a 2×2 Markovian chain was presented in Sect. 6.1.1, from this result study the case when the transition matrix $\mathbf{T}_1$ is

$$\mathbf{T}_1 = \begin{pmatrix} \beta & (1 - \beta) \\ (1 - \beta) & \beta \end{pmatrix}, \quad 0 < \beta < 1$$

and the values of the process are $\pm\Delta$. Then, the recurrence relation (6.10) for this *discrete-time* dichotomic "noise" is

$$P(s, n+1) = \sum_{j=1}^{2} \mathbf{T}_1\left(s, s_j\right) P\left(s_j, n\right), \ n = 0, 1, 2, \cdots$$

i.e., for example: $P(\Delta, n+1) = \beta P(\Delta, n) + (1-\beta) P(-\Delta, n)$. The case $\beta \to 1/2$ corresponds to a situation when all transition sequences $\pm\Delta$ are equally weighted. On the other hand, note that cases $\beta = \{0, 1\}$ correspond to deterministic regimes. Calculate the *discrete-time* evolution of the conditional matrix $\mathbf{T}_n$; i.e., study elements $\mathbf{T}_n\left(s \mid s'\right) = P(s, n \mid s', 0)$. Show that the correlation function is $\langle\langle s(n) s(0)\rangle\rangle_{\text{st}} = \Delta^2 (2\beta - 1)^n$. This correlation is alternating and decreasing for $0 < \beta < \frac{1}{2}$; monotonically decreasing for $\frac{1}{2} < \beta < 1$, and it vanishes for $\beta = \frac{1}{2}$.

6.7.3   Kramers-Moyal and van Kampen Ω Expansions

A general Master Equation (ME) can be written in the (continuous) form[51]:

$$\frac{\partial P(y, t)}{\partial t} = \int W(y-r; r) P(y-r, t) dr - P(y, t) \int W(y; -r) dr, \qquad (6.123)$$

where

$$W\left(y \mid y'\right) = W(y-r; r); \ r = y - y'.$$

In Sect. 3.18 we presented the derivation of the **F-P** equation from the Chapman-Kolmogorov equation. In particular if we included all moments we found that the propagator fulfills an equation like this:

$$\frac{\partial}{\partial t} P(y, t) = \sum_{n=1}^{\infty} \frac{(-1)^n}{n!} \left(\frac{\partial}{\partial y}\right)^n \left[a_n(y) P(y, t)\right]; \quad \forall t \geq t_0. \qquad (6.124)$$

Show that if we introduce a Taylor expansion in (6.123) and define moments:

$$a_n(y) = \int r^n W(y; r) dr$$

---

[51] Here we use the short notation: $P(y, t) = P(y, t \mid y_0, t_0)$

we obtain (6.124), i.e., the Kramers-Moyal equation.[52] Formally this equation is identical to the ME and neither is easy to deal with, but it suggests a possible approach to work out the problem. From (6.124) we can calculate any moment $\langle y^q \rangle$, in terms of a coupled hierarchy of equations, for example, show that

$$\frac{d}{dt} \langle y \rangle = \langle a_1 (y) \rangle \tag{6.125}$$

$$\frac{d}{dt} \langle y^2 \rangle = 2 \langle ya_1 (y) \rangle + \langle a_2 (y) \rangle , \text{ etc.} \tag{6.126}$$

Hint: integrate by parts and use that the surface terms go to zero.

In 1961 van Kampen[53] introduced a similar expansion based on the fundamental innovation that the pdf should be scaled down with some size parameter Ω; this might be particle number, volume, or any other suitable macroscopic parameter. Therefore he assumed that the transition probability had the *canonical* form

$$W (y|y') = W (y';r) = f (\Omega) \left[\Phi_0 \left(\frac{y'}{\Omega};r \right) + \Omega^{-1} \Phi_1 \left(\frac{y'}{\Omega};r \right) + \Omega^{-2} \Phi_2 + \cdots \right], \tag{6.127}$$

here y'/Ω is the time dependent density, r the step size, and $f (\Omega)$ an arbitrary function of the size parameter only, which can later be absorbed scaling the time. If the transition rate W does not have this *form* the Ω-expansion does not apply.

Substitute this canonical form in the ME and show that to the lowest order it results in

$$\frac{\partial}{\partial t} P(y,t) \simeq f (\Omega) \left[\int \Phi_0 \left(\frac{y-r}{\Omega};r \right) P (y-r,t) \, dr - P (y,t) \int \Phi_0 \left(\frac{y}{\Omega};-r \right) dr \right] + \cdots . \tag{6.128}$$

The next essential step in the Ω-expansion comes introducing the *ansatz* that—at later times after the initial condition: $P (y,t_0) = \delta (y - y_0)$—the pdf $P (y,t)$ has a sharp peak at some position of the order $\mathcal{O} (\Omega)$, while its width will be of order $\mathcal{O} \left(\sqrt{\Omega} \right)$. This ansatz can be justified at posteriori, and is mathematically expressed setting

$$y (t) = \Omega \phi (t) + \sqrt{\Omega} \xi (t) , \tag{6.129}$$

---

[52]J.E. Moyal, J. Roy. Statist. Soc. B 11, 150 (1949); H.A. Kramers, Physica A 7, 284 (1940).

[53]Kramers-Moyal's expansion is by no means essential for the derivation of van Kampen's Ω-expansion, [Can. J. Phys. 39, 551 (1961)]. A similar *size*-expansion was also presented in the context of path integral method to get the propagator; see R. Kubo, K. Matsuo and K. Kitahara, J. Stat. Phys. 9, 51, (1973).

the first term is macroscopic and $\phi(t)$ has to be adjusted to follow the motion of the peak of the pdf, the process $\xi(t)$ characterizes the fluctuations of the system. Thus the pdf $P(y, t)$ becomes the function $\Pi(\xi, t)$ according to

$$P(y, t) = P\left(\Omega\phi(t) + \sqrt{\Omega}\xi, t\right) = \Pi(\xi, t).$$
(6.130)

Show, using (6.129) and (6.130) that the transformation of the derivatives are

$$\frac{\partial^n \Pi}{\partial \xi^n} = \Omega^{n/2}\frac{\partial^n P}{\partial y^n}$$

$$\frac{\partial \Pi}{\partial t} = \frac{\partial P}{\partial t} + \sqrt{\Omega}\frac{d\phi}{dt}\frac{\partial \Pi}{\partial \xi}.$$

Introducing (6.127) and the transformations (6.129), (6.130) in the ME allows a systematic treatment in terms of the size parameter. This is the starting point of the analysis of the Ω-expansion. It should be noted that this expansion determines consistently the evolution equation of $\phi(t)$, which is the function to be used in the transformation (6.129). Thus $\phi(t)$ determines the macroscopic part of y in such a way that the fluctuations are of $\mathcal{O}\left(\sqrt{\Omega}\right)$ (proving the previous ansatz). The stability analysis of the evolution equation

$$\frac{d\phi(t)}{dt} = \alpha_{1,0}(\phi) \quad \text{with} \quad \alpha_{n,q}(x) = \int r^n \Phi_q(x; r)\, dr,$$
(6.131)

allows a concrete classification of what type of ME we are dealing with, therefore giving rise to a rigorous classification of the system. Compare the macroscopic evolution (6.131) against (6.125). The full analysis of the Ω-expansion is beyond the scope of this introduction and can be seen in detail in van Kampen's book.

6.7.4   Enlarged Master Equation (Stochastic Liouville Equation)

Consider a differential equation for the n-dimensional vector $\vec{u} = (u_1, u_2, \cdots, u_n)$ of the form

$$\frac{d\vec{u}}{dt} = F\left(\vec{u}, y\right),$$
(6.132)

here $F\left(\vec{u}, y\right)$ is a vector function of $\vec{u}$ and y; in particular when $y = \eta(t)$ is a noise with arbitrary correlation (in general a nonwhite **sp**); Eq. (6.132) turns out to be an **sde** for the non-Markovian process $\vec{u}$. If we restrict ourselves to the case when the noise $\eta(t)$ is discrete and Markovian; that is, its conditional pdf Π obeys an ME of the form:

$$\frac{d}{dt}\Pi = \mathbf{H} \cdot \Pi. \tag{6.133}$$

We can then formally solve the problem (6.132) defining a new process U with $n+1$ components: $U = (u_1, u_2, \cdots, u_n, y)$. Then, its probability density $\mathcal{P} \equiv \mathcal{P}(\vec{u}, y, t)$ varies in time owing to the flow in $\vec{u}$-space and the jumps of y. The corresponding enlarged ME is found combining the continuity equation for the pdf of $\vec{u}$ with the ME (6.133) for y, this is the so-called stochastic Liouville equation, originally introduced by Kubo[54]

$$\partial_t \mathcal{P} = -\sum_{i=1}^{n} \frac{\partial}{\partial u_i} \left[F(\vec{u}, y) \mathcal{P} \right] + \mathbf{H} \cdot \mathcal{P}. \tag{6.134}$$

The initial condition of (6.134) should be: $\mathcal{P}(\vec{u}, y, 0) = \delta(\vec{u} - \vec{a}) \delta(y - y_0)$. But frequently, we assume that the **sp** $\eta(t)$ that appears in the **sde** (6.132) is in its stationary state. Then, as initial condition we take $\mathcal{P}(\vec{u}, y, 0) = \delta(\vec{u} - \vec{a}) \Pi^{st}(y)$. Here the stationary distribution of y is determined by the zero eigenvalue of the ME; that is, $\mathbf{H} \cdot \Pi^{st} = 0$. Defining the marginal distribution:

$$P(\vec{u}, t) = \sum_{y_n} \mathcal{P}(\vec{u}, y_n, t),$$

we can calculate all moments of $\vec{u}$. Unfortunately Eq. (6.134) is very hard to solve; but it is greatly simplified when $F(\vec{u}, y)$ is linear, in this case it is possible to work out the problem analytically.

Exercise.   Consider the case when u is a scalar, $F(u, y)$ is linear in u and y, and $\eta(t) = \pm 1$ is a symmetric dichotomic **sp**, so the matrix $\mathbf{H}$ in (6.133) is

$$\mathbf{H} = \begin{pmatrix} -\gamma & \gamma \\ \gamma & -\gamma \end{pmatrix}.$$

Write the enlarged ME (6.134) for the pdf $\mathcal{P}(u, \pm 1, t)$. Using Kubo's oscillator model: $F(u, y) = -i(\omega_0 + \alpha \eta(t)) u$ where ω_0, α are constants, and defining the quantities

$$m_+(t) = \int u \mathcal{P}(u, +1, t) \, du$$

$$m_-(t) = \int u \mathcal{P}(u, -1, t) \, du.$$

[54]R. Kubo, J. Math. Phys. Soc. Japan 4, 174 (1963). See also R.C. Bourret, Can. J. Phys. 44, 2519 (1966); and N. van Kampen, Phys. Rep. 24, 171 (1976).

Show that $m_{\pm}(t)$ obeys[55]

$$\frac{dm_+(t)}{dt} = -i\left(\omega_0 + \alpha\right) m_+(t) - \gamma m_+(t) + \gamma m_-(t) \tag{6.135}$$

$$\frac{dm_-(t)}{dt} = -i\left(\omega_0 - \alpha\right) m_-(t) - \gamma m_-(t) + \gamma m_+(t), \tag{6.136}$$

then, from these equations, solutions $m_{\pm}(t)$ can be found[56], and from them the moment $\langle u \rangle$ follows as

$$\langle u \rangle = m_+(t) + m_-(t)$$

$$= ae^{-i(\omega_0 + \gamma)t} \left\{ \cos\left(t\sqrt{\alpha^2 - \gamma^2}\right) + \frac{\gamma \sin\left(t\sqrt{\alpha^2 - \gamma^2}\right)}{\sqrt{\alpha^2 - \gamma^2}} \right\}.$$

Study the limits $\gamma \ll \alpha$ and $\gamma \gg \alpha$. Compare the **mv** of u with the result when $\eta(t)$ is a GWN; see advanced exercise 3.20.11. The pdf for the additive case $F(\vec{u}, y) = -\gamma u + \eta(t)$ with $\eta(t)$ a dichotomic noise can be solved using the Green function technique; see M.O. Cáceres, Phys. Rev. E **67**, 016102 (2003), and also in Sect. 3.14. Related models when $F(\vec{u}, y)$ is linear and $\eta(t)$ are, in general, non-Markovian noises can also be solved using the characteristic functional technique; see M.O. Cáceres and A.A Budini, J. Phys. A Math. Gen. **30**, 8427 (1997).

6.7.5 Enlarged Markovian Chain (Noisy Map)

A map driven by a discrete noise can be worked out considering an enlarged Markovian chain; that is, a Markovian chain with internal states. In this case, the discrete nature of the time introduces new scenarios in the bifurcation analysis of the map. For example, consider the logistic map perturbed by a multiplicative two-level noise in the form

$$x_{n+1} = \mu\left(1 + \xi_n\right) x_n \left(1 - x_n\right), \ n = 0, 1, 2, \cdots \tag{6.137}$$

here ξ is a discrete-time dichotomic "noise" taking two possible values $\pm\Delta$ with a probability $\beta \in [0, 1]$ of repeating the same value in the following iteration. For $\beta > \frac{1}{2}$ the most probable sequence is formed by repetitive values of $+\Delta$ and $-\Delta$, for $\beta = \frac{1}{2}$ all sequences are equally weighted, while for $\beta < \frac{1}{2}$ the most probable

---

[55]Hint: multiply (6.132) by u and integrate over u (integrating by parts the surface term is assumed to vanish).

[56]In this case the initial condition is $\mathcal{P}(u, y, 0) = \delta(u - a) \Pi^{\text{st}}(y) = \delta(u - a)/2$, for $y = \pm1$.

sequence is formed by alternating values of $+\Delta$ and $-\Delta$. The parameters Δ and β completely define the process ξ_n, here Δ gives the intensity of the noise, and β characterizes the correlation of the dichotomic process; see also exercise 6.7.2.

In order to analyze the evolution of the noisy map (6.137) it is necessary to define the joint probability $P_n(x, \pm\Delta)$ of finding the map and the noise with values x and $\pm\Delta$ after n iterations. Due to the discrete nature of the time, this joint probability follows an enlarged recurrence relation[57]

$$P_{n+1}(x, \pm\Delta) = \beta \int \delta(x - f_\pm(y)) P_n(y, \pm\Delta) \, dy \qquad (6.138)$$

$$+ (1 - \beta) \int \delta(x - f_\mp(y)) P_n(y, \mp\Delta) \, dy,$$

where for the present logistic model $f_\pm(y) = \mu(1 \pm \Delta) y(1 - y)$. Show that the invariant solution of Eq. (6.138) obeys:

$$P(x, \pm\Delta) = \beta \int \delta(x - f_\pm(y)) P(y, \pm\Delta) \, dy \qquad (6.139)$$

$$+ (1 - \beta) \int \delta(x - f_\mp(y)) P(y, \mp\Delta) \, dy,$$

then the stationary distribution for the noisy map $\rho(x)$ is given by

$$\rho(x) = P(x, +\Delta) + P(x, -\Delta).$$

Equation (6.139) can be analyzed by perturbation techniques, therefore interesting predictions for the noisy map can be obtained; see J.M. Gutierrez, A. Iglesias, M.A. Rodriguez, Phys. Rev. E, 48, 2507 (1993). More general models of enlarged non-Markovian chains can also be studied in the context of the CTRW theory with internal states; see Sect. 7.4.

References

1. N.G. van Kampen, *Stochastic Process in Physics and Chemistry*, 2nd edn. (North-Holland, Amsterdam, 1992)
2. E.W. Montroll, B.J. West, *Fluctuation Phenomena*, ed. by E.W. Montroll, J.L. Lebowitz (North-Holland, Amsterdam, 1987)
3. M.O. Cáceres, C.E. Budde, Physica A **153**, 315 (1988)
4. M. Shlesinger, J. Stat. Phys. **10**, 421, (1974).
5. B.D. Hughes, E.W. Montroll, M.F. Shlesinger, J. Stat. Phys. **30**, 273 (1983)

---

[57]A Markovian chain with internal states will be treated in Chap. 7; see also C.B. Briozzo, C.E. Budde, O. Osenda, M.O. Cáceres, J. Stat. Phys. 65, 167 (1991).

6. N.S. Goel, N. Richter-Dyn, *Stochastic Models in Biology*(Academic Press, New York, 1974)
7. S. Bustingorry, M.O. Cáceres, E.R. Reyes, Phys. Rev. **B 65**, 165205 (2002)
8. P.L. Taylor, *A Quantum Approach to the Solids State* (Prentice-Hall, Englewood Cliffs, New York, 1970)
9. S. Alexander, J. Bernasconi, W.R. Schneider, R. Orbach Rev. Mod. Phys. **53**, 175 (1981)
10. S. Chandrasekhar, Rev. Mod. Phys. **15**, 1 (1943)
11. E. Hernández-García, M.O. Cáceres, Phys. Rev. A **42**, 4503 (1990)
12. M.O. Cáceres, H. Matsuda, T. Odagaki, D.P. Prato, W. Lamberti, Phys. Rev. B **56**, 5897 (1997)
13. P.A. Pury, M.O. Cáceres, Phys. Rev. E **66**, 021112 (2002)
14. A. Papoulis, *Probability Random Variables and Stochastic Process*, 3rd edn. (McHGraw Hill, New York, 1991)
15. A.A. Budini, M.O. Cáceres, J. Phys. A **32**, 4005 (1999)
16. P. Reimann, G.J. Schmid, P. Hanggi, Phys. Rev. E **60**, 1 (1999)
17. P.A. Pury, M.O. Cáceres, E. Hernández-García, Phys. Rev. A **49**, R967 (1994)
18. Z. Schuss, *Theory and Applications of Stochastic Differential Equations* (Wiley, New York, 1980)

Chapter 7
Diffusion in Disordered Media

There are circumstances under which the diffusion approximation is no longer valid. Problems like this arise when such transport is studied in disordered media, the reason being that in this case the long-range order is broken by the presence of impurities and diffusion becomes anomalous. Since these phenomena are of great theoretical and experimental interest in this chapter we will give an introduction to diffusion in disordered systems. Now, this problem could be studied based on the **F-P** equation with disorder; however, from the practical point of view it is easier to attack the problem with discrete stochastic models. On the other hand, continuous diffusion models always can be related to diffusion models on the lattice [1].

7.1 Disorder in the Master Equation

In this chapter we will discuss, in detail, the master equation (transport in a lattice) and its relation to anomalous diffusion in disordered systems. In order to simplify the presentation of the problem, consider the $1D$ case. Let $P(s, t) \equiv P(s, t \mid s_0, t_0)$ be the conditional probability of finding the RW[1] at site s and at time t. If the ME is characterized by the transition matrix (master Hamiltonian)

$$\mathbf{H}_{s,s'} = W_{s,s'} - \delta_{s,s'} \sum_{s_2} W_{s_2,s},$$

(7.1)

then in the presence of disorder matrix $\mathbf{H}$ acquires a random nature, either because the network is not regular or because the transition probabilities $W_{s,s'}$, between different sites, have themselves a disordered character. One of the easiest ways of addressing the problem of diffusion in disordered media is to consider that the

---

[1]Note that here we are using a shorthand notation without specifying the initial condition (s_0, t_0).

© Springer International Publishing AG 2017
M.O. Cáceres, *Non-equilibrium Statistical Physics with Application to Disordered Systems*, DOI 10.1007/978-3-319-51553-3_7

transition elements $W_{s,s\pm1}$ (on models with transitions to first next neighbors) are functions of random variables ξ_s, then every element $W_{s,s\pm1}$ may depend on each particular site s. It should be noted that for a fixed configuration of the disorder $[\xi]$ the evolution equation for $P(s, t)$ is a ME. This means that for each configuration the RW will be described by a Markovian stochastic process; that is to say, in general we have

$$\partial_t P(s, t \mid [\xi]) = \sum_{s'} \mathbf{H}_{s,s'}([\xi], t) P(s', t \mid [\xi]). \tag{7.2}$$

The explicit $[\xi]$ dependence in the argument of the master Hamiltonian $\mathbf{H}$ indicates that the solution of (7.2) depends on the settings of the disorder $[\xi] \equiv \{\cdots\xi_{s-1}, \xi_s, \xi_{s+1}, \cdots\}$, and the explicit argument t in $\mathbf{H}$ indicates any possible time dependence of the transition probabilities. In principle, associated with each site s there is a random variable that can influence the value taken by the transition matrix element $W_{s,s\pm1}$. The problem thus posed is of formidable complexity.[2] Therefore, to address this problem of disorder it is necessary to introduce further simplifications in the model. However, from the formal point of view for the stationary case, i.e., when $\mathbf{H}$ is not explicitly time dependent, and if we introduce the Laplace transform in (7.2) we obtain in matrix notation

$$u\mathbf{P}(u \mid [\xi]) - \mathbf{P}(t = 0) = \mathbf{H}([\xi]) \cdot \mathbf{P}(u \mid [\xi]), \tag{7.3}$$

where $\mathbf{P}(t = 0)$ is the initial condition and we assume that it does not depend on the disorder configuration $[\xi]$. Then what we are trying to do is to assess the **mv** of the solution (7.3), i.e., to evaluate the average considering all configurations of disorder $[\xi]$. For $\mathbf{P}(t = 0) = \mathbf{1}$, where $\mathbf{1}$ is the identity matrix of the same dimension as the matrix $\mathbf{H}([\xi])$, we have

$$\langle \mathbf{P}(u \mid [\xi]) \rangle_{\text{Disorder}} = \left\langle [u\mathbf{1} - \mathbf{H}([\xi])]^{-1} \right\rangle_{\text{Disorder}}. \tag{7.4}$$

Compare this situation with cases where the system is finite and with disordered transition elements [see, for example, Sect. 6.4.]. When working with an infinite system, the strategy is to calculate a homogeneous *effective* master Hamiltonian. For example, in 1D without drift, $\mathbf{H}_{\text{eff}}$ depends on the Laplace variable through a generalized diffusion coefficient $\mu(u)$; that is, $\mathbf{H}_{\text{eff}}(u) = \mu(u)\left(E^+ + E^- - 2\right)$. Thus the average resolvent of (7.3) can be approximated asymptotically by an effective Green function:

$$\left\langle [u\mathbf{1} - \mathbf{H}([\xi])]^{-1} \right\rangle_{\text{Disorder}} \simeq [u\mathbf{1} - \mathbf{H}_{\text{eff}}(u)]^{-1}. \tag{7.5}$$

---

[2]Even though we are only considering the case where $\mathbf{H}_{s,s'}([\xi])$ does not depend explicitly on time.

This approach is broadly applicable and allows to find several universal properties of the RW (i.e., to characterize the **mv** properties of the diffusion in disordered media). We will use this type of approach when we present the Effective Medium Approximation (EMA) and the continuous-time generalized non-Markovian *Random Walk* (CTRW.[3]) In the following sections we will develop these issues in detail.

Excursus (**Disordered Mechanical Systems**). Another important motivation for studying the ME with disorder lies in the fact that Eq. (7.3) can be related to models of vibrations in harmonic networks with a disordered distribution of masses, or in circuits with random components [2]. Thus the calculation of DOS and/or disordered vibration properties in mechanical systems is mathematically equivalent to the study of classical diffusion in disordered media.[4]

Excursus (**Random First Passage Times**). Extremal probability densities are those in which there appears a typical length-scale of the finite system, for example, the first passage time probability density across a certain border of the domain, see Sect. 6.6. If we are interested in a probability density of this type, the problem in the presence of disorder can be addressed using a technique similar to EMA. This is because, for each configuration of the disorder, the calculation of the survival probability satisfies an evolution equation analogous to the ME.[5] The use of projection operators [see Sect. 1.14] allows the calculation of the **mv** of the evolution equation for the accumulative function, and thus we can introduce an approach for solving the problem of anomalous diffusion in finite systems and/or the calculation of extremal densities in disordered media [3–5].

7.2 Effective Medium Approximation

In this section we present the outlines of the EMA technique, while in Appendix C some of the mathematical details are developed.

---

[3]The approach *Continuous Time Random Walk* was introduced by E.W. Montroll and G.H. Weiss, in J. Math. Phys. **6**, 178 (1965).

[4]The density of vibrational states (DOS) of harmonic chains can be calculated using the Green function formalism, in particular chains with impurities. Also in the same way a linear ferromagnetic lattice with nearest-neighbor interactions can be studied; see D. Prato and C.A. Condat, Am. J. Phys. 51(2), 140 (1983).

[5]The evolution of the accumulation function is governed by the "backward" equation or the adjoint operator. This equation is discussed in the context of the Fokker–Planck process in Chaps. 3, 4 and 9.

7.2.1   The Problem of an Impurity

The method of EMA may be presented in different ways[6] [2, 6–8]. Here we will proceed as when considering the problem of an impurity.

Consider an RW and assume that there is only *one bond* with characteristics different from the rest of the lattice.[7] In this case the matrix associated with the ME, i.e., H_i, will depend on an extra variable: *the impurity bond*. To fix this idea, assume, in 1D, that the impurity is the bond between sites $s = a$ and $s + 1 = b$, then the master Hamiltonian with the impurity can be written as

$$H_i = H_0 + V, \tag{7.6}$$

where $H_0 = \mu(E^+ + E^- - 2)$ is the ordered part and V represents the contribution coming from the presence of an impurity between sites a, b. That is, for a symmetrical model with nearest neighbor transitions we can write the *ordered* master Hamiltonian as

$$H_0 = \sum_s [|s> \mu <s+1| + |s> \mu <s-1|] - z \sum_s |s> \mu <s|, \tag{7.7}$$

where $z = 2$ is the coordination number in 1D; and the impurity term V adopts a simple form[8]

$$V = (\mu - W_{ba}) \, |a><a| + (\mu - W_{ab}) \, |b><b| \\ + (W_{ab} - \mu) \, |a><b| + (W_{ba} - \mu) \, |b><a|. \tag{7.8}$$

Exercise. Show from (7.6) that the different elements of the matrix H_i are

$$<a\,|\,H_i\,|\,a> = -\mu - W_{ba}, \quad <b\,|\,H_i\,|\,b> = -\mu - W_{ab} \tag{7.9}$$

$$<a\,|\,H_i\,|\,b> = W_{ab}, \quad <b\,|\,H_i\,|\,a> = W_{ba}.$$

---

[6]The time dependent perturbation approach necessarily leads to the concept of Terwiel's cumulant. These cumulants [ordered temporarily] were presented in Chap. 1. If the disturbance of disorder has Gaussian characteristics, Novikov's theorem allows closed-form expressions for calculating average values; see for example Sect. 1.14.

[7]For site-disorder, the problem can be analyzed analogously [9].

[8]Consider the ordered case of Fig. 6.4. An impurity between sites a and b can be placed anywhere on the lattice and in that location we must replace $\mu = W_{ba}$ and $\nu = W_{ab}$. For the symmetric case we have $\mu = \nu$ and so $W_{ab} = W_{ba}$. Note that this expression for the contribution of the impurity is also valid in the isotropic and symmetrical n-dimensional case.

Exercise. Show that in the asymmetric case (i.e., with drift: $\mu \neq \nu$ and $W_{ba} \neq W_{ab}$) the impurity term **V** takes the form

$$\mathbf{V} = (\mu - W_{ba}) \mid a >< a \mid + (\nu - W_{ab}) \mid b >< b \mid$$
$$+ (W_{ab} - \nu) \mid a >< b \mid + (W_{ba} - \mu) \mid b >< a \mid.$$

Exercise (**Impurity in** nD). In the case of an isotropic symmetric nD lattice, show that the elements of the matrix $\mathbf{H_i}$ for the problem of an impurity in arbitrary dimensions remain the same as in (7.9). Hint: Use again the perturbation **V** given in (7.8) and the homogeneous master Hamiltonian with transitions to first next neighbors characterized by

$$\mathbf{H}_0 = \sum_{s,s'(\neq)}^{N.N.} \mid s > \mu < s' \mid -z \sum_s \mid s > \mu < s \mid,$$

where $z = 2n$ for a hypercubic lattice.

Guided Exercise (**Asymmetric Anisotropic** $2D$ **Case**). Show that in a 2D lattice with transitions to first next neighbors the impurity term can be written in the form

$$\mathbf{V}_{\alpha\beta} = b_1 \delta_{\alpha a} \delta_{\beta a} + b_4 \delta_{\alpha b} \delta_{\beta b} + b_2 \delta_{\alpha a} \delta_{\beta b} + b_3 \delta_{\alpha b} \delta_{\beta a}. \tag{7.10}$$

Here indices α, β run over a simple cubic lattice, and we have defined $b_1 = A - W_{ba}$, $b_2 = W_{ab} - B$, $b_3 = W_{ba} - A$, $b_4 = B - W_{ab}$ if the sites a and b are horizontal; and $b_1 = C - W_{ba}$, $b_2 = W_{ab} - D$, $b_3 = W_{ba} - C$, $b_4 = D - W_{ab}$ if the sites a and b are vertical. In Fig. 7.1 a schematic diagram of a $2D$ ordered anisotropic and asymmetric lattice is shown. Note that in the ordered $2D$ case the ME can be written as follows[9]:

$$\partial_t \mathbf{P}_{m,n}(t) = A\,\mathbf{P}_{m-1,n}(t) + B\,\mathbf{P}_{m+1,n}(t) + C\,\mathbf{P}_{m,n-1}(t) + D\,\mathbf{P}_{m,n+1}(t)$$
$$- (A + B + C + D)\,\mathbf{P}_{m,n}(t).$$

That is, the ordered master Hamiltonian is

$$\mathbf{H}_0 = \sum_{m,n} \{ B \mid m - 1, n >< m, n \mid + A \mid m + 1, n >< m, n \mid$$

$$+ D \mid m, n - 1 >< m, n \mid + C \mid m, n + 1 >< m, n \mid \}$$

$$- \sum_{m,n} \{ (A + B + C + D) \mid m, n >< m, n \mid \}.$$

---

[9]Here we have omitted, for simplicity, the initial condition.

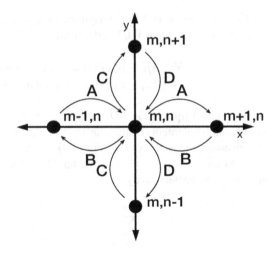

7.2.2 Calculation of the Green Function with an Impurity

The Green function corresponding to the *one* impurity problem can be calculated
exactly and for arbitrary dimension using Dyson's technique. For example, the
Green function (in *2D*) for the problem of a bond impurity characterized by (7.10)
is $G^i(u) = (u\mathbf{1} - \mathbf{H}_i)^{-1}$, which in turn is a function that can be expressed in terms
of the Green function associated with a homogeneous ordered lattice $G^0(u) =
(u\mathbf{1} - \mathbf{H}_0)^{-1}$. The mathematical details of this proof can be found in Appendix C.
Using the homogeneous anisotropic and asymmetric Green function we obtain for
element $G^i_{ab} \equiv < a \mid G^i(u) \mid b >$ the expression[10]:

$$G^i_{ab} = < a \mid (u\mathbf{1} - \mathbf{H}_i)^{-1} \mid b > = \frac{G^0_{ab} + b_4 \left(G^0_{ab}G^0_{ba} - G^0_{aa}G^0_{bb}\right)}{1 + b_1 \left(G^0_{ab} - G^0_{aa}\right) + b_4 \left(G^0_{ba} - G^0_{bb}\right)}. \tag{7.11}$$

In the *2D* isotropic case without drift, we note that $W_{ba} = W_{ab} = W$ is the
impurity bond between sites a and b (horizontal or vertical sites). In this case we
have, then, a single parameter for the homogeneous lattice: $A = B = C = D = \mu$,
and the Green function for *one* impurity adopts a simpler form:

$$G^i_{ab} = \frac{G^0_{ab} + \mathcal{B}\left(G^0_{ab}G^0_{ba} - G^0_{aa}G^0_{bb}\right)}{1 + \mathcal{B}\left(G^0_{ab} + G^0_{ba} - G^0_{bb} - G^0_{aa}\right)}, \tag{7.12}$$

where $\mathcal{B} = \mu - W$.

---

[10]This problem corresponds to a *2D* anisotropic and asymmetric lattice with a drift in any direction,
[10].

Note that in the $1D$ case without drift $W_{ba} = W_{ab} = W$ and so $A = B = \mu$, whence it follows that $b_1 = b_4 = -b_3 = -b_2$. Then, defining as before the quantity $\mathcal{B} = \mu - W$, we can see that the same expression is obtained (7.12); the only difference with the $2D$ case is in the structure itself of the homogeneous Green function: G_{ab}^0. In a $1D$ lattice, elements of the homogeneous Green function can be obtained from the analysis presented in Sect. 6.3.3; therefore we get

$$
G_{ab}^0 = G_{ba}^0 = \left[\frac{2\mu}{\left(u + 2\mu + \sqrt{u^2 + 4\mu u} \right)} \right]^{|a-b|} \frac{1}{\sqrt{u^2 + 4\mu u}} \tag{7.13}
$$

$$
G_{aa}^0 = G_{bb}^0 = \frac{1}{\sqrt{u^2 + 4\mu u}}.
$$

Optional Exercise. Show that in the isotropic nD case without drift (symmetrical), the Green function of one impurity is again given by expression (7.12). Then, the only difference is that G_{ab}^0 is now the Green function corresponding to the homogeneous isotropic and symmetrical nD lattice (see Appendix C).

Guided Exercise. Show that for the isotropic and symmetric case, in nD, the homogeneous Green function meets the identity

$$
G_{00}^0 - G_{10}^0 = \left(1 - uG_{00}^0 \right) / z\mu. \tag{7.14}
$$

The coordination number z shows the number of nearest neighbors in a simple hypercubic nD lattice. To prove (7.14) simply use the definition of Green's function or, equivalently, its evolution equation: $uG^0 - 1 = H_0 G^0$.

$$
< 0 \,|\, (uG^0 - 1) \,|\, 0 > = < 0 \,|\, H_0 G^0 \,|\, 0 > .
$$

That is, using the notation $< 0 \,|\, H_0 \,|\, l > \equiv H_{0l}$, etc., we can write

$$
uG_{00}^0 - 1 = \sum_l H_{0l} \, G_{l0}^0 = H_{00} \, G_{00}^0 + zH_{01} \, G_{10}^0 ,
$$

where we have used that transitions are to first neighbors only. Noting that $H_{00} = -z\mu$ and $H_{01} = \mu$ the proof follows.

7.2.3 Effective Medium

When trying to solve the problem of a disordered system where the transition probability W (for each bond in the lattice) is a random variable, the equation that provides an approximation for the effective medium arises from imposing that the difference between the propagator with *one* impurity (i.e., with a random character) and the homogeneous propagator (in the effective medium) is zero in **mv**. For example, in $1D$, we write, first, the homogeneous propagator in the form $G^0(u, \Gamma) \equiv (u\mathbf{1} - \mathbf{H}_0)^{-1}$, where $\mathbf{H}_0 = \Gamma(E^+ + E^- - 2)$ and $\Gamma \equiv \mu(u)$ is an unknown function. That is, in the homogeneous Green function we have replaced the constant μ by an unknown function of the Laplace variable, which we call *effective medium* (or frequency dependent diffusion coefficient). Second, we replace the Green function of *one* impurity by a Green's function of an impurity but with a diffusion coefficient Γ, i.e., $G^i(u) \rightarrow G^i(u, \Gamma) = (u\mathbf{1} - \mathbf{H}_i)^{-1}$, where $\mathbf{H}_i = \Gamma(E^+ + E^- - 2) + \mathbf{V}$ and the perturbation $\mathbf{V}$ is given as before, for example for bond disorder (7.8), but substituting[11] $\mu \rightarrow \Gamma$. If we now consider that the impurity is a random variable characterized by a certain probability distribution $\mathcal{P}$, the self-consistent condition is given by

$$\left\langle G^i(u, \Gamma) \right\rangle_{\mathcal{P}} = G^0(u, \Gamma), \quad \forall u. \tag{7.15}$$

This equation allows for the $\Gamma = \Gamma(u)$ unknown function. Figure 7.2 shows, for a $1D$ lattice, the procedure involved in the scheme of EMA. Finally, once the effective medium $\Gamma = \Gamma(u)$ is known, the effective Green function $G^{\text{eff}} = [u\mathbf{1} - \mathbf{H}_{\text{eff}}(u)]^{-1}$ is obtained by replacing μ by Γ in the homogeneous solution.

In general, in an isotropic nD system without drift, where there is only one random variable W, the average involved in (7.15) will be characterized by a single probability distribution $\mathcal{P}(W)$. Averaging over the elements of the Green function between sites $\{a = 0, b = 1\}$, from (7.15) we obtain

$$\left\langle < 0 \mid G^i(u, \Gamma) \mid 1 > \right\rangle_{\mathcal{P}(W)} = < 0 \mid G^0(u, \Gamma) \mid 1 > .$$

From which it follows, using (7.12) with $\mu = \Gamma$, that the self-consistent equation for solving the effective medium Γ is

$$\left\langle \frac{G^0_{01} + \mathcal{B} \left(G^0_{01} G^0_{10} - G^0_{00} G^0_{11} \right)}{1 + \mathcal{B} \left(G^0_{10} + G^0_{01} - G^0_{11} - G^0_{00} \right)} \right\rangle = G^0_{01}. \tag{7.16}$$

Given that the random variable W appears only in the quantity $\mathcal{B}$, the expression (7.16) can be simplified further, using (7.14); we finally get:

$$\left\langle \frac{\mathcal{B}}{1 + 2\mathcal{B} \left(G^0_{10} - G^0_{00} \right)} \right\rangle = \int \frac{(\Gamma - W)}{1 - 2(\Gamma - W)\left(1 - u G^0_{00} \right)/z\Gamma} \mathcal{P}(W) \, dW = 0. \tag{7.17}$$

---

[11] For an isotropic and symmetric nD lattice, the procedure is the same as for the (symmetrical) $1D$ case. But not if the lattice has, for example, asymmetry in one direction because of the application of a field. If so, there is no single effective medium. In this case, when considering a lattice with drift, the effective medium depends on both sense and direction.

(a)

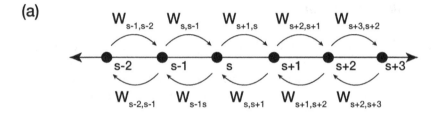

(b)

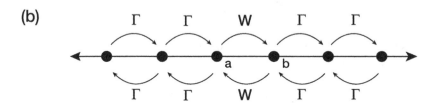

(c)

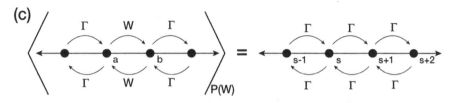

Fig. 7.2 (**a**) Diagrammatic representation in a 1D disordered lattice. (**b**) Representation of a bound impurity (symmetrical) embedded in a homogeneous medium. (**c**) Representation of the procedure of calculating the effective medium, $\Gamma = \Gamma(u)$, using a self-consistent approach

Exercise. Prove for a 1D lattice without drift and with bond disorder, that expression (7.17) which is valid for any probability distribution $\mathcal{P}(W)$ takes the form:

$$\int \frac{(\Gamma - W)}{W + (\Gamma - W)uG_{00}^0} \mathcal{P}(W)\, dW = 0, \quad \forall u. \tag{7.18}$$

Excursus. The effective medium approximation for a n-dimensional anisotropic and asymmetric lattice can also be solved in closed form; i.e., in terms of the Green function of the corresponding (homogeneous) problem. In this case we must solve $2n$ self-consistent effective media [10].

Exercise (**Second Moment**). Using the Fourier–Laplace representation of the effective solution $G^{\mathrm{eff}}(k, u) = [u + 2\Gamma(u)(1 - \cos k)]^{-1}$, for the 1D case, show that the second moment is given by $\langle s(u)^2 \rangle = 2\Gamma(u)/u^2$.

7.2.4 Short Time Limit

Here we will use the results of the EMA technique to study the effective diffusion coefficient in the short time regime. In particular, in this section we will restrict ourselves to the $1D$ case without drift. Given that the short time limit is equivalent to the analysis of the limit $u \to \infty$ in the Laplace representation [11], from expression (7.13) it follows that

$$G_{00}^0 \sim \frac{1}{u}, \quad u \to \infty.$$

From this fact we can approximate, in the short time regime, the quantity $u G_{00}^0 \sim 1$. Then, from (7.18) it follows that

$$\int \frac{(\Gamma - W)}{\Gamma} \mathcal{P}(W) \, dW = 0, \quad u \to \infty,$$

This implies that, for all types of probability distribution $\mathcal{P}(W)$, the effective medium is given by

$$\Gamma = \langle W \rangle, \quad u \to \infty.$$

This result is consistent with our short times prediction of Chap. 6.

7.2.5 The Long Time Limit

Now we use the EMA, the self-consistent equation (7.17), for studying the effective diffusion coefficient in the long time regime. Again, in this section we shall treat only the $1D$ case without drift. In this case the limit of interest of the Laplace variable is $u \to 0$. Then, from (7.13) it follows that

$$G_{00}^0 \sim 1/\sqrt{4\Gamma u}, \quad u \to 0,$$

from where we can extrapolate that for long times: $u G_{00}^0 \sim \sqrt{u/4\Gamma}$. Then, from (7.18) and taking the limit $u \to 0$ we obtain

$$\int \frac{(\Gamma - W)}{W + (\Gamma - W) u G_{00}^0} \mathcal{P}(W) \, dW \longrightarrow \int \frac{(\Gamma - W)}{W + (\Gamma - W) \sqrt{u/4\Gamma}} \mathcal{P}(W) \, dW,$$

and thus in particular

$$\int \frac{(\Gamma - W)}{W} \mathcal{P}(W) \, dW = 0, \quad u = 0.$$

From this expression, calculating the effective diffusion coefficient is straightforward. Finally, applying the normalization condition $\int \mathcal{P}(W) \, dW = 1$ we obtain

$$\Gamma \left\langle \frac{1}{W} \right\rangle = 1, \quad u = 0. \tag{7.19}$$

This expression is consistent with our previous result for the effective diffusion coefficient of the system in the long time regime [see Sect. 6.4.3]. Note that this expression is valid for any distribution $\mathcal{P}(W)$ as long as the quantity $\langle W^{-1} \rangle$ is defined; thus we get $\Gamma \neq 0$ in the limit $u \to 0$.

Guided Exercise (**Strong Disorder**).   A model of strong disorder is characterized by a finite probability of finding a cut bond, i.e., $\langle W^{-1} \rangle = \infty$. For example, we can use $\mathcal{P}(W) = (1 - \alpha)W^{-\alpha}$ with $W \in (0, 1)$ and $\alpha \in (0, 1)$. If we try to solve (7.17) with this distribution $\mathcal{P}(W)$, we cannot obtain an analytical expression $\Gamma(u), \forall u$. However, it is possible to study $\Gamma(u)$ asymptotically for $u \sim 0$. To make a rigorous perturbative analysis of $\Gamma(u)$ around $u \sim 0$, it is necessary to use the Mellin transform.[12] As this transformation is not well known, we present here a simple way to get the dominant order of the effective medium $\Gamma(u)$ using the Laplace transform; however, as we have emphasized, this technique does not allow to study higher order corrections when $\mathcal{P}(W) = (1 - \alpha)W^{-\alpha}$. We have already commented that $G_{00}^0 - G_{10}^0 = \left(1 - uG_{00}^0\right) / z\Gamma(u)$, and because for strong disorder $\lim_{u \to 0} \Gamma(u) \to 0$, we see that $\left(G_{00}^0 - G_{10}^0\right)$ diverges. Then, the self-consistent equation should be written in a way to avoid a division by zero. If we define $R \equiv \left[2 \left(G_{00}^0 - G_{10}^0\right)\right]^{-1} - \Gamma$, we see immediately that this quantity [nonrandom] is small and tends to zero for $u \to 0$. Then, we can write (7.17) for $1D$ in the form $\langle (\Gamma - W) / (R + W) \rangle = 0$; that is to say:

$$\Gamma \left\langle \frac{1}{R + W} \right\rangle = \left\langle \frac{W}{R + W} \right\rangle, \quad \forall u. \tag{7.20}$$

Now we have to study this equation in the regime when $R \sim 0$, using $\mathcal{P}(W) = (1 - \alpha)W^{-\alpha}$. It is easy to see that $\lim_{R \to 0} \left\langle \frac{W}{R+W} \right\rangle \to 1$ and, on the other hand, $\lim_{R \to 0} \left\langle \frac{1}{R+W} \right\rangle \to \infty$. Then we can approximate (7.20) by

$$\Gamma \left\langle \frac{1}{R + W} \right\rangle = 1, \quad u \to 0. \tag{7.21}$$

---

[12]Both Mellin and the Laplace transforms can be found at: *Transformations Intégrales et Calcul Opérational,* V. Ditkine et A. Proudnikov, Moscou, Ed. Mir (1978).

Now we need to calculate how much does $\langle (R + W)^{-1} \rangle$ diverge for $R \to 0$; for this purpose we introduce the change of variable $W = R e^t$, and we get:

$$\left\langle \frac{1}{R+W} \right\rangle = (1-\alpha) \int_0^1 \frac{W^{-\alpha}\, dW}{W(1+R/W)} = \frac{(1-\alpha)}{R^\alpha} \int_{-\infty}^{-\ln R} \frac{e^{-\alpha t}\, dt}{1+e^{-t}}$$

$$= \frac{(1-\alpha)}{R^\alpha} \left\{ \int_{-\infty}^0 \frac{e^{-\alpha t}\, dt}{1+e^{-t}} + \int_0^{-\ln R} \frac{e^{-\alpha t}\, dt}{1+e^{-t}} \right\}.$$

The first integral is simply a Laplace transform, that is,

$$\int_{-\infty}^0 \frac{e^{-\alpha t}\, dt}{1+e^{-t}} = \int_0^\infty \frac{e^{\alpha t}\, dt}{1+e^t} = \int_0^\infty \frac{e^{-(1-\alpha)t}\, dt}{1+e^{-t}}$$

$$= \mathcal{L}_s \left[\frac{1}{1+e^{-t}} \right] \Big|_{s=1-\alpha} = \frac{1}{2} \left(\psi(\frac{1-\alpha+1}{2}) - \psi(\frac{1-\alpha}{2}) \right),$$

where $\psi(z)$ is the Digamma function.[13] The second integral can be approached extending the upper limit to infinity; then we can again use the Laplace transform:

$$\int_0^{-\ln R} \frac{e^{-\alpha t} dt}{1+e^{-t}} \approx \int_0^\infty \frac{e^{-\alpha t} dt}{1+e^{-t}} = \mathcal{L}_s \left[\frac{1}{1+e^{-t}} \right] \Big|_{s=\alpha} = \frac{1}{2} \left(\psi(\frac{\alpha+1}{2}) - \psi(\frac{\alpha}{2}) \right).$$

Whereupon, gathering all the expressions and using $\psi(\frac{1}{2} - z) = \psi(\frac{1}{2} + z) - \pi \tan(\pi z)$ and $\psi(1 - z) = \psi(z) + \pi/\tan(\pi z)$, we finally obtain:

$$\left\langle \frac{1}{R+W} \right\rangle \approx \frac{(1-\alpha)\pi}{R^\alpha \sin(\pi\alpha)}, \quad R \to 0.$$

Applying this asymptotic expression in (7.21), show that in 1D [using $R \sim \sqrt{u\Gamma/4}$] we get

$$\Gamma(u \sim 0) \approx \left(\frac{\sin(\pi\alpha)}{2\pi(1-\alpha)} \right)^{2/(2-\alpha)} u^{\alpha/(2-\alpha)}, \quad \alpha \in (0,1), \tag{7.22}$$

a result that matches the dominant order when it is calculated using the Mellin transform.

Excursus (**Electric Conductivity**). In general, the frequency behavior of $\Gamma(u), \forall u$, cannot be obtained analytically from the self-consistent equation. However, for $u \to 0$ and $u \to \infty$ they can be obtained from perturbation expansions,

---

[13]The Digamma function $\psi(z)$ is defined as the derivative of the logarithm of the Gamma function $\Gamma(z)$ [not to be confused with our notation for effective medium]; see, for example, *An Atlas of Functions*, J. Spanier and K.B. Oldham, Berlin, Springer-Verlag (1987).

which are also useful for comparing with numerical results of the above equation. From the calculation of $\Gamma(u)$ it is possible to determine the frequency dependence of the electrical conductivity, by using the generalized Einstein's relation [12]: $\sigma_{ac}(\omega) \propto \Gamma(u = -i\omega)/k_BT$; see Chap. 8.

Optional Exercise (**Localization in Disordered Systems**). In a $1D$ lattice with bond disorder,[14] characterized by $\mathcal{P}(W) = (1 - \alpha)W^{-\alpha}$ with $W \in (0, 1)$ and $\alpha \in (0, 1)$, we have rigorously proved—in the context of the EMA—that, asymptotically,

$$\Gamma(u \sim 0) \sim u^{\alpha/(2-\alpha)}$$

is the dominant term, as shown in (7.22). Note that for this type of "strong" disorder $\Gamma(u \rightarrow 0) \rightarrow 0$. Show that the second moment of displacement of an RW asymptotically behaves like

$$\langle s(u \sim 0)^2 \rangle \sim \Gamma(u)/u^2 \sim u^{(3\alpha-4)/(2-\alpha)}.$$

This implies a long times behavior like[15]

$$\langle s(t)^2 \rangle \sim t^{2(1-\alpha)/(2-\alpha)}, \quad 0 < \alpha < 1, t \rightarrow \infty,$$

that is, subdiffusive. Here it is noted that the "critical" exponent $2(1 - \alpha)/(2 - \alpha)$ is not universal and depends on the intensity of the disorder. On the other hand, in Sect. 6.3.4 we already noted that the analysis of the DOS can establish criteria to characterize the localization of a classical particle diffusing in a disordered medium. From the asymptotic behavior of the Green function $G^{\text{eff}} = [u\mathbf{1} - \mathbf{H}_{\text{eff}}(u)]^{-1}$, show that

$$G_{00}^{\text{eff}} \sim 1/\sqrt{4\Gamma(u)u} \sim u^{-1/(2-\alpha)}, \quad u \rightarrow 0,$$

then it is possible to observe—taking the inverse Laplace transform—that in the EMA

$$P(0, t) \sim t^{-(1-\alpha)/(2-\alpha)}, \quad t \rightarrow \infty.$$

That is, in a system with strong disorder, the relaxation of the initial condition is slower than in the diffusive case: $P(0, t) \sim t^{-1/2}$.

---

[14]In reference [2] this type of disorder is called W-disorder, while the site-disorder is called C-disorder since it corresponds to an electrical circuit with random capacitances. The asymptotic behavior of $\langle s(t)^2 \rangle$ is the same for both W and C disorder types, see [7].

[15]Taking the inverse Laplace transform. Indeed, Tauberian theorem shows that if $f(u) \sim u^{-\beta}$, for $u \rightarrow 0$, with $\beta \geq 0$; then asymptotically $f(t) \sim t^{\beta-1}$, for $t \rightarrow \infty$, see advanced exercise 7.5.6, or reference [13].

Exercise (**Static Limit** $u = 0$). Using that

$$uG_{00}^0(u)\big|_{u=0} = P(0, t = \infty \mid 0, 0) = 0.$$

Show, for an isotropic nD lattice without drift and bond disorder, that the equation for calculating the self-consistent diffusion coefficient (effective) is

$$\int \frac{(\Gamma - W)}{z\Gamma - 2(\Gamma - W)} P(W)\, dW = 0, \quad u = 0.$$

Hint: use (7.14), (7.17) and the fact that the static diffusion coefficient is given by $\Gamma\big|_{u=0}$; that is, the effective medium evaluated at zero frequency. For example, in $2D$ the self-consistent equation takes the form

$$\left\langle \frac{\Gamma - W}{\Gamma + W} \right\rangle = 0, \quad u = 0. \tag{7.23}$$

Enter now a distribution $\mathcal{P}(W)$ such that it allows—with nonzero probability—the possibility of finding a "broken" bond ($W = 0$). Interpret and compare this result with that obtained in a $1D$ lattice [i.e., using (7.19)] with the same distribution $\mathcal{P}(W)$.

Optional Exercise (**Percolation of "Unbroken" Bond**). Consider a nD lattice with bond disorder of the percolation type: $\mathcal{P}(W) = p\delta(W - W_0) + (1 - p)\delta(W)$, where $p \in [0, 1]$; i.e., with probability p to find a bond W_0 between two arbitrary sites. From the self-consistent equation for $u = 0$, show that the static diffusion coefficient has a threshold value $p_c = 2/z = 1/n$ (z is the coordination number in hypercubic lattice) below which the diffusion coefficient vanishes [6]. Figure 7.3 shows a $2D$ lattice with a concentration of "healthy" (unbroken) bond $p = 0.6$.

Excursus (**Transport with Site-Disorder**). For a nD lattice with site-disorder [9], show that the self-consistent equation for calculating the effective medium $\Gamma = \Gamma(u)$ is

$$\frac{1}{\Omega + \Gamma} = \left\langle \frac{1}{\Omega + W} \right\rangle, \quad \forall u, \tag{7.24}$$

Fig. 7.3 Random configuration of a $2D$ square lattice with broken bonds, represented by *dotted lines*. In this realization of the disorder the "healthy bond" concentration is $p = 0.6$

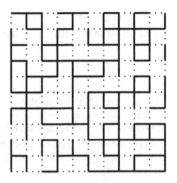

where $\Omega = \Gamma / \left(1 - uG_{00}^0\right) - \Gamma$. Hint: obtain $G^i = (u\mathbf{1} - \mathbf{H}_0 - \mathbf{V})^{-1}$ for a hypercubic lattice with one impurity at site a, where $\mathbf{V}$ is now given by the expression:

$$\mathbf{V} = z\,(\mu - W_a)\,|\,a><a\,| + (W_a - \mu)\,|\,a{+}1><a\,| + (W_a - \mu)\,|\,a{-}1><a\,|.$$

Compare with (7.8).

Note It should be emphasized that the EMA for models of weak disorder [site or bond] reproduces exactly, to order $\mathcal{O}(\sqrt{u})$, the same results we have obtained perturbatively for long times using Dyson's expansion. These results were presented in Chap. 6 in terms of quantities $\langle W^{-p} \rangle \neq \infty, \forall p$. For example, from the self-consistent equation, using weak disorder in (7.24), it is possible to calculate the diffusion coefficient for site-disorder up to the first correction of order $\mathcal{O}(\sqrt{u})$. Alternatively, this calculation can be performed more easily using the procedure presented in Sect. 6.4.3. In this way, the following exact perturbative expansion is obtained:

$$\Gamma(u) = \langle W^{-1} \rangle^{-1} \left\{ 1 + \langle W^{-1} \rangle^{-3/2} \frac{\sqrt{u}}{2} \langle Q^2 \rangle + \mathcal{O}(u) \right\}, \quad u \to 0. \tag{7.25}$$

Here we have defined the quantity $\langle Q^2 \rangle \equiv \langle W^{-2} \rangle - \langle W^{-1} \rangle^2$, i.e., the "dispersion" of the inverse moment. It is possible to prove that the results of higher orders calculated by EMA [7] must be corrected diagrammatically to match the exact results from perturbative calculation. On the other hand, it is interesting to note that the result (7.25) is also valid for weak bond-disorder. This is so because we have assumed that the initial condition of the RW is δ_{s,s_0}. Finally, it is noteworthy that in the case of strong disorder, i.e., when the inverse moments are not defined, it is not possible to perform a perturbative calculation as presented in (7.25). In this case the EMA is of inestimable value, because it does allow to obtain an asymptotic expansion for $\Gamma(u \sim 0)$ [see Eq. (7.22)]. However, it is important to remark that the EMA for strong disorder gives the correct "critical" exponents, but does not give the correct coefficients. That is, it is possible to introduce a *singular* perturbation theory in terms of Terwiel's cumulants with which a correct perturbation expansion is obtained; whence we can prove that EMA is the first contribution. In particular, it can be shown that the critical exponents calculated from EMA are correct to the dominant order; however, in order to get the correct coefficients it is necessary to add other diagrams that go beyond the EMA. These assertions have been proved, for 1D lattices at least, for systems with and without boundary conditions [3, 5, 7].

Excursus (**Oscillatory Boundary Condition**). The response of a system to an external field is standard information used regularly in the study of condensed matter. Under a periodic forced boundary condition at one of the confines of the system (1D), the output flow from the other boundary is proportional to the perturbation (in the linear regime) and the proportionality constant is called the admittance.[16] The frequency dependence of the admittance contains varied information on the dynamics of the responding carriers. The boundary perturbation method when the carrier follows a stochastic dynamics has been studied from a generic ME in 1D, and then the EMA has been used to analyze the long time behavior of the probability distribution at the confines of the system for different types of the random distribution of jump rates in the ME [5].

7.3 Anomalous Diffusion and the CTRW Approach

This section presents the general outline of the Continuous Time Random Walk (CTRW) technique [13, 14]; that is, the generalization to a non-Markovian RW, in Appendix D some of its mathematical details are presented. Consider a recurrence relation of the RW type, but, unlike the case of Markov chains now let time be a continuous variable. Then we can write for the conditional probability *density* of reaching site s *precisely* at time t

$$R(s,t \mid s_0,0) = \int_0^t \sum_{s'} \psi(s-s',t-\tau)R(s',\tau \mid s_0,0) \, d\tau + \delta_{s,s_0}\delta(t). \qquad (7.26)$$

Here $\psi(s-s',t-\tau)\,d\tau$ represents the probability that the RW (non-Markovian) jumps from site s' to s in a regular lattice[17] having been trapped, since arriving at s' during a time interval $t-\tau$. Normalization requires:

$$\int_0^\infty \sum_s \psi(s-s',\tau) \, d\tau = 1$$

The recurrence relation (7.26) generalizes a Markov chain, because the conditional pdf of being at site s just at time t; that is, $R(s,t \mid s_0,0)$ is given in terms of a temporal convolution likely to be at different sites s' in previous instants $R(s',\tau \mid s_0,0)$ with $\tau \le t$. As can be seen from recurrence relation (7.26) and since $R(s,t \mid s_0,0)$ is a probability density, in order to find the conditional probability $P(s,t \mid s_0,0)$ we must integrate the solution $R(s,t' \mid s_0,0)$ multiplied

---

[16]The experimental techniques of "the frequency response method" and "Intensity modulated photocurrent spectroscopy" can be considered to belong to the generic boundary perturbation method; see: T. Odagaki, M. Kawassaki, M.O. Cáceres, and H. Matsuda, p. 171, in: *Statistical Physics, Experiments, Theories and Computer Simulation*, Proc. 2nd Tohwa University International Meeting; Eds. M. Tokuyama and I. Oppenheim, World Scientific, Singapore, 1997.

[17]In this section we will not focus on the dimension of the lattice, because the expressions can easily be generalized to arbitrary dimension. Furthermore, in the context of CTRW theory, although the lattice dimension is important, the crucial aspect is characterizing the temporal behavior of pauses or entrapments of the RW at an arbitrary site.

by the probability of remaining at site s without jumping during the interval of time $t - t'$, i.e.:

$$P(s, t \mid s_0, 0) = \int_0^t R(s, t' \mid s_0, 0)\phi(t - t') \, dt'. \tag{7.27}$$

Here we have used that the chance to stay on site s without jumping during the time interval $[0, t]$ is independent of the site[18] and is given by

$$\phi(t) = 1 - \int_0^t \sum_{s'} \psi(s' - s, t') \, dt' \equiv 1 - \int_0^t \psi(t') \, dt'. \tag{7.28}$$

This equation defines the probability density for waiting times between successive jumps $\psi(t)$.[19] On the other hand, it is easier to find the solution of the system if we first transform all CTRW equations to Laplace representation, i.e.:

$$R(s, u \mid s_0, 0) = \sum_{s'} \psi(s - s', u)R(s', u \mid s_0, 0) + \delta_{s,s_0} \tag{7.29}$$

$$P(s, u \mid s_0, 0) = R(s, u \mid s_0, 0)\phi(u) \tag{7.30}$$

$$\phi(u) = \frac{1 - \psi(u)}{u}. \tag{7.31}$$

Now, as we did when we solved an RW in a translationally invariant lattice, we should move to the Fourier representation. Since the recurrence relation (7.29) involves a spatial convolution (in the infinite lattice), after defining the Fourier transform,

$$P(k, u \mid s_0, 0) = \sum_{s'} e^{iks'} P(s', u \mid s_0, 0), \tag{7.32}$$

and calculating the corresponding functions $\psi(k, u)$ and $R(k, u \mid s_0, 0)$, we get immediately the solution of the CTRW as

$$P(k, u \mid s_0, 0) = \frac{1 - \psi(u)}{u} \frac{e^{iks_0}}{1 - \psi(k, u)} \tag{7.33}$$

$$\equiv \phi(u) R(k, u \mid s_0, 0)$$

---

[18]Due to the translational invariance in the lattice.

[19]This density is called the *waiting-time function*. It is also possible to present the CTRW scheme considering that the density for the "first pause" $\psi^{(1)}(t)$ is different from the density $\psi(t)$; see advanced exercises in Sect. 7.5. This generalization is useful in models of disordered systems where the initial preparation of the system is not in equilibrium; see, J.W. Haus and K.W. Kehr, Phys. Rep. **150**, 263 (1987); M.A. Ré and C.E. Budde, Phys. Rev. E **61**, 1110 (2000).

In general, the CTRW propagator is characterized by the function $\psi(k, u)$, which in turn determines univocally the waiting-time function

$$\psi(u) = \psi(k = 0, u) = \sum_{s'} e^{ik(s'-s)} \psi(s' - s, u)\Big|_{k=0}.$$

The simplest transport model in the CTRW context corresponds to the case where the function $\psi(k, u)$ is separable in the following sense[20]:

$$\psi(k, u) = \lambda(k)\,\psi(u). \tag{7.34}$$

Then $\lambda(k)$ characterizes the hopping structure in the regular network [in general we will use transitions to first neighbors; compare with function $T_1(k)$ of Sect. 6.2] and $\psi(u)$ is the waiting-time function that characterizes the random time intervals when the non-Markovian RW is (stuck) fixed at a lattice site. The Green function or CTRW propagator corresponds to the case when the initial condition is at the origin of coordinates:

$$G_{CTRW}(k, u) = \frac{1 - \psi(u)}{u\,[1 - \psi(k, u)]}. \tag{7.35}$$

Exercise (**Rate of Probability**). From (7.27) it is clear that $R(s, t \mid s_0, 0)$ is a probability per unit time, because $\phi(t)$ is the probability of staying at site s during the time interval $[0, t]$ following the walkers arrival at the site. Show that in the Markovian case, that is, when $\psi(u) = 1/(1 + \langle t \rangle u)$, and in the long time limit we get asymptotically:

$$P(s, t \to \infty \mid s_0, 0) = \langle t \rangle R(s, t \to \infty \mid s_0, 0).$$

For future notation it is convenient to clarify that $R(s, t \mid s_0 = 0, 0) \equiv R(s, t)$, and in the Fourier–Laplace representation we simply write $R(k, u)$. If the model is separable as in (7.34) we get the identity $R(k, u) \equiv R(k, \psi)$ with $\psi = \psi(u)$, occasionally we also use the notation $R(k, z)$ when coming from the discrete time representation of an RW; see Sects. 7.3.2 and 7.5.

Exercise (**The Green Function**). Show that $G_{CTRW}(k, u)$ is the Green function in the Fourier–Laplace representation of the problem proposed by Eqs. (7.29)–(7.31).

Transport properties of a CTRW can be characterized by studying the moments of the Green function $G_{CTRW}(k, u)$. In particular, when there is no drift the second

---

[20]The notion of "separability" can equivalently be introduced in the original space-time variables, i.e.: $\psi(s - s', t - \tau) = \lambda(s - s')\,\psi(t - \tau)$.

moment is given by

$$\langle s(u)^2 \rangle = \frac{1 - \psi(u)}{u\left[1 - \psi(k, u)\right]^2} \frac{d^2}{dk^2} \psi(k, u)\bigg|_{k=0} \tag{7.36}$$

$$= \frac{1}{u\left[1 - \psi(u)\right]} \frac{d^2}{dk^2} \psi(k, u)\bigg|_{k=0},$$

here we have used parity $\psi(k, u) = \psi(-k, u)$ and space normalization $\psi(k = 0, u) = \psi(u)$. Then, in the separable case, the (asymptotic) temporal behavior of the second moment is characterized (in Laplace's representation) by the function

$$\frac{1}{u\left[1 - \psi(u)\right]}, \quad u \to 0. \tag{7.37}$$

Consider the separable case and assume that the probability $\lambda(s - s')$, which characterizes the jumps in the regular lattice has all its moments well defined. To study the temporal behavior of the second moment of a CTRW we need to compute only the inverse Laplace transform of the function (7.37). To this end suppose that for long times (i.e., in Laplace $u \to 0$) we can approximate

$$\psi(u) = 1 - \langle t \rangle u + \mathcal{O}(u), \quad \text{with} \quad \langle t \rangle = \int_0^\infty t \psi(t) \, dt, \tag{7.38}$$

here $\mathcal{O}(u)$ indicates a higher than linear order. Then the behavior for long times of the second moment of the displacement of a CTRW is given by

$$\langle s(u)^2 \rangle \to \frac{1}{\langle t \rangle u^2} \lambda''(0), \quad u \to 0.$$

Whence we observe that the model is diffusive, because from the inverse Laplace transform it follows that $\langle s(t)^2 \rangle \to \frac{\lambda''(0)}{2\langle t \rangle} t$ for $t \to \infty$. Here $\lambda''(0)$ is a constant characterizing the length scale of jumps in a regular lattice, and $\langle t \rangle$ is the mean waiting time at each lattice site. Other models of CTRW may also be analyzed similarly [14].

Exercise. Consider a CTRW model in which there is a drift, i.e.:

$$\frac{d\psi(k, u)}{dk}\bigg|_{k=0} \neq 0.$$

Calculate the first and second moments of the displacement of a CTRW. Hint: Use a model with separable CTRW and exponential waiting-time function: $\psi(u) = (1 + \langle t \rangle u)^{-1}$ [compare with the asymptotic result (7.38)]; to model $\lambda(k)$ use *characteristic functions* as those presented in Sect. 1.4. Compare with the results obtained using the scheme of discrete time, presented in Sect. 6.2.2.

Exercise (**Diffusive Behavior**). A n-dimensional CTRW model becomes asymptotically diffusive only if the hopping structure function $\lambda(k)$ and the waiting-time function $\psi(u)$ are regular, that is, if they have a Taylor series expansion around $k \sim 0$ and $u \sim 0$, respectively. Show, for the case without drift, that if the second moment of the one-step probability distribution σ^2 and the first moment of the waiting time pdf $\langle t \rangle$ are finite, we have $\lambda(k) = 1 - \frac{\sigma^2}{2n} \mid k \mid^2 + \cdots$, and $\psi(u) = 1 - \langle t \rangle \, u + \cdots$. Then the CTRW propagator takes the form

$$G_{\text{CTRW}}(k, u) = \frac{1 - \psi(u)}{u \, [1 - \psi(k, u)]} \approx \frac{\langle t \rangle}{\frac{\sigma^2}{2n} \mid k \mid^2 + \langle t \rangle \, u}, \quad k \to 0, u \to 0.$$

That is, it reaches a Gaussian regime with a diffusion coefficient $d = \sigma^2 / (2n \langle t \rangle)$.

7.3.1 Relationship Between the CTRW and the Generalized ME

Consider a separable CTRW model. Depending on the form of the waiting-time function $\psi(t)$, different possibilities for the evolution of a CTRW can be characterized. That is, given the propagator $G_{\text{CTRW}}(k, u)$, the following question arises: What will be its evolution equation? To answer this question we reconstruct its evolution equation from the Fourier–Laplace representation of the CTRW propagator in the following manner. First we multiply $G_{\text{CTRW}}(k, u)$ times u and subtract one

$$u G_{\text{CTRW}}(k, u) - 1 = \frac{1 - \psi(u)}{[1 - \lambda(k)\psi(u)]} - 1 \tag{7.39}$$

$$= -[1 - \lambda(k)] \frac{\psi(u)}{[1 - \lambda(k)\psi(u)]}$$

$$= -[1 - \lambda(k)] \frac{1 - \psi(u)}{u \, [1 - \lambda(k)\psi(u)]} \frac{u\psi(u)}{[1 - \psi(u)]}.$$

If we now define the function

$$\Phi(u) = \frac{u\psi(u)}{[1 - \psi(u)]}, \tag{7.40}$$

from (7.39) we obtain the identity

$$u G_{\text{CTRW}}(k, u) - 1 = -\Phi(u) [1 - \lambda(k)] G_{\text{CTRW}}(k, u), \tag{7.41}$$

which is the evolution equation we sought. To recognize this fact we must return to the variable t. To do this we rely on $u G_{\text{CTRW}}(k, u) - 1 = \mathcal{L}_u [dG_{\text{CTRW}}(k, t) / dt]$, and that multiplication of functions in the Laplace variable corresponds to a convolution in time variable t, i.e.:

$$\frac{dG_{CTRW}(k, t)}{dt} = \int_0^t \Phi(t - \tau) [\lambda(k) - 1] G_{CTRW}(k, \tau) \, d\tau, \tag{7.42}$$

note that the memory kernel $\Phi(t)$ is not a pdf, and

$$\left. \frac{dG_{CTRW}(k, t)}{dt} \right|_{t=0} = 0.$$

This equation is of ME type, but in (7.42) there is a memory kernel $\Phi(t - \tau)$ representing the non-Markovian character of the CTRW. The term $[\lambda(k) - 1]$ represents the gain and loss in the ME, and it is characterized, as expected, by the jump structure in Fourier space $\lambda(k)$ [14, 15]. We note that $\Phi(t)$ need not to be positive for all times even when $\psi(t)$ is, as required by the positivity of a pdf; see advanced exercises in Sect. 8.8, and Appendix E.

Exercise (**Deltiform Memory Kernel**). Show that if the waiting-time function is an exponential $\psi(t) = \frac{1}{\langle t \rangle} e^{-t/\langle t \rangle}$, then the memory of the CTRW disappears and a conventional ME for the conditional probability is obtained.

Exercise (**Non-separable Kernel**). Show that if the structure of jumps is not separable, the kernel of the generalized ME is given by

$$\Phi(k, u) = \frac{u \, [\psi(k, u) - \psi(u)]}{1 - \psi(u)},$$

i.e., the evolution equation in the Fourier representation is

$$\frac{dG_{CTRW}(k, t)}{dt} = \int_0^t \Phi(k, t - \tau) \, G_{CTRW}(k, \tau) \, d\tau.$$

Optional Exercise (**Telegrapher's Equation**). The Telegrapher equation is an equation that characterizes for long times a diffusive behavior; however, at short time it takes into account that the particle motion is performed at a finite speed c [11, 16]. Hence it follows that the evolution equation is a partial differential equation of second order in time:

$$\frac{\partial^2}{\partial t^2} P(x, t) + \frac{c^2}{D} \frac{\partial}{\partial t} P(x, t) = c^2 \frac{\partial^2}{\partial x^2} P(x, t).$$

Show that this evolution equation can be written in the form of a generalized ME (with the condition $\frac{\partial}{\partial t} P(x, t)|_{t=0} = 0$) if the memory kernel $\Phi(t)$ is an exponential $\Phi(t) = c^2 \exp\left(-\frac{c^2}{D} t\right)$. Then, the waiting-time function $\psi(t)$ of the

CTRW associated with Telegrapher's equation is characterized by the function (see also Appendix E)

$$\psi(u) = \frac{a}{a + u(u + \lambda)},$$

where $a = c^2 > 0$ and $\lambda = c^2/D > 0$; with the restriction that $\lambda^2 > 4a$, such that $\psi(t)$ will be a probability density for all t [14, 17].

7.3.2 Return to the Origin*

The study of the probability of return to the origin is an old problem originally formulated and solved—for Markovian processes—by G. Polya.[21] In particular, this problem is related to the analysis of the first passage time probability distribution (FPTD), which was already mentioned in Sect. 6.6. However, we present here the solution of this problem in an alternative way, which is useful for studying *approximately* the problem of return to the origin in the context of non-Markovian processes. In Sect. 6.6, we showed that the probability distribution of the first passage time could be calculated properly (for Markov processes[22]) using absorbing boundary conditions. The method we present now is equivalent but makes it easier to visualize the need for a Markovian condition.

Let $f(s^*, t)$ be the FPTD through the site s^* [having left the origin of coordinates at $t_0 = 0$], then it is clear that the density $f(s^*, t)$ satisfies the following equation

$$P(s^*, t \mid 0, 0) = \phi(t)\,\delta_{s^*, 0} + \int_0^t f(s^*, \tau)P(s^*, t \mid s^*, \tau, \{s(\bullet)\})d\tau, \qquad (7.43)$$

where $\phi(t)$ is the probability of staying (up to time t) at the origin, $P(s^*, t \mid 0, 0)$ is the conditional probability of being at s^* at time t and $P(s^*, t \mid s^*, \tau, \{s(\bullet)\})$ is an analogous conditional probability but depends on the previous history path, represented here by the object $\{s(\bullet)\}$. From this expression it is clear that it is only in the Markovian case, that the FPTD $f(s^*, t)$ is related to the conditional probability of the system. Then, under the constraint of Markovian process

$$(P(s^*, t \mid s^*, \tau, \{s(\bullet)\}) \to P(s^*, t \mid s^*, \tau))$$

and using the spatial translational invariance of $P(s^*, t \mid s^*, \tau)$, it follows that

$$P(s^*, t \mid 0, 0) = \phi(t)\,\delta_{s^*, 0} + \int_0^t f(s^*, \tau)P(0, t \mid 0, \tau)\,d\tau. \qquad (7.44)$$

---

[21] See: G. Polya, Math. Ann. **84**, 149 (1921).

[22] See the excursus "Reflection principle" in Sect. 6.5.6.

Taking the Laplace transform on (7.44) we finally get

$$f(s^*, u) = \frac{P(s^*, u) - \phi(u)\delta_{s^*,0}}{P(0, u)}$$

(7.45)

$$= \frac{R(s^*, u)}{R(0, u)} - \frac{\delta_{s^*,0}}{R(0, u)},$$

where we have used a compact notation $P(s^*, t \mid 0, \tau) \equiv P(s^*, t - \tau)$, etc. Furthermore the last equality is given in terms of the function $R(s, t \mid s', t')$. Expression (7.45) completely solves the first passage time problem if $R(s, u)$ is known. This quantity is given in terms of the structure function $\lambda(k)$ and the waiting-time function $\psi(u)$ of the CTRW. From (7.29) and for a separable model it follows that $R(k, u) = [1 - \psi(u)\lambda(k)]^{-1}$, then

$$R(s, \psi) = \frac{1}{(2\pi)^D} \int_{-\pi}^{\pi} dk_1 \cdots \int_{-\pi}^{\pi} dk_D \, e^{-ik \cdot s} \frac{1}{1 - \psi\lambda(k)}, \quad \text{with} \quad \psi \equiv \psi(u),$$

(7.46)

where $s = (s_1, \cdots, s_D)$ and $k = (k_1, \cdots, k_D)$. Therefore, the problem is reduced to calculating an inverse Fourier transform (in D dimension).

It is important to recall that for the study of extremal properties of an RW, such as the return to the origin, or the number of different sites visited on a walk, the dimension of the lattice plays a crucial role in assessing those quantities. For this reason we will calculate here some structure functions $\lambda(k)$ for different lattices.

Exercise.   Using the definition of structure function:

$$\lambda(k) = \sum_{s_1 = -\infty}^{\infty} \cdots \sum_{s_D = -\infty}^{\infty} e^{ik \cdot s} \lambda(s),$$

prove, using transitions to first neighbors, that for $D = 2$ it follows

$$\lambda(k) = \begin{cases} [\cos k_1 + \cos k_2]/2 & \text{for a square lattice} \\ [\cos k_1 + \cos k_2 + \cos(k_1 + k_2)]/3 & \text{for a triangular lattice} \\ [e^{ik_1} + 2\cos k_2]/3 & \text{for a hexagonal lattice,} \end{cases}$$

while for $D = 3$ we obtain

$$\lambda(k) = \begin{cases} [\cos k_1 + \cos k_2 + \cos k_3]/3 & \text{for a sc lattice} \\ [\cos(k_1)\cos(k_2)\cos(k_3)] & \text{for a bcc lattice} \\ [\cos(k_1)\cos(k_2) + \cos(k_1)\cos(k_3) + \cos(k_2)\cos(k_3)]/3 & \text{for a fcc lattice} \end{cases}$$

Exercise. For a simple hypercube lattice show that

$$R(0, \psi) = \left(1 - \psi^2\right)^{-1/2} \to (1 - \psi)^{-1/2}, \quad \text{for} \quad \psi \to 1, \text{ (in 1D)}$$

$$R(0, \psi) = \frac{2}{\pi} K(\psi) \to \ln(1 - \psi)^{-1}, \quad \text{for} \quad \psi \to 1, \text{ (in 2D)}, \qquad (7.47)$$

where $K(z)$ is the complete elliptic integral of first type. In general, for a simple hypercubic lattice with $D \geq 3$ we have the expression

$$R(s, \psi) = \int_0^\infty dx \, e^{-x} I_{s_1}(x\psi/D) \cdots I_{s_D}(x\psi/D), \quad \text{with} \quad s = (s_1, \cdots, s_D), \psi \equiv \psi(u),$$

where $I_{s_1}(z)$ is the hyperbolic Bessel's function; see [11].

In the previous section we said that depending on the waiting-time function $\psi(t)$ it is possible to characterize different diffusion models; in particular, only if $\psi(t)$ is exponential will the CTRW be a Markov process. It is important to note that only in the Markovian case is the relation (7.45) correct.[23] From (7.45) it is easy to understand that the probability of return to origin—at any time—is given by the quantity

$$\int_0^\infty f(0, t) \, dt = f(0, u = 0),$$

an expression that depends on the dimension of the lattice through the relaxation of the initial condition [compare with Sect. 6.3.4]; that is to say:

$$f(0, u = 0) = 1 - \frac{1}{R(0, \psi = 1)}. \qquad (7.48)$$

From this expression we note that if the quantity $R(0, \psi = 1)$ diverges, the RW returns with certainty to the origin of coordinates. In particular, it is easy to recognize that for $D = \{1, 2\}$ the quantity $R(0, \psi = 1)$ is infinite,[24] but not for $D = 3$ and its value depends on the type of lattice structure.[25] This means that the distribution $f(0, t)$ is normalized only in $D = 1$ and 2, or equivalently that for dimensions $D \geq 3$ the RW can escape to infinity and may never return to the origin from which the walk started.

---

[23]See for example appendix A in: "*Passage times of asymmetric anomalous walks with multiple paths*", M.O. Cáceres and G.L. Insua, J. Phys. A: Math. Gen. **38**, 3711 (2005).

[24]See the asymptotic behavior in (7.47).

[25]For example in 3D and for a sc lattice we get $R(0, \psi = 1) = 1, 516386 \cdots$; for a bcc lattice we get $R(0, \psi = 1) = 1, 393203 \cdots$; and for a fcc it is obtained $R(0, \psi = 1) = 1, 344661 \cdots$, see advanced exercise 7.5.5.

Guided Exercise (**Return to the Origin in the Diffusive Case**). Let us calculate the asymptotic time behavior of the return to the origin. Taking the Laplace transform of (7.27) we get (7.30), then using (7.47) for $1D$ in the Markovian case, i.e., when using $\psi(u) = (1 + \langle t \rangle u)^{-1}$, we get $R(s = 0, u \to 0) \to 1/\sqrt{\langle t \rangle u}$; and from (7.31) we know that $\phi(u) = (1 - \psi(u))/u \to \langle t \rangle$. Then, introducing all these results in (7.30) we get finally the expression

$$\lim_{u \to 0} P(s = 0, u) \to \sqrt{\frac{\langle t \rangle}{u}}.$$

The asymptotic time behavior of the return to the origin can be calculated using the Tauberian theorem[26]; see advanced exercise 7.5.6.

$$P(s = 0, t) \sim \frac{\sqrt{\langle t \rangle}}{\Gamma(3/2)} t^{-1/2}, \quad \text{for} \quad t \to \infty. \tag{7.49}$$

Compare this result with relaxation of the initial condition presented in Sect. 6.3.4. Note that although in $1D$ the probability of return to the origin is certain (this is so in 1 and 2 dimensions because $R(0, \psi = 1) = \infty$), the average time until the RW returns to the origin for the first time diverges. The time dependence of the (density) of the first passage time at the *initial site*, $f(0, t)$, at large time behaves like

$$f(0, t) \sim \begin{cases} 1/t^{2 - D/2}, & D < 2 \\ 1/\ln t, & D = 2 \end{cases}$$

Optional Exercise (**The FPTD for the Wiener sp**). For continuous Markovian process (such as the Wiener process), we have shown that the FPTD could be calculated if we used the method of images [see Sect. 6.6]. Show that if we place an absorbing boundary condition at point x in the line, the FPTD $f(x, t)$ [having left the origin of coordinate at the instant $t_0 = 0$] is given by expression:

$$f(x, t) = x (4\pi\epsilon)^{-1/2} t^{-3/2} \exp\left(\frac{-x^2}{4\epsilon t}\right), \quad \text{with} \quad \{x, t\} > 0, \tag{7.50}$$

where ϵ is the diffusion coefficient; i.e., its asymptotic behavior is $\sim xt^{-3/2}$. Compare (7.50) with the expression obtained from (7.45) if Wiener's propagator is used (in $1D$):

$$P(x, t) = (4\pi\epsilon t)^{-1/2} \exp\left(-x^2/4\epsilon t\right),$$

which in Laplace's representation takes the form:

$$P(x, u) = \frac{1}{2} |u\epsilon|^{-1/2} \exp\left(-|x| |\frac{u}{\epsilon}|^{1/2}\right).$$

---

[26]Tauberian theorem states that: if $f(u) \sim u^{-(\theta+1)}$, for $u \to 0$, then $f(t) \sim t^{\theta}$, for $t \to \infty$.

Hint: using that

$$\int_0^\infty e^{izk}\, dz = \pi\delta(k) + i\mathcal{P}\left(\frac{1}{ik}\right),$$

it is possible to prove the relationship:

$$\frac{-i}{k}\frac{d}{dt}\int_{-\infty}^\infty \frac{dk}{k}\,\exp\left(-ikx - k^2\epsilon t\right) = \frac{-\epsilon}{\pi}\frac{d}{dx}\int_{-\infty}^\infty dk\,\exp\left(-ikx - k^2\epsilon t\right),$$

from which (7.50) can be inferred. Calculate the first moment of the FPTD given in (7.50) and interpret the result.

Optional Exercise (**Distinct Sites Visited on a Walk**). Interestingly if the FPTD is known, also other extremal properties can be calculated from $f(s^*, t)$; for example, the average number of different sites visited on a walk, etc. For the purposes of analyzing this quantity, let us introduce here the analogous function but in the case of discrete time $t \to n$; that is, $f(s^*, t) \to f_n(s^*)$. Thus, the discrete version of (7.44) would be

$$P_n(s^*) = \sum_{m=1}^n f_m(s^*)\, P(s^*, n\,|\, s^*, m)$$

$$= \sum_{m=1}^n f_m(s^*)\, P_{n-m}(0), \tag{7.51}$$

this formula accounts for the fact that the walker must have visited site s^* for the first time at some step $m \le n$, and we have used translational invariance. Interpret graphically the recurrence relation (7.51). Noting that the initial condition is $P_n(s) = \delta_{s,0}$, we can write $P_n(0) = \delta_{s,0} + \sum_{m=1}^n f_m(0)\, P_{n-m}(0)$. So the probability of first return to the origin at step n is $f_n(0)$. The recurrence relation (7.51) can be solved using the generating functions of $f_n(s^*)$ and $P_n(s^*)$; that is: $f(s^*, z) = \sum_{n=0}^\infty f_n(s^*)z^n$ and $R(s^*, z) = \sum_{n=0}^\infty P_n(s^*)z^n$ respectively. Then the final expression equivalent to (7.45) is

$$f(s^*, z) = \frac{R(s^*, z)}{R(s=0, z)} - \frac{\delta_{s^*,0}}{R(s=0, z)},$$

where $R(k, z) = [1 - z\mathbf{T}_1(k)]^{-1}$ is the generating function of a discrete-time RW [see Sect. 6.2.1], compare with the expression $R(k, u) = [1 - \psi(u)\lambda(k)]^{-1}$. From this result many interesting properties can be studied, for example the probability that the walker will *eventually* reach the origin

$$Q(0) = \sum_{n=0}^\infty f_n(0) = f(0, z = 1). \tag{7.52}$$

The *average* return time:

$$\langle n \rangle = \sum_{n=0}^{\infty} n f_n(0) = \left. \frac{df(0,z)}{dz} \right|_{z=1}. \tag{7.53}$$

The average number of distinct sites *visited* by a n-steps walk is given by the relationship:

$$S_n = \sum_s \{f_1(s) + f_2(s) + \cdots + f_n(s)\}, \tag{7.54}$$

where $f_n(s)$ is the probability that site s is visited for the first time at step n. From (7.54) it is possible to obtain, using the generating function technique, an expression for calculating the asymptotic behavior of S_n in the limit as $n \to \infty$; see advanced exercise 7.5.5. We can also find other extremal properties in a similar way; for example, the number of visits (on a n steps walk) that an RW makes to a given lattice site; the number of visited sites in the lattice—exactly m-times—in a n-steps walk, or Polya's problem for many walkers, see for example [13, 14].

7.3.3 Relationship Between Waiting-Time and Disorder*

Depending on the nonanalyticity (around $u \sim 0$) of the waiting-time function $\psi(u)$ it is possible to characterize different models of anomalous diffusion [12–14]. That is, we can study situations where the second moment of the displacement is no longer linear in time. There is a direct relationship between a model of disorder in a ME and the waiting-time function for a CTRW (see Appendix D). This relationship is characterized by the identification

$$\psi(s - s', t) = \left\langle W_{s,s'}\, e^{-t \sum_s W_{s,s'}} \right\rangle_{\text{Disorder}}. \tag{7.55}$$

If we consider a 1D model with site-disorder and transitions to first neighbors on a regular lattice, we have $W_{s,s'} = w_{s'} \delta_{s \pm 1, s'}$ (and zero in any other case). Then, to represent the disorder it is assumed that the quantity $w_{s'}$ is random. From (7.55) we can show that $\lambda(s - s') = \frac{1}{2}\left(\delta_{s+1,s'} + \delta_{s-1,s'}\right)$ and so the average (7.55) involves only the **sirvs** $w_{s'}$.

In general (in arbitrary dimension), if $w_{s'}$ are characterized by a probability distribution—equally distributed in all lattice sites—of the form

$$P(w_s) = P(w) = \begin{cases} \frac{\theta}{w_0}\,(w/w_0)^{\theta-1}, & 0 \le w \le w_0,\ \theta \in (0,1) \\ 0, & w > w_0, \end{cases} \tag{7.56}$$

from (7.55) we conclude that $\psi(s - s', t) = \lambda(s - s')\psi(t)$ where[27]

$$\psi(t) = \frac{z\theta w_0}{\tilde{t}^{\theta+1}} \gamma(\theta + 1, \tilde{t}), \quad \theta \in (0, 1) \quad \text{con} \quad \tilde{t} = zw_0t. \tag{7.57}$$

Here $\gamma(\theta + 1, \tilde{t})$ is the incomplete Gamma function,[28] then for $\tilde{t} \to \infty$ we can use the asymptotic result: $\gamma(\theta + 1, \tilde{t}) \to \Gamma(\theta + 1)$, and thus in the asymptotic long-time limit the waiting-time function $\psi(t)$ is approximated by

$$\psi(t) \sim \frac{z\theta\Gamma(\theta + 1)}{w_0^\theta (zt)^{\theta+1}}, \quad \theta \in (0, 1), t \to \infty. \tag{7.58}$$

The Laplace transform of $\psi(t)$, in this case, has the asymptotic behavior

$$\psi(u) \sim 1 - \text{const.}\, u^\theta, \quad \theta \in (0, 1), \text{ as } u \to 0. \tag{7.59}$$

From which it follows, for $\theta \in (0, 1)$, a nonanalytic behavior around $u = 0$. This result says that the second moment of the displacement of the CTRW; i.e., $\langle s(u)^2 \rangle \propto (u[1 - \psi(u)])^{-1}$ for $u \to 0$, is given by the expression

$$\langle s(u)^2 \rangle \sim u^{-(\theta+1)}, \quad \theta \in (0, 1), u \to 0. \tag{7.60}$$

Applying the inverse Laplace transform,[29] it follows that the second moment of displacement shows anomalous behavior in the long time limit:

$$\langle s(t)^2 \rangle \sim t^\theta, \quad \theta \in (0, 1), t \to \infty, \tag{7.61}$$

that is, subdiffusive.

Furthermore, the analysis of the generalized ME in the context of CTRW theory allows to establish that the effective medium is given by $\Phi(u) = u\psi(u)/[1 - \psi(u)]$. Then, we see that in the regime $u \sim 0$ and when the waiting-time function has a nonanalytical behavior of the type given in (7.59), the effective medium, in the CTRW approach, is given by

$$\Phi(u \to 0) \sim u^{1-\theta}; \quad \text{then} \quad \Phi(t \to \infty) \sim t^{-(2-\theta)}. \tag{7.62}$$

---

[27]Here $\lambda(s - s')$ represents the structure of jumps in a regular lattice of dimension n. Assuming hypercubic lattices the coordination number is given by $z = 2n$.

[28]See for example, *An Atlas of Functions*, J. Spanier and K.B. Oldham, Berlin, Springer-Verlag (1987).

[29]Using Tauberian theorem: if $f(u) \sim u^{-(\theta+1)}$, for $u \to 0$, then $f(t) \sim t^\theta$, for $t \to \infty$; see advanced exercise 7.5.6(((; also "*Random walks and random environments*", Vol.1, B.D. Hughes, Oxford Science Publ. 1995.

Note From this result it is possible to determine the behavior of the *ac* electrical conductivity at low frequency. That is, using the generalized Einstein's relation:

$$\sigma_{ac}(\omega) \propto \Phi(u = -i\omega)/k_B T,$$

between conductivity and the diffusion coefficient (frequency dependent), we can study the transport properties of electric charges in amorphous systems in the context of CTRW [12]; see Sect. 8.4.

Optional Exercise (**Classical Localization**). From the expression for the generalized diffusion coefficient (7.62) the relaxation of the initial condition can be calculated in the context of CTRW. For example, in $1D$ we have

$$< 0 \mid G_{CTRW}(u) \mid 0 > \sim 1/\sqrt{4\Phi(u)u} \sim u^{-(2-\theta)/2}, \quad u \to 0.$$

Taking the inverse Laplace's transform, show that the behavior at long times is given by

$$P(0, t) \sim t^{-\theta/2}, \quad t \to \infty.$$

Using $\theta = 1 - \alpha$, show that the anomalous exponent obtained by the CTRW approach differs from the one obtained with the EMA technique [2].

Excursus (**Internal States**). If when describing a transport diffusing model, in a disordered medium, the particle has—in addition—another degree of freedom to be taken into account (i.e., speed, energy, etc.) it is possible to generalize the CTRW theory to a system with internal states [13–15, 18]. A simple model with internal states is a $1D$ Lorentz gas or persistent random walk [17]; the $2D$ case will be discussed in detail in Sect. 7.4.1. The derivation of the Lorentz gas model on a lattice from the linear Boltzmann equation is presented in Sect. 5.5.1. Furthermore, the Telegrapher equation (or Goldstein's equation) is also related to a Lorentz gas model in the lattice. In reference [19] the study of Goldstein's equation and certain irreversible thermodynamic processes are presented.

Excursus (**Dynamic Disorder**). It is possible to construct models with dynamic disorder which emulate "complex materials" [20] where there are mechanisms or on/off traps that fluctuate over time; that is, temperature-dependent traps. The difference between static disorder[30] and dynamic (time dependent) disorder lies then in that impurities are treated as stochastic processes, rather than as random variables. Dynamic models of disorder can be studied in the context of the CTRW theory with internal states, or alternatively using generalizations of the EMA [21]. Dynamic impurities in quantum systems can also be studied in a similar way [22].

---

[30]Sometimes called *Quenched disorder*.

Excursus (**Absorption Probability Density**). The absorption probability density of the walker by a dynamic trap, may not coincides with the first passage time density in presence of a static trap. This model is of interest in the study of the problem of a gate which opens and closes at random times. Note that in the dynamic trap model the absorption probability density does not coincide with the first passage time density because even if the trap position lies on the path it does not mean that the walker will be absorbed. The dynamic problem can be solved using the multistate CTRW technique in a way similar to that used for the problem of non-Markovian global dynamic disorder [21]. This approach is achieved by generalizing the effect of traps on a lattice walk [14]; that is, introducing a dynamic impurity into the lattice; then, the absorption probability density can be solved in Laplace's representation. The results are analytic for every switching-time probability density of the trap [23].

Excursus (**Anomalous Diffusion with Drift**). If the transport description is made in the continuous spatial coordinate (i.e., x) taking into account any external force that may be present and produces a drift term in the stream of particles, the evolution equation for the conditional probability is the **F-P** equation [see Smoluchowski's equation in Sect. 3.18], where the "linear" term in $\partial/\partial x$ characterizes the drift, convection or drag in the diffusive transport. As we have noted already, this can also be modeled by ME[31] considering asymmetric transition probabilities $W_{s,s'}$ or, in the context of the CTRW[32] process considering an asymmetric jump structure $\psi(s - s', t - \tau)$ for the transition $s \leftrightarrow s'$.

Excursus (**Diffusion-Advection Equation**). If the medium itself is in a state of motion described by a velocity field $\vec{V}(\vec{r}, t)$, a test particle diffusing in that medium also moves together with the medium. Then the $\partial/\partial t$ operation in the diffusion equation[33] must be replaced by a material derivative, leading to the diffusion-advection equation

$$\frac{\partial n(\vec{r}, t)}{\partial t} = -\nabla \cdot \left(n(\vec{r}, t)\, \vec{V}(\vec{r}, t) \right) + d\, \nabla^2 n(\vec{r}, t). \tag{7.63}$$

If the velocity field $\vec{V}(\vec{r}, t)$ is turbulent, then it can be modeled by a stochastic process[34] in both variables $\{\vec{r}, t\}$, and Eq. (7.63) becomes a stochastic partial differential equation, which can be studied using perturbation theory. Another problem is generalizing the diffusion term (7.63) in order to describe a process of anomalous transport with advection.

---

[31]The effective medium approximation with asymmetric transition rates in 2D, was solved in: S. Bustingorry, M.O. Cáceres, and E.R. Reyes, Phys. Rev. B, 65, 165205 (2002).

[32]The more general case of asymmetric anomalous CTRW with multiple paths, was solved in: M.O. Cáceres, and G.L. Insua, J. Phys. A: Math. Gen., 38, 3711 (2005).

[33]To study Fick's method in order to get the diffusion equation, see Sect. 8.5.2.

[34]Actually here appears the concept of stochastic field; see, for example, J. García-Ojalvo and J.M. Sancho, in *Noise in Spatially Extended Systems*, Berlin, Springer (1999); and references therein.

Excursus (**Anomalous Diffusion with Random Advection**). Consider a transport process in a disordered medium with an advection term characterized by a static vector field $\vec{V}(\vec{r})$. It is possible to prove that the CTRW process can be generalized to consider anomalous diffusion in the presence of advection; moreover if the velocity field $\vec{V}(\vec{r})$ is random, this system can be studied using a perturbation approach [24], see Sect. 8.5.4.

7.3.4 Superdiffusion*

The anomalous behavior of a diffusion process is characterized by the second moment increasing with time as $\langle s(t)^2 \rangle \sim t^\eta$, where $\eta \neq 1$. If the exponent is $\eta < 1$, it is said that we are in presence of subdiffusion, while if $\eta > 1$, it is called superdiffusion. The CTRW theory allows to study both cases in a unified manner.[35] We have already seen that subdiffusion appears when disorder is strong enough to localize the particle somewhere in the net. The opposite case, superdiffusion appears when the dispersion is amplified,[36] a situation that can occur when the structure function of the CTRW $\psi(s, t)$ is not separable, so that, for example, long jumps become likely at the expense however of the *penalty* of long waiting times.

The probability $\psi(s, t)dt$ completely characterizes the propagator of the CTRW. Note that if the structure function $\psi(s, t)$ is coupled in space and time, it is given by the product of probabilities, where one of them is conditioned in space or in time.

First consider the factorization $\psi(s, t) = \Lambda(s \mid t)\psi(t)$, where obviously the normalization is given by

$$\int_0^\infty \psi(t)\, dt = 1; \quad \sum_s \Lambda(s \mid t) = 1, \forall t.$$

For example, if in the Fourier representation one has

$$\psi(k, t) = \psi(t) \exp\left(-D(t)k^2\right), \tag{7.64}$$

a *simple* conditional coupled model is obtained. On the other hand, from (7.64) it is clear that $\psi(k, t)$ is a probability density for any positive function of time $D(t)$. Then $\psi(k, t)$ has the same meaning as above, that is, $\psi(t) = \psi(k = 0, t) = \sum_s \psi(s, t)$ is the waiting-time function, while $\exp\left(-D(t)k^2\right)$ is the Fourier representation of the

---

[35]It is important to note that the EMA allows the study of subdiffusion for any kind of disorder. While CTRW theory gives the "correct" anomalous exponents, only in some particular cases; see for example: G.H. Weiss and R. J. Rubin, in *Random Walk: Theory and Selected Applications*, Adv. Chem. Phys. **52**, 363, (1983); E. Hernández-García, M.O. Cáceres and M. San Miguel, Phys. Rev. **A 41**, 4562, (1990), and [7].

[36]For example: a passive scalar field has enhanced diffusion in a fluctuating fluid (the medium).

conditional probability $\Lambda(s \mid t)$ of taking a jump of length s given a pause t.[37] It should be noted that the asymptotic analysis of a non-separable CTRW is a complex matter because the limits $k \to 0$ and $u \to 0$ do not necessarily commute. In general (and without drift), if there is a Taylor series expansion around $k = 0$, we can write (note that we have agreed to first take the limit $k \to 0$):

$$\psi(k, u) = \psi(0, u) + \frac{1}{2}\psi_{kk}(0, u)k^2 + \cdots . \tag{7.65}$$

In calculating the second moment of the walk, we are interested in the second derivative with respect to k of the structure function $\psi(k, u)$ evaluated at $k = 0$. It follows, then, that in this case it is only necessary to know $\psi(0, u)$ and $\psi_{kk}(0, u)$, where

$$\left.\frac{\partial^2 \psi(k, u)}{\partial k^2}\right|_{k=0} \equiv \psi_{kk}(0, u), \tag{7.66}$$

to characterize the dispersion of the displacement of a CTRW [see Eq. (7.36)]. It is possible to make asymptotic expansions of the function $\psi(k, u)$ around $k \sim 0$ and $u \sim 0$ to characterize different anomalous exponents in superdiffusion.

Excursus (**Asymptotic CTRW Propagator for Superdiffusion**).   Consider the following asymptotic model for the structure function: $\psi(k, u) = 1 - C_1 u^\phi - C_2 k^2 u^{\phi-\eta} + \mathcal{O}(k^4)$, with $0 < \phi \leq 1$. In this case the CTRW propagator adopts the asymptotic form:

$$G_{\text{CTRW}}(k, u) = \frac{1 - \psi(u)}{u[1 - \psi(k, u)]} \approx \frac{C_1 u^\phi}{u[C_1 u^\phi + C_2 k^2 u^{\phi-\eta}]} \tag{7.67}$$

$$\approx \frac{C_1 u^\eta}{u[C_1 u^\eta + C_2 k^2]}, \qquad \text{with} \qquad k \to 0, u \to 0.$$

Show that, at long times the second moment of the displacement has the behavior $\langle s(t)^2 \rangle \approx t^\eta$, i.e., superdiffusive if $\eta > 1$. Note that for $\eta = 1$ and $\phi < 1$ we get $\psi(k, u) = 1 - C_1 u^\phi - C_2 k^2 u^{\phi-1} + \mathcal{O}(k^4)$ and so we obtain the mean waiting time: $-\left.\partial_u \psi(k = 0, u)\right|_{u=0} = \infty$, and the mean *one-step* squared displacement: $-\left.\partial_k^2 \psi(k, u)\right|_{k=0, u=0} = \infty$. Interestingly, a coupled effect could offset each other to yield a Gaussian distribution for $G_{\text{CTRW}}(s, t)$. On the other hand, in the case $\langle t \rangle = -\left.\partial_u \psi(k = 0, u)\right|_{u=0} < \infty$ but with $\langle t^2 \rangle = \left.\partial_u^2 \psi(k = 0, u)\right|_{u=0} = \infty$; that is, when $\psi(k = 0, u \sim 0) \simeq 1 - \langle t \rangle u - B u^{1+\gamma} + \cdots$, $\gamma \in [0, 1]$, which means $\psi(t \to \infty) \propto t^{-2-\gamma}$. Then, taking $D(t) \sim t^\eta$ in Eq. (7.64), this choice yields $\psi(k, u) = 1 - \langle t \rangle u - C_2 k^2 u^{1+\gamma-\eta} + \mathcal{O}(k^4)$. In the case $(\eta - \gamma) > 1$, and introducing this $\psi(k, u)$ in (7.67) we get

$$G_{\text{CTRW}}(k, u) \approx \frac{(\text{const.})u^{\eta-1-\gamma}}{u^{\eta-\gamma} + (\text{const.})k^2},$$

---

[37] In this case a Gaussian coupling for the jumps.

and so $\left\langle s\,(t)^2 \right\rangle \approx$ (const.)$u^{\eta-\gamma}$. Therefore we see that the profile is non-Gaussian and superdiffusive. In the case $\eta - \gamma \leq 1$ it can be seen that an asymptotic Gaussian behavior results because $\psi\,(t)\,D\,(t)$ decays rapidly enough for the leading behavior of its Laplace transform to be a constant [25].

Consider now the opposite factorization, i.e.: $\psi(s,t) = \Lambda(t \mid s)p(s)$. In this case the normalization is given by

$$\int_0^\infty \Lambda(t \mid s)\,dt = 1, \forall s; \quad \sum_s p(s) = 1,$$

that is, $\Lambda(t \mid s)$ is the conditional probability density of waiting a time interval t given that the jump has a long s.

Example (**Superdiffusion**). Take, in dimensionless units, $\Lambda(t \mid s) = \delta(t- \mid s \mid)$ and, asymptotically, the expression $p(s) \sim \mid s \mid^{-2-\beta}$, with $-1 < \beta$, for the probability of a jump of size s. In this case we get superdiffusion if $\beta \in (-1,1)$ [26]. To prove this result we use the following corollary of the Abelian theorem. Consider the asymptotic behavior

$$f(t) \sim C\,t^{-1-\gamma}, \quad C = \text{constant}, \; \gamma > 0, t \to \infty, \tag{7.68}$$

if $n - 1 < \gamma < n$, with n a natural number, the first $n - 1$ moments are finite:

$$\langle t^m \rangle = \int_0^\infty t^m f(t)\,dt \propto \frac{1}{(\gamma - 1)(\gamma - 2)\cdots(\gamma - (m-1))},$$

whereas moments of order greater than or equal to n diverge. So in this case there is an asymptotic expansion for the Laplace transform of the form:

$$f(u) \equiv \mathcal{L}_u\left[f(t)\right] = 1 + \sum_{m=1}^{n-1} \frac{(-1)^m}{m} \langle t^m \rangle\, u^m - \frac{C\Gamma(1-\gamma)}{\gamma} u^\gamma + \cdots, \quad u \to 0. \tag{7.69}$$

Now we calculate $\psi(0, u)$ and $\psi_{kk}(0, u)$. Indeed we can see that asymptotically

$$\psi(k,u) = \int_0^\infty dt\, e^{-ut} \sum_s e^{iks}\Lambda(t \mid s)p(s) \sim \sum_s e^{iks} \mid s \mid^{-2-\beta} \int_0^\infty dt\, e^{-ut} \delta(t- \mid s \mid)$$

$$= \sum_s e^{iks} \mid s \mid^{-2-\beta} e^{-u\mid s\mid} \sim \mathcal{R}e \left[\int_0^\infty s^{-2-\beta} e^{-(u-ik)s}\,ds \right]$$

$$\approx \mathcal{R}e \left[\mathcal{L}_{(u-ik)}[s^{-2-\beta}] \right],$$

where we have approximated $\sum_s \cdots$ by $\int_0^\infty \cdots ds$; now, using (7.68) and (7.69), we get

$$\psi(k,u) \sim \begin{cases} \mathcal{R}_e \left[1 - \text{const.}\,(u-ik)^{\beta+1} \right], & \text{if } 0 < \beta + 1 < 1 \\ \mathcal{R}_e \left[1 - \text{const.}\,(u-ik) - \text{const.}\,(u-ik)^{\beta+1} \right], & \text{if } 1 < \beta + 1 < 2 \\ \mathcal{R}_e \left[1 - \text{const.}\,(u-ik) + \text{const.}\,(u-ik)^2 - \text{const.}\,(u-ik)^{\beta+1} \right], & \text{if } 2 < \beta + 1. \end{cases}$$

This expansion presents some difficulty in taking the limits $\{k, u\} \to 0$. However, it is possible to see that $\psi_{kk}(0,u) \sim u^{\beta-1}$, if $-1 < \beta < 1$, and that $\psi_{kk}(0,u) \sim$ const., if $1 < \beta$. Then, using the corresponding expressions for $\psi(0,u)$ and $\psi_{kk}(0,u)$ in (7.36), and applying Abelian theorem[38] we get

$$\langle s(t)^2 \rangle \sim \begin{cases} t^2, & -1 < \beta \leq 0 \\ t^{2-\beta}, & 0 \leq \beta < 1 \\ t, & 1 < \beta, \end{cases}$$

that is, a superdiffusive behavior if $\beta \in (-1, 1)$. With this technique turbulent models and chaotic maps presenting superdiffusion can be studied [25–27].

Excursus (**Turbulence**). Consider a coupled jump structure function so that in the time interval $[t, t + dt]$ only a thin "layer" of points on the sphere of radius $\mathbf{r}$ [in D-dimension] is accessible to the walker. That is, if at time $t + dt$ the walker is at the origin of coordinates only infinitesimally close points to that layer are accessible, but not points further away, in that same time interval. Show that in this case the structure function is not separable and can be written in the form

$$\psi(\mathbf{r}, t) = \text{const.}\, r^{-\mu}\, \delta(r - t^{\nu}), \qquad \mu > 0, \nu > 0, r > 0,$$

where $r \equiv |\mathbf{r}|$. Defining the quantity: $\mu^* = \mu - D + 1$ it is possible to find asymptotically anomalous transport: $\langle r(t)^2 \rangle \sim t^{\eta}$, $\eta \neq 1$. That is, if $\nu\mu^* > 2$:

$$\langle r(t)^2 \rangle \sim \begin{cases} t^{2-\nu\mu^*+2\nu}, & \nu\mu^* < 1 + 2\nu \\ t & \nu\mu^* > 1 + 2\nu \end{cases},$$

and if $1 < \nu\mu^* < 2$:

$$\langle r(t)^2 \rangle \sim \begin{cases} t^{\nu\mu^*-1}, & \nu\mu^* > 1 + 2\nu \\ t^{2\nu} & \nu\mu^* < 1 + 2\nu \end{cases}.$$

Difficulties for the numerical simulations, in the transient regime, in order to test these predictions have been reported but the general conclusion remains;

---

[38] See advanced exercise 7.5.6.

showing also the agreement with the theoretical predictions for the exponent of
the mean number of distinct sites visited at time t, which is also another important
characterization of anomalous transport [28].

7.4 Diffusion with Internal States

7.4.1 The Ordered Case

In this section we present the outline of the technique of Markov chains with internal
states in ordered lattices [1, 14, 29]. Its generalization to the disordered case can
easily be studied in the context of the factorized CTRW theory.[39] To fix ideas, we
present the analysis of internal states considering a particular case: the Lorentz gas
model.[40] It is a system of noninteracting classical particles moving in a fixed array of
scattering centers.[41] In this model the diffusing particle has two degrees of freedom:
"position" and "speed." Then, we use internal states to represent the direction in
which the RW moves in the lattice.

Let us begin by describing a Markov chain with internal states associated with a
Lorentz gas model in 2D. In this case $P_n^l(x, y)$ represents the probability of being on
lattice site (x, y), with internal state l in the n-th step. Then, the recurrence relations
will be

$$P_{n+1}^1(x, y) = p\, P_n^1(x-1, y) + q\, P_n^2(x-1, y) + r\, P_n^3(x-1, y) + r'\, P_n^4(x-1, y)$$

$$P_{n+1}^2(x, y) = p\, P_n^1(x+1, y) + q\, P_n^2(x+1, y) + r'\, P_n^3(x+1, y) + r\, P_n^4(x+1, y)$$

$$P_{n+1}^3(x, y) = r'\, P_n^1(x, y-1) + r\, P_n^2(x, y-1) + p\, P_n^3(x, y-1) + q\, P_n^4(x, y-1)$$

$$P_{n+1}^4(x, y) = r\, P_n^1(x, y+1) + r'\, P_n^2(x, y+1) + q\, P_n^3(x, y+1) + p\, P_n^4(x, y+1).$$

$$(7.70)$$

---

[39]That is, a non-Markovian random walk with spatio-temporal separable structure, and with a
unique waiting-time function $\psi(u)$ which is independent of internal states.

[40]A Lorentz gas can be used to model different problems: (a) Neutron diffusion in a medium in
which the dispersion is anisotropic; (b) Coil-globule transition in physics of polymers, identifying a
polymer chain with an RW on a network where the position of each monomer depends on the prior
position; (c) Frequency response of superionic conductors, where the ion-ion interaction introduces
correlations between subsequent steps in an RW scheme, etc.

[41]Note that for a lattice Lorentz gas, the *Scattering Centers* are the same lattice sites, and RW
transitions are to first neighbors.

That is, the internal states $l = 1, 2$ represent "movement" on the x-axis in the positive or negative direction respectively; and states $l = 3, 4$ represent "movement" on the y axis in the positive or negative direction respectively. Then p (q) is the jump probability "forward" ("backward"), and r (r') is the probability of scattering by $\pm\pi/2$ from its previous direction. Since probability is conserved by normalization we get $p + q + r + r' = 1$.

In general—as in the Lorentz gas (7.70)—for an RW with internal states we can write a recurrence relation defining a structure matrix Ψ in the form[42]

$$P^l_{n+1}(\mathbf{s}) = \sum_{\mathbf{s}'} \sum_j \Psi_{l,j}(\mathbf{s}, \mathbf{s}') P^j_n(\mathbf{s}'); \quad 1 = \sum_{\mathbf{s}'} \sum_j \Psi_{j,l}(\mathbf{s}', \mathbf{s}). \tag{7.71}$$

Here $\mathbf{s} \equiv (x, y)$ represents a vector in the $2D$ square lattice, but in general it could be an arbitrary lattice in nD; see advanced exercise 7.5.3. If there is translational invariance,[43] we can immediately use the Fourier representation,[44] then from (7.71) we obtain

$$P^l_{n+1}(\mathbf{k}) = \sum_j \Psi_{l,j}(\mathbf{k}) P^j_n(\mathbf{k}). \tag{7.72}$$

In particular, for the Lorentz gas model in $2D$, characterized by (7.70), the structure matrix takes the form [30]

$$\Psi(\mathbf{k}) = \begin{pmatrix} p\,e^{ik_x} & q\,e^{ik_x} & r\,e^{ik_x} & r'\,e^{ik_x} \\ q\,e^{-ik_x} & p\,e^{-ik_x} & r'\,e^{-ik_x} & r\,e^{-ik_x} \\ r'\,e^{ik_y} & r\,e^{ik_y} & p\,e^{ik_y} & q\,e^{ik_y} \\ r\,e^{-ik_y} & r'\,e^{-ik_y} & q\,e^{-ik_y} & p\,e^{-ik_y} \end{pmatrix}. \tag{7.73}$$

Exercise. From (7.70) get the structure matrix $\Psi(\mathbf{k})$ shown in (7.73). Note that $\Psi(\mathbf{k} = 0)$ is a Markov matrix.

The solution of the *matrix* recurrence relation (7.72) is obtained using a procedure analogous to that employed when solving the Markov chain [see Sect. 6.2.1]; then, using a generating function, we obtain

$$\mathbf{R}(\mathbf{k}, Z) = [\mathbf{1} - z\Psi(\mathbf{k})]^{-1} \mathbf{P}(\mathbf{k}, n = 0). \tag{7.74}$$

Here $\mathbf{P}(\mathbf{k}, n = 0) \equiv (P^1_{n=0}(\mathbf{k}), P^2_{n=0}(\mathbf{k}), \cdots)$ is the "vector" initial condition, where each component represents the probability with internal state l, and $\mathbf{1}$ indicates the identity matrix. The solution in discrete time, $\mathbf{P}(\mathbf{k}, n)$, is obtained from the vector

---

[42]In this section, for convenience, we use the vector notation in "bold" letters $\mathbf{s}, \mathbf{k}$.

[43]That is, it holds that all elements (l, j) of the matrix structure Ψ satisfy: $\Psi_{l,j}(\mathbf{s}, \mathbf{s}') = \Psi_{l,j}(\mathbf{s} - \mathbf{s}')$.

[44]Discrete Fourier transform, where $\mathbf{k} = (k_x, k_y, \cdots)$.

$R(k, z)$, expanding the expression (7.74) in power series of z. That is, from the following relationship:

$$P(k, n) = \frac{1}{n!} \left[\frac{\partial^n}{\partial z^n} R(k, z) \right]_{z=0}. \tag{7.75}$$

Exercise. Find $R(k, z) \equiv \sum_{n=0}^{\infty} z^n P(k, n)$ from (7.72).

Exercise. Obtain an expression (in Fourier space) for the probability with internal state l in the n-th step, i.e.: $P_n^l(k)$.

7.4.2 The Disordered Case

If we want to consider a Lorentz gas model in a *disordered* medium, we can use the solution for discrete time (7.74) and introduce the concept of random waiting-time between different steps in an RW; see advanced exercise 7.5.4. That is, we simply replace z by the waiting-time function $\psi(u)$ in the expression for $R(k, z)$. Then, in the CTRW context,[45] the probability $P^l(k, t)$ is given by the convolution in time of $R(k, t)$ with the function $\phi(t)$; i.e., in Fourier–Laplace

$$P(k, u) = \phi(u) R(k, z = \psi);$$

see (7.74) and (7.33).

Then, the general solution for the disordered problem is given in the Fourier–Laplace representation by using the matrix Green function [with internal states]; that is:

$$G(k, u) = \frac{1 - \psi(u)}{u} [1 - \psi(u)\Psi(k)]^{-1}. \tag{7.76}$$

Obviously, as we can see from (7.76), this solution represents a factorized model.[46] Non-factorized models with internal states can also be studied analogously; in the next section we will present the outlines of that approach with a non-factorized waiting-time matrix [14, 15]. In particular, we will show a model in which the walker inside a "channel" has a spatiotemporal structure which is not separable, thereby allowing to represent superdiffusion when the walker enters into that particular channel.

---

[45] In Sect. 7.3.3 we found the relationship between the waiting-time function $\psi(u)$ and a model of site-disorder in the ME. Furthermore, the definition of "effective medium" $\Phi(u)$ in the context of the CTRW theory was also established there.

[46] Note that this expression is general and represents the solution of an arbitrary model of non-Markovian process with internal states in a disordered medium. Indeed, the site-disorder approach is taken into account considering a factorizable waiting-time function $\psi(u)$, i.e., independent of internal states.

Exercise (**Green's Function with Internal States**). Using the CTRW approach [see Sect. 7.3] along with the solution for discrete time (7.74), show that the matrix Green's function (7.76) is the solution of a CTRW (separable) with internal states [18].

Optional Exercise (**Disordered Lorentz Gas**). Using (7.73) and (7.76), consider different waiting-time functions[47] $\psi(u)$, to study the moments of the marginal probability $\mathcal{P}(\mathbf{s}, t) = \sum_l P^l(\mathbf{s}, t)$ for a Lorentz gas model in $2D$ [30].

Excursus (**Non-Markovian Chain**). It should be mentioned that if we are only interested in knowing the marginal probability $\mathcal{P}_n(\mathbf{s}) = \sum_l P_n^l(\mathbf{s})$, this probability satisfies a recurrence relation with memory; i.e., it represents a non-Markovian chain [31, 32]; see the first exercise of Sect. 8.5.1.

Excursus (**Fractured Porous Media**). A master equation with internal states may also be used to describe transport in porous geological systems in the presence of fractures. In this case the internal states are used to represent different alternative "paths" generated by the cracks in the porous rocks [33–35].

Excursus (**Return to the Origin**). Other interesting properties of an RW with *internal states*, such as the average number of different sites visited or the probability of return to the origin [see Sect. 7.3.2], can also be analyzed in the representation of discrete times from the solution with internal states $\mathbf{R}(\mathbf{k}, z)$, or in the continuous time representation from solution $\mathbf{G}(\mathbf{k}, u)$; see, for example, reference [14].

Excursus (**Diffusion Limited by Annihilation**). To describe transport of particles which decay in time, as in the case of positron diffusion in matter, models of ME in the context of EMA and stochastic non-Markovian processes of the CTRW type with internal states can be used [36, 37].

7.4.3   Non-factorized Case*

It usually happens that when trying to describe the transport of "complex" entities in a given medium it is necessary to consider that the medium fluctuates over time, thus producing different interactions with the particle or entity that diffuses [1]. These temporal fluctuations may produce anomalous transport in describing the particle; for example, the walker could find in its path dynamic traps that are activated or deactivated stochastically. Another example might be to try to describe the diffusion of a tracer in a fractured porous medium. In this case, when the particle (or tracer) diffuses into the porous medium itself, we expect subdiffusive behavior, whereas if the tracer diffuses through a crack we can expect superdiffusion. There are some

---

[47]That is, using different models of disorder.

examples of "complex systems" in which in order to describe the *true* transport of entities, it is imperative to use internal states in a non-factorized context [14, 20, 25]. For this reason we now present briefly a generalization of the CTRW approach with internal states.

The starting point is again Eq. (7.26), but now considering another degree of freedom that we call internal state l [or l-th channel.[48]] Then, in the presence of internal states, we write:

$$R_l(\mathbf{s}, t \mid \mathbf{s}_0, 0) = \int_0^t \sum_{l'} \sum_{\mathbf{s}'} \eta_{ll'}(\mathbf{s} - \mathbf{s}', t - \tau) R_{l'}(\mathbf{s}', \tau \mid \mathbf{s}_0, 0) \, d\tau + \delta_{\mathbf{s}, \mathbf{s}_0} \delta(t) c_l.$$

(7.77)

Here $\eta_{ll'}(\mathbf{s} - \mathbf{s}', t - \tau) d\tau$ represents the probability that the RW (non-Markovian) makes a jump from $\mathbf{s}'$ to $\mathbf{s}$ in a regular lattice and the transition from the internal state l' to l, after being trapped since coming to the site $\mathbf{s}'$ with internal state l' during a time interval $t - \tau$. Here we have assumed that the initial condition for the CTRW is $\delta_{\mathbf{s}, \mathbf{s}_0} c_l$; that is, c_l characterizes the probability that at the initial time $t = 0$ the RW is with internal state l [or in the l-th "channel"]. Another important quantity that characterizes the evolution of a CTRW with internal states is the likelihood of staying in place $\mathbf{s}$ and internal state l without changing place or state during the interval $[0, t]$, that is:

$$\phi_l(t) = 1 - \int_0^t \sum_{l'} \sum_{\mathbf{s}'} \eta_{l'l}(\mathbf{s}' - \mathbf{s}, t') \, dt'.$$

(7.78)

We should, then, define the diagonal matrix, which will soon be useful

$$\Xi(t) = \text{diagonal}(\cdots, \phi_{l-1}(t), \phi_l(t), \phi_{l+1}(t), \cdots).$$

In a similar way as we did in previous sections, the solution of a CTRW with internal states can be obtained immediately in the Fourier–Laplace representation:

$$\mathbf{G}(\mathbf{k}, u) = \Xi(u) \cdot [\mathbf{1} - \eta(\mathbf{k}, u)]^{-1}.$$

(7.79)

Exercise. Taking the Fourier–Laplace transform in (7.77) and (7.78), show that (7.79) is the corresponding Green's function where the dimension of the matrix $\mathbf{G}(\mathbf{k}, u)$ is given by the number of internal states of the CTRW.

Thus (7.79) generalizes solution (7.76) for the case where the waiting-time function does not factorize. That is, the matrix $\eta(\mathbf{k}, u)$ replaces the structure $\psi(u) \Psi(\mathbf{k})$; whence it is easy to see that the probability of staying depends on the

---

[48]It is important to note that internal states are not necessarily degrees of freedom of the particle, they may also describe a particular state of the medium in which the particle diffuses. As in Sect. 7.4.1, here it is helpful to use the notation $\mathbf{s}$ to indicate a vector of arbitrary dimension.

internal state,[49]

$$\Xi(u)_{ll} \equiv \phi_l(u) = \frac{1 - \sum_{l'} \eta_{l'l}(\mathbf{k} = 0, u)}{u},$$

the normalization condition is now (note that $\boldsymbol{\eta}(\mathbf{k} = 0, u = 0)$ is a Markov matrix)

$$1 = \sum_{l'} \eta_{l'l}(\mathbf{k} = 0, u = 0).$$

In general, the matrix $\boldsymbol{\eta}(\mathbf{k}, u)$ contains all the information characterizing the model. Many complex systems, such as the motion of dimers over surfaces, radioactive ion transport on a membrane, chemical kinetics cycles, multienergetic neutron transport, polymer relaxation, etc., have been modeled using this technique; see, example [13, 14, 18, 20], and references cited therein.

Note It is important to mention that if we just want to know the marginal probability

$$\mathcal{P}(\mathbf{s}, t \mid \mathbf{s}_0, 0) = \sum_l P^l(\mathbf{s}, t \mid \mathbf{s}_0, 0),$$

its calculation is equivalent to adding components of the matrix Green function (7.79), appropriately weighted with initial conditions $e^{i\mathbf{k}\cdot\mathbf{s}_0} c_{l'}$. That is, using that

$$P^l(\mathbf{k}, u) = \Xi(u)_{ll} \sum_{l'} [1 - \boldsymbol{\eta}(\mathbf{k}, u)]_{ll'}^{-1} c_{l'}.$$

Optional Exercise (**Subdiffusion Versus Superdiffusion**). One possible way of studying a model showing the inhibition of subdiffusion[50] is considering the presence of alternative "paths" without obstacles; that is, a CTRW with two internal states such that when the RW diffuses into the disordered medium [internal state 1] it is characterized by the structure matrix element $\eta_{11}(\mathbf{s} - \mathbf{s}', t)$, whereas if the RW diffuses through the alternative path [internal state 2] it is characterized by the element $\eta_{22}(\mathbf{s} - \mathbf{s}', t)$. Suppose further that the transition from one internal state to another is instantaneous and without changing the position of RW in space, i.e., it is characterized by off diagonal elements:

$$\eta_{12}(\mathbf{s} - \mathbf{s}', t) = (1 - p)\,\delta(t)\,\delta_{\mathbf{s},\mathbf{s}'}$$

$$\eta_{21}(\mathbf{s} - \mathbf{s}', t) = p\,\delta(t)\,\delta_{\mathbf{s},\mathbf{s}'},$$

---

[49]In the Laplace representation.
[50]Conversely: the inhibition of superdiffusion by the presence of disordered "roads."

with $p \in [0, 1]$. Then, consider, for example the non-factorized matrix structure (in 1D space[51]):

$$\eta(k, u) = \begin{pmatrix} (1-p)\,\psi_1(u)\cos k & (1-p) \\ p & p\,\psi_2(k, u) \end{pmatrix},$$

where p is the concentration of alternative paths (crack). Characterize asymptotically the transport in a disordered medium by $\psi_1(u) \sim 1 - Cu^\theta$, with $\theta \in (0, 1)$, and the alternative path or "crack" by $\psi_2(k, u) \sim 1 - C_1 u^\phi - C_2 u^{\phi-\eta} k^2 + \mathcal{O}(k^4)$, with $0 < \phi \le 1$, $\eta > 1$. Calculate the long time behavior of the second moment of the CTRW, regardless of its internal state, i.e.: $\langle s(t)^2 \rangle = \sum_s s^2 \sum_{l=1}^2 P^l(s, t)$. Show using Tauberian theorem the following asymptotic results

$$\langle s(t)^2 \rangle \sim \begin{cases} \dfrac{2C_2}{C} \dfrac{p^2}{(1-p)^2}\, t^{\eta+\theta-\phi}, & \text{if } \theta < \phi,\ 0 \le p < 1 \quad \text{for } t \to \infty, \\[2ex] \dfrac{2C_2}{C_1}\, t^\eta, & \text{if } \theta > \phi,\ 0 < p \le 1 \quad \text{for } t \to \infty. \end{cases}$$

Physically interpret the meaning of the sign $(\theta - \phi)$ studying the probability of remaining $\Xi(t)_{ll}$.

7.5 Additional Exercises with Their Solutions

7.5.1 Power-Law Jump and Waiting-Time from an Entropic Viewpoint

We have shown that the second moment of an RW may become anomalous: $\langle X(t)^2 \rangle \sim t^\nu$, $\nu \neq 1$. In particular in the presence of disorder it is possible to get subdiffusion $\nu < 1$, while in a turbulent regime we may get superdiffusion $\nu > 1$. In a Markov chain it is clear that this behavior is mainly controlled by the one-step jump $p(x)$; while in the context of the CTRW approach the waiting-time distribution $\psi(t)$ plays also an important role in the anomalous behavior.

(a) Let us focus now on the one-step jump $p(x)$, where Gaussian distributions represent the *minimum* of information of the random variable. This fact can be seen from the viewpoint of an entropic maximization principle[52] using: $S \propto \int_{-\infty}^{+\infty} p(x) \ln p(x) dx$. In this spirit, it is interesting to find that power-law distributions emerge also under a maximization pseudo-entropic principle.

---

[51]Obviously, a 1D lattice is not a realistic model, but the present conclusion is by itself interesting. Furthermore take, for example, the initial condition $P^l(s, t = 0 \mid 0, 0) = \delta_{s,0}/2$, $\forall l$.

[52]See: "*Maximum Entropic Formalism, Fractals, Scaling Phenomena, and 1/f Noise: a Tale of Tails*", by E. W. Montroll and M.S. Shlesinger, J. Stat. Phys. 32, 209, (1983).

Consider Tsallis' generalized entropy[53]

$$S_q[p(x)] = k_B \frac{1 - \int_{-\infty}^{+\infty} \frac{dx}{\sigma} [\sigma p(x)]^q}{q - 1},$$

here x represents the **rv** and σ is a characteristic length-scale. In the limit $q \to 1$ the standard entropy is recovered. Under the constraint $\int_{-\infty}^{+\infty} p(x)dx = 1$ and

$$[x^2]_q = \frac{\int_{-\infty}^{+\infty} x^2 [p(x)]^q \, dx}{\int_{-\infty}^{+\infty} [p(x)]^q \, dx} = \sigma^2,$$

and applying extremal techniques to $S_q[p(x)]$ (maximize if $q > 0$ and minimize if $q < 0$, in the $\lim q \to 1^{\pm}$ the Gaussian pdf is recovered), find the distribution $p_q(x)$ for cases $q > 1$ and $q < 1$. In particular, show that the support of $p_q(x)$ is compact if $q \in (-\infty, 1)$, while $p_q(x)$ becomes a power-law distribution for $q > 1$ with[54]

$$p_q(x) \sim \left(\frac{\sigma}{x}\right)^{2/(q-1)}, \quad \text{for } |x|/\sigma \to \infty.$$

Therefore $\langle x^2 \rangle = [x^2]_1$ is finite if $q < 5/3$ and diverges if $5/3 \le q \le 3$.

(b) Consider now the same extremal entropic technique applied to the waiting-time $\psi(t)$ pdf of the CTRW theory. Define Boltzmann's entropy as $S[\psi(t)] = -\int_0^{+\infty} \psi(t) \ln \psi(t) dt$, under constraints $\int_0^{+\infty} \psi(t) dt = 1$ and $\langle t \rangle = \int_0^{+\infty} t\psi(t) dt = 1/\lambda$. Show that an exponential waiting-time is obtained: $\psi(t) = \lambda e^{-\lambda t}$. This corresponds to a Poisson process for the number of steps in a CTRW.

(c) A power-law waiting-time function can be obtained from the Weierstrass probability that we have introduced in Chap. 1.

$$\psi(t) = \lambda \frac{1-a}{a} \sum_{s=1}^{\infty} (ab)^s \exp(-\lambda b^s t), \quad 0 < b; \quad 0 < a < 1. \qquad (7.80)$$

This expression fulfills $\int_0^{\infty} \psi(t) dt = 1$ and represents an infinite superposition of Poisson processes each one characterized by $\lambda_s \exp(-\lambda_s t)$, where λ_s is a characteristic waiting time scale $1/\lambda_s = 1/(\lambda b^s)$. In a similar way as we did in Chap. 1, show that Eq. (7.80) can be written in the form

$$\psi(bt) = \frac{1}{(ab)} [\psi(t) - (1-a)\lambda b \exp(-\lambda bt)],$$

[53] This function was originally introduced by C. Tsallis in: J. Stat. Phys. 52, 479, 1988.

[54] See: *"Nonextensive foundation of Lévy distributions"*, D. Prato and C. Tsallis, Phys. Rev. E 60, 2398 (1999).

this formula clearly shows a characteristic self similarity in the waiting-time. Then, for long times we can write

$$\psi\,(bt) \sim \psi\,(t)\,/\,(ab)\,,$$

which says that $\psi\,(t)$ is a homogeneous function; from this equation show that the asymptotic solution is

$$\psi\,(t) \propto t^{-1-\gamma},\ \ \text{with}\ \ \gamma = \ln\,(a)\,/\,\ln\,(b)\,.$$

Note that if $b^m < a$, with $m - 1 < \gamma < m$ all moments $\langle t^n \rangle = \infty$, $\forall n \geq m$.

(d) The waiting-time pdf $\psi\,(t)$ is also an important ingredient in the renewal theory.[55] Consider the continuous-time recurrence relation for the *probability density* $\psi_{(n)}\,(t)$

$$\psi_{(n)}\,(t) = \int_0^t \psi\,(t-\tau)\,\psi_{(n-1)}\,(\tau)\,d\tau, \quad \int_0^\infty \psi_{(n)}\,(t)\,dt = 1.$$

The probability that up to time t there have been exactly n events is

$$P_{(n)}\,(t) = \int_0^t \phi\,(t-\tau)\,\psi_{(n)}\,(\tau)\,d\tau,$$

where $\phi\,(t) = 1 - \int_0^t \psi\,(\tau)\,d\tau$ is the probability that there has been no renewal after the last one until time t. Show that the mean number of events till time t, that is $\langle n(t) \rangle = \sum_{n=1}^\infty n P_{(n)}\,(t)$, is given in terms of functions $\psi\,(u)$ and $\psi_{(1)}\,(u)$ (the waiting-time for the first event) in the form:

$$\langle n(t) \rangle = \mathcal{L}_u^{-1} \left[\frac{1}{u} \frac{\psi_{(1)}\,(u)}{1 - \psi\,(u)} \right], \tag{7.81}$$

where $\mathcal{L}_u^{-1}\,[\cdots]$ represents the inverse Laplace transform. Hint: formula (7.81) can be obtained using the Laplace representation and the generating function technique (see the next exercise 7.5.4). Note that if we take $\psi_1\,(u) = \psi\,(u)$ this assumption is equivalent to staying that $\psi_{(0)}\,(u) = \delta\,(t - 0_+)$. A renewal process with a fractal time structure can be modeled using a power-law waiting-time function.[56]

[55]See D.R. Cox in: "Renewal Process", Monographs on Statistical and Applied Probability". General eds. D.R. Cox and D.W. Hinkley, Chapman and Hall, London 1962 (reprinted 1982).

[56]See: "*Fractal time in condensed matter*", M.F. Shlesinger, Ann. Rev. Phys. Chem. 39, 269 (1988).

7.5.2 Telegrapher's Equation

The equation:

$$\partial_{tt}P(x,t) + \frac{c^2}{D}\partial_t P(x,t) = c^2\partial_{xx}P(x,t)$$

with the initial condition $\partial_t P(x,t)|_{t=0}$ was presented in Sect. 7.3.1, and the connection of Telegrapher's equation with the CTRW approach was established by using the waiting-time function

$$\psi(u) = \frac{a}{a + u(u + \lambda)}, \tag{7.82}$$

where $a = c^2 > 0$ and $\lambda = c^2/D > 0$; with $\lambda^2 > 4a$, such that $\psi(t) \geq 0$ for all t. Calculate the probability that the walker remains at site without making a jump during the time interval $(0, \tau)$. Hint: applying the Laplace transform to $\phi(\tau) = 1 - \int_0^\tau \psi(\tau')\,d\tau'$ we get $\phi(u) = [1 - \psi(u)]/u$. Then introducing (7.82) in this expression and after inversion of the Laplace it follows that

$$\phi(t) = e^{-\lambda t/2}\left\{\cosh\left(\frac{t}{2}\sqrt{\lambda^2 - 4a}\right) + \frac{\lambda}{\sqrt{\lambda^2 - 4a}}\sinh\left(\frac{t}{2}\sqrt{\lambda^2 - 4a}\right)\right\}.$$

Show also that the waiting-time function in real time is given by

$$\psi(t) = \frac{2ae^{-\lambda t/2}}{\sqrt{\lambda^2 - 4a}}\sinh\left(\frac{t}{2}\sqrt{\lambda^2 - 4a}\right),$$

while the memory kernel of the associated ME is $\Phi(t) = a\exp(-\lambda t)$. Then, interpret a "flight-time" scale as $t_f \sim 1/\sqrt{a}$, and a "residence-time" scale as $t_r \sim \lambda/a$.

7.5.3 RW with Internal States for Modeling Superionic Conductors

An accurate model of ionic transport in β-alumina takes into account the true dimensionality of the material and the structure of the interstitial sites lattice. Na β-alumina is composed of parallel oxygen planes with a triangular lattice structure, separated by intervening spinel structure blocks.[57] Na ions move in these

---

[57]See, C.R. Peters, M. Bettman, J.W. Moore, and M.D. Glick, in: Acta Crystallogr. B **27**, 1826 (1971).

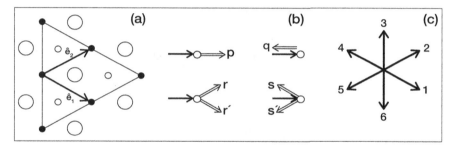

Fig. 7.4 (a) Interstitial sites structure of Na β-alumina. *Large open circles* represent oxygen atoms, *small circles* the interstitial sites (*solid circles* are Beevers-Ross sites). $\Delta = 5.595$ Å. Basis vectors $\hat{e}_1$ and $\hat{e}_2$ are shown. (b) Definition of parameters p,q,r,r',s,s'. *Solid arrows* represent incoming direction, open ones the outgoing direction. (c) Numbering of the six possible flight directions for a Na ion on an instertitial site

planes avoiding the oxygen ions by hopping between interstitial sites which form a hexagonal, two-dimensional lattice.[58] As only half of these sites are minima of the ionic potential, the effective lattice for the ionic motion is triangular and has the same structure factor as the oxygen one [see Fig. 7.4a]; then, we can take as basis vectors $\hat{e}_1 = \frac{1}{2}\Delta\left(\sqrt{3}\hat{x} - \hat{y}\right)$ and $\hat{e}_2 = \frac{1}{2}\Delta\left(\sqrt{3}\hat{x} + \hat{y}\right)$, where $\Delta = 5.595$ Å is the lattice parameter. For a Na ion hopping on this triangular lattice, correlation effects make the probability of moving to a given neighboring site to depend on which site does the particle come from. We assign probabilities p for the particle to go right through, q for turning back, r (s) for making a turn of $60°$ $(120°)$ to the left, and r' (s') for a turn of $60°$ $(120°)$ to the right [see Fig. 7.4b]; and probability conservation requires $p + q + r + s + r' + s' = 1$. Different assignments for these parameters allows for modeling correlations from the caterpillar ($p \simeq 1$) to the bounce-back ($q \simeq 1$) mechanisms. For the totally symmetrical case $p = q = r = s = r' = s'$ we recover the uncorrelated Markovian RW model. The present correlated model can be written in terms of a Markovian chain with internal states, where the probability $P_n^i(\mathbf{s})$ $(i = 1, \cdots, 6)$ for the ion to be at site $\mathbf{s} = s_1\hat{e}_1 + s_2\hat{e}_2$ (s_1 and s_2 integers) at the n-th step having come from the i-th direction is

$$P_n^i(\mathbf{s}) = \sum_{\mathbf{s}'}\sum_{j=1}^{6}\psi_{ij}\left(\mathbf{s} - \mathbf{s}'\right)P_n^j(\mathbf{s}), \quad (i = 1, \cdots, 6),\qquad (7.83)$$

where $\psi_{ij}(\mathbf{s})$ are the elements of a 6×6 matrix $\Psi(\mathbf{s})$ characterizing the transitions, and assuming translational invariance. The flight directions are numbered as shown in Fig. 7.4c.

---

[58] See A.S. Barker, Jr., J.A. Ditzenberger, and J.P. Remeika, in Phys. Rev. B **14**, 4254 (1976).

Introducing the Fourier transform in (7.83) we get

$$P_n^i(\mathbf{k}) = \sum_{s'} \sum_{j=1}^{6} \psi_{ij}(\mathbf{k}) P_n^j(\mathbf{k}),$$ (7.84)

where

$$P_n^j(\mathbf{k}) = \sum_{s'} e^{i\mathbf{k}\cdot\mathbf{s}} P_n^j(\mathbf{s}), \quad (j = 1, \cdots, 6),$$ (7.85)

and similarly for $\psi_{ij}(\mathbf{k})$, $\mathbf{k}$ are the vectors of the reciprocal lattice. Defining $\epsilon_1 = \exp(i\mathbf{k} \cdot \hat{e}_1)$, $\epsilon_2 = \exp(i\mathbf{k} \cdot \hat{e}_2)$, and $\epsilon_3 = \exp(i\mathbf{k} \cdot [\hat{e}_2 - \hat{e}_1])$, show that the matrix $\Psi(\mathbf{k})$ takes the form:

$$\Psi(\mathbf{k}) = \begin{pmatrix} p\epsilon_1 & r\epsilon_1 & s\epsilon_1 & q\epsilon_1 & s'\epsilon_1 & r'\epsilon_1 \\ r'\epsilon_2 & p\epsilon_2 & r\epsilon_2 & s\epsilon_2 & q\epsilon_2 & s'\epsilon_2 \\ s'\epsilon_3 & r'\epsilon_3 & p\epsilon_3 & r\epsilon_3 & s\epsilon_3 & q\epsilon_3 \\ q\epsilon_1^* & s'\epsilon_1^* & r'\epsilon_1^* & p\epsilon_1^* & r\epsilon_1^* & s\epsilon_1^* \\ s\epsilon_2^* & q\epsilon_2^* & s'\epsilon_2^* & r'\epsilon_2^* & p\epsilon_2^* & r\epsilon_2^* \\ r\epsilon_3^* & s\epsilon_3^* & q\epsilon_3^* & s'\epsilon_3^* & r'\epsilon_3^* & p\epsilon_3^* \end{pmatrix},$$

where the asterisk denotes complex conjugation. Equation (7.84) can be solved using the discrete-time generating function, as we did in Sect. 7.4.1. Of particular interest is the marginal probability to be at site $\mathbf{s}$ independently of the arrival direction, which is

$$P_n(\mathbf{s}) = \sum_{j=1}^{6} P_n^j(\mathbf{s}),$$

this probability can also be obtained from the discrete-time generating function [38]. Then, all moments of the RW can be calculated, in particular from the second moment the calculation of the *ac* conductivity $\sigma(\omega)$ can be obtained from the generalized Einstein's relation, see Sect. 8.5.

7.5.4 Alternative n-Steps Representation of the CTRW

Here we introduce the CTRW theory using an alternative presentation,[59] this may be of interest if we want to consider the number of steps, n, that the walker takes in

---

[59]This presentation is similar to the renewal theory, see previous Exercise 7.5.1 (d).

his continuous-time evolution. Calling $P_n(\mathbf{s}, t)\, dt$ the probability of being on site $\mathbf{s}$ at time t having taken n steps, we can write the recurrence relation:

$$P_{n+1}(\mathbf{s}, t) = \sum_{\mathbf{s}'} \int_0^t P_n(\mathbf{s}', \tau)\, \psi(\mathbf{s} - \mathbf{s}', t - \tau)\, d\tau, \tag{7.86}$$

where $\psi(\mathbf{s}, \mathbf{s}', \tau)\, d\tau$ is the probability that after arriving at position $\mathbf{s}'$ (a lattice site in arbitrary dimension) the walker moves to position $\mathbf{s}$ at instant τ. In (7.86) we have assumed translational invariance in the lattice, which is reflected by the $\mathbf{s} - \mathbf{s}'$ dependence of ψ. The normalization condition requires

$$\sum_{\mathbf{s}} \int_0^\infty \psi(\mathbf{s} - \mathbf{s}', \tau)\, d\tau = 1. \tag{7.87}$$

Here the initial condition that we shall use for the recurrence relation (7.86) is: $P_0(\mathbf{s}, t) = \delta_{\mathbf{s},0}\delta(t - 0_+)$. In order to solve (7.86) we introduce the generating function

$$\mathcal{R}(\mathbf{s}, t, z) = \sum_{n=0}^{\infty} z^n P_n(\mathbf{s}, t). \tag{7.88}$$

Replacing (7.88) in (7.86) and introducing the Fourier and Laplace transforms, show that the solution of the function $\mathcal{R}(\mathbf{k}, u, z)$ is

$$\mathcal{R}(\mathbf{k}, u, z) = \frac{1}{1 - z\psi(\mathbf{k}, u)}. \tag{7.89}$$

Introducing in (7.89) a series expansion is z and comparing with (7.88), show that the solution $P_n(\mathbf{s}, t)$ is given by

$$P_n(\mathbf{s}, t) = \mathcal{F}_{\mathbf{k}}^{-1} \mathcal{L}_u^{-1} [\psi(\mathbf{k}, u)^n]. \tag{7.90}$$

If we are interested in the probability $P(\mathbf{s}, t)$ of being at position $\mathbf{s}$ at time t, independently of the number of steps taken by the walker, we have to take into account the possibility that the walker arrived at $\mathbf{s}$ at an earlier time and stayed there until t; then, this probability is given by the expression:

$$\boxed{P(\mathbf{s}, t) = \int_0^t \mathcal{R}(\mathbf{s}, t - \tau, z = 1)\, \phi(\tau)\, d\tau,}$$

where $\phi(\tau) = 1 - \int_0^\tau \psi(\mathbf{k} = 0, \tau') d\tau'$ is the probability that after arrival, the walker remains at the site during the interval of time $[0, \tau]$. Show that the Fourier–Laplace transform of $P(\mathbf{s}, t)$ is the same as that found in Eq. (7.33).[60]

7.5.5 Distinct Visited Sites: Discrete and Continuous Time Results

In Sect. 7.3.2 we defined S_n as the expected number of distinct points visited (in the lattice) during the course of a n-steps walk:

$$S_n = \sum_s \{f_1(s) + f_2(s) + \cdots + f_n(s)\}, \quad S_0 = 0,$$

here $A_j = \sum_s f_j(s)$ is the expected number of *new* points visited on a j-steps walk (remember that $f_j(s)$ is the probability of first passage through site s at the j-th step of a walk which started at the origin at $n = 0$). To study the asymptotic behavior of S_n, we first calculate the generating function of A_n, i.e.,

$$A(z) = \sum_{n=0}^\infty z^n A_n = \sum_s \sum_{n=0}^\infty z^n f_n(s)$$

$$= \sum_s f(s, z) = \sum_s \frac{R(s, z) - \delta_{s,0}}{R(0, z)},$$

then, noting that $\sum_s R(s, z) = (1 - z)^{-1}$ we get

$$A(z) = \frac{z}{(1 - z) R(0, z)}.$$

In order to calculate the generating function $S(z) = \sum_{n=0}^\infty z^n S_n$ we use that $S_n - S_{n-1} = A_n$, and that $A_0 = 0$, $S_0 = 0$, thus

$$\sum_{n=1}^\infty z^n S_n = \sum_{n=1}^\infty z^n S_{n-1} + \sum_{n=1}^\infty z^n A_n$$

$$= z \sum_{m=0}^\infty z^m S_m + A(z),$$

---

[60]In the separable case $P(k, t)$ is solution of (7.42). The CTRW with degrees of freedom can be presented in a similar way; this approach has been used to model transport of multienergetic neutrons, [18].

then we get

$$S(z) = A(z) / (1 - z)$$

$$= \frac{z}{(1 - z)^2 R(0, z)}, \tag{7.91}$$

this result is the starting point for studying the asymptotic long-time behavior of the number of distinct sites visited by a walk. This analysis can be done in discrete or continuous time by using the Tauberian theorem [13]. Here we propose to do this analysis in continuous time, that is replacing $z \to \psi(u)$, then (7.91) adopts the form[61]

$$S(u) = \frac{\psi(u)}{(1 - \psi(u))^2 R(0, \psi(u))}. \tag{7.92}$$

The asymptotic long-time behavior follows introducing an expression for $\psi(u \sim 0)$. Show that in the Markovian diffusive case, i.e., using

$$\psi(u) \simeq 1 - \langle t \rangle u + \cdots$$

$$\lambda(\mathbf{k}) \simeq 1 - \frac{\sigma^2}{2D} \mathbf{k}^2 + \cdots,$$

the asymptotic behavior of $S(t \to \infty)$ for square lattices is

$$S(t) \sim \begin{cases} \sqrt{\dfrac{8\sigma^2 t}{\langle t \rangle \pi}}, & \text{for } D = 1 \\[3mm] \dfrac{\pi \sigma^2 t}{\langle t \rangle \ln\left(\frac{t}{\langle t \rangle}\right)}, & \text{for } D = 2 \\[3mm] \dfrac{t}{\langle t \rangle R(s=0, z=1)}, & \text{for } D \geq 3, \end{cases}$$

where $R(\mathbf{s} = 0, z)$ depends on the type of lattice and the dimension $D \geq 3$, for example for symmetric transitions using $\lambda(\mathbf{k})$ from Sect. 7.3.2 we get

$$R(\mathbf{s} = 0, z = 1) = \frac{1}{\pi^D} \int_0^\pi dk_1 \cdots \int_0^\pi dk_D \frac{1}{1 - \lambda(\mathbf{k})}$$

$$= \begin{cases} 1.516386959, & \text{for a } sc \text{ 3D-lattice} \\[2mm] 1.393203929, & \text{for a } bcc \text{ 3D-lattice} \\[2mm] 1.344661073, & \text{for a } fcc \text{ 3D-lattice} \end{cases}$$

---

[61]To realize the replacement $z \to \psi(u)$, see previous Exercise 7.5.4.

7.5.6 Tauberian Theorem

This theorem plays an important role in the analysis of anomalous diffusion, so we will summarize some of the more important results, the proof of this theorem is rather involved and can be found in Feller's book [see reference in Chap. 1].

Consider $f(t)$ to be real, and $\mathcal{L}[f(t)] = f(u)$ its Laplace transform. Let $L(x)$ be a "slowly varying function," i.e., continuous and positive for $x > 0$ fulfilling the property:

$$\lim_{x \to \infty} \frac{L(cx)}{L(x)} \to 1,$$

for example $L(x) = \ln x$ is a *slowly varying function*, while $L(x) = x^\nu$ is not. If

$$\lim_{u \to 0} f(u) \sim C\frac{L(1/u)}{u^\gamma}, \quad \gamma \geq 0, \tag{7.93}$$

and $f(t) \geq 0$; then it is true that

$$F(t) = \int_0^t f(\tau) \, d\tau \sim \frac{CL(t)}{\Gamma(\gamma + 1)} t^\gamma, \quad \text{for } t \to \infty. \tag{7.94}$$

If in addition we assume that $f(t)$ is asymptotically a monotonic function of t, we can then formally differentiate equation (7.94), reaching the asymptotic conclusion

$$f(u) \sim C\frac{L(1/u)}{u^\gamma} \ (\text{for } u \to 0) \Rightarrow \frac{C}{\Gamma(\gamma + 1)} t^{\gamma-1} \ (\text{for } t \to \infty).$$

Incidentally it is also true that the theorem just stated is valid in the opposite limit [13], that is

$$f(u) \sim C\frac{L(1/u)}{u^\gamma} \ (\text{for } u \to \infty) \Rightarrow \frac{C}{\Gamma(\gamma + 1)} t^{\gamma-1} \ (\text{for } t \to 0).$$

The Abelian Theorem

This theorem concerns the inverse problem; that is, finding which types of asymptotic behavior in the function $f(t)$, for $t \to \infty$ lead to a specific singularity in $f(u)$; see Weiss' book [13]; i.e.:

$$\lim_{t \to \infty} f(t) \sim t^m \ (\text{for } m > 0) \Rightarrow \frac{\Gamma(m + 1)}{u^{m+1}} \ (\text{for } u \to 0).$$

References

1. N.G. van Kampen, *Stochastic Process in Physics and Chemistry*, 2nd edn. (North-Holland, Amsterdam, 1992)
2. S. Alexander, J. Bernasconi, W.R. Schneider, R. Orbach, Rev. Mod. Phys. **53**, 175 (1981)
3. E. Hernández-García, M.O. Cáceres, Phys. Rev. A **42**, 4503 (1990)
4. P.A. Pury, M.O. Cáceres, E. Hernández-García, Phys. Rev. A **49**, R967 (1994)
5. M.O. Cáceres, H. Matsuda, T. Odagaki, D.P. Prato, W.P. Lamberti, Phys. Rev. B **56**, 5897 (1997)
6. T. Odagaki, M. Lax, Phys. Rev. **24**, 5284 (1981)
7. E. Hernández-García, M.A. Rodríguez, L. Pesquera, M. San Miguel, Phys. Rev. B **42**, 10653 (1990)
8. D.P. Prato, W.P. Lamberti, H. Matsuda, J. Phys. A **32**, 4027 (1999)
9. T. Odagaki, J. Phys. A **20**, 6455 (1987)
10. E.R. Reyes, M.O. Cáceres, P.A. Pury, Phys. Rev. B **61**, 308 (2000)
11. P.M. Morse, H. Feshbach, *Method of Theoretical Physics* (McGraw-Hill, New York, 1953)
12. H. Scher, M. Lax, Phys. Rev. B **7**, 4491 (1973)
13. G.H. Weiss, *Aspects and Applications of the Random Walk* (North-Holland, Amsterdam, 1994)
14. E.W. Montroll, B.J. West, *Fluctuation Phenomena*, ed. by E.W. Montroll, J.L. Lebowitz (North-Holland, Amsterdam, 1987)
15. M.O. Cáceres, Phys. Rev. A **33**, 647 (1986)
16. S. Goldstein, Quart. J. Mech. Appl. Math. **4**, 129 (1951)
17. M.O. Cáceres, Phys. Scripta **37**, 214 (1988)
18. M.O. Cáceres, H.S. Wio, Z. Phys. B. Condens. Matt. **54**, 175 (1984)
19. R.C. Bourret, Can. J. Phys. **39**, 133 (1961); **38**, 665 (1960)
20. M.O. Cáceres, H. Schnörer, A. Blumen, Phys. Rev. A **42**, 4462 (1990)
21. A. Harrison, R. Zwanzig, Phys. Rev. A **32**, 1072 (1985); A. Brissaud, U. Frisch, J. Math. Phys. **15**, 524 (1974); C.E. Budde, M.O. Cáceres, Phys. Rev. Lett. **60**, 2712 (1988)
22. M.O. Cáceres, A.K. Chattah, Phys. Lett. A **276**, 272 (2000); A.K. Chattah, M.O. Cáceres, J. Phys. A Math. Gen. **34**, 5795 (2001)
23. M.O. Cáceres, C.E. Budde, M.A. Ré, Phys. Rev. E **52**, 3462 (1995)
24. A. Compte, Phys. Rev. E **55**, 6821 (1997); A. Compte, M.O. Cáceres, Phys. Rev. Lett. **81**, 3140 (1998)
25. E.W. Montroll, M.F. Shlesinger, *Nonequilibrium Phenomena II, from Stochastic to Hydrodynamic*. Studies in Statistical Mechanics, ed. by J.L. Lebowitz, E.W. Montroll, vol. XI (Elsevier Science Publisher B.V., Amsterdam, 1984), pág. 1
26. M.F. Shlesinger, J. Klafter, Phys. Rev. Lett. **54**, 2551 (1985)
27. M.F. Shlesinger, B.J. West, J. Klafter, Phys. Rev. Lett. **58**, 1100 (1987)
28. A. Blumen, G. Zumofen, J. Klafter, Phys. Rev. A **40**, 3964 (1989)
29. C.B. Briozzo, C.E. Budde, O. Osenda, M.O. Cáceres, J. Stat. Phys. **65**, 167 (1991)
30. C.B. Briozzo, C.E. Budde, M.O. Cáceres, Physica A **159**, 225 (1989)
31. M.O. Cáceres, C.E. Budde, Phys. Lett. A **125**, 369 (1987)
32. C.E Budde, M.O. Cáceres, Physica A **153**, 315 (1988)
33. B.D. Hughes, M. Sahimi, Phys. Rev. E **48**, 2776 (1993)
34. M.O. Cáceres, Phys. Rev. A **69**, 036302 (2004)
35. M.O. Cáceres, G.L. Insua, J. Phys. A Math. Gen. **38**, 3711 (2005)
36. W.E. Frieze, K.G. Lynn, D.O. Welch, Phys. Rev. B **31**, 15 (1985)
37. P.A. Pury, P.D. Prato, Z. Phys. B Cond. Matt. **85**, 117 (1991)
38. C.B. Briozzo, C.E. Budde, M.O. Cáceres, Phys. Rev. A **39**, 6010 (1989)

Chapter 8
Nonequilibrium Statistical Mechanics

8.1 Fluctuations and Quantum Mechanics

All along this book we have studied fluctuations in a classical physical system by implementing its description in terms of an ME and/or Fokker-Planck dynamics; in particular its relaxation to the thermal equilibrium distribution demands a constraint that nowadays is well known as the detailed balance condition. A short cut technique for studying fluctuations, which is very popular, is the so-called Langevin approach which has been generalized in many ways, leading sometimes to controversial results in nonlinear systems due to the white noises nature. The general conclusion concerning the Langevin or Fokker-Planck approach is that in order to describe nonequilibrium fluctuations of a (nonlinear) classical system it is necessary to start the analysis from a more fundamental description (master equation) in order to associate a proper model for the noise term in the Langevin differential equation.

An extension to these questions would be to know how to incorporate fluctuations in a quantum system. Many attempts have been made on this subject, which goes back to 1957 with the pioneer work of Redfield[1] for writing an evolution equation for the density matrix. A natural question, on this issue, is to know the structure for a quantum Markov process, this question was solved by Kossakowski and Lindblad independently, see references in [1]. On the other hand, getting a quantum Markov process from a given physical system is also another interesting problem to be solved. One way would be to introduce *a la Langevin* noise and dissipation terms in the Schrödinger equation, but soon we realize that a damping term in the Schrödinger equation would dissipate probability rather than energy. Another possibility is to write the quantum Langevin equation in the Heisenberg representation, but this leads to the unsolved problem of formulating the required

---

[1] A.G. Redfield, IBM J. Research Devel, 1, 19 (1957).

© Springer International Publishing AG 2017

M.O. Cáceres, *Non-equilibrium Statistical Physics with Application to Disordered Systems*, DOI 10.1007/978-3-319-51553-3_8

stochastic properties for a random force operator.[2] A suitable shortcut option is to write a Schrödinger-Langevin equation *a la van Kampen,* but this merely leads to the formula for the Kossakowski-Lindblad general structure of a Markov semigroup. In Appendix I we present a brief summary on quantum open systems and the problem of quantum noise, see [31].[3]

The conclusion is that in quantum mechanics one cannot avoid adding explicitly a description of the mechanism that is the physical cause of the fluctuation and the damping on the system of interest. To do this we have to write the total Hamiltonian, i.e., for the system $\mathcal{S}$, the thermal bath $\mathcal{B}$, and the interaction between $\mathcal{S}$ and $\mathcal{B}$. Then, using von Neumann's equation of motion for the density matrix of the *"Universe"*

$$\partial_t \rho^{\mathrm{T}} = -i \left[H_{\mathrm{T}}, \rho^{\mathrm{T}} \right] / \hbar,$$

we trace out the degree of freedom of the bath to get the *reduced* density matrix

$$\rho(t) = Tr \left[\rho^{\mathrm{T}}(t) \right]_{\mathrm{Bath}}.$$

The problem arises when we try to get an evolution equation for $\rho(t)$ because generally speaking this evolution equation does not have the required Kossakowski-Lindblad algebraic structure [32]. So more approximations must be introduced to get a proper Kossakowski-Lindblad infinitesimal generator for a quantum Markov process. In conclusion, considering an *open* quantum system coupled to a thermal bath $\mathcal{B}$ of temperature T, the nonequilibrium evolution of the reduced density matrix can only be approximated by a quantum Markovian stochastic process under several restrictions.[4]

The diagonal elements of the density matrix are indeed probabilities (a summary of the density matrix and its evolution equation is presented in Appendices F and I); and in the semi-classical regime it is possible, under certain restrictions, to see that the temporal evolution of the diagonal elements of the density matrix are governed by the Pauli ME[5] [24]. In addition the steady state of this equation coincides with

---

[2]See for example one of the pioneer works by: J.R. Klauder and E.C.G. Sudarshan, *Fundamentals of Quantum Optics* (Benjamin, New York, 1968); C.W. Gardiner, *Quantum Noise* (Springer, Berlin, 1991).

[3]Actually in quantum mechanics the generator of a Markov process is called the Kossakowski-Lindblad infinitesimal generator (of a noncommutative semigroup). In a physical context this evolution equation is often called the quantum master equation, as it meets all the requisites to evolve a density matrix, as well as to fulfill all commutation relations (uncertainties) required by quantum mechanics. In the remainder of this text we always clarify if we are talking about the ME or the quantum master equation.

[4]See: *Langevin and Master Equation in Quantum Mechanics*, N.G. van Kampen and I. Oppenheim, J. Stat. Phys. 87, 1325, (1997).

[5]The Pauli ME has the same structure as the ME (in classical mechanics) for discrete Markovian processes and continuous time. Transition elements W_{ij} are characterized by Fermi's golden rule, from which we can see that the model meets the principle of detailed balance. Then, and as expected, the stationary solution of the Pauli ME matches the equilibrium solution of quantum

the Gibbs distribution of statistical mechanics [1, 31]. Under these conditions it is plausible to claim modeling transport phenomena in disordered media by using a classical ME.

8.2 Transport and Quantum Mechanics

In general, the loss of long-range structural order is the central feature of disordered systems. This loss of order in the material generates a random potential that causes the phenomenon of localization (quantum mechanics) of electrons and/or elementary excitations in solid state physics. Thus, in the study of anomalous transport phenomena (diffusion in disordered materials) it is necessary to resort to physical models in which the charged carriers move by *hopping* between the localization centers. This means that now it is the atomic orbitals around each site which will be the relevant quantum number for characterizing the charged carriers, instead of the $\mathbf{k}$ wave vectors of the delocalized case.[6] This raises the concern of studying the problem of transport in disordered media in terms of the probability

$$P(\mathbf{r}, t \mid \mathbf{r}_0, 0) = \mid \left\langle \mathbf{r} \mid e^{-itH/\hbar} \mid \mathbf{r}_0 \right\rangle \mid^2,$$

rather than in terms of complex quantum probability amplitude $\langle \mathbf{r} \mid e^{-itH/\hbar} \mid \mathbf{r}_0 \rangle$. On the other hand the study of $P(\mathbf{r}, t \mid \mathbf{r}_0, 0)$ is a simpler problem than using the density matrix.

In previous chapters various models of (classical) transport in disordered media have been defined, these models were presented in the context of the ME in regular networks with untidy transitions to first neighbors (see Chaps. 6 and 7). A different, but equivalent analysis, when $\mathbf{k}$ is a good quantum number, is focused on the study of kinetic equations in semi-classical approximations. Among the best known is the Boltzmann equation for the distribution of the number of electrons with quantum state $\mathbf{k}$. Under certain restrictions transport coefficients can be obtained using this type of kinetic theories. We shall not develop into kinetic methods but will focus instead on the linear response approximation.

The alternative way to study transport using linear response theory is a powerful technique [see Chap. 5]. In this context and against small perturbations caused by an external force (the system can be placed out of equilibrium), we can then study

---

statistical mechanics [31]. That is, in Pauli's ME the dynamics of off-diagonal elements of the density matrix "separate" from the dynamics of the diagonal elements; see also N.G. van Kampen, Physica A 20, 603 (1954).

[6]The wave vector $\mathbf{k}$ appears quite naturally when a free particle is studied in quantum mechanics; that is, when the Hamiltonian of the system is $H = -(\hbar^2/2m)\nabla^2$. In this case the solution of the eigenstates problem $H \mid \mathbf{k} > = E_\mathbf{k} \mid \mathbf{k} >$, is characterized by the eigenvalue: $E_\mathbf{k} = (\hbar^2/2m) \mid \mathbf{k} \mid^2$, and the eigenfunction $< \mathbf{r} \mid \mathbf{k} >$ is a plane wave.

the *response* of the system needed in order to dissipate the delivered energy (for example studying the electric conductivity). It is possible to see when calculating the diffusion coefficient that only fluctuations around equilibrium are necessary, whether from the velocity correlation function or from the dispersion of the carrier displacement. This is the basic idea that allows to study all transport coefficients (nonequilibrium) by appropriate calculations taking into account only fluctuations around equilibrium. It should be noted that although the kinetic theory can go beyond linear response, one needs linear response theory in order to address the problem of transport in disordered media.

The history of the evolution of this issue, so exciting, is long and goes back to the pioneering work on fluctuation and dissipation presented by Green [15], Kubo [18], and Lax [20]. In the following sections we briefly discuss the basic ideas of this topic. Some of its mathematical details are presented in Appendix G and in advanced exercises in Sect. 8.8. The much more involved problem of the description of open quantum systems (quantum stochastic models) will be briefly commented in Appendix I.

8.3   Transport and Kubo's Formula*

In 1957 Kubo found from microscopic principles an expression for the linear response function[7] in terms of correlations of the system around thermodynamic equilibrium. Thus he concluded and generalized Einstein's and Onsager's famous works, deriving the famous Fluctuation-Dissipation theorem [19]. A macroscopic scheme of linear response theory has already been analyzed in detail in previous chapters of this text (see Sect. 5.4 and Appendix G). The following discussion is due to Kubo, Toda, and Hashitsume [19]; on the other hand, in Appendix G.1 an alternative proof of this theorem is presented (without the use of super-operator algebra[8] [9]).

Consider that the system of interest S is in thermal equilibrium with a bath at temperature T. Assume that a monochromatic external force $F(t) = E_0 \cos(\omega_o t)$ interacts with the system via a certain operator (observable) A, thus producing a perturbation $-F(t)A$ in the total Hamiltonian of the system S, namely: $H_T = H - F(t)A$.

In this case we say that the force has a mechanical character; while if thermal gradients, or chemical gradients, etc., appear, they are related to *thermal forces* (nonmechanical) and will not be discussed in this text.[9] Obviously, due to this

---

[7]Sometimes called: *The after effect function.*

[8]A superoperator is a linear operator acting on the space of density matrices, see Appendices F, G, and I.

[9]In this case the analysis must be made in terms of an **sde** or using the **F-P** equation; this presentation due to Green and independently by Mori can be seen, for example in Kubo's text [19].

external time-dependent perturbation the $\mathcal{S}$ system is no longer in thermodynamic equilibrium and the density matrix will depend on time. That is, ρ is no more the canonical equilibrium distribution:

$$\rho^{\text{eq}} = \mathcal{Z}^{-1} e^{-H\beta}, \quad \text{where} \quad \mathcal{Z} = Tr[e^{-H\beta}], \beta \equiv \frac{1}{k_B T}.$$

Suppose that we characterize the response of the system $\mathcal{S}$ by measuring the temporal change, relative to thermodynamic equilibrium, of a certain observable B. We define

$$\langle \delta B(t) \rangle \equiv Tr[\rho(t)B] - Tr[\rho^{\text{eq}}B]. \tag{8.1}$$

This variation characterizes the statistical average of the quantum observable B at time t, in the presence of the disturbance. In the case of working in the classical limit (8.1) is simply a statistical mean value, where operators ρ and ρ^{eq} become ensemble distribution functions f and f^{eq} of the canonical variables $\{p, q\}$ (see Appendices F and G).

In the linear response approximation we can write the temporal variation of the observable B, due to the disturbance caused by the term $-A F(t)$ in the total Hamiltonian, as follows[10]:

$$\mathcal{F}_\omega \left[\langle \Delta B(t) \rangle_\rho \right] = \mathcal{F}_\omega \left[\phi_{BA}(t) \right] \mathcal{F}_\omega \left[F(t) \right]$$

$$= \phi_{BA}(\omega) F(\omega),$$

where $\langle \delta B(t) \rangle = Tr \left[\rho(t) \left(B - \langle B \rangle_{\text{eq}} \right) \right] = Tr[\rho(t) \Delta B] \equiv \langle \Delta B(t) \rangle_\rho$.

Here, as in the above formulas, ρ represents the density matrix of $\mathcal{S}$, which is the solution of the nonequilibrium problem (Liouville-Neumann equation in the presence of external disturbance, see Appendix F):

$$\partial_t \rho(t) = i \left[\mathcal{L}_0 + \mathcal{L}_{\text{ext}}(t) \right] \rho(t).$$

This operational equation is solved iteratively up to the first order in the external interaction, i.e., to order $\mathcal{O}(E_0)$, then with this approximate solution the **mv** $\langle \Delta B(t) \rangle_\rho$ is calculated. In what follows in this section we formulate Kubo's theorem leaving Appendix G for details of its proof.

In Green's derivations, the assumption is that variables are Markovian, while in Mori's derivation the assumption is the local equilibrium distribution. An alternative "mechanical" derivation of thermal transport coefficients (viscosity, heat conductivity, etc.) made in a *"Hamiltonian formalism"* can be seen in: J.M. Luttinger, Phys. Rev. 135, A1505 (1964).

[10]This expression is the Fourier transform of the after effect function (linear response function), see Sect. 5.2.

8.3.1 Theorem III (Kubo)

In the context of Kubo's linear response theory (microscopic), the susceptibility $\phi_{BA}(\omega)$ is expressed in terms of the system's fluctuations around thermodynamic equilibrium. In the Fourier representation one has

$$\phi_{BA}(\omega) = \int_0^\infty \phi_{BA}(t) \exp(i\omega t) \, dt, \tag{8.2}$$

where the response function (classical or quantum) is given by

$$\phi_{BA}(t) = Tr\left[\{\rho^{eq}, \Delta A\} \, \Delta B(t)\right] = Tr\left[\rho^{eq} \{\Delta A, \Delta B(t)\}\right]. \tag{8.3}$$

Here $\Delta A = A - \langle A \rangle_{eq}$ and $\Delta B(t) = B(t) - \langle B \rangle_{eq}$; indeed $\{\cdot, \cdot\}$ represents Poisson's brackets in the classical or quantum case (see appendix G); that is, the Liouville-Neumann operator:

$$i\mathcal{L}\rho = \{H, \rho\} = \frac{1}{i\hbar}\left[H, \rho\right].$$

8.3.2 Kubo's Formula

In this section we present different ways of writing the response function $\phi_{BA}(t)$, i.e., Kubo's theorem (8.3). In particular, for a start we prove a very useful result, see also advanced exercise 8.8.1.

Guided Exercise (**Kubo's Identity**). Prove the identity

$$\{e^{-\beta H}, Q\} = e^{-\beta H} \int_0^\beta \dot{Q}(-i\hbar\lambda) \, d\lambda. \tag{8.4}$$

The first member of (8.4) is the quantum commutator, i.e.:

$$\{e^{-\beta H}, Q\} = \frac{1}{i\hbar}\left(e^{-\beta H}Q - Qe^{-\beta H}\right).$$

Also, by changing variable $\tau = -i\hbar\lambda$, the second member of (8.4) can be written as:

$$e^{-\beta H} \int_0^\beta \dot{Q}(-i\hbar\lambda) \, d\lambda = \frac{e^{-\beta H}}{-i\hbar} \int_0^{-i\hbar\beta} \dot{Q}(\tau) \, d\tau$$

$$= \frac{e^{-\beta H}}{-i\hbar}\left(Q(-i\hbar\beta) - Q\right),$$

where the second line comes from integrating by parts. Using the equation of motion $dQ(t)/dt = \frac{1}{i\hbar}[Q, H]$, we conclude that $Q(-i\hbar\beta) = e^{\beta H}Qe^{-\beta H}$, then we can finally write

$$e^{-\beta H}\int_0^\beta \dot{Q}(-i\hbar\lambda)\,d\lambda = \frac{e^{-\beta H}}{-i\hbar}\left(e^{\beta H}Qe^{-\beta H} - Q\right) = \frac{1}{-i\hbar}\left(Qe^{-\beta H} - e^{-\beta H}Q\right),$$

which is what we wanted to prove.

Now we use Kubo's identity (8.4) to show that in quantum mechanics formula (8.3) can be written in the form

$$\phi_{BA}(t) = \int_0^\beta Tr\left[\dot{A}(-i\hbar\lambda)\Delta B(t)\rho^{eq}\right]\,d\lambda, \tag{8.5}$$

where

$$\Delta B(t) \equiv e^{-tH/i\hbar}\Delta B e^{+tH/i\hbar}$$

$$A(-i\hbar\lambda) = e^{\lambda H}Ae^{-\lambda H}.$$

From (8.3) we obtain immediately

$$Tr\left[\{\rho^{eq}, \Delta A\}\Delta B(t)\right] = Tr\left[\mathcal{Z}^{-1}\{e^{-\beta H}, A\}\Delta B(t)\right]$$

$$= Tr\left[\mathcal{Z}^{-1}e^{-\beta H}\int_0^\beta \dot{A}(-i\hbar\lambda)\Delta B(t)\,d\lambda\right]$$

$$= \int_0^\beta Tr\left[\mathcal{Z}^{-1}e^{-\beta H}\dot{A}(-i\hbar\lambda)\Delta B(t)\right]\,d\lambda,$$

which is what we wanted to prove. Note that we have used $\Delta A = A - \langle A\rangle_{eq}$, then $\{\rho^{eq}, \Delta A\} = \{\rho^{eq}, A\}$.

It is also possible to obtain alternative expressions for the response function $\phi_{BA}(t)$. Again, using Kubo's identity in (8.3) we can write $\phi_{BA}(t)$ in the form:

$$\phi_{BA}(t) = Tr\left[\mathcal{Z}^{-1}e^{-H\beta}\int_0^\beta d\lambda\,\Delta\dot{A}(-i\hbar\lambda)\Delta B(t)\right], \tag{8.6}$$

then we can explicitly express the operator $\Delta\dot{A}$ in imaginary time $-i\hbar\lambda$ as

$$\phi_{BA}(t) = \beta\,Tr\left[\frac{\rho^{eq}}{\beta}\int_0^\beta d\lambda\,e^{\lambda H}\Delta\dot{A}e^{-\lambda H}\Delta B(t)\right] \equiv \beta\langle\Delta\dot{A};\Delta B(t)\rangle. \tag{8.7}$$

The latter expression defines the canonical Kubo correlation $\langle \Delta\dot{A}; \Delta B(t)\rangle$, which in the classical limit corresponds to the standard correlation between classical objects $\Delta\dot{A}$ and $\Delta B(t)$.

Optional Exercise (**Classical Limit**). In a similar way as was done for proving (8.7) and using

$$\{e^{-H\beta}, Q\} = \beta e^{-H\beta}\dot{Q}, \tag{8.8}$$

show that in classical mechanics the response function can be written as:

$$\phi_{BA}(t) \rightarrow \beta \int \mathcal{D}p \int \mathcal{D}q\, f_{eq}\, \Delta\dot{A}\, \Delta B(t),$$

where we have replaced $Tr\, [\rho^{eq}B] \rightarrow \int \mathcal{D}p \int \mathcal{D}q\, f_{eq}(p,q)\, B(p,q)$.[11]

Relationship (8.8) is trivial to obtain as the classical limit ($\hbar \rightarrow 0$) of (8.4). Instead, given an arbitrary function Q which depends on the generalized coordinates q_i, conjugate moments p_i and time t, and using the equations of motion of functions $q_i(t)$ and $p_i(t)$ we can calculate the total time derivative in the form:

$$\frac{dQ}{dt} = \{Q, H\} + \frac{\partial Q}{\partial t}, \tag{8.9}$$

where, in general, for any arbitrary function f and g we define the operation:

$$\{f, g\} \equiv \sum_i \left(\frac{\partial f}{\partial q_i}\frac{\partial g}{\partial p_i} - \frac{\partial f}{\partial p_i}\frac{\partial g}{\partial q_i} \right),$$

as the Poisson bracket (see Appendix G). Using (8.9) show (8.8).

Guided Exercise. Prove that the response function $\phi_{BA}(t)$ given in (8.7) can be written alternatively as follows:

$$\phi_{BA}(t) = -\beta\langle \Delta A; \Delta\dot{B}(t)\rangle.$$

This expression follows immediately since

$$Tr\, [\rho^{eq}\{\Delta A, \Delta B(t)\}] = Tr\, [\rho^{eq}\Delta A\Delta B(t) - \rho^{eq}\Delta B(t)\Delta A]$$
$$= Tr\, [\Delta A\Delta B(t)\rho^{eq} - \rho^{eq}\Delta B(t)\Delta A]$$
$$= Tr\, [\Delta B(t)\rho^{eq}\Delta A - \rho^{eq}\Delta B(t)\Delta A]$$

---

[11]In the classical limit, $\int \mathcal{D}p \int \mathcal{D}q$ indicates an integration in phase space of canonical variables $\{p, q\}$ of N-body.

$$= Tr\left[\{\Delta B(t), \rho^{eq}\}\, \Delta A\,\right]$$

$$= Tr\left[\Delta A\; \{\Delta B(t), \rho^{eq}\}\right]$$

$$= -Tr\left[\Delta A\; \{\rho^{eq}, \Delta B(t)\}\right].$$

Excursus (**Wannier States**). On the other hand, it is also possible to write (8.7) in the form

$$\phi_{BA}(t) = \frac{-\partial}{\partial t} \int_0^\beta Tr\left[\Delta B(t) A(i\hbar\lambda)\rho_{eq}\right]\, d\lambda. \tag{8.10}$$

This equality follows cyclic permutation within the trace and using the fact that Kubo's canonical correlation $\langle\Delta A; \Delta\dot{B}(t)\rangle$ is invariant under temporal translation, i.e.:

$$\langle\Delta A(t_1); \Delta B(t_2)\rangle = \langle\Delta A(t_1 + \tau); \Delta B(t_2 + \tau)\rangle, \quad \forall\tau,$$

because the probability distribution ρ^{eq} is stationary.

The expression (8.10) is the starting point for the study of the generalization of Einstein's relation for localized particles in a semi-classical context [21]. That is, we proceed performing all quantum mechanical calculations involved in (8.10) but choosing a basis $\{| s >\}$ where the position operator $\hat{s}$ is diagonal, i.e.: $\hat{s} \,| s > = s \,| s >$.

8.3.3   Application to the Electrical Conductivity

If we particularize Theorem III to the calculation of the electrical conductivity we must consider that the observable A is proportional to the "displacement" of the particles $A = \sum_{\alpha=1}^{N} Q_\alpha \hat{s}_\alpha$, as the perturbation in the Hamiltonian is simply the electric potential (the force has mechanical character). Furthermore $B = \sum_{\alpha=1}^{N} Q_\alpha \, d\hat{s}_\alpha/dt$, because the electric current is proportional to the velocity of charged carriers. Whereupon from (8.7) and (8.2) we can identify the susceptibility $\phi_{BA}(\omega)$ with (*ac*) electric conductivity in frequency; that is, by defining a matrix $\sigma_{jl}(\omega)$ [the static conductivity in the *j*-direction will be characterized by the matrix element $\sigma_{jj}(0)$]. In general, we can write[12]

$$\sigma_{jl}(\omega) = \frac{1}{k_B T} \int_0^\infty \langle J_j; J_l(t)\rangle_0\, e^{i\omega t}\, dt, \tag{8.11}$$

---

[12]This expression is often called Green-Kubo formula, in tribute to those who deduced this formula independently (M.S. Green [15] and R. Kubo [18]).

where $\langle J_j; J_l(t)\rangle_0$ indicates Kubo's canonical correlation[13] (it is generally an N-body quantum correlation function). This expression is a widely applicable formula and it reduces to Einstein's expression for the dc-conductivity (static) in the case of a classical free electron gas scattered by impurity centers.

8.4 Conductivity in the Classical Limit

As mentioned above, the transport of excitations (electrons, phonons, etc.) is profoundly modified whenever spatial translational invariance is broken (impurities, faults, dislocations, etc.). If the degree of disorder is large enough, it gives rise to the phenomenon of localization of excitations. In this situation quantum transport cannot be described by equations of the Boltzmann-Bloch type. It is necessary to represent transport by *hopping* mechanisms, so that Wannier's states are more natural that the plane wave $\mathbf{k}$ representation [30].

To fix ideas, consider the problem of electrical conductivity in amorphous materials. It should be emphasized that we must first study the disordered system (around equilibrium), then analyze nonequilibrium disturbances (caused by the application of an electric field). This work plan is of formidable mathematical complexity, which is why the use of the Fluctuation-Dissipation theorem and semiclassical approximations are of vital importance.

The Fluctuation-Dissipation theorem establishes the relationship between susceptibility (linear system response) and fluctuations around thermodynamic equilibrium. In particular, for a system of independent particles[14] and in the classical limit,[15] the frequency response (8.11) can be written in $1D$, in the form[16]

$$\chi(\omega) = \beta \int_0^\infty \exp(i\omega t) \langle V(0)V(t)\rangle_0 \ dt, \quad \beta = \frac{1}{k_B T}. \tag{8.12}$$

---

[13]If a symmetrized correlation is defined instead of using Kubo's canonical correlation, expression (8.11) can be written as an integral from $-\infty$ to $+\infty$ in time. In this case, an explicit factor appears: $(1 - \exp(\beta\omega))/2\omega$, which it is directly related to the fluctuation-dissipation theorem [14, 19]. A critical comparison between Greenwood-Peierls dc-conductivity against the Green-Kubo formula is presented in [9].

[14]That is, we do not take into account the interaction between the charged carriers.

[15]The condition of validity for a classical approach is related to the analysis of the partition function $Tr\{e^{-\beta H}\}$ of the N-bodies system. For example, if we consider a quantum gas of independent particles, it is possible to show that if $\rho\lambda_T^3 \ll 1$ [where $\rho = N/\Omega$ is the density of particles and $\lambda_T = h(2\pi m k_B T)^{-1/2}$ the thermal wave length] the classical approach is acceptable; see, for example, D.N. Zubarev, in: *Nonequilibrium Statistical Thermodynamics*, New York, Plenum (1974).

[16]Note that in the static case ($\omega = 0$) we recover Einstein's relationship. A purely classical derivation can be found in Sect. 3.13; see also linear response theory in Sect. 5.4.

Here $\langle V(0)V(t)\rangle_0$ is the *stationary* velocity correlation function of a charged carrier (note that in equilibrium and with no external force $\langle V(t)\rangle_0 = 0$). Then the *ac*-electric conductivity is given by the general relationship

$$\sigma(\omega) = NQ^2 \chi(\omega). \tag{8.13}$$

Note that in this expression we have used the independent particles approximation, this is why the electric current is proportional to the number of charged carriers N; here Q represents the electric charge of the carrier. The fact that the charge appears squared is because the force on the carrier is also proportional to Q.

8.4.1   Conductivity Using an Exponential Relaxation Model

If the velocity correlation function is modeled by an exponential decay

$$\langle V(0)V(t)\rangle_0 = V_0^2 \exp(-|t|/\tau_r). \tag{8.14}$$

The susceptibility $\chi(\omega)$ can easily be calculated using the Laplace transform of the correlation (8.14), i.e.,

$$\mathcal{L}_u\left[\langle V(0)V(t)\rangle_0\right] = \frac{V_0^2}{1/\tau_r + u},$$

whence we conclude that

$$\chi(\omega) = \beta\mathcal{L}_u\left[\langle V(0)V(t)\rangle_0\right]_{u=-i\omega} = \frac{V_0^2 \beta \tau_r}{1 - i\omega\tau_r}. \tag{8.15}$$

From this expression for the susceptibility, it is possible to calculate the electric conductivity (frequency dependent) as the real part of $NQ^2\chi(\omega)$

$$\mathcal{R}_e\left[\sigma(\omega)\right] = \mathcal{R}_e\left[NQ^2 V_0^2 \beta \tau_r\left(\frac{1 + i\omega\tau_r}{1 + \omega^2\tau_r^2}\right)\right]. \tag{8.16}$$

In particular, the static conductivity σ_{dc} is obtained by taking the limit $\omega \to 0$ in (8.16)

$$\sigma_{\text{dc}} = \frac{NQ^2 V_0^2}{k_B T}\tau_r. \tag{8.17}$$

Here τ_r^{-1} represents the scattering cross section of the charged carriers with *impurities* and *phonons*.

From this expression it follows that the conductivity (8.17) has the expected T^{-1} behavior for semiconductors or metals at high temperatures, as well as the expected proportionality of the conductivity and the mean free path $l = V_0 \tau_r$.[17] The relationship (8.17) coincides with Einstein's formula for the electrical conductivity[18]

$$\sigma_{dc} = \frac{NQ^2}{k_B T} d, \tag{8.18}$$

where d is the diffusion coefficient of the material.

In amorphous systems the T^{-1} law is invalid, and the frequency dependence does not follow a Lorentzian form as predicted in (8.16). It is therefore necessary to have alternative models for the velocity correlation function. In general, in disordered or amorphous materials electrons are localized. For this reason, a model by conduction bands is invalid and we must introduce the concept of transport by jumps (*hopping*) between localized sites. In this case the Scher and Lax formula [29] for electric conductivity allows efficiently, to study the susceptibility $\chi(\omega)$ as a function of temperature through the analysis of fluctuations in the movement of charged carriers.

Excursus (**Kinetic Metal Theory**). If the temperature is low, the conductivity of metals can be studied using Boltzmann's kinetic theory and electron scattering by impurities and phonons[19] (in this case the wave vector **k** is a good quantum number for describing an electron in a crystal lattice). In particular, at low temperatures only low energy phonons are important, then it is possible to obtain the temperature dependence T^{-5} as it is observed in simple metals. However, in transition metals such as nickel, platinum, palladium, etc., the electron-electron scattering plays a role in limiting the current, then the observed conductivity behaves as $\sigma \sim T^{-2}$, instead of T^{-5}.

---

[17]If Boltzmann kinetic theory is used and we only take into account dispersion by impurities, the scattering will be elastic (the electron conserves energy). Moreover, at very low temperature, using $kT \ll \mathcal{E}_f$ [Fermi's energy], and the approximation that the relaxation time τ is isotropic (that is, τ is independent of the direction of the Bloch wave vector **k**), elementary kinetic theory predicts that the conductivity behaves as $\sigma = NQ^2 \tau / \Omega m$ (where Ω is the volume of the sample). That is, τ^{-1} is the probability—per unit time—that in a collision an electron will lose the momentum gained, $QE\tau$, due to the application of the external field E.

[18]If the Boltzmann kinetic theory is used [30] and we take into account phonon dispersion, the scattering will be inelastic (the electron does not conserve energy due to the emission or absorption of a phonon of energy $\hbar\omega_q$ from the lattice). Then, at low temperatures kinetic theory predicts that the conductivity behaves as $\sigma \sim T^{-5}$. Furthermore, at high temperature when the phonon distribution can be approximated by $P_q \sim kT/\hbar\omega_q$, Boltzmann's kinetic theory predicts the behavior $\sigma \sim T^{-1}$.

[19]The fact that we can "add separately" the contribution of scattering by impurities and phonons (i.e., the two processes can be considered incoherent) is known as Matthiessen's law, and has proved to be a good approximation [30].

8.5   Scher and Lax Formula for the Electric Conductivity

In the classical regime the starting point is again Eq. (8.12). If we define the function $d(\omega) = \chi(\omega)/\beta$,[20] using the cosine transform in (8.12) we have

$$\mathcal{R}_e\,[d(\omega)] = \int_0^\infty \cos(\omega t)\,\langle V(0)V(t)\rangle_0\,dt, \tag{8.19}$$

and, taking the inverse transform,

$$\langle V(0)V(t)\rangle_0 = \frac{2}{\pi}\int_0^\infty \cos(\omega t)\mathcal{R}_e\,[d(\omega)]\,d\omega.$$

Since $\mathcal{R}_e\,[d(\omega)]$ is an even function, we can also write (8.19) in the form

$$\langle V(0)V(t)\rangle_0 = \frac{1}{\pi}\int_{-\infty}^\infty \exp(i\omega t)\mathcal{R}_e\,[d(\omega)]\,d\omega. \tag{8.20}$$

At thermal equilibrium, $\langle V(0)V(t)\rangle_0$ is a stationary correlation, then it follows

$$\langle V(t')V(t)\rangle_0 = \frac{1}{\pi}\int_{-\infty}^\infty \exp(i\omega\,|\,t-t'\,|)\mathcal{R}_e\,[d(\omega)]\,d\omega. \tag{8.21}$$

Given the relation

$$X(t) - X(0) = \int_0^t V(s)\,ds,$$

we can write

$$\left\langle (X(t) - X(0))^2 \right\rangle = \left\langle \int_0^t ds \int_0^t ds'\, V(s')V(s) \right\rangle_0 = \int_0^t ds \int_0^t ds'\, \langle V(s')V(s)\rangle_0\,.$$

Then by using (8.21) and integrating twice we can obtain a relation for the variance of the displacement of the charged carrier:

$$\left\langle (X(t) - X(0))^2 \right\rangle = \int_0^t ds \int_0^t ds'\, \left\{ \frac{1}{\pi}\int_{-\infty}^\infty \exp(i\omega\,|\,s-s'\,|)\mathcal{R}_e\,[d(\omega)]\,d\omega \right\}$$

$$= \frac{1}{\pi}\int_{-\infty}^\infty \mathcal{R}_e\,[d(\omega)] \left\{ \int_0^t ds \int_0^t ds'\, \exp(i\omega\,|\,s-s'\,|) \right\}\,d\omega. \tag{8.22}$$

Now using

$$\int_0^t ds \int_0^t ds'\, \exp(i\omega\,|\,s-s'\,|) = \frac{|\,e^{i\omega t} - 1\,|^2}{\omega^2} = \frac{2[1 - \cos(\omega t)]}{\omega^2},$$

---

[20]Where $\beta \equiv 1/k_B T$.

in (8.22) we get

$$\left\langle (X(t) - X(0))^2 \right\rangle = \frac{2}{\pi} \int_{-\infty}^{\infty} \mathcal{R}_e \left[d(\omega) \right] \frac{\left[1 - \cos(\omega t) \right]}{\omega^2} \, d\omega,$$

and, since this whole argument is even, we can write

$$\left\langle (X(t) - X(0))^2 \right\rangle = \frac{4}{\pi} \int_{0}^{\infty} \mathcal{R}_e \left[d(\omega) \right] \frac{\left[1 - \cos(\omega t) \right]}{\omega^2} \, d\omega.$$

Defining the quantity $\frac{d}{dt} \left\langle (X(t) - X(0))^2 \right\rangle$, that is:

$$\frac{d}{dt} \left\langle (X(t) - X(0))^2 \right\rangle = \frac{4}{\pi} \int_{0}^{\infty} \mathcal{R}_e \left[d(\omega) \right] \frac{\sin(\omega t)}{\omega} \, d\omega,$$

we can invert in the sine transform the quantity $\mathcal{R}_e \left[d(\omega) \right] / \omega$:

$$\frac{2}{\omega} Re \left[d(\omega) \right] = \int_{0}^{\infty} \left\{ \frac{d}{dt} \left\langle (X(t) - X(0))^2 \right\rangle \right\} \sin(\omega t) dt . \tag{8.23}$$

Integrating by parts (8.23) and introducing a convergence factor e^{-at} for $a \to 0$ we finally obtain:

$$\mathcal{R}_e \left[d(\omega) \right] = -\frac{\omega^2}{2} \int_{0}^{\infty} \left\langle (X(t) - X(0))^2 \right\rangle \cos(\omega t) \, dt. \tag{8.24}$$

On the other hand, Kramers-Kronig's relations (see Chap. 5) require

$$\mathcal{I}_m \left[d(\omega) \right] = \frac{-1}{\pi} P \int_{-\infty}^{\infty} d\omega' \frac{\mathcal{R}_e \left[d(\omega') \right]}{(\omega' - \omega)} \tag{8.25}$$

$$\mathcal{R}_e \left[d(\omega) \right] = \frac{1}{\pi} P \int_{-\infty}^{\infty} d\omega' \frac{\mathcal{I}_m \left[d(\omega') \right]}{(\omega' - \omega)}. \tag{8.26}$$

Then

$$\mathcal{I}_m \left[d(\omega) \right] = \frac{-1}{\pi} P \int_{-\infty}^{\infty} d\omega' \frac{1}{(\omega' - \omega)} \left\{ -\frac{(\omega')^2}{2} \int_{0}^{\infty} \left\langle (X(t) - X(0))^2 \right\rangle \cos(\omega' t) \, dt \right\}$$

$$= \frac{1}{2\pi} \int_{0}^{\infty} dt \left\langle (X(t) - X(0))^2 \right\rangle \left\{ P \int_{-\infty}^{\infty} d\omega' \frac{(\omega')^2 \cos(\omega' t)}{(\omega' - \omega)} \right\}, \tag{8.27}$$

and since

$$P \int_{-\infty}^{\infty} d\omega' \frac{(\omega')^2 \cos(\omega' t)}{(\omega' - \omega)} = (-\frac{d^2}{dt^2}) P \int_{-\infty}^{\infty} d\omega' \frac{\cos(\omega' t)}{(\omega' - \omega)}$$

$$= (-\frac{d^2}{dt^2})[-\pi \sin(\omega t)] = -\pi \omega^2 \sin(\omega t),$$

we can rewrite (8.27) in the form:

$$\mathcal{I}_m [d(\omega)] = \frac{1}{2\pi} \int_0^{\infty} \left\langle (X(t) - X(0))^2 \right\rangle \{ -\pi \omega^2 \sin(\omega t) \} \; dt$$

$$= \frac{-\omega^2}{2} \int_0^{\infty} \left\langle (X(t) - X(0))^2 \right\rangle \sin(\omega t) \; dt,$$

which together with (8.24) can be grouped into the final expression:

$$d(\omega) = -\frac{\omega^2}{2} \int_0^{\infty} \left\langle (X(t) - X(0))^2 \right\rangle \exp(i\omega t) \; dt. \tag{8.28}$$

That is, if we introduce the Laplace transform of the variance of the displacement, the generalized diffusion coefficient, in $1D$, is written in the form

$$d(\omega) = \left[\frac{u^2}{2} \mathcal{L}_u \left\langle (X(t) - X(0))^2 \right\rangle \right]_{u=-i\omega}. \tag{8.29}$$

In general, in n-dimensions and for the isotropic case the expression in the Laplace variable is:

$$d(u) = \frac{1}{2n} \left[u^2 \mathcal{L}_u \left\langle (X(t) - X(0))^2 \right\rangle \right].$$

Then, using formula (8.29), we can express the ac-electrical conductivity as a generalized Einstein relation [29]:

$$\sigma(\omega) = NQ^2 \chi(\omega) = \frac{NQ^2}{k_B T} d(u)|_{u=-i\omega}. \tag{8.30}$$

Using this formula to calculate the electrical conductivity in amorphous materials, the task is reduced to the study of the generalized diffusion coefficient $d(u)$, which, for example, can be analyzed for different models of disorder by using the ME [16].

Optional Exercise (**Velocity-Velocity Correlation**). Suppose the Telegrapher equation describes transport in a given physical system. Using the relationship between CTRW theory and the generalized ME, calculate by using the Scher and Lax formula the velocity correlation function associated with this model. Hint: use the relation between the kernel of the generalized ME and the waiting-time function of the CTRW; see Sect. 7.3.1.

8.5.1 Susceptibility of a Lorentz Gas

As an example of application of the generalized Einstein relation (8.30), we calculate here the frequency dependent electric conductivity (or susceptibility) for a $1D$ Lorentz gas model. A Lorentz gas in the $1D$ lattice is a persistent RW [16]). It is called "persistent" because, depending on the correlation parameter c the direction of an RW jump in an arbitrary transition will be with high probability in the same direction as for the preceding jump if $c \in (0, 1)$. In the opposite case if $c \in (0, -1)$, successive transitions show a peculiar "bouncing back and forth" between two sites with little chance of moving (advancing) in any direction. To see this more clearly, we transcribe here the recurrence relation of a persistent RW $1D$.

From the recurrence relation for Markov chain with internal states, presented in Sect. 7.4.1, it follows that the $1D$ recurrence relations for a Lorentz gas are

$$P_{n+1}^1(x) = pP_n^1(x - 1) + qP_n^2(x - 1)$$

$$\text{(8.31)}$$

$$P_{n+1}^2(x) = qP_n^1(x + 1) + pP_n^2(x + 1).$$

Where, as before, the internal states $l = 1, 2$ of the RW represent "movement" on the x-axis in the positive or negative direction respectively. Here p (q) is the probability of spread "forward" ("backward") and also normalization $p + q = 1$ is satisfied. Moreover, the correlation parameter is defined as $c = p - q$.

Guided Exercise (**Boltzmann-Lorentz Equation**). In the first excursus of Sect. 7.3.3, we mentioned that the Lorentz gas model in the lattice can be obtained from the kinetic Boltzmann equation [11]. Using the linear Boltzmann equation for noninteracting classical particles in a fixed regular lattice of scattering centers, the distribution function of a particle $f(\mathbf{r}, \mathbf{v}, t)$, satisfies the kinetic equation (in $3D$):

$$\left\{\frac{\partial}{\partial t} + \mathbf{v} \cdot \nabla\right\} f(\mathbf{r}, \mathbf{v}, t) = -W(\mathbf{r}, \mathbf{v}) f(\mathbf{r}, \mathbf{v}, t) + \int W(\mathbf{r}, \mathbf{v}', \mathbf{v}) f(\mathbf{r}, \mathbf{v}', t) \, d\mathbf{v}', \qquad \text{(8.32)}$$

where $W(\mathbf{r}, \mathbf{v}) = \int W(\mathbf{r}, \mathbf{v}, \mathbf{v}') \, d\mathbf{v}'$ is the probability that the particle has a collision at the site $\mathbf{r}$, regardless of its final velocity. Using the dispersion model $W(\mathbf{r}, \mathbf{v}', \mathbf{v}) = q_0 \delta(\mathbf{v}' + \mathbf{v})$,[21] and introducing a discrete expression for the collision operator

$$\left\{\frac{\partial}{\partial t} + \mathbf{v} \cdot \nabla\right\} f(\mathbf{r}, \mathbf{v}, t) = \frac{1}{\Delta t} [f(\mathbf{r} + \mathbf{v}\Delta t, \mathbf{v}, t + \Delta t) - f(\mathbf{r}, \mathbf{v}, t)],$$

show that in $1D$ the Boltzmann-Lorentz equation is reduced to

$$f(r, v, t) = pf(r - v\Delta t, v, t - \Delta t) + qf(r - v\Delta t, -v, t - \Delta t). \qquad \text{(8.33)}$$

---

[21]That is, the scattering is independent of the spatial position $\mathbf{r}$.

Where we have used $q \equiv q_0 \Delta t$ and the normalization condition $(p = 1 - q)$. On the other hand, introducing a spatial and temporal discretization in the form

$$r_i + v_i \Delta t = r_{i+1}; \ t_i + \Delta t = t_{i+1},$$

from (8.33) follows (8.31).

Exercise (2-**Steps Non-Markovian Memory**). From the two recurrence relations (8.31), show that the Lorentz gas model in the $1D$ lattice can also be written as an RW with memory. That is, defining the marginal probability[22] $\mathcal{P}_n(x) = P_n^1(x) + P_n^2(x)$, this probability satisfies the recursion relation

$$\mathcal{P}_n(x) = p \left[\mathcal{P}_{n-1}(x-1) + \mathcal{P}_{n-1}(x+1) \right] - c \, \mathcal{P}_{n-2}(x), \quad c = (p-q) \qquad (8.34)$$

This means that (8.34) is a non-Markovian chain with a two-step memory. Note that if the correlation parameter c vanishes, it becomes the usual (symmetrical) RW.

Exercise (**Mesoscopic Preparation**). There is an interesting result that proves that for every Markov chain with internal states

$$P_{n+1}^j(\mathbf{s}) = \sum_l \sum_{s'} \Psi_{j,l}(\mathbf{s} - \mathbf{s}') P_n^l(\mathbf{s}'), \quad \text{with} \quad \mathbf{s} = (s_1, \cdots, s_D),$$

if the structure matrix $\Psi(\mathbf{k})$ is of dimension $m \times m$, the marginal probability $\mathcal{P}_n(\mathbf{s}) = \sum_l P_n^l(\mathbf{s})$ depends on m parameters, $\{\mathcal{P}_0(\mathbf{s}), \mathcal{P}_1(\mathbf{s}), \cdots, \mathcal{P}_{m-1}(\mathbf{s})\}$, which represent the mesoscopic preparation of the non-Markovian chain. It is also proved that the asymptotic behavior of the marginal probability $\mathcal{P}_n(\mathbf{s})$, for $n \to \infty$, does not depend on this preparation. Show that the solution of the recurrence relation (8.34) depends on two parameters [8].

Since in order to use the generalized Einstein's relation we need to calculate the dispersion $\langle x(t)^2 \rangle$ of the RW regardless its internal state, we are now only interested in the solution of (8.34). We can obtain this solution by the generating function technique $R(k, z) = \sum_{n=0}^{\infty} z^n \, \mathcal{P}_n(k)$,[23] where $\mathcal{P}_n(k)$ is the discrete Fourier transform of $\mathcal{P}_n(x)$.

Exercise (**Marginal Solution of Lorentz' Gas in** $1D$). From the memory recurrence relation (8.34) shows that the marginal solution of the persistent RW is given in terms of the generating function:

$$R(k, z) = \frac{1 - zc \cos(k)}{\left[1 - z(1+c) \cos(k) + cz^2 \right]}, \qquad (8.35)$$

---

[22]Independent of the internal states.
[23]See Sect. 6.2.1.

where we have used the following *marginal* initial conditions:

$$\mathcal{P}_{n=0}(k) = \sum_x e^{ikx} \delta_{x,0}$$

$$\mathcal{P}_{n=1}(k) = \frac{1}{2} \sum_x e^{ikx} (\delta_{x,1} + \delta_{x,-1}) = \cos(k),$$

which are equivalent to specifying the initial conditions with "internal states":

$$P_{n=0}^1(x) = P_{n=0}^2(x) = \frac{\delta_{x,0}}{2}.$$

The expression (8.35) is the solution we sought in the discrete time formulation.

From expression (8.35) and as explained in Sect. 7.3., the solution in the continuous-time formulation is obtained by replacing $z \to \psi(u)$, and taking the temporal convolution of $R(k, \psi(t))$ with the probability $\phi(t)$ to stay on the site without jumping during the time interval $[0, t]$. That is, in the Laplace representation $P(k, u)$ is given by the expression:

$$P(k, u) = \frac{1 - \psi(u)}{u} R(k, z = \psi(u)),$$

then from (8.35) we get

$$P(k, u) = \frac{1 - \psi(u)}{u} \frac{(1 - c\,\psi(u)\cos(k))}{[1 - (1 + c)\,\psi(u)\cos(k) + c\,\psi(u)^2]}. \tag{8.36}$$

From this it is immediate to calculate the expression for the second moment of the displacement of the persistent RW, i.e., in the Laplace representation we obtain

$$\langle x(u)^2 \rangle = \frac{-\partial^2}{\partial k^2} P(k, u) \Big|_{k=0} \tag{8.37}$$

$$= \frac{\psi(u)\,(1 + c\,\psi(u))}{u\,(-1 + \psi(u))\,(-1 + c\,\psi(u))}.$$

From here we can easily see that if the correlation parameter is zero, $c = 0$, the second moment coincides with the expression for the dispersion of a usual CTRW [note that in (8.37) we have taken the lattice parameter to be unity].

To use the generalized Einstein's relation (8.30) it is necessary to specify a model for the waiting-time function $\psi(u)$. Assume here an exponential model for the probability density of pausing times between successive steps[24]

$$\psi(u) = \frac{1}{1 + \tau u},\tag{8.38}$$

then τ is the average waiting-time between each jump in the CTRW.

Substituting (8.38) in (8.37)' and taking the inverse Laplace transform, the following time behavior for the dispersion of a persistent CTRW is obtained:

$$\langle x(t)^2 \rangle = \frac{(1+c)}{(1-c)} \frac{t}{\tau} + \frac{2c}{(1-c)^2} \left[\exp\left(-(1-c)\frac{t}{\tau} \right) - 1 \right].\tag{8.39}$$

Exercise (**Non-diffusive Limits**). From the dispersion (8.39) show that the limits $c \to \pm 1$ are not diffusive situations. That is, $c \to 1$ is the ballistic limit; whereas $c \to -1$ corresponds to a collapse of the realizations. After a transient of the order $\tau/(1-c)$ the variance approaches a linear diffusive regime.

From the same substitution it is easily seen that the generalized diffusion coefficient $d(u)$ is given by

$$d(u) = \frac{u^2}{2} \langle x(u)^2 \rangle = \frac{1 + c + \tau u}{2\tau (1 - c + \tau u)}.\tag{8.40}$$

This result shows that the response of a persistent RW; that is, its susceptibility is frequency dependent $\propto d(u = -i\omega)$. This marks a noticeable difference with the "white" answer for the usual RW model (or Wiener process in continuous space). This structure, in the susceptibility is the result of including the degree of freedom corresponding to the *velocity* in the description of an RW.[25] This is clearly seen when the persistent RW is derived from the kinetic Boltzmann-Lorentz model; see (8.33). From (8.40) we find that the real part of the susceptibility behaves as

$$\mathcal{R}_e[\chi(\omega)] = \beta \, \mathcal{R}_e[d(u = -i\omega)].$$

Then the normalized conductivity to zero frequency has the expression:

$$\frac{\chi_1(\omega)}{\chi_1(0)} = \left(\frac{1-c}{1+c} \right) \frac{\omega^2 + \tau^{-2}(1-c^2)}{\omega^2 + \tau^{-2}(1-c)^2}.$$

---

[24]Then, the CTRW behaves as a Markov process; see the analysis of the associated ME [Sect. 7.3.1]. This is equivalent to saying that the model corresponds to a discrete-time RW in an "orderly" lattice, where the probability of making n jumps in the time interval $[0, t]$ is given by the Poisson probability $P(n) = \frac{1}{n!}(\frac{t}{\tau})^n \exp(-t/\tau)$; see advanced exercises 7.5.1(d) and 7.5.4, or references [16, 31].

[25]That is, the statistical description of an RW with internal states has a nontrivial structure that is present even if we only are interested in knowing the marginal probability: $\mathcal{P}(k, u) = \sum_l P^l(k, u)$.

Fig. 8.1 Normalized
conductivity $\chi(\omega)/\chi(0)$ as a
function of ω. Continuous
lines correspond to the mean
waiting-time $\tau = 1$, while the
dashed lines correspond to
$\tau = 1/3$

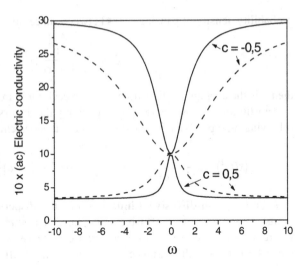

The coefficient of static diffusion is related to the high frequency limit by a
correlation factor:

$$\chi_1(0) = \left(\frac{1+c}{1-c}\right)\chi_1(\infty).$$

In Fig. 8.1 the real part of the (normalized) susceptibility is shown as a function
of the frequency,[26] for different values of the correlation parameter c and mean
time τ. From this graph we see that when the RW is persistent; that is, $c > 0$,
the conductivity increases as the frequency approaches zero. This is so because
the realizations of the RW are less random, they are more directionally "stretched"
when the persistent effect dominates. In the anti-persistent case, i.e., when $c < 0$,
the effect is the opposite, for the realizations tend to "collapse" and thus the static
coefficient of diffusion decreases. Conversely, for high frequencies, the conductivity
shows saturation behavior for both cases. Dependence on τ is reflected in the mean
"width" of the structure (peak): the shorter the mean waiting-time τ between jumps
the wider the "nonwhite" structure response of the system.

On the other hand, the imaginary part of the susceptibility is given by

$$\chi_2(\omega) = \frac{2c\beta\omega\tau}{(\tau\omega)^2 + (1-c)^2},$$

that is, dissipation is proportional to the correlation coefficient c.

Figure 8.2 shows $\chi_2(\omega)/\beta$ as a function of ω, for different values of the
correlation parameter c and mean time τ.

---

[26]For reasons of elegance only the part corresponding to $\omega < 0$ is represented.

Fig. 8.2 Normalized imaginary part of the susceptibility $\chi(\omega)/\beta$ as a function of ω. *Continuous lines* correspond to the mean time $\tau = 1$, while *dashed lines* correspond to $\tau = 1/3$

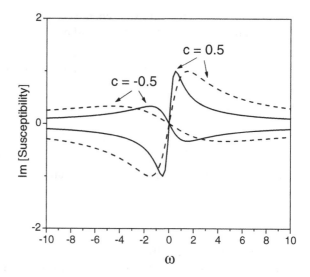

8.5.2   Fick's Law (Static Limit)

We have already mentioned that in the limit of zero frequency, Scher & Lax's formula reduces to an elegant relationship between the diffusion coefficient (static) and the velocity correlation function. This relationship is merely the static limit of Kubo's formula for electrical conductivity, obtained from the linear response theory (8.12).

Furthermore, defining the static diffusion coefficient, d, as the coefficient of proportionality between the particle flow $j(\mathbf{x}, t)$ and the gradient of concentration $\nabla n(\mathbf{x}, t)$ (Fick's law):

$$j(\mathbf{x}, t) = -d\,\nabla n(\mathbf{x}, t), \tag{8.41}$$

if the medium is homogeneous, and combining Fick's law with particle conservation:

$$\frac{\partial n(\mathbf{x}, t)}{\partial t} = -\nabla \cdot j(\mathbf{x}, t),$$

the diffusion equation for concentration follows:

$$\frac{\partial n(\mathbf{x}, t)}{\partial t} = d\,\nabla^2 n(\mathbf{x}, t). \tag{8.42}$$

Then, using (8.42), the diffusion coefficient can be calculated by a functional relationship[27] (for example, for the x coordinate)[28]:

$$\mathcal{D}_x(t) = \frac{1}{2}\frac{d}{dt}\left\langle [x(t) - \langle x(t)\rangle]^2\right\rangle. \tag{8.43}$$

In general, the diffusion coefficient is defined as the asymptotic limit of $\mathcal{D}_x(t)$, i.e.:

$$d_x = \lim_{t\to\infty}\mathcal{D}_x(t),$$

which is equivalent to our previous demonstration in Sect. 3.13.

$$d_x = \lim_{t\to\infty}\int_0^t \langle V_x(0)V_x(\tau)\rangle_0\, d\tau, \tag{8.44}$$

that is, the static limit (8.19).

Depending on the velocity correlation function the integral (8.44) will converge or not. In particular, if d_l (for $l = x, y, z$) vanishes, we said that the process is subdiffusive,[29] whereas if the integral diverges the process is called superdiffusive. Only if the integral converges and also $d_l \neq 0$ will we be asymptotically in the presence of a diffusive process.

Exercise. Taking the Laplace transform of (8.43) and (8.44) [without taking the limit $t \to \infty$], show that the velocity correlation function is given by

$$\mathcal{L}_u\left[\langle V_x(0)V_x(t)\rangle_0\right] = \frac{u^2}{2}\mathcal{L}_u\left[\left\langle [x(t) - \langle x(t)\rangle]^2\right\rangle\right].$$

Guided Exercise (**Particles in the Presence of a Magnetic Field**). The diffusion of charged particles in the presence of a constant magnetic field **B** is an issue of interest in plasma physics [2]. In particular, the velocity correlation function $\langle V_j(0)V_j(\tau)\rangle_0$ can be modeled considering an **sde** which includes the Lorentz force (oriented in the direction perpendicular to **B**), a dissipative term (proportional to the velocity), and the influence of stochastic accelerations from collisions with particles of the medium (Langevin's forces). From expressions[30] for the velocity correlations with components perpendicular and parallel to **B**, we calculate here the

---

[27]It should be noted that if the process is subdiffusive, the temporal behavior of this function does not match that obtained from the formula $d_x = \lim_{t\to\infty}\langle x(t)^2\rangle/2t$. However, one hopes that both coincide asymptotically, but the precise duration of this transient regime is an unknown factor.

[28]Integrating by parts twice and taking the surface terms at infinity as zero. In the anisotropic case we have to study separately the dispersion for each coordinate axis. For the nonhomogeneous case, see advanced Exercise 4.9.5.

[29]Note that for the integral to vanish the correlation function must take positive and negative values so that the area is zero.

[30]See Sect. 5.3.2.

perpendicular and *parallel* static diffusion coefficients using (8.19). If the field **B** is oriented in the $\hat{z}$ direction, we get that $d_\perp = d_x = d_y$, and the perpendicular diffusion coefficient is given by

$$
\begin{aligned}
d_\perp &= \lim_{t\to\infty} \int_0^t \langle V_\perp(0)V_\perp(\tau)\rangle_0 \; d\tau \\
&= \lim_{t\to\infty} \int_0^t \frac{V_T^2}{2} \exp(-v\,|\,\tau\,|)\,\cos\,(\Omega\tau)\;d\tau \\
&= \lim_{t\to\infty} \left[\frac{V_T^2}{2}\frac{v + (\Omega \sin \Omega t - v \cos \Omega t)\,e^{-vt}}{v^2 + \Omega^2}\right] \\
&= \frac{V_T^2}{2}\frac{v}{v^2 + \Omega^2}.
\end{aligned}
$$

While the expression for the diffusion coefficient in the parallel direction is

$$
\begin{aligned}
d_z &= \lim_{t\to\infty} \int_0^t \langle V_z(0)V_z(\tau)\rangle_0 \; d\tau \\
&= \lim_{t\to\infty} \int_0^t \frac{V_T^2}{2} \exp(-v\,|\,\tau\,|)\;d\tau \\
&= \lim_{t\to\infty} \left[\frac{V_T^2}{2}\frac{1 - e^{-vt}}{v}\right] \\
&= \frac{V_T^2}{2v},
\end{aligned}
$$

where $V_T^2 = 2k_BT/m$. The difference in behavior of the diffusion coefficients as a function of dissipative parameter v is precisely due to the helical dynamics of the realization of the particles along the field direction $\hat{z}$ in the presence of the magnetic field **B**.

Exercise. Study the diffusion coefficient $d_\perp$ in the regime of strong magnetic field; that is, $\Omega/v \gg 1$.

Excursus (**Lorentz' Model with Multiplicative Noise**). In the advanced exercise 3.20.10 we have introduced the dynamics of a Lorentz' model in the presence of a *random charge*. Consider a particle with velocity $\vec{V}$ in a magnetic field in the $\hat{z}$ direction, using Lorentz' force and assuming that the charge is $\varepsilon(t) = q_0 + q(t)$ where $q(t)$ is a discrete stochastic process, show that the perpendicular velocity satisfies the **sde**:

$$
\dot{V}_x(t) = \varepsilon(t)\,V_y(t), \qquad \dot{V}_y(t) = -\varepsilon(t)\,V_x(t).
$$

Calculate the correlation function $\langle\langle V_\perp(t)\,V_\perp(t')\rangle\rangle_0$ and the perpendicular diffusion coefficient $d_\perp = \lim_{t\to\infty}\int_0^t \langle\langle V_\perp(0)V_\perp(\tau)\rangle\rangle_0 \; d\tau$.

8.5.3 Stratified Diffusion (the Comb Lattice)

The study of anomalous diffusion in non-Euclidean networks has been a continuous and inexhaustible source of inspiration to create different models of transport. In particular, the comb lattice model has proven to be very useful in the study of transport networks having dead ends (*Dead fingers*). In a "comb" lattice it is postulated, for example, that the particle stream in the $\hat{x}$ direction is only possible if $y = 0$; that is, using the notation $(\mathbf{x}, t) \equiv (x, y, t)$ we have for each component of the current $j(\mathbf{x}, t)$ that[31]

$$j_x(\mathbf{x}, t) = -d_x\, \delta(y)\, \frac{\partial n(\mathbf{x}, t)}{\partial x}, \quad j_y(\mathbf{x}, t) = -d_y \frac{\partial n(\mathbf{x}, t)}{\partial y}.$$

Then, according to Fick's scheme [see (8.41) and (8.42)], it follows that the concentration satisfies

$$\left[\frac{\partial}{\partial t} - d_x\, \delta(y)\, \frac{\partial^2}{\partial x^2} - d_y \frac{\partial^2}{\partial y^2} \right] n(\mathbf{x}, t) = 0.$$

Exercise. By using the corresponding Green function for diffusion on the y-axis, that is:

$$\frac{\partial n(\mathbf{x}, t)}{\partial t} = d_y \frac{\partial^2 n(\mathbf{x}, t)}{\partial y^2},$$

consider the term $d_x\, \delta(y)\, \partial_x^2 n(\mathbf{x}, t)$ as a "source"; then, obtain an integral equation for the concentration $n(\mathbf{x}, t)$. Show that the nonlocal equation for the concentration $n(x, y = 0, t)$ is

$$n(x, 0, t) = n(x, 0, 0) + \frac{d_x}{\sqrt{\pi d_y}} \frac{\partial^2}{\partial x^2} \int_0^t \frac{n(x, 0, t')}{\sqrt{t - t'}} dt'.$$

Calculate the second moment of the displacement and show that $\langle x(t)^2 \rangle = d_x \sqrt{t/d_y}$, i.e., subdiffusion.[32] Hint: to simplify this integral equation use the Laplace transform of the convolution, and the results: $\mathcal{L}_u \left[1/\sqrt{t} \right] = \sqrt{\pi/u}$ and $\mathcal{L}_u \left[\partial_t f(t) \right] = u f(u) - f(t = 0)$, see also advanced exercise 8.8.5.

Exercise. In relation to the previous exercise on the comb network, we heuristically can check the aforementioned subdiffusion behavior $\langle x(t)^2 \rangle \sim t^{1/2}$ if we consider a $1D$ model in the CTRW approach [i.e., diffusion on the axis x] in which the waiting-time function $\psi(t)$ [at each integer site x] is characterized by

---

[31]This problem can also be studied in a lattice, see last optional exercise in Appendix H.2.
[32]Note that dimensions of coefficients d_x and d_y are different.

Fig. 8.3 Scheme representing a comb network. The *circle* represents an RW making an excursion in the direction *y* before returning to the origin (intersection with the *x* axis)

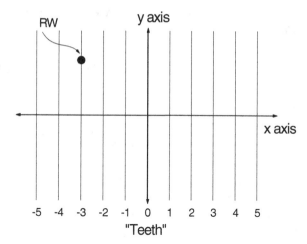

the probability density for return to the "origin" [each time the RW leaves the x axis for an excursion in the y direction]. In Fig. 8.3 a schematic representation of a comb network is shown. Show that using $\psi(t) \sim t^{-3/2}$ [see Sect. 7.3.2] this implies (asymptotically) a CTRW with $\psi(u) \sim 1 - u^{1/2}$, from which we can deduce the mentioned subdiffusive behavior.

Excursus (**Disordered Comb Lattice**).   The study of transport in a comb lattice with teeth of different lengths leads to the problem of a random comb lattice. This problem can also be studied in the context of an ME, then EMA or equivalent approaches can easily be applied; see Appendix H.2.

8.5.4   Diffusion-Advection and the CTRW Approach

The study of standard diffusion in a shear flow is of great applicability, in particular the analysis of anomalous diffusion-advection motion is an interesting problem that can also be tackled in the context of the CTRW approach. We start with the standard diffusion-advection equation for diffusion in a fluid moving along the $\hat{x}$ direction with constant velocity υ:

$$\frac{\partial \rho(\mathbf{x}, t)}{\partial t} + \upsilon \frac{\partial \rho(\mathbf{x}, t)}{\partial x} = d \, \nabla^2 \rho(\mathbf{x}, t), \qquad (8.45)$$

in Fourier representation this equation becomes

$$\frac{\partial \rho(\mathbf{k}, t)}{\partial t} = -\left(ik_x \upsilon + dk^2 \right) \rho(\mathbf{k}, t). \qquad (8.46)$$

In the context of the generalized ME for a CTRW process we have shown in Chap. 7 that

$$\frac{dG_{CTRW}(k,t)}{dt} = \int_0^t \Phi(k, t-\tau) \, G_{CTRW}(k,\tau) \, d\tau. \tag{8.47}$$

with

$$\Phi(k,u) = \frac{u\left[\psi(k,u) - \psi(u)\right]}{1 - \psi(u)}, \tag{8.48}$$

Thus Eq. (8.46) might be understood in the framework of the CTRW if the function $\psi(k,u)$ is written in terms of a "resting" waiting-time function $\psi_0(k,u)$ times the advection factor, that is

$$\psi(k,u) = \exp\left(-i\tau_a \mathbf{k} \cdot \mathbf{v}\right) \psi_0(k,u). \tag{8.49}$$

Therefore, consider in particular a Markovian model for the resting system [12]

$$\psi_0(k,u) = \frac{\exp\left(-\sigma^2 k^2\right)}{(1 + u\tau)},$$

the kernel (8.48) adopts the form

$$\Phi(k,u) = \frac{1}{\tau}\left[\exp\left(-i\tau_a \mathbf{k} \cdot \mathbf{v} - \sigma^2 k^2\right) - 1\right].$$

To retain just the essential properties of the diffusion-advection equation which manifest macroscopically, we take the limits $\tau \to 0, \tau_a \to 0, \sigma \to 0$ keeping σ^2/τ =constant and τ_a/τ =constant. After taking these limits and rescaling of time and space, we get the diffusion-advection equation (8.46). Non-Markovian models using $\psi(u) = 1/(1+u^\gamma)$, as well as Levy-like models (for the resting waiting-time) $\psi_0(k,u) = \exp\left(-\sigma^2 k^{2\beta}\right)/(1 + u\tau)$—leading to a generalized Laplacian term like $\nabla^{2\beta}$ for $\beta \in (0, 1)$—can also be studied using this approach.

In the presence of a nonhomogeneous velocity field $\mathbf{v}(x)$ the diffusion-advection equation can also be studied in the context of the CTRW considering a generalization of (8.49) in the form

$$\psi(k,u) = \psi_0(k,u) \int dx' e^{-i\mathbf{k}\cdot\mathbf{x}'} \exp\left(-i\tau_a \mathbf{k} \cdot \mathbf{v}\left(x'\right)\right),$$

then many interesting flow motions can be studied [12], in particular the situation when the flow has a random characteristic can be analyzed by perturbation theory [13].

8.6 Anomalous Diffusive Transport (Concluded)

8.6.1 The CTRW Approach

Transport models that include the disordered nature of a lattice can be considered in the context of the CTRW theory using different waiting-time functions $\psi(u)$; in particular, those which are not analytical, around $u = 0$, in the case of strong disorder.[33] Overall, given a CTRW propagator, the inclusion of strong disorder[34] can be done parametrically using a waiting-time function with "long tail". Asymptotically, for u small, a function with "long tail" in the Laplace representation behaves as $\psi(u \sim 0) \approx 1 - C u^{\theta}$, $0 < \theta < 1$ [the constant C should not to be confused with mean time between jumps, as in the case $\theta = 1$; see Sect. 7.3]. Introducing the behavior of $\psi(u \sim 0)$ into (8.30), we can analytically obtain the electric conductivity of the disordered system at low frequencies.

We can classify the anomalous behavior of the CTRW propagator considering the regime of small $\{k, u\}$. For a separable CTRW model in arbitrary dimension $\psi(k, u) = \lambda(k)\psi(u)$, if the jumps structure behaves for $k \to 0$ as

$$\lambda(k \sim 0) \approx 1 - \text{const.} \mid k \mid^{\mu}, \quad 0 < \mu < 2, \tag{8.50}$$

a non-Gaussian profile is asymptotically obtained. This profile is related to the Lévy probability distribution if $\psi(u)$ does not have a long-tail [see excursus in Sects. 3.14, 6.2.3 and Appendix H.2]. Remember that a Lévy distribution does not have a defined second moment; often CTRW models based on jump structures like (8.50) are called superdiffusive processes. However, note that throughout this text we have reserved the classification *superdiffusion* for the temporal analysis of the second moment of the RW.[35]

On the other hand, if the waiting-time function behaves for $u \to 0$ as

$$\psi(u \sim 0) \approx 1 - \text{const.} u^{\theta}, \quad 0 < \theta < 1, \tag{8.51}$$

we get for long times a subdiffusive profile, as long as $\lambda(k)$ is analytic around a $k = 0$; i.e., when the one-step distribution does not have a long-tail as in (8.50). In Chap. 7, we noted that the evolution equation of this subdiffusive profile is controlled by a non-Markovian character depending on the parameter θ. The memory kernel $\Phi(t)$ in the separable case is identified (in the Laplace representation) with the generalized diffusion coefficient $\Phi(u)$.

---

[33]The CTRW theory allows in a very simple way to introduce internal states that are useful in describing transport of complex systems.

[34]We said that the disorder is strong when the anomalous behavior of the variance is subdiffusive.

[35]It is often desirable to classify stochastic processes in which the second moment is not defined in terms of fractional moments; see Appendix H.2.

Finally, we mention, in Sect. 7.3.4, that some CTRW models with non-separable structure $\psi(k, u)$, can lead to stochastic processes that have superdiffusion [i.e., $\langle s(t)^2 \rangle \sim t^\eta$, for $t \to \infty$, with $\eta > 1$]. In addition diffusion-advection models can be studied in the framework of the CTRW approach generalizing the structure function $\psi(k, u)$ with an *advection factor*; see previous section.

Exercise (**Scaling Relations**). A subdiffusion model can be represented by a CTRW propagator if the jump structure is regular $\lambda(k \sim 0) \approx 1 - C_2 \mid k \mid^2$, and the waiting-time function has an asymptotic behavior of the form (8.51). Then, the $G_{\text{CTRW}}(k, u)$ takes the asymptotic expression

$$G_{\text{CTRW}}(k, u) = \frac{1 - \psi(u)}{u\left[1 - \lambda(k)\psi(u)\right]} \tag{8.52}$$

$$\approx \frac{C_1 u^\theta}{u\left[C_1 u^\theta + C_2 \mid k \mid^2\right]}$$

$$= \frac{1}{u\left[1 + C_2 \mid k \mid^2 (C_1 u^\theta)^{-1}\right]}, \quad k \to 0, u \to 0, \theta \in (0, 1).$$

Show that the CTRW propagator in Fourier-Laplace representation satisfies the scale relationship

$$\frac{1}{\Lambda} G_{\text{CTRW}}(\frac{k}{\Lambda^H}, \frac{u}{\Lambda}) = G_{\text{CTRW}}(k, u), \quad \text{with } H = \theta/2. \tag{8.53}$$

Compare with the scale relationship introduced in Sect. 3.11 using the $\{k, t\}$ representation. Then, we can define the following fractal dimensions for this **sp** [for a derivation of these quantities see Appendix H].

$$D_B = 2 - \theta/2, \quad D_D = 2/\theta, \quad \text{with } \theta \in (0, 1).$$

Remember that in the present case, the CTRW represents a *mean* **sp** over the disorder. In particular, D_D matches the definition of the fractal dimension d_w, introduced in order to study anomalous diffusion in non-Euclidean lattices by analyzing the behavior of the second moment of an RW on a fractal network [see Appendix H.2].

Alternatively, the CTRW theory allows to study the transient of the propagator if we have a closed expression for $\psi(u), \forall u$.

Exercise (**Transient**). Consider an arbitrary solution of a separable CTRW and study the transient of its propagator if the jump structure $\lambda(k)$ is symmetrical and the waiting-time function is $\psi(u) = a/[a + u(u + \lambda)]$. Note that using this model we are considering that "the residence time" at an arbitrary site of the lattice is the order $\tau_r \approx \lambda/a$; while "the flight-time" is the order $\tau_f \approx a^{-1/2}$. Hint: see advanced exercise 7.5.2.

8.6.2 The Self-Consistent Technique (EMA)

In the context of the theory of effective medium approximation the starting point is the ME, in which we can explicitly introduce different models of disorder. Then, in the EMA scheme the problem reduces to the calculation of a self-consistent effective medium $\Gamma(u)$.[36] From this result and applying the generalized Einstein's relation (8.30), an expression for the electric conductivity in different models of disordered systems is obtained. In general, in the context of EMA the effective propagator $G^{\mathrm{eff}} = [u\mathbf{1} - \mathbf{H}_{\mathrm{eff}}(u)]^{-1}$ is simply the homogeneous propagator where we have replaced the diffusion constant μ by the effective medium $\Gamma(u)$. In this text we have discussed, but not explicitly developed, that the EMA technique can also be used in the presence of drift; in this case and due to the asymmetry in some coordinate axis two effective media [one in favor and one against the external field] will appear. If the system presents anisotropy and asymmetry, the problem becomes much more complex, but in principle, a self-consistent system of coupled equations needs to be solved [6, 27].

Note that throughout the text we have introduced two distinct notations to represent the same object $d(u)$, depending on the type of approach that we are using. In the CTRW approximation we used $\Phi(u)$, whereas when we use the EMA technique we talk about $\Gamma(u)$; see Sects. 7.2 and 7.3.

Exercise. In the context of the EMA, anomalous diffusion can be represented by the effective propagator G^{eff}; for example, in 1D, without bias, and in the Fourier-Laplace representation we get:

$$G^{\mathrm{eff}}(k, u) = [u + \Gamma(u)\,(1 - \cos k)]^{-1}. \tag{8.54}$$

Consider a model of strong (bond) disorder in which $\rho(W) = (1 - \alpha)W^{-\alpha}$; then, study asymptotically the scale relationship

$$\frac{1}{\Lambda} G^{\mathrm{eff}}(\frac{k}{\Lambda^H}, \frac{u}{\Lambda}) = G^{\mathrm{eff}}(k, u).$$

Show that the fractal dimension D_D matches the expression $d_w = (2 - \alpha)/(1 - \alpha)$; see Appendix H.2.

Excursus (**Asymptotic Profile**). In general, the inverse transform of any of the propagators $G^{\mathrm{eff}}(k, u)$ or $G_{\mathrm{CTRW}}(k, u)$ [see expressions (8.54) or (8.52) respectively] is not a simple task; however, using concepts of scale variables it is possible to prove that asymptotically the CTRW profile of concentration behaves like a *stretched exponential,* and only in the limit of normal diffusion is a Gaussian regime for the profile obtained [2].

---

[36]Or the frequency dependent diffusion coefficient.

Throughout this text we have presented alternately, using the techniques of CTRW or AME, several conclusions on transport for disordered media. It is worth mentioning here that each of these techniques has its advantages and disadvantages, which we have clearly exposed in several of our discussions. In general, we can say that the EMA technique gives the correct critical exponents that characterize anomalous transport in different models of disorder. Moreover, it has been discussed [Chap. 7] that the EMA is the first contribution of a singular perturbation theory. The CTRW approach originally "conceived" in a different context does not reproduce the correct mean-square-displacement exponent, nevertheless this approach can be used to describe more than just subdiffusive anomalous transport. In particular, we have shown how to extend this approach to characterize superdiffusion, the transport of a complex system with internal states, and disordered diffusion-advection models.

In conclusion it should be noted that in the context of a theory of diffusion in disordered media, in the limit of long times, two very different results appear, depending on the strength of the disorder. If the disorder is strong, the generalized diffusion coefficient is zero: $d(u) \rightarrow 0$ for $u \rightarrow 0$; while this does not happen if the disorder is weak, i.e.: $\lim_{u \rightarrow 0} d(u) \rightarrow$ const. In the latter case, for sufficiently long times, all information on the disorder appears in redefining a new diffusion constant. In general, for weak disorder, we can say that the transport coefficients are finite and must be "renormalized" depending on the disorder parameters; but if the disorder is strong, the "temporal" laws are substantially different, as is the case for subdiffusion. The universal (or not) time behavior of the transport processes in the presence of disorder is a fascinating field of research and continuous study.

Excursus (**Conductivity in Superionic Materials**). The study of electric conductivity in superionic materials has shown that ion-ion interaction plays a crucial role for a correct description of transport in these materials [28]. In particular, the use of the generalized Einstein's relation and the theory of anomalous diffusion [in the CTRW scheme] allows modeling the aforementioned interaction [4, 7].

Excursus (**Conductivity in Granular Metal Materials**). Granular metal materials which exhibit the phenomenon of reentrance of superconductivity can be modeled by disordered systems in which hopping transport competes with tunneling. Thus, using the EMA and the generalized Einstein relation this problem can be analyzed as a function of the temperature and the random size of the metal grains, to study the phenomenon of reentrant superconductivity in thin granular films [25].

Excursus (**Ionic Conductivity in Granular Materials**). The characteristics of ionic conductivity in amorphous solid electrolyte materials are of industrial interest. In this context it should be noted that the stochastic transport scheme is very useful for understanding some of the important mechanisms for the description of these materials [5]. Using the generalized Einstein's relation it is possible to study the electric conductivity in granular ionic materials; in particular, these materials exhibit an anomalous transport due to the high conductance interface which appears between the conducting material and insulating particles dispersed in a second phase. This system can be studied in the context of EMA, considering a mixture of two "phases" (conductor and insulator) and placing special emphasis on the analysis of the different configurations of paths of high conductance, which occur due to such interface [10].

Excursus (**Transport with Drift**). The characteristics of electric conductivity beyond the linear regime present problems of theoretical and experimental interest. The transport coefficients for disordered systems that we have studied throughout this text are strictly limited to the linear response regime. In the presence of a strong external field it is necessary to study these transport coefficients explicitly considering an asymmetric character for the transition probabilities in the field direction. That is, we have to study an ME with biased transition probabilities. This problem leads to new and interesting questions about the universal behavior of an RW for different types of disorder and intensity of external field [3, 6, 34].

8.6.3 Fractional Derivatives*

Consider now the diffusion equation (8.42), but written in spherical coordinates. In this case the quantity of interest is the conditional probability:

$$P(r, \theta, \phi, t \mid r_0, \theta_0, \phi_0, t_0).$$

If we assume that there is rotational symmetry (θ, ϕ), the probability only depends on the modulus $r \in [0, \infty]$. Writing the Laplacian ∇^2 in spherical coordinates we obtain after removal of the angular variables, the diffusion equation for the particle concentration $n(r, t)$ in the variable r

$$\frac{\partial n(r, t)}{\partial t} = \left[\frac{d}{r^{D-1}} \frac{\partial}{\partial r} r^{D-1} \frac{\partial}{\partial r} \right] n(r, t), \quad \forall D = 1, 2, 3. \tag{8.55}$$

Note that if $D = 1$ we recover the diffusion equation in one-dimension (for $r > 0$). Furthermore, we note that the probability of finding the particle within $[r, r + dr]$ is proportional to the quantity $r^{D-1} n(r, t)$; for example, in 3-dimensions it is given by $4\pi r^2 n(r, t)$.

Exercise (**Transformation of Coordinate**). Using the transformation theorem for rvs we know that $\Pi(x, y, z) = J \Pi(r, \theta, \phi)$, where J is the Jacobian of the transformation [see Sect. 1.13], then interpret (8.55) in terms of the **rv** r.

Optional Exercise (**Moments of the Modulus**). Show that the moments of the modulus r in an Euclidean space of dimension D behave as $\langle r^{2q}(t) \rangle \propto t^q, \forall q > 0$, independently of the dimension.

Excursus (**Diffusion on Fractal Lattices**). The interesting point in using spherical coordinates (8.55) is that it allows to approach the study of diffusion in a fractal geometry. In this context we mention the work of B. O'Shaughnessy and I. Procaccia [23], in which the diffusion equation on a non-Euclidean space is generalized to fractional dimension [see advanced exercise 8.8.6 and Appendix H]. On the other hand, it is important to note that the CTRW approach allows to reinterpret its propagator solution, so that it is possible to establish a connection with the analysis of diffusion on a fractal structure [2, 3, 16, 17].

Optional Exercise (**Fractional Structures**). We mentioned that the CTRW propagator (8.52) is completely characterized if we know the function $\psi(k, u)$. Consider the following separable models that asymptotically are given by

$$\psi(k, u) \sim 1 - C_1 u^\theta - \frac{\sigma^2}{2} \mid k \mid^2, \quad \text{with} \quad 0 < \theta < 1 \quad \text{(subdiffusion } nD)$$

$$\tag{8.56}$$

$$\psi(k, u) \sim 1 - \tau u - C_2 \mid k \mid^\mu, \quad \text{with} \quad 0 < \mu < 2 \quad \text{(Lévy RW } nD), \tag{8.57}$$

it is clear that both models lead to non-Gaussian profiles. The concern now is to determine the relationship between these structure models and the concept of fractional derivative. Given the relationship between the memory kernel of the generalized ME and the function $\psi(k, u)$ [see Sect. 7.3.1] show that asymptotically:

$$\Phi(k, u) \sim -\frac{\sigma^2}{2\tau} u^{1-\theta} \mid k \mid^2, \quad \text{(subdiffusion)}$$

$$\Phi(k, u) \sim -\frac{C_2}{\tau} \mid k \mid^\mu, \quad \text{(Lévy RW)},$$

and so the evolution equation can be put in the form:

$$u G_{\text{CTRW}}(k, u) - G_{\text{CTRW}}(k, t = 0) \approx \Phi(k, u) G_{\text{CTRW}}(k, u).$$

Noting that in Fourier and Laplace representations, we have the following results

$$\mathcal{L}_u \left[\frac{\partial^\alpha f(t)}{\partial t^\alpha} \right] = u^\alpha f(u)$$

$$\mathcal{F}_k [\nabla^\beta f(r)] = \mid k \mid^\beta f(k),$$

see for example: S.G. Samko, A.A. Kilbas, and O.I. Marichev, *Fractional Integrals and Derivatives, Theory and Applications*, New York, Gordon; and also Appendix E.1. Prove that an evolution equation with fractional derivative for the CTRW can asymptotically be written in the form:

$$\frac{\partial^\theta G_{\text{CTRW}}}{\partial t^\theta} \propto \nabla^2 G_{\text{CTRW}}, \quad \text{with} \quad 0 < \theta < 1 \quad \text{(subdiffusion)}$$

$$\frac{\partial G_{\text{CTRW}}}{\partial t} \propto \nabla^\mu G_{\text{CTRW}}, \quad \text{with} \quad 0 < \mu < 2 \quad \text{(Lévy RW)}.$$

8.7 Transport and Mean-Value Over the Disorder

In an attempt to describe transport on non-Euclidean lattices, in the last 40 years different approaches have been introduced. Some of these techniques have been presented in the second part of this text, as possible methods of addressing the problem of transport in disordered media. However, although we have not discussed many other valuable tools such as scaling techniques, renormalization group approach, Monte Carlo dynamics, etc., we cannot fail to mention here that they have been extremely effective in tackling the problem of disorder; see, for example, references cited thereon [3, 17].

Having a closed expression for the **mv** over the disorder of the propagator allows us also to study other responses besides the electric conductivity in disordered media. For example, relaxation, transients of the system, non-Markovian effects, spatial structure function,[37] etc., and in general, many other important properties for the understanding of transport in amorphous systems. Some of these details are beyond the scope of this text; however, they can be understood using the different mathematical techniques presented here.[38]

Moreover, the various generating and characteristic functions that we have widely used along several chapters facilitate the study of other physical problems in which the quantity of interest also involves taking averages over disorder. A particular case is the study of the structure factor involved when one is interested in calculating the intensity of X-rays scattered by a disordered material.[39]

Excursus (**X-Ray Scattering in Disordered Superlattices**). The structural characteristics of solid state matter can also be studied by the X-ray scattering technique, it is interesting to note that the theoretical predictions about the average intensity of the diffracted X-rays by a disordered lattice are only consistent with experimental results if models and averages involved are adequately considered.[40] In this context the generating functions technique has proved to be very useful in obtaining analytical expressions for calculating X-ray scattering in disordered superlattices [26].

Finally, it should be emphasized that when studying transport in quantum disordered systems, the average over the disorder is made on a physical observable; i.e., over the expectation value of some observable. This involves the use of the

---

[37]The incoherent quasielastic dynamic structure function is related to the RW propagator $P(k, u)$ through the expression $S_{\text{inc}}(k, \omega) = \frac{1}{\pi} \mathcal{R}_e [P(k, u = i\omega)]$.

[38]See for example references: [3, 16, 33].

[39]Note that the electric field of an incident plane wave in free space recalls the shape of a characteristic function: $E(\mathbf{r}) = E_0 \, e^{i\mathbf{k}\cdot\mathbf{r}}$.

[40]See, for example, E.E. Fullerton, I.K. Schuller, H. Vanderstraeten and Y. Bruynseraede, Phys. Rev. B 45, 9292 (1992).

density matrix[41] [which, ultimately, entails averaging simultaneously over two Green's functions]. Therefore this mathematical problem is even more complex than the one we have studied over the last chapters. However, the EMA technique— presented in Chap. 7—provides us with a first approach towards understanding this fundamental quantum problem.[42]

8.8   Additional Exercises with Their Solutions

8.8.1   Quantum Notation*

Suppose we have two wave functions Φ and Ψ and that we want to work with some operator Q, the operator is said to be Hermitian when

$$\int \Phi^* Q \Psi dr = \int (Q\Phi)^* \Psi dr, \tag{8.58}$$

here the integral is over some region of the space where the system is defined, and we also have to specify the boundary conditions fulfilled by functions Φ and Ψ. In general the operator Q may involve space derivatives (as in the case of momentum operator: $\mathbf{p} = -i\hbar\nabla$). Show that $\mathbf{p}$ is a Hermitian operator.

In quantum mechanics when working with an *observable* operator Q that is not explicitly time dependent, it is usual to write the *symbol dQ/dt*, but which we should interpret as is the operational formula:

$$\frac{dQ}{dt} = \frac{i}{\hbar}(HQ - QH) \equiv \frac{i}{\hbar}[H, Q]. \tag{8.59}$$

To prove this formula we proceed considering Schrödingers equation valid for both wave functions; then, for example

$$\frac{\partial \Psi}{\partial t} = \frac{-i}{\hbar}H\Psi \tag{8.60}$$

$$\frac{\partial \Phi^*}{\partial t} = \frac{i}{\hbar}(H\Phi)^*, \tag{8.61}$$

---

[41]In the linear response approximation, Kubo's formula is the starting point for addressing this problem. In Appendix G we present in detail the relationship between this formula and the equilibrium density matrix.

[42]The perturbation introduced by various impurities in a Tight-Binding Hamiltonian can be studied using the effective medium technique. In solid state theory this technique is called Coherent Potential Approximation (CPA) and it is used to calculate the density of states [the CPA can be seen in: E.N. Economou, *Green Functions in Quantum Physics*, 2nd ed., Berlin, Springer Verlag (1983); A. Gonis, *Green Functions for Ordered and Disordered Systems*, Amsterdam, North Holland (1992)]. In the case when the impurities vary with time, for example for temperature effects, this problem can also be addressed with similar mathematical techniques to the one presented in this text; see, for example: A.K. Chattah and M.O. Cáceres, J. Phys. A Math. and Gen. **34**, 5795 (2001) and references therein.

here as usual $(\cdots)^*$ means the complex conjugate, and H is the Hermitian Hamiltonian of the system. Let us have some Hermitian operator Q (with no explicit time dependence); then, using (8.58), (8.60), and (8.61) we get

$$\frac{d}{dt}\int \Phi^* Q\Psi dr = \int \frac{\partial \Phi^*}{\partial t} Q\Psi dr + \int \Phi^* Q \frac{\partial \Psi}{\partial t} dr$$

$$= \int \frac{i}{\hbar} (H\Phi)^* Q\Psi dr + \int \Phi^* Q \left(\frac{-i}{\hbar} H\Psi\right) dr$$

$$= \frac{i}{\hbar} \int \Phi^* HQ\Psi dr - \frac{i}{\hbar} \int \Phi^* QH\Psi dr$$

$$= \frac{i}{\hbar} \int \Phi^* (HQ - QH) \Psi dr.$$

Then defining the operator dQ/dt in the form

$$\int \Phi^* \frac{dQ}{dt} \Psi dr = \frac{d}{dt} \int \Phi^* Q\Psi dr,$$

the operational relation (8.59) follows.

Show by explicit substitution that $Q(t) = e^{+itH/\hbar} Q e^{-itH/\hbar}$ is a solution of equation $\dot{Q} = \frac{i}{\hbar} [H, Q]$. Therefore, evolution in imaginary time gives $Q(-i\hbar\beta) = e^{+\beta H} Q e^{-\beta H}$.

By using the series expansion of the exponential, show that if Q is Hermitian, e^{iQ} is not.

Let A, B be two operators, using series expansion show that

$$Ae^{iB} = e^{iB} A$$

$$e^{i(A+B)} = e^{iA} e^{iB},$$

only if A and B commute. If A and B do not commute; that is, $[A, B] = AB - BA = C$, and if C commutes with A and B we can use the following formula

$$e^{(A+B)} = e^{C/2} e^B e^A = e^A e^B e^{-C/2}$$

All these expressions are useful for going through Sect. 8.3, in particular to get the Pauli ME.[43]

---

[43] The proof of the exponential relation $e^{(A+B)}$, as well as Pauli's evolution equation for the diagonal elements of the density matrix (by using time-dependent perturbation theory) can be found in many books of quantum mechanics. Concerning time-dependent perturbation theory see Appendix G.1. An extended analysis of these subjects can be seen in the book: *Principles of Magnetic Resonance*, by C.P. Slichter, Springer-Verlag, Berlin, 1978.

8.8.2 Classical Diffusion with Weak Disorder

The description of a diffusion process in the presence of weak disorder can easily
be considered in the context of the CTRW approach using a waiting-time function
analytic around $u \sim 0$. A two-parameters pdf that fits this condition is the
Gamma pdf

$$\psi\,(t) = \frac{1}{\tau}\left(\frac{t}{\tau}\right)^{b-1}\frac{\exp\,(-t/\tau)}{\Gamma\,(b)}, \quad \tau > 0, b > 0. \tag{8.62}$$

Note that this density becomes sharp in the limit $b\tau^2 \to 0$, and in the limit $b \to \infty$,
$\tau \to 0$ with $\tau b \to$const. it becomes singular $\psi\,(t) \to \delta\,(t - \tau b)$. For values $b < 1$
the waiting-time (8.62) diverges in the limit $t \to 0$, on the other hand for $b > 1$ the
diffusion coefficient is renormalized by the disorder and turns out to be smaller that
in the Markovian case. For $b = 1$ we recover the Markovian description for an RW
(characterized by a Poisson number of steps n during the time interval $[0, t]$).

The Laplace transform of the pdf (8.62) is

$$\psi\,(u) = \frac{1}{(1 + \tau u)^b}. \tag{8.63}$$

Show that all moments are well defined, in particular

$$\langle t \rangle = \tau b$$
$$\langle t^2 \rangle = (1 + b)\,b\tau^2, \text{ etc.}$$

The kernel of the generalized ME associated with the CTRW process is

$$\Phi\,(u) = \frac{u}{(1 + \tau u)^b - 1}. \tag{8.64}$$

In Fig. 8.4 we show the memory kernel in the time representation, for a weak
disorder model (8.62) through a numerical integration of the inverse Laplace
transform of (8.64). Show that the generalized diffusion coefficient, in the static limit
$\Phi\,(u = 0)$, vanishes for $b \to \infty$ keeping τ fixed. Using the waiting-time (8.63) in
the context of *renewal* theory, calculate the asymptotic long-time limit of the mean
number of events (steps) n; see advanced exercise 7.5.1(d).

8.8.3 Fractal Waiting-Time in a Persistent RW

From the second moment of an RW, calculate the asymptotic time behavior of
the variance of a persistent RW when the waiting-time function represents strong
disorder and has a fractal characteristic:

$$\psi\,(u) = \frac{1}{1 + u^\gamma}, \quad \gamma \in (0, 1), \tag{8.65}$$

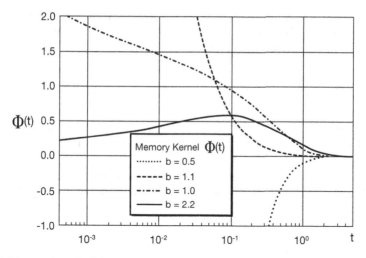

Fig. 8.4 Memory kernel of the generalized ME of a CTRW when the waiting-time function is a Gamma dpf. Typical time behavior, in dimensionless time, of $\Phi(t)$ for a waiting-time corresponding to weak disorder (8.62) with $\tau = 1$ and for different values of positive b. For the case $b < 1$ the memory kernel is not always positive, and it is not plotted for $t < 0.3$ due to numerical problems in the integration. From the initial value theorem: $\Phi(t \to 0^+) \geq 0$

note that from the asymptotic behavior $\psi(u \sim 0) \approx 1 - u^\gamma$, from Tauberian theorem it follows that $\psi(t \to \infty) \sim t^{-1-\gamma}$; see advanced exercise 7.5.1(c). Introducing $\psi(u)$ from (8.65) in the Laplace representation of the second moment, (8.37), we get

$$\langle x(u)^2 \rangle = \frac{\psi(u)\,(1 + c\,\psi(u))}{u\,(-1 + \psi(u))\,(-1 + c\,\psi(u))}$$

$$= \frac{1}{u^{\gamma+1}} \left(\frac{c + 1 + u^\gamma}{-c + 1 + u^\gamma} \right),$$

then, once again using Tauberian theorem it is possible to study asymptotically the behavior of $\langle x(t \to \infty)^2 \rangle$, interpret the results in terms of the physical parameters of the model. Calculate for a persistent RW model, the susceptibility at low frequency assuming the presence of a strong disorder like the one modeled in (8.65).

8.8.4 Abel's Waiting-Time Probability Distribution

Abel was probably the first to give an application of fractional calculus.[44] He used derivatives of arbitrary order to solve the isochrone problem in classical mechanics,

---

[44]N.H. Abel, *"Solution de quelques problems á l'aide díntegrales défínies"*, Werke **1**, 10, (1823). Fractional derivatives can be seen in [22]. Applications of the Abel's waiting-time can be seen in: A.A. Budini and M.O. Cáceres, J. Phys A Math. and Gen. **37**, 5959, (2004).

and the integral equation he worked out was precisely the one Riemann used to define fractional derivatives. It is interesting to note that the particular class of normalized one-side stable (Lévy) type of probabilities

$$\psi(t) = \frac{\tau^{\theta}}{\Gamma(\theta)} t^{-1-\theta} \exp(-\tau/t), \text{ with } \tau > 0, t > 0, \theta > 0, \tag{8.66}$$

are solutions of a class of fractional differential equations of the Abel type. In particular it is easy to see that the integers q–moments of the random variable t are finite if $\theta > q$.

From now on let us refer to an Abel distribution in honor of the great mathematician. The Laplace transform of the probability density (8.66) is straightforward to calculate:

$$\psi(u) = \frac{2}{\Gamma(\theta)} \left(\sqrt{u\tau}\right)^{\theta} K_{\theta}\left(2\sqrt{u\tau}\right), \tag{8.67}$$

where $K_{\theta}(z)$ is the Basset function [22]. Therefore using,

$$K_{\theta}(x) \simeq \frac{\Gamma(\theta) x^{-\theta}}{2^{1-\theta}} + \frac{\Gamma(-\theta) x^{\theta}}{2^{1+\theta}}, \ 0 < \theta < 1, \ x \sim 0,$$

it follows that the asymptotic behavior of $\psi(u \sim 0)$ is given by

$$\psi(u \sim 0) \simeq 1 - \frac{\pi \csc(\pi\theta)}{\theta \, \Gamma(\theta)^2} (\tau u)^{\theta} + \cdots, \text{ with } \theta \in (0, 1),$$

which is of the form used for strong disorder (8.51). The interesting point to mention is that (8.67) allows us to study, for all u, waiting-time models of strong disorder, and in addition we can "reduce" the disorder strength by increasing θ, i.e., depending on the value of θ we can obtain non-divergent integer moments of the random waiting times

$$\langle t^q \rangle = \int_0^{\infty} t^q \, \psi(t) \, dt = \tau^q \frac{\Gamma(\theta - q)}{\Gamma(\theta)}, \text{ if } \theta > q. \tag{8.68}$$

Using Abel's model of waiting-time density $\psi(t)$, the memory kernel in the Laplace representation becomes [using Eq. (8.48) for a separable model]

$$\Phi(u) = \frac{u\left(\sqrt{\tau u}\right)^{\theta} K_{\theta}(2\sqrt{\tau u})}{\frac{\Gamma(\theta)}{2} - \left(\sqrt{\tau u}\right)^{\theta} K_{\theta}(2\sqrt{\tau u})}, \text{ with } \theta > 0. \tag{8.69}$$

From the properties of the Laplace transform it is possible to check that for this model of strong disorder the area of the memory kernel is null, i.e., the static

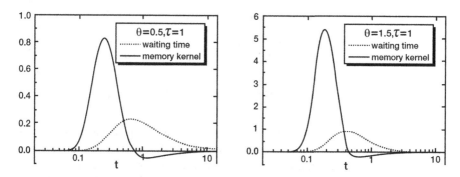

Fig. 8.5 Typical time behavior (dimensionless) of the memory kernel $\Phi(t)$ using the Abel probability distribution (8.66) to characterize strong disorder for different θ and $\tau = 1$. The memory kernel is not always positive and shows a maximum; note that in the case $1 < \theta < 2$ the kernel shows a sharper peak leading to a well defined first moment. The plot also shows the corresponding waiting-time probability densities $\psi(t)$

diffusion coefficient is zero $\Phi(u = 0) = 0$. Show also that the short and long time limits are easily obtained

$$\Phi(t = 0) = \lim_{u \to \infty} u\Phi(u) \to 0^+,$$

and

$$\Phi(t = \infty) = \lim_{u \to 0^+} u\Phi(u) \to 0^+.$$

After taking numerically the inverse Laplace transform of (8.69), in Fig. 8.5 we have shown some typical cases for the memory kernel $\Phi(t)$, for $\theta = 1/2$ (all integer moments diverge) and $\theta = 3/2$ (only the first moment is finite). In addition we have also plotted the corresponding Abel's densities $\psi(t)$.

8.8.5   Nonhomogeneous Diffusion-Like Equation

A nonhomogeneous partial differential equation can be written in its integral representation using the associated Green function. Consider a diffusion-like equation with an extra linear annihilation term (the pdf initial condition is given: $n(x, t = 0) \geq 0, \ \forall x \in [-\infty, \infty]$)

$$\partial_t n(x, t) = D\partial_x^2 n(x, t) - f(x, t) n(x, t), \ D > 0, \ n(x, t) \geq 0, \ f(x, t) \geq 0, \ \forall t \geq 0.$$
$$(8.70)$$

Let $G(x, t | x', t')$ be solution of the Green problem:

$$\left[\partial_t - D\partial_x^2\right] G(x, t | x', t') = \delta(x - x') \delta(t - t').$$
$$(8.71)$$

Show that an integral representation of Eq. (8.70) can be written in the form

$$n(x,t) = \int_0^\infty dt' \int_{-\infty}^\infty G\left(x,t\,|\,x',t'\right) n\left(x',t'\right) \delta\left(t'\right) dx' \tag{8.72}$$

$$- \int_0^\infty dt' \int_{-\infty}^\infty G\left(x,t\,|\,x',t'\right) f\left(x',t'\right) n\left(x',t'\right) dx',$$

here $G\left(x,t\,|\,x',t'\right) = \Theta\left(t-t'\right) P\left(x,t\,|\,x',t'\right)$, where $\Theta\left(z\right)$ is the step function (causality) and $P\left(x,t\,|\,x',t'\right)$ the Wiener propagator. By direct substitution prove that $G\left(x,t\,|\,x',t'\right)$ is a solution of (8.71). Using the integral representation (8.72) write an iterative solution for $n\left(x,t\right)$ up to second order in $f\left(x,t\right)$. Consider now the case when $f\left(x,t\right)$ is a space-time stochastic field, a simple one could be: $f\left(x,t\right) = \delta\left(x - W\left(t\right)\right)$ where $W\left(t\right)$ is some **sp**, it may be a Wiener or a dichotomic **sp**, etc. A much more involved model for $f\left(x,t\right)$ could be the *Random Rectifier* process of Sect. 3.20.13, but the problem is hard to handle. Write an approximation for $\langle n\left(x,t\right)\rangle$, here the *average* means the **mv** over all realizations of the stochastic space-time process: $f\left(x,t\right)$. Similar situations have been worked out considering disorder in diffusion models.[45]

8.8.6 Diffusion with Fractional Derivative

The interesting point in the alternative description of diffusion using spherical coordinates is that it allows to reinterpret equation (8.55) so that it is possible to approach the study of diffusion in a fractal geometry. Prove that (8.55) can be written in the equivalent form:

$$\partial_t^{1/2} n(r,t) = \left[\frac{\partial}{\partial r} n(r,t) + \frac{D-1}{2r} n(r,t) \right], \quad \forall D = 1,3 \tag{8.73}$$

Hint: apply the operator $\partial_t^{1/2}$ to both sides of Eq. (8.73) to recover (8.55); the fractional operator $\partial_t^{1/2}$ can be seen for example in [22]. Note that for dimension $D = 2$ the Eq. (8.73) is only an asymptotic approximation for $r^2/\langle r(t)^2\rangle \to \infty$.

---

[45]See: *"Diffusion in a continuous medium with space-correlated disorder"*, A. Valle, M.A. Rodríguez, and L. Pesquera, Phys. Rev. A, 43, 984, (1991); and also related references using the Terwiell expansion technique for the perturbation in references [3–5, 7] of Chap. 7.

8.8.7 Anomalous Diffusion-Advection Equation

In the context of the CTRW approach we have shown that the diffusion-advection equation can be written asymptotically in the form

$$\frac{dG_{\text{CTRW}}(k, t)}{dt} = \int_0^t \Phi(k, t - \tau) \, G_{\text{CTRW}}(k, \tau) \, d\tau. \tag{8.74}$$

with

$$\Phi(k, u) = \frac{u \left[\psi(k, u) - \psi(u)\right]}{1 - \psi(u)}$$

$$\psi(k, u) = \exp\left(-i\tau_a \mathbf{k} \cdot \mathbf{v}\right) \psi_0(k, u)$$

where $\psi(k, u)$ is written in terms of a "resting" waiting-time $\psi_0(k, u)$ times the advection factor (here the external *velocity* vector is $\mathbf{v}$).

Show that in the case of using a long-tail waiting-time function $\psi(u) \simeq 1 - u^\theta$, the evolution equation for (sub) diffusion-advection problems in a velocity *vector field* $\mathbf{v}(\mathbf{x})$ can be written asymptotically in the form [12]:

$$\frac{\partial G_{\text{CTRW}}(\mathbf{x}, t)}{\partial t} + \nabla \cdot \left[\mathbf{v}(\mathbf{x}) \frac{\partial^{1-\theta} G_{\text{CTRW}}(\mathbf{x}, t)}{\partial t^{1-\theta}}\right] = d \, \nabla^2 \left[\frac{\partial^{1-\theta} G_{\text{CTRW}}(\mathbf{x}, t)}{\partial t^{1-\theta}}\right], \ \theta \in (0, 1),$$
$$\tag{8.75}$$

Using the result of Appendix E.1 show that Eq. (8.75) corresponds asymptotically to a fractional time-derivative diffusion-advection equation of the form

$$\frac{\partial^\theta G_{\text{CTRW}}(\mathbf{x}, t)}{\partial t^\theta} \simeq -\nabla \cdot \left[\mathbf{v}(\mathbf{x}) G_{\text{CTRW}}(\mathbf{x}, t)\right] + d \, \nabla^2 G_{\text{CTRW}}(\mathbf{x}, t).$$

References

1. R. Aliki, K. Lendi, *Quantum Dynamical Semigroups and Applications*. Lectures Notes in Physics, vol. 286 (Springer, Berlin, 1987)
2. R. Balescu, *Statistical Dynamics* (Imperial College Press, London, 1997)
3. J.P. Bouchaud, A. Georges, Anomalous diffusion in disordered media: statistical mechanisms, models physical applications. Phys. Rep. **195**, 127 (1990)
4. C.B. Briozzo, C.E. Budde, M.O. Cáceres, Continuous-time random-walk model for superionic conductors. Phys. Rev. A **39**, 6010 (1989)
5. A. Bunde, W. Dieterich, E. Roman, Dispersed ionic conductors and percolation theory. Phys. Rev. Lett. **55**, 5 (1985)
6. S. Bustingorry, E.R. Reyes, M.O. Cáceres, Biased diffusion in anisotropic disordered systems. Phys. Rev. E **62**, 7664 (2000); S. Bustingorry, M.O. Cáceres, E.R. Reyes, Effective-medium approximation with asymmetric transition rates. Phys. Rev. B **65**, 165205 (2002)
7. M.O. Cáceres, On the problem of free-jump diffusion with sublattice disorder. Phys. Scripta **37**, 214 (1988)

8. M.O. Cáceres, C.E. Budde, The continuous-time resolvent matrix for non-markovian chains. Phys Lett. A **125**, 369 (1987)
9. M.O. Cáceres, S. Grigera, Concerning the microscopic linear response theory. Phys. A**291**, 317 (2001)
10. M.O. Cáceres, E.R. Reyes, AC conductivity in dispersed ionic conductors: the effective medium approximation. Phys. A **227**, 277 (1996)
11. M.O. Cáceres, H.S. Wio, Non-Markovian diffusion-like equation for transport processes with anisotropic scattering. Phys. A**142**, 563 (1987)
12. A. Compte, Stochastic foundations of fractional dynamics. Phys. Rev. E **53**, 4191 (1996); A. Compte, Continuous time random walks on moving fluids. Phys. Rev. E **55**, 6821 (1997)
13. A. Compte, and M.O. Cáceres, Fractional dynamics in random velocity fields. Phys. Rev. Lett. **81**, 3140 (1998)
14. R.P. Feynman, A.R. Hibbs, *Quantum Mechanics and Path Integrals* (Mc Graw-Hill Book Company, New York, 1965)
15. M.S.J. Green, Brownian motion in a gas of noninteracting molecules. Chem. Phys. **19**, 1036 (1951)
16. J.W. Haus, K.W. Kehr, Diffusion in regular and disordered lattices. Phys. Rep. **150**, 141 (1987)
17. S. Havlin, D. Ben-Avraham, Diffusion in disordered media. Adv. Phys. **36**, 695 (1987)
18. R. Kubo, Statistical-mechanical theory of irreversible processes. I. General theory and simple applications to magnetic and conduction problems. J. Phys. Soc. Jpn. **12**, 570 (1957)
19. R. Kubo, M. Toda, N. Hashitsume, *Statistical Physics II: Nonequilibrium Statistical Mechanics* (Springer, Berlin, 1985)
20. M. Lax, Generalized mobility theory. Phys. Rev. **109**, 1921 (1958); idem, Rev. Mod. Phys. **32**, 25 (1960)
21. T. Odagaki, M. Lax, Coherent-medium approximation in the stochastic transport theory of random media. Phys. Rev. **24**, 5284 (1981)
22. K.B. Oldham, J. Spanier, *The Fractional Calculus* (Academic, New York, 1974); J. Sapnier, K.B. Oldham, *An Atlas of Functions* (Springer, Berlin/Heildelberg, 1987)
23. B. O'Shaughnessy, I. Procaccia, Phys. Rev. A **32**, 3073 (1985)
24. W. Pauli, *Collected Scientific Papers by Wolfgang Pauli*, vol. 1, ed. by R. Kronig, V.F. Wiesskopf (Interscience, New York, 1964)
25. P.A. Pury, M.O. Cáceres, Tunneling percolation model for granular metal films. Phys. Rev. B **55**, 3841 (1997), and reference therein
26. C.A. Ramos, M.O. Cáceres, D. Lederman, X-ray scattering in disordered superlattices: theory and application to FeF2/ZnF2 superlattices. Phys. Rev. B **53**, 7890 (1996)
27. E.R. Reyes, M.O. Cáceres, P.A. Pury, The nonisotropic effective medium approximation for diffusion problems in random media. Phys. Rev. B **61**, 308 (2000)
28. H. Sato, R. Kikuchi, Cation diffusion and conductivity in solid electrolytes. I. J. Chem. Phys. **55**, 677 (1971)
29. H. Scher, M. Lax, Stochastic transport in a disordered solid. I. theory. Phys. Rev. B **7**, 4491 (1973)
30. P.L. Taylor, *A Quantum Approach to the Solids State* (Prentice-Hall, Englewood Cliffs/New York, 1970)
31. N.G. van Kampen, *Stochastic Process in Physics and Chemistry*. 2nd edn. (North-Holland, Amsterdam, 1992)
32. N.G. van Kampen, A new approach to noise in quantum mechanics. J. Stat. Phys. **115**, 1057 (2004)
33. G.H. Weiss, *Aspects and Applications of the Random Walk* (North-Holland, Amsterdam, 1994)
34. K.W. Yu, R. Orbach, Frequency-dependent hopping conductivity in disordered networks in the presence of a biased electric field. Phys. Rev B **31**, 6337 (1985)

Chapter 9
Metastable and Unstable States

Nonlinear systems away from equilibrium exhibit a variety of instabilities when the appropriate control parameters are changed. Due to such changes of control parameters the system may reach a stationary state which is not globally stable. One of the phenomena in which statistical fluctuations play a crucial role in nonequilibrium descriptions is the transient dynamics associated with the relaxation from states that have lost their global stability due to changes of appropriate control parameters. A relevant quantity in the characterization of the relaxation dynamics is the lifetime of such states, i.e., the random time that the system takes to leave the vicinity of the initial state.

Historically, this kind of problem was first studied by Kramers [1]. He was interested in computing the rate of escape of a particle across a barrier due to thermal fluctuations. To tackle this problem he first derived a two-dimensional **F–P** equation associated with the phase-space mechanical description of a particle in interaction with a thermal bath. Thus the stochastic variables of the problem: position X and velocity V were characterized by the conditional probability distribution that we discussed in Chap. 4 (see also Sect. 4.7.3). By introducing a large viscosity approximation he was able to derive a one-dimensional **F–P** equation for the position stochastic variable X (the Smoluchowski equation for the reduced problem). Kramers' treatment is based on an ansatz that is valid when the barrier height of the mechanical problem is larger than the thermal energy. Applications of Kramers' approach can be used in the dynamics of an equilibrium or nonequilibrium phase transition, where the variable X would be the order parameter while the potential would be the free energy-like functional [2]. Explicit examples of Kramers' approach can also be found in the analysis of the rotational Brownian motion of a single domain magnetic particle in an anisotropic potential [3], and in general chemical reactions [4].

Kramers' method is based on Einstein's theory of Brownian motion, so we will first focus on his pioneer classic work [1]. An alternative approach in computing the rate of escape is to estimate the mean time for a stochastic process, within a

© Springer International Publishing AG 2017
M.O. Cáceres, *Non-equilibrium Statistical Physics with Application to Disordered Systems*, DOI 10.1007/978-3-319-51553-3_9

given region, to first reach the boundary of that region. The statistics of these times is described by the First Passage Time Distribution (FPTD) that we have already presented in Sect. 6.6, and the Mean First Passage Time (MFPT) is identified with the lifetime of the initial state as we will show below. There are standard techniques for calculating the MFPT for Markov processes [4, 5]. Another alternative route to these techniques focuses on the individual stochastic path of the process and extracts the FPTD from some approximations of these paths [6, 7]. This is the so-called stochastic path perturbation approach which can also be generalized to tackle non-Markov processes [8], non-Gaussian noises [9], and stochastic differential equations with distributed time-delay [10]. From a practical point of view the stochastic path perturbation approach is useful in the calculation of the MFPT in situations in which standard techniques cannot be applied in a straightforward way, such as, for example in extended dynamically systems, as well as in the analysis of the MFPT in stochastic partial integro-differential equations (or nonlocal models). These last topics can be worked out by introducing a stochastic multiple scale expansion [11], see advanced exercise 9.3.3.

9.1 Metastable States

9.1.1 Decay Rates in the Small Noise Approximation

In the large viscosity approximation we can neglect the time evolution of the velocity in the phase space of the macroscopic system, therefore, the (two-dimensional) F–P equation can be reduced to the Smoluchowski equation for the space stochastic variable X. Then the F–P equation for the conditional pdf $P(x, t | x_0, t_0)$ describing the system of interest can be written in the form

$$\frac{\partial}{\partial t} P(x, t | x_0, t_0) = \left[-\frac{\partial}{\partial x} \bar{K}(x, t) + \frac{\partial^2}{\partial x^2} \bar{D}(x, t) \right] P(x, t | x_0, t_0). \tag{9.1}$$

The structure of the state functions $\bar{K}(x, t), \bar{D}(x, t)$ (see Sect. 3.18) depends on the stochastic calculus and the Langevin equation that we have used to model the system. We shall restrict ourselves to the case when the space is one dimensional and the time is homogeneous, therefore $\bar{K}(x, t) = K(x)$ is independent of time and we can take $\bar{D}(x)$ =constant, this constant being proportional to the intensity of the noise and in mechanical problems proportional to the temperature of the bath. Note that, as we mentioned in Sect. 3.18, the classification as: *multiplicative* Langevin equations is superfluous in one dimension.

If the boundary condition is natural the stationary solution of (9.1) is obtained by setting the stationary current to zero, hence

$$J_{st}(x) = \left[K(x) - \frac{D}{2} \frac{\partial}{\partial x} \right] P_{st}(x) = 0, \tag{9.2}$$

in this last expression we have taken the diffusion $\bar{D}(x)$ to be the constant[1] $D/2$ in (9.1), and its generalization to the case when it is a function of x is simple to calculate, see next exercise. We use Eq. (9.2) to find the stationary pdf

$$P_{st}(x) = \frac{\mathcal{N}}{D} \exp\left(2 \int^x \frac{K(x')}{D} dx' \right), \tag{9.3}$$

see also Sect. 4.7.1. Therefore, we can easily introduce a *potential* $U(x)$ such as

$$-\frac{\partial U(x)}{\partial x} = K(x) \tag{9.4}$$

and then the stationary pdf can be written as

$$P_{st}(x) \propto \exp\left[\frac{-U(x)}{(D/2)} \right]. \tag{9.5}$$

Guided Exercise (**Nonzero Current**). Consider the one-variable **F–P** equation written in the form

$$\frac{\partial}{\partial t} P(x, t| x_0, 0) + \frac{\partial}{\partial x} J(x, t) = 0, \tag{9.6}$$

with

$$J(x, t| x_0, 0) = \left[K(x) - \frac{\partial}{\partial x} \frac{D(x)}{2} \right] P(x, t| x_0, 0). \tag{9.7}$$

In general $J(x, t| x_0, 0)$ can be interpreted as the probability current, and (9.6) as the equation of conservation of probability. The general solution of (9.6) for $t = \infty$ is obtained by writing

$$P(x, t = \infty| x_0, 0) = v(x)/D(x), \tag{9.8}$$

so that Eq. (9.7) becomes

$$\frac{dv}{dx} - 2\frac{K(x)}{D(x)}v = -2J(\infty).$$

---

[1] Here the factor $\frac{1}{2}$ is for convenience only, it is not associated with the derivation from the Langevin equation, see Sect. 3.18.

This linear equation in $v(x)$ admits the solution

$$v(x) = -2J(\infty) \int^x \exp\left\{2\int_{x'}^x \frac{K(\xi)}{D(\xi)}d\xi\right\} dx' + C \exp 2\int^x \frac{K(\xi)}{D(\xi)}d\xi, \qquad (9.9)$$

where C is an arbitrary constant of integration. Substituting this expression for $v(x)$ into (9.8) we get the solution for $P(x,\infty|x_0,0)$ in terms of two constants $J(\infty)$ and C the normalization constant. Here $J(\infty)$ characterizes the nature of the processes, if there is no flow of probability into the state space from outside we have $J(x,t|x_0,0) \geq 0$. When $J(x,t|x_0,0)$ vanishes at both boundaries for all $t \geq 0, J(\infty) = 0$ by (9.6) and so the stationary pdf has the form (9.3).

In the context of an *equilibrium* model, (9.5) is the familiar pdf with D being proportional to the thermal energy k_BT, see Sect. 4.7.3. To illustrate this idea let us focus on a bistable potential as indicated in Fig. 9.1. In this figure the point x_M is a maximum of the potential, the point x_1 is a metastable minimum, while x_2 is a stable minimum, therefore x_M is an unstable point (or a saddle point in a more general multidimensional context, see the last section). The present analysis will be restricted to the high barrier/weak-noise limit which implies that $U(x_M)/D$ is a large number.

The **F–P** equation is parabolic, and the time evolution of the conditional pdf can therefore be written as an eigenfunction expansion, with the eigenvalues appearing in exponential decay factors associated with each eigenfunction (see Sect. 4.7.5. and the next Sect. 9.1.4). The zero eigenvalue corresponds to the stationary pdf, and the nontrivial eigenvalues determine the decay rates for metastable states. The question that we want to answer is: starting from an arbitrary initial pdf $P(x_0, t_0)$, how long does one have to wait for the pdf to evolve into $P_{st}(x)$. Kramers' argument was to separate time scales in the regime $t_{qs} \leq t \leq t_{st}$. Here t_{qs} is the time scale to reach a quasi-stationary pdf, and t_{st} the time-scale to reach the stationary pdf (9.5). Intuitively, in the small noise approximation, we can expect t_{qs} to be much shorter than the time t_{st}, so the interesting question is to study the *slow* time evolution of the pdf between t_{qs} and t_{st}.

Fig. 9.1 Representation of the potential $U(x)$

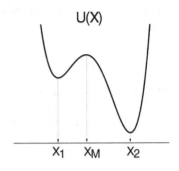

9.1.2 The Kramers Slow Diffusion Approach

In the regime $t_{qs} \leq t \leq t_{st}$ the pdf $P(x, t)$ is varying so slowly that we can neglect its time derivative, i.e., we can consider $P(x, t) \approx P_{qs}(x)$, so accordingly the probability current $J(x, t) \approx J_{qs}$ should be a nonzero constant. Hence, from the **P–F** equation we can write

$$J_{qs} = \left[K(x) - \frac{D}{2} \frac{\partial}{\partial x} \right] P_{qs}(x). \tag{9.10}$$

Introducing an integration factor this last equation can be written in the form

$$J_{qs} = -\frac{D}{2} e^{-U(x)/(D/2)} \frac{\partial}{\partial x} \left(P_{qs}(x) e^{U(x)/(D/2)} \right), \tag{9.11}$$

which can immediately be integrated from a generic point x_0 to x:

$$J_{qs} = \frac{(D/2) \left[P_{qs}(z) e^{U(z)/(D/2)} \right]_x^{x_0}}{\int_{x_0}^x \exp \left[\frac{U(z)}{(D/2)} \right] dz}. \tag{9.12}$$

Thus in the time domain $t_{qs} \leq t \leq t_{st}$, the population of the state x_1 is depleted while that of the state x_2 is increased at an almost steady state across the state x_M. So in this scheme of separation of time scales a reaction rate can be approximated as $r \approx J_{qs}/N_1$, where N_1 is the number of particles near x_1 (which can be written using $P_{qs}(x)$). In order to do the calculation we shall adopt an initial condition for the quasi-stationary pdf, i.e., we assume that in this regime no particles have yet arrived at x_2, while near x_1 thermal equilibrium has practically been established, therefore, using (9.12) we obtain

$$J_{qs} \simeq \frac{(D/2) P_{qs}(x_1) \exp \left[U(x_1)/(D/2) \right]}{\int_{x_1}^{x_2} \exp \left[\frac{U(z)}{(D/2)} \right] dz}. \tag{9.13}$$

The number N_1 of particles near x_1 can be calculated approximating $P_{qs}(x)$ near the point x_1 in the form

$$P_{qs}(x) \approx P_{qs}(x_1) \exp \left[-\frac{(U(x) - U(x_1))}{(D/2)} \right].$$

Therefore, if we assume that $U(x)$ near x_1 can be represented by a harmonic oscillator we can use $U(x) \approx U(x_1) + \frac{1}{2} U''(x_1)(x - x_1)^2$ and we can write for the number of particles

$$N_1 \approx \int_{\mathcal{D}_1} P_{qs}(x) dx = P_{qs}(x_1) \int_{\mathcal{D}_1} e^{-\frac{1}{2} U''(x_1)(x-x_1)^2/(D/2)} dx. \tag{9.14}$$

In the limit of small noise we can extend the limit of the integral to $\pm\infty$, and finally get

$$N_1 = P_{qs}(x_1)\sqrt{\frac{2\pi(D/2)}{U''(x_1)}}.$$ (9.15)

Therefore, using (9.13) and (9.15) the rate $r \approx J_{qs}/N_1$ is given by

$$\frac{J_{qs}}{N_1} = \frac{(D/2)P_{qs}(x_1)e^{U(x_1)/(D/2)}}{\int_{x_1}^{x_2} \exp\left[\frac{U(z)}{(D/2)}\right]dz} \frac{1}{P_{qs}(x_1)\sqrt{\frac{2\pi(D/2)}{U''(x_1)}}},$$

so

$$r = \frac{(D/2)e^{U(x_1)/(D/2)}}{\sqrt{\frac{2\pi(D/2)}{U''(x_1)}}\int_{x_1}^{x_2}\exp\left[\frac{U(z)}{(D/2)}\right]dz}.$$ (9.16)

The main contribution to the integral in (9.16) is due to a small region near x_M. Assuming, as would be consistent with the drawing of $U(x)$ in Fig. 9.1, that there is no sudden jump in the curvature, we may write near x_M

$$U(x) \approx \Delta - \frac{1}{2}\left|U''(x_M)\right|(x-x_M)^2 + U(x_1),$$

where $\Delta \equiv U(x_M) - U(x_1)$, and extending the limit of integration to $\pm\infty$, we get

$$\int_{x_1}^{x_2}\exp\left[\frac{U(z)}{(D/2)}\right]dz \approx e^{\Delta/(D/2)+U(x_1)/(D/2)}\int_{-\infty}^{+\infty}e^{-\frac{1}{2}|U''(x_M)|(x-x_M)^2}dx$$

$$= e^{[\Delta+U(x_1)]/(D/2)}\sqrt{\frac{2\pi(D/2)}{|U''(x_M)|}},$$

so for the reaction velocity we get

$$r \simeq \frac{1}{2\pi}\sqrt{|U''(x_M)|\,U''(x_1)}\,e^{-\Delta/(D/2)}.$$ (9.17)

From this result we see that the reaction rate depends only on the temperature (noise) through the Arrhenius factor: $e^{-\Delta/(D/2)}$. This would not be the case for other shapes of the potential barrier. In that case we need to use different techniques, such as the first passage time theory and the use of stretching transformations [12].

Guided Exercise. Consider a two-level system characterized by the states n_1 and n_2. The ME associated with this system is

$$\dot{n}_1 = \omega_{12}n_2 - \omega_{21}n_1$$ (9.18)

$$\dot{n}_2 = \omega_{21}n_1 - \omega_{12}n_2.$$

We can associate the states n_1 and n_2 with the number of particles around the spatial sites x_1 and x_2 of the previous one-dimensional problem, and then use the same Kramers' arguments to derive the rates ω_{21} and ω_{12} and the familiar rate equations (9.18). After a long time of order $\tau_{12} \equiv \omega_{12}^{-1}$ or $\tau_{21} \equiv \omega_{21}^{-1}$ the two-level system reaches the stationary values

$$n_1(\infty) = \frac{\tau_{12}}{\tau_{12} + \tau_{21}}$$

$$n_2(\infty) = \frac{\tau_{21}}{\tau_{12} + \tau_{21}}.$$

Then, in the small noise approximation we can write a stationary solution for the continuous system as a distribution with two peaks near x_1 and x_2:

$$P_{st}(x) \simeq n_1(\infty) \sqrt{\frac{U''(x_1)}{2\pi(D/2)}} \exp\left[-U''(x_1)\frac{(x-x_1)^2}{(D/2)}\right]$$

$$+ n_2(\infty) \sqrt{\frac{U''(x_2)}{2\pi(D/2)}} \exp\left[-U''(x_2)\frac{(x-x_2)^2}{(D/2)}\right].$$

9.1.3 Kramers' Activation Rates and the Mean First Passage Time

Through use of the theory of the first passage time it is possible to introduce an exact method for the calculation of the activation rate. We have already presented in Sects. 7.3.2 and 6.6 the crossing level theory (for discrete and/or continuous Markov systems). Let us focus on the introduction of the MFPT in a slightly different way. The system point is assumed to be initially at x_0 at time $t = 0$, where $-\infty \leq x_0 \leq x_L$. We imagine that an absorbing boundary is erected at x_L such that once the trajectory point reaches x_L it is removed from any further consideration (the absorbing boundary condition can only be used for diffusive processes, why is it so? [13]). We denote by $\tau(x_L | x_0)$ the time taken by the system point to reach x_L for the first time, having started from x_0 at $t = 0$. This is the so-called first passage time and it is clearly a random variable which varies from realization to realization. We shall later identify the MFPT $\langle \tau(x_L | x_0) \rangle$ with τ_{12} of the previous section.

Exercise (**Continuous Processes with Prob. 1**). Justify the usefulness of the absorbing boundary condition for calculating the first passage time in continuous Markov processes. Hint: use the concept of continuous stochastic processes (see Sect. 3.8.3, or the pedagogical example given in Appendix of [12]).

Starting at x_0 and after a short time Δt the particle arrives at point x' with probability $T_{\Delta t}(x' | x_0)$. Hence, as the system is assumed to be Markovian we can write

$$\langle \tau \left(x_L | x_0 \right) \rangle = \Delta t + \int \langle \tau \left(x_L | x' \right) \rangle T_{\Delta t} \left(x' | x_0 \right) \, dx'. \tag{9.19}$$

Where the conditional probability $T_{\Delta t} \left(x' | x_0 \right)$ is given for short times by the usual stationary expression (fulfilling the initial condition and normalization, see Sect. 6.3)

$$T_{\Delta t} \left(x_2 | x_1 \right) = \delta \left(x_2 - x_1 \right) + \Delta t \, \mathbf{H} \left(x_2 | x_1 \right) + \mathcal{O} \left(\Delta t \right). \tag{9.20}$$

Introducing this expression in (9.19) and taking the limit of $\Delta t \to 0$ we get

$$\int \langle \tau \left(x_L | x' \right) \rangle \mathbf{H} \left(x' | x_0 \right) dx' = -1, \tag{9.21}$$

compare with the result of Sect. 6.6. Thus, $\tau(x_0) \equiv \langle \tau \left(x_L | x_0 \right) \rangle$ as a function of the starting point x_0 obeys an equation in which the transpose of the kernel $\mathbf{H} \left(x' | x_0 \right)$ appears.

In particular for an **F–P** evolution the integral operator in (9.21) is a differential form, then using the adjoint operator (see Sect. 3.18), in the case of $D(x) = D/2 =$constant, from (9.21) we get the Dynkin equation:

$$\left[-U'(x_0) \frac{d}{dx_0} + (D/2) \frac{d^2}{dx_0^2} \right] \tau(x_0) = -1. \tag{9.22}$$

This equation has to be solved for $x_0 < x_L$ with two boundary conditions. Using $\tau(x_0) = 0$ at $x_0 = x_L$ and $d\tau(x_0)/dx_0 = 0$ at $x_0 = -\infty$ (reflecting boundary condition, see advanced Exercises 3.20.14 and 9.3.2) the general solution of (9.22) can be written as

$$\tau(x_0) = \frac{1}{(D/2)} \int_{x_0}^{x_L} \exp \left(+ \frac{U(\xi)}{(D/2)} \right) d\xi \int_{-\infty}^{\xi} \exp \left(- \frac{U(\eta)}{(D/2)} \right) d\eta. \tag{9.23}$$

Exercise. For the potential $U(x)$ given in Fig. 9.1, calculate the MFPT from $x_0 = x_1$ to $x_L = x_M$ using Eq. (9.23). Taking into account that the first exponential has a sharp maximum at x_M and the second at x_1, show in the small noise limit that we can approximate the MFPT by

$$2 \langle \tau \left(x_M | x_1 \right) \rangle \simeq \frac{2\pi}{\sqrt{U''(x_1) | U''(x_M) |}} \exp \left(\frac{U(x_M) - U(x_1)}{(D/2)} \right). \tag{9.24}$$

This result is in agreement with Kramers' activation rate $r = \tau_{21}^{-1}$ (9.17). Note that $\tau_{21} = 2 \langle \tau \left(x_M | x_1 \right) \rangle$, because once the particle has arrived at x_M it has equal chances of escaping or returning to x_1.

Guided Exercise (**Survival Probability**). Consider an **sp** $x(t)$ defined in the domain $x \in [-\infty, \infty]$ for all $t \geq t_0$. The probability that at time t the system point is still within the interval $-\infty$ to x_L, not having reached x_L even once, is given by

$$S(x_0, t) = \int_{-\infty}^{x_L} \bar{P}(x, t | x_0, t_0) dx, \tag{9.25}$$

here $\bar{P}(x, t | x_0, t_0)$ states specifically that the point representing the system never reached x_L in the time interval t_0 to t. Then the quantity $S(x_0, t)$ equals the probability that $\tau(x_L | x_0) > t$. Using the initial condition (at $t_0 = 0$) of the propagator $\bar{P}(x, t_0 | x_0, 0) = \delta(x - x_0)$ in (9.25) we get

$$S(x_0, 0) = \begin{cases} 1, & \text{if } -\infty \leq x_0 < x_L \\ 0, & \text{elsewhere} \end{cases}. \tag{9.26}$$

On the other hand, if x_0 happens to be equal to x_L the process *dies* immediately, hence

$$S(x_L, t) = 0,$$

in addition because the particle will ultimately reach x_L we get

$$S(x_0, \infty) = 0.$$

Since $S(x_0, t)$ is the probability that the first passage has not occurred, $1 - S(x_0, t)$ is the probability that it has. Show that the MFPT is given by (here we use $t_0 = 0$)

$$\langle \tau(x_L | x_0) \rangle = - \int_0^\infty t \, \dot{S}(x_0, t) \, dt = \int_0^\infty S(x_0, t) \, dt. \tag{9.27}$$

The adjoint **F–P** operator was introduced in Sect. 4.8, in the context of the Kolmogorov operator, see also first optional exercise in Sect. 3.18. For a one-variable **F–P** equation as in (9.6) the adjoint operator is

$$\mathcal{L}(x, \partial_x)^\dagger = \left[K(x) \frac{\partial}{\partial x} + \frac{D(x)}{2} \frac{\partial^2}{\partial x^2} \right]. \tag{9.28}$$

Applying the operator $\mathcal{L}(x_0, \partial_{x_0})^\dagger$ on (9.27) we can write

$$\mathcal{L}(x_0, \partial_{x_0})^\dagger \langle \tau(x_L | x_0) \rangle = \int_0^\infty \mathcal{L}(x_0, \partial_{x_0})^\dagger S(x_0, t) \, dt.$$

Now from $\partial_t \bar{P}(x, t | x_0, 0) = \mathcal{L}(x_0, \partial_{x_0})^\dagger \bar{P}(x, t | x_0, 0)$, integrating by parts and noting that $\bar{P}(x, \infty | x_0, 0) = 0$ for $t_0 = 0$; show that the MFPT satisfies the equation

$$\mathcal{L}(x_0, \partial_{x_0})^\dagger \langle \tau(x_L | x_0) \rangle = -1, \tag{9.29}$$

which is the Dynking equation for a general continuous diffusion process. Solving this (one-dimensional) equation imposes one more boundary condition besides

$$\langle \tau (x_L | x_L) \rangle = 0,$$

this extra boundary condition can be inferred from the backward **F–P** equation, see advanced exercise 9.3.2; typically it could be reflecting or absorbing [4, 5, 14].

9.1.4 Variational Treatment for Estimating the Relaxation Time

The **F–P** equation is a second order partial differential equation in the space derivatives and of first order in the time derivative, therefore the time-dependent conditional pdf can be written as an eigenfunction expansion. The eigenvalues of the **F–P** operator have negative real part leading to decay factors associated with each eigenfunction and thus approaching the stationary state, if it exists. This last fact has been proved using the Liapunov function in Sect. 4.6 for a general multidimensional **F–P** equation.

As in the previous section we focus here on the one-variable case, therefore, the conditional pdf of the F–P equation can be written as

$$P(x, t | x_0, t_0) = a_0 \phi_0(x) + \sum_{n>0} a_n(x_0)\phi_n(x) \exp\left(-\lambda_n (t - t_0)\right), \qquad (9.30)$$

here we are assuming the completeness of the eigenfunctions, an issue that is not possible to prove in general for any **F–P** operator. The first term corresponds to the eigenvalue zero which yields the stationary pdf $\phi_0(x)$ with a_0 the normalization constant (as $t \rightarrow \infty$ and assuming that λ_0 is a single, not a multiple root). Introducing this expansion in (9.6) leads to the elliptic partial differential equation

$$\left[-\frac{\partial}{\partial x} K(x) + \frac{\partial^2}{\partial x^2} \frac{D(x)}{2} \right] \phi_n(x) = -\lambda_n \phi_n(x). \qquad (9.31)$$

The substitution of $\Psi_n(x) = \phi_n(x)/\phi_0(x)$ transforms (9.31) into a self-adjoint eigenvalue equation of the Sturm–Liouville type

$$\frac{\partial}{\partial x} \left[\frac{D(x)}{2} \phi_0(x) \frac{\partial}{\partial x} \right] \Psi_n(x) = -\lambda_n \phi_0(x)\Psi_n(x), \qquad (9.32)$$

here we have assumed that the current is zero in the stationary state. Equation (9.32) can be considered as the Euler–Lagrange equation for the functional

$$F\left[\Psi_n(x)\right] = \int_{-\infty}^{+\infty} dx \, \phi_0(x) \left[\frac{1}{2} D(x) \left(\frac{\partial \Psi_n(x)}{\partial x} \right)^2 - \lambda_n \Psi_n(x)^2 \right]. \qquad (9.33)$$

Exercise. Obtain (9.32) under the condition $\delta F[\Psi_n]/\delta \Psi_n(z) = 0$. Hint: use the exercise in Sect. 3.1.2., here the stationary pdf $\phi_0(x)$ enters the functional $F[\Psi_n(x)]$ as a weight density.

Following standard procedure [15], the eigenvalues λ_n of (9.31) can be shown to obey the Rayleigh–Ritz inequality

$$\lambda_n \leq \frac{\Omega[\Psi_n(x)]}{\Xi[\Psi_n(x)]}, \tag{9.34}$$

where

$$\Omega[\Psi_n(x)] = \int_{-\infty}^{+\infty} dx \, \phi_0(x)\frac{1}{2}D(x)\left(\frac{\partial\Psi_n(x)}{\partial x}\right)^2$$

$$\Xi[\Psi_n(x)] = \int_{-\infty}^{+\infty} dx \, \phi_0(x)\Psi_n(x)^2.$$

and $\Psi_n(x)$ is the trial function.

We have mentioned before, in the Kramers problem, that we are only interested in the lowest eigenvalue λ_1 for the description of the passage from the quasistationary state into the stationary state. Hence the minimizing procedure is enormously simplified. We can assume a particular form for the trial function $\Psi_1(x)$ in terms of certain variational parameters α in such a way that it minimizes the right-hand side of (9.34) with respect to these parameters α. The variational choice of the eigenfunction $\Psi_1(x)$ must be normalized and orthogonal to the lowest eigenfunction $\Psi_0(x) = 1$, i.e., $\int_{-\infty}^{+\infty} dx \, \Psi_0(x)\Psi_1(x) = 0$. This implies that the trial function $\Psi_1(x)$ must change its sign as the system point moves from the interval $x_1 \leq x \leq x_M$ to $x_M \leq x \leq x_2$. The study of the upper bound for the eigenvalues of an **F–P** has been an important motivation of analysis with applications in many areas of research [16, 17].

Excursus. The thermal relaxation behavior of single-domain magnetic particles can be investigated in the context of **F–P** dynamics [3], in particular the decay of the metastable states has been calculated using a variational approach. An interesting trial function providing Kramers' activation rate result, in the large barrier limit, and also the lower bound to the **F–P** eigenvalue was presented in [18].

9.1.5 Genesis of the First Passage Time in Higher Dimensions*

The decay of a metastable state is a key element in the analysis of nonequilibrium phenomena. In general, the state of these systems is assumed to be determined by an **F–P** equation, and under the condition of weak noise important results can be obtained. As we mentioned before in the limit of weak noise, the pioneer work for the case of one-variable was presented by Kramers [1], and for higher-dimensional systems by Landauer and Swanson [19] and Langer [20]. However, their approaches were restricted to systems with detailed balance and with essentially one single transition point (saddle). These restrictions can be dropped in the framework of first passage time theory. In one-dimension the MFPT can be solved analytically [13], and in the multidimensional case with a gradient drift field and with a scalar constant diffusion it was solved, in the small noise approximation, by Schuss and Matkowsky [21] and Schuss [22]. For a more general situation: several variables without potential, the problem was tackled by Talkner and Ryter [23], van Kampen [12], and Gardiner [4].

Here we will present the result for the MFPT in a two-dimensional gradient drift field and with a scalar constant diffusion θ. This **F–P** model describes the overdamped motion of a Brownian particle in an external field. If the motion is not overdamped, its velocity must be included in the dynamics, and the problem becomes essentially more difficult to work out.

Consider an n-component stochastic process $\vec{x} = (x_1, x_2, \cdots)$ taking an initial value $\vec{x}_0$ somewhere within a volume Ω. The first passage time is now defined, for a given realization of $\vec{x}$ by first reaching the boundary $\partial\Omega$ (an $n - 1$ dimensional surface). As in the one-dimensional case (9.29) the MFPT satisfies the Dynking equation in n-dimensions (the adjoint **F–P** operator was given in Sect. 4.8)

$$\left[\sum_j K_j(\vec{x}) \frac{\partial}{\partial x_j} + \sum_{ij} \frac{D(\vec{x})_{ij}}{2} \frac{\partial^2}{\partial x_j \partial x_i} \right] \tau(\vec{x}) = -1. \tag{9.35}$$

In the two-dimensional case with $\vec{K}$ given by the gradient of a potential, see Fig. 9.2, and the diffusion matrix given by a scalar independent of $\vec{x}$ we can write

$$- \vec{\nabla} U(\vec{r}) \cdot \vec{\nabla} \tau(\vec{r}) + \theta \nabla^2 \tau(\vec{r}) = -1, \ (\vec{r} \in \Omega); \ \theta > 0, \tag{9.36}$$

with the boundary condition

$$\tau(\vec{r}) = 0, \ (\vec{r} \in \partial\Omega). \tag{9.37}$$

Fig. 9.2 A crater-like potential in two dimensions, the crater ridge is the boundary $\partial\Omega$, and it has the lowest point at $\vec{b}$ of height W

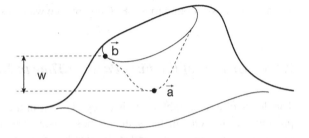

We ask for the escape time from the trough around $\vec{a}$, then $\tau(\vec{a})$ will be identified—as before—with the MFPT of the problem by $2\tau(\vec{a}) = \tau_{21}$. The crater ridge has a lowest point $\vec{b}$ of height $W = U(\vec{b}) - U(\vec{a})$. The escape time will include the Arrhenius factor $e^{W/\theta}$ (escape across any higher point of the ridge will be exponentially less likely). In addition the MFPT should contain a factor carrying the information on the surface near the point $\vec{b}$ (the saddle). This is the factor that was tackled by Schuss and Matkowsky [21] introducing a perturbation expansion in the small parameter θ. The basic idea is to isolate the singular part of $\tau(\vec{r})$ on θ and assume the rest to be regular in θ expandable in a power series. Thus, take

$$\tau(\vec{r}) = v(\vec{r})e^{W/\theta} \tag{9.38}$$

$$v(\vec{r}) = v^{(0)}(\vec{r}) + \theta v^{(1)}(\vec{r}) + \cdots,$$

introducing this expression into

$$\left[-\vec{\nabla}U(\vec{r}) \cdot \vec{\nabla} + \theta\nabla^2 \right] v(\vec{r}) = -e^{-W/\theta} \approx 0, \tag{9.39}$$

to each order θ we get

$$\mathcal{O}\left(\theta^0\right) \Longrightarrow -\vec{\nabla}U(\vec{r}) \cdot \vec{\nabla}v^{(0)}(\vec{r}) \approx 0 \tag{9.40}$$

$$\mathcal{O}\left(\theta^1\right) \Longrightarrow -\vec{\nabla}U(\vec{r}) \cdot \vec{\nabla}v^{(1)}(\vec{r}) + \nabla^2 v^{(0)}(\vec{r}) \approx 0 \tag{9.41}$$

$$\text{etc.}$$

Equation (9.40) would imply that $v^{(0)}(\vec{r})$ is a nonzero constant in the region where

$$-\vec{\nabla}U(\vec{r}) \neq 0,$$

but in view of (9.38) this solution would contradict the boundary condition (9.37). This issue can be solved if there is, at least, one direction along which $v^{(0)}(\vec{r})$ varies rapidly near the surface $\partial\Omega$ over a scale $\sqrt{\theta}$ (this is the stretching transformation introduced by van Kampen [12]) in such a way that its second derivative times θ ends being independent of θ. This special direction, which we call z, is expected to lie along the path saddle point $\vec{b}$.

In Fig. 9.3 we show the local set of axes $(\vec{\rho}, \vec{z})$ centered on the point $\vec{b}$. The unit vector $\vec{z}$ points inward from $\partial\Omega$ and the other direction $\vec{\rho}$ is normal to $\vec{z}$. Taking all this into account we can reinterpret (9.39) and write to $\mathcal{O}\left(\theta^0\right)$

$$\left[-\left(\frac{\partial}{\partial z} U(\vec{\rho}, \vec{z}) \right) \frac{\partial}{\partial z} + \theta \frac{\partial^2}{\partial z^2} \right] v^{(0)}(\vec{\rho}, \vec{z}) \approx 0 \tag{9.42}$$

$$\left[-\frac{\partial}{\partial \rho} U(\vec{\rho}, \vec{z}) \frac{\partial}{\partial \rho} \right] v^{(0)}(\vec{\rho}, \vec{z}) \approx 0. \tag{9.43}$$

Fig. 9.3 Local coordinates
near the mountain pass
around the point $\vec{b}$

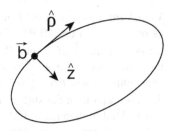

Introducing an expansion for $U(\rho, z)$ around the saddle point $\vec{b}$ in the form
$U(\rho, z) \simeq U(\vec{b}) + \frac{1}{2}\partial_{\rho\rho}U(\vec{b})\rho^2 - \frac{1}{2}\left|\partial_{zz}U(\vec{b})\right|z^2$ Eq. (9.42) can be integrated.
From (9.43) we get $v^{(0)} = v^{(0)}(\vec{z})$ independent of $\vec{\rho}$. On the other hand, from
the (9.37) and (9.38) we get $v^{(0)}(0) = 0$ so we end up with the equation:

$$\left|\partial_{zz}U(\vec{b})\right|z\partial_z v^{(0)}(z) + \theta\partial_z^2 v^{(0)}(z) = 0,$$

with can be integrated giving the result

$$v^{(0)}(z) = C\int_0^z \exp\left(\frac{-1}{2\theta}\left|\partial_{zz}U(\vec{b})\right|\xi^2\right)d\xi. \qquad (9.44)$$

The constant C can be solved multiplying Dynking's equation (9.36) by $\exp\left(-\frac{U(\vec{r})}{\theta}\right)$
and integrating over Ω

$$-\int_\Omega dV \exp\left(\frac{-U(\vec{r})}{\theta}\right) = \int_\Omega dV \exp\left(\frac{-U(\vec{r})}{\theta}\right)\left[-\vec{\nabla}U(\vec{r})\cdot\vec{\nabla} + \theta\nabla^2\right]\tau(\vec{r})$$

$$= \int_\Omega dV \vec{\nabla}\cdot\left[\exp\left(\frac{-U(\vec{r})}{\theta}\right)\vec{\nabla}\tau(\vec{r})\right]$$

$$= -\oint_{\partial\Omega} ds \exp\left(\frac{-U(\vec{r})}{\theta}\right)\vec{\nabla}\tau(\vec{r}). \qquad (9.45)$$

In the last line we have used the Gauss theorem and noted that the normal to the
surface $\partial\Omega$ (at $\vec{b}$) lies inward along $\vec{z}$. The main contribution to the right-hand side
integral comes from the minimum of the potential (at $\vec{b}$), therefore, we can use (9.44)
to approximate

$$\left[\vec{\nabla}\tau(\vec{r})\right]_{\text{near } \vec{b}} \simeq C e^{W/\theta}$$

in (9.45) and so solve for the constant C as

$$C = \frac{\int_\Omega dV \exp\left(\frac{-U(\vec{r})}{\theta}\right)}{e^{W/\theta}\oint_{\partial\Omega} ds \exp\left(\frac{-U(\vec{r})}{\theta}\right)}.$$

Then from (9.38) we get a closed formula for the escape time

$$\tau(\vec{r}) \simeq \frac{1}{\theta} \sqrt{\frac{\pi\theta}{2\left|\partial_{zz}U(\vec{b})\right|}} \frac{\int_\Omega dV \, \exp\left(\frac{-U(\vec{r})}{\theta}\right)}{\oint_{\partial\Omega} ds \left[\exp\left(\frac{-U(\vec{r})}{\theta}\right)\right]_{\text{near } \vec{b}}}, \qquad (9.46)$$

here we have used (9.44) and taken the integral to infinity (small noise approximation), then we get that $v^{(0)}(z)$ is asymptotically constant

$$v^{(0)}(\vec{z}) \approx \frac{C}{2} \sqrt{\frac{2\pi\theta}{\left|\partial_{zz}U(\vec{b})\right|}}.$$

Each of the integrals in formula (9.46) can be calculated using sharp peaking arguments. For the present two-dimensional model and using a harmonic approximation
for the potential $U(\vec{r})$ near $\vec{a}$ we can write from the numerator

$$\int_\Omega dV \, \exp\left(\frac{-U(\vec{r})}{\theta}\right) \simeq \int\int dx \, dy \, \exp\left(\frac{-1}{\theta}\left[U(\vec{a}) + \frac{1}{2}U_{xx}\,x^2 + \frac{1}{2}U_{yy}\,y^2 + U_{xy}\,xy\right]\right)$$

$$= 2\pi \exp\left(\frac{-U(\vec{a})}{\theta}\right) [\partial_{xx}U(\vec{a})\partial_{yy}U(\vec{a}) - \partial_{xy}U(\vec{a})]^{-1/2},$$

(to calculate this integral see Sect. 1.12). From the denominator using a harmonic
approximation for $U(\vec{r})$ near $\vec{b}$ we get

$$\int_{\partial\Omega} ds \, \exp\left(\frac{-U(\vec{r})}{\theta}\right) \simeq \int d\rho \, \exp\left(\frac{-1}{\theta}\left[U(\vec{b}) + \frac{1}{2}U_{\rho\rho}\,\rho^2\right]\right)$$

$$= \exp\left(\frac{-U(\vec{b})}{\theta}\right) \sqrt{2\pi\theta/\partial_{\rho\rho}U(\vec{b})}.$$

Collecting the results for the integrals, from (9.46) we get a final expression for the
MFPT from the metastable state 1 (state a in the two-dimensional model) to state 2
(the possible stable state outside the valley of attraction of a)

$$\tau_{21} = 2\tau(\vec{a}) = \frac{2\pi e^{W/\theta} \left[\partial_{\rho\rho}U(\vec{r})\right]_{\vec{b}}^{1/2}}{\left[\left|\partial_{zz}U(\vec{r})\right|\right]_{\vec{b}}^{1/2} \left[\partial_{xx}U(\vec{r})\partial_{yy}U(\vec{r}) - \partial_{xy}U(\vec{r})\right]_{\vec{a}}^{1/2}} \qquad (9.47)$$

This formula generalizes the MFPT (9.24) when the system is bistable in a
two-dimensional configurational space. Cases when the F–P has several variables
and when there is *no potential present* can also be treated by using a stretching
transformation [4, 12], similar to what we have presented here. Of course there
are also other instabilities, for example limit cycles, models with several saddles

and critical points. In principle all these problems could be tackled using the F–P approach, but their presentation is outside the scope of this introduction. Another interesting situation to be studied is the relaxation from an unstable point. This topic will be presented in the following section.

9.2 Unstable States

9.2.1 *Relaxation in the Small Noise Approximation*

The interesting problem of finding the pdf $P(x, t | x_0, t_0)$ for an initial point x_0 close to an unstable state, like point X_M in Fig. 9.1, has also been subject of much research, in particular using scaling transformations [24], path integrals [25], and other techniques [26, 27]. The basic idea is, again, to consider the time-evolution near the instability in linear approximation and then smoothly join the result with the solution for the nonlinear region. This approach is very useful when the instability can be linearized.

Specifically the scaling theory [24] entails dividing the whole time range into three regions: the *initial region*, in which the linear approximation is valid, the *scaling region*, in which the scaling law holds, and the *final region*, in which the system approaches the equilibrium state. The scaling region is specified by the time regime in which the scaling-time variable is of order unity. The interesting point is that this time scaling can be associated with the escape time for leaving the unstable point. In the present section we will introduce the stochastic path perturbation approach to calculate this characteristic escape time. In addition, this small noise perturbation can also be applied when there is not a dominant linear regime at the unstable point, when the system is non-Markovian, or when the model is multivariable.

For simplicity, we introduce here a one-dimensional model, take the unstable point at the origin of coordinates and adjust the time scale in such a way that we can approximate the potential (near the origin) by

$$U(x) = -\frac{1}{2}x^2 + \mathcal{O}\left(x^3\right), \tag{9.48}$$

and use a small constant diffusion θ in the F–P (9.1). With the change of variable $x = \sqrt{\theta}\eta$ we get to the lowest order

$$\frac{\partial P(\eta, t | \eta_0, t_0)}{\partial t} = \left[-\frac{\partial}{\partial \eta}\eta + \theta\frac{\partial^2}{\partial \eta^2}\right] P(\eta, t | \eta_0, t_0), \tag{9.49}$$

with $x_0 \sim \mathcal{O}\left(\sqrt{\theta}\right)$. The solution in the old variable x is

$$P(x,t|x_0,0) = \frac{\exp\left[-\frac{\left(x-\sqrt{\theta}\eta_0 e^t\right)^2}{2\theta\left(e^{2t}-1\right)}\right]}{\sqrt{2\pi\theta\left(e^{2t}-1\right)}}, \tag{9.50}$$

then moments $\langle x(t)^m \rangle$ can be calculated, in particular: $\langle x(t) \rangle = \sqrt{\theta}\eta_0 e^t$ and $\langle x(t)^2 \rangle = \theta\left(e^{2t}-1\right) + \langle x(t) \rangle^2$.

Exercise. From the solution (9.50) calculate the total probability on the right (splitting of the probability) $\int_0^\infty P(x,t|x_0,0)dx$. Note that in the limit $e^t \gg 1$ the integration limit is constant in time and then we can write

$$\text{Prob}\left[x \in (0,\infty); \forall t > 0, x(0) = x_0\right] = \frac{1}{\sqrt{2\pi}} \int_{-x_0/\sqrt{\theta}}^{\infty} e^{-z^2/2} dz.$$

Also, the total probability on the left will be

$$\text{Prob}\left[x \in (-\infty,0); \forall t > 0, x(0) = x_0\right] = \frac{1}{\sqrt{2\pi}} \int_{x_0/\sqrt{\theta}}^{\infty} e^{-z^2/2} dz.$$

From these results we see that in the limit $\theta \to 0$ we regain the deterministic result.

In order to find the pdf for longer times, when the nonlinear terms in $U'(x)$ are important, we could use the Chapman–Kolmogorov equation to join smoothly the result from the linear and nonlinear regimes:

$$P(x,t|x_0,0) = \int_{-\infty}^{\infty} P(x,t|x_1,t_1)P(x_1,t_1|\sqrt{\theta}\eta_0,0)dx_1. \tag{9.51}$$

Here $P(x_1,t_1|\sqrt{\theta}\eta_0,0)$ can be taken to be the solution in the linear regime (9.50), and for $P(x,t|x_1,t_1)$ several approximations can be used. For example: we can assume $P(x,t|x_1,t_1)$ to be a Gaussian approximation with a time-dependent mean value following deterministic evolution and a consistent time-dependent variance; or $P(x,t|x_1,t_1)$ can represent a pure deterministic evolution from the nonlinear regime. This last approximation means that we only take into account fluctuations in the vicinity of the unstable point.

The value of the matching time t_1 should be of the order of that for which the fluctuations are anomalously large i.e., Suzuki's scaling-time. But for a true Markovian process the validity of the Chapman–Kolmogorov equation is independent of t_1. In the next section we will show that the matching time is of the order of the MFPT from the unstable point. Denoting by $\varphi\left(t - t_1|x_1\right) = x(t)$ the solution of the nonlinear deterministic evolution, the pdf for the stochastic problem can be approximated by:

$$P(x,t|x_0,0) \simeq \int_{-\infty}^{\infty} \delta(x - \varphi\left(t - t_1|x_1\right))P(x_1,t_1|\sqrt{\theta}\eta_0,0)dx_1 \tag{9.52}$$

Exercise. From (9.52), using the deterministic evolution $\dot{x} = -U'(x)$ and the inverse transformation $x_1 = \varphi\,(t_1 - t|\,x)$ show that the pdf can be written in the form

$$P(x,t|\,x_0,0) \simeq \frac{e^{-t_1}}{\sqrt{2\pi\theta}} \frac{\partial \varphi\,(t_1 - t|\,x)}{\partial x} \exp\left[-\frac{(\varphi\,(t_1 - t|\,x) - x_0 e^{t_1})^2}{2\theta e^{2t_1}} \right].$$

Note that this solution is not independent of the precise choice of the time t_1. In the limit of small noise a possible selection for t_1 is to use the value of the MFPT (to leave the unstable state). Another alternative is to introduce a self-consistent approximation [12, 26, 27].

9.2.2   The First Passage Time Approach

The distribution of the first passage time emphasizes the role of the individual realizations of the process. It is particularly well suited to answer questions related to time scales of evolution. For example, the lifetime of a given state can be defined as the MFPT required for leaving the vicinity of that state. The associated variance of the first passage time distribution determines if that lifetime is a meaningful quantity.

An example of a situation in which a passage time description is useful is the study of the transient relaxation processes triggered by noise, as when a system leaves a state which is not globally stable (Kramers' problem). In the previous sections we have shown that this is the case of metastability, in which the lifetime of the metastable state, identified as the MFPT, is $\tau_K \propto \exp(W/\theta)$, where W is an activation energy and θ measures the intensity of the fluctuations (temperature). A second case is the relaxation from the state that becomes unstable in a supercritical pitchforks bifurcation [24, 26, 27]. In a deterministic treatment $dX/dt \sim aX$, so the order parameter $X(t)$ starts to grow exponentially as $X(t)^2 \sim e^{2at}$. However, the lifetime of the state is not a^{-1} because the lifetime is determined by fluctuations. It turns out to be Suzuki's time-scale $\tau_S \sim \frac{1}{2a} \ln \theta^{-1}$, here θ is proportional to the small fluctuations. The relaxation from an unstable state in a supercritical pitchfork bifurcation initially follows Gaussian statistics for $X(t)$. But more complicated is the description of the relaxation from the state that loses its stability in a saddle-node bifurcation [6] or in states of marginal stability, for which Gaussian statistics, or equivalently, linear theory, do not hold at any time of the relaxation process.

9.2.3   Suzuki's Scaling-Time in the Linear Theory

Consider the following stochastic dynamics (pitchfork bifurcation)

$$\frac{dX}{dt} = aX(t) - bX(t)^3 + \sqrt{\epsilon}\,\xi(t); \quad a > 0,\ b > 0,\ X(0) = 0, \tag{9.53}$$

here the last term represents a GWN of zero mean, with small intensity ϵ, and correlation

$$\langle \xi(t) \xi(t') \rangle = \delta(t - t'), \tag{9.54}$$

therefore $\xi(t)dt = dW(t)$ is a Wiener differential.

For short times any realization is dominated by the linear instability term in (9.53), then it can be approximated to $\mathcal{O}(\sqrt{\epsilon})$ by

$$X(t) \simeq \sqrt{\epsilon}\, e^{at} \int_0^t e^{-as} dW(s) \equiv \sqrt{\epsilon}\, e^{at} h(t), \quad t \geq 0. \tag{9.55}$$

Where we have defined a new stochastic process $h(t) \equiv \int_0^t e^{-as} dW(s)$, which is the solution of the **sde**

$$\frac{dh(t)}{dt} = e^{-at}\xi(t), \quad h(0) = 0, \quad t \geq 0.$$

It is important to note that even when the correlation function of the process $\xi(t)$ is white, the stochastic process $h(t)$ saturates at long times $(t \gg a^{-1})$. In particular it is possible to prove that $\langle h(\infty)^2 \rangle = (2a)^{-1}$, and then the stationary pdf is

$$P(h, \infty) \equiv P(\Omega) = \frac{e^{-a\Omega^2}}{\sqrt{\pi/a}}, \quad \Omega \in (-\infty, +\infty). \tag{9.56}$$

Therefore we can approximate (9.55) in the form:

$$X(t) \simeq \sqrt{\epsilon}\, \Omega \exp(at). \tag{9.57}$$

Formula (9.57) gives t as a map from the random number Ω.

Realizations (9.55) give an accurate representation of the paths for short and intermediate times, except for the small fluctuations around the final steady state $X(\infty)^2 = a/b$, i.e., in the long-time limit $t \to \infty$. At *intermediate* times, the random escape times t_e (when the stochastic paths leave the initial domain $\mathcal{O}(\sqrt{\epsilon})$ and fall into the attractor of the saturation valley) can be obtained by inverting t_e from (9.57) in the form $X(t_e)^2 \simeq \epsilon \Omega^2 \exp(2at_e) = X_f^2$. Therefore to $\mathcal{O}(\sqrt{\epsilon})$ we get the transformation law

$$t_e = \frac{1}{2a} \log\left(\frac{X_f^2}{\epsilon \Omega^2}\right). \tag{9.58}$$

This formula says that the MFPT is just given by the mean value $\langle t_e \rangle$ over the distribution of the random variable Ω. Therefore using (9.56) we get Suzuki's scaling-time

$$\tau_S = \langle t_e \rangle = \frac{1}{2a} \int_{-\infty}^{\infty} \log \left(\frac{X_f^2}{\epsilon \Omega^2} \right) P(\Omega) \, d\Omega$$

$$= \frac{1}{2a} \left(\ln \frac{X_f^2 a}{\epsilon} - \psi \left(\frac{1}{2} \right) \right), \tag{9.59}$$

here we have used $\int_{-\infty}^{\infty} \log (\Omega^2) \sqrt{\frac{a}{\pi}} e^{-a\Omega^2} \, d\Omega = -\left(\log a - \psi \left(\frac{1}{2} \right) \right)$, where $\psi \left(\frac{1}{2} \right) \simeq -1.9351$ is the Digamma function.

Exercise. Show that Suzuki's scaling time for a bistable flux is the same as the one obtained for the noisy Logistic equation with reflecting BC at the origin [9].

Exercise. From the formula for the MFPT (see, Sect. 9.1.3) to reach the threshold X_f, show that corrections from the nonlinear saturation term are given by

$$\langle \tau (X_f | 0) \rangle = \frac{1}{2a} \left(\ln \frac{X_f^2 a}{\epsilon} - \psi \left(\frac{1}{2} \right) + \frac{X_f}{K} - \frac{\sqrt{\epsilon/a}}{K} D \left[\frac{X_f}{\sqrt{\epsilon/a}} \right] + \cdots \right), \tag{9.60}$$

here we have used the one-dimensional potential $U(x) = -\frac{1}{2}ax^2 + \frac{1}{3}\frac{a}{K}x^3$. To obtain this result we have imposed reflecting boundary conditions at $x = 0$ and used the Dawson function $D[z]$ [28]. Formula (9.60) means that asymptotically for small noise Suzuki's scaling-time is equivalent to the dominant term in the MFPT. Using $\lim_{z \to \infty} D[z] \to \frac{1}{2z} + \frac{1}{4z^2} \cdots$, we see that the first correction from the nonlinear saturation term is the constant X_f / K

In order to obtain the FPTD $\mathcal{P}(t_e)$ (i.e., the probability that amplitude $X(t)$ reaches, for the first time, a given threshold $\pm X_f$ between t_e and $t_e + dt_e$), we go back to the relation between Ω and t_e in Eq. (9.58). Then the FPTD $\mathcal{P}(t_e)$ follows as:

$$\mathcal{P}(t_e) = \int \delta (t_e - t_e [\Omega]) P(\Omega) d\Omega.$$

Using the Jacobian of the transformation $\left| \frac{d\Omega}{dt_e} \right| = a |\Omega|$, we get

$$\mathcal{P}(t_e) = \mathcal{N} \frac{aX_f}{\sqrt{\epsilon}} \exp(-at_e) P \left(\Omega = \frac{X_f}{\sqrt{\epsilon}} \exp(-at_e) \right), \quad t_e \in (0, \infty), \tag{9.61}$$

where $P(\Omega)$ is given in Eq. (9.56) and the normalization constant is $\mathcal{N} = 2/ \operatorname{erf} \left(X_f \sqrt{a/\epsilon} \right)$.

Excursus (**Non-Markov Effects**). It is interesting to note that the FPTD $\mathcal{P}(t_e)$ for the escape process for an exponential distributed time-delay stochastic bistable flux:

$$\dot{X}(t) = rX_G(t) - X(t)^3 + \sqrt{\epsilon}\,\xi(t) \qquad (9.62)$$

$$X_G(t) = \lambda \int_0^\infty e^{-\lambda s} X(t-s)\,ds,$$

can be read from Eq. (9.61) replacing $X_f \to \mathcal{N}_{non-A}$ and $a \to \mathcal{A}_{non-A}$. Here the nonadiabatic constants $\{\mathcal{N}_{non-A}, \mathcal{A}_{non-A}\}$ take into account non-Markov effects [10].

9.2.4 Anomalous Fluctuations

From the FPTD (9.61) it is possible to study the dynamics of the nonlinear process (9.53) (bistable flux) by introducing an instanton-like approximation

$$X(t) \simeq \pm\Theta\left(t - t_e\right), \qquad (9.63)$$

here $\Theta\left(z\right)$ is the step function and t_e the random escape times characterized by $\mathcal{P}(t_e)$, this approximation gives a good description for the analysis of the anomalous fluctuations of the process. To calculate the anomalous fluctuation *analytically*, we approximate the transient toward the global attracting solution using (9.63) which is for $t > t_e$ the $\mathcal{O}(1)$ macroscopic amplitude size. Then, the transient anomalous fluctuation is given by

$$\sigma_X^2(t) = \langle\Theta(t - t_e)\rangle - \langle\Theta(t - t_e)\rangle^2, \quad t \geq 0, \qquad (9.64)$$

where

$$\langle\Theta(t - t_e)\rangle = \int_0^\infty \Theta(t - t_e)\,\mathcal{P}(t_e)\,dt_e$$

$$= \int_0^t \mathcal{P}(t_e)\,dt_e,$$

In this instanton-like approximation the maximum of the function $\sigma_X^2(t)$ is at the most probable escape value. The function $\sigma_X^2(t)$ depicts the qualitative behavior of the anomalous fluctuation [24]. From the behavior of $\sigma_X^2(t)$ we see that in the transient regime the initial fluctuations are amplified and give rise to the transient anomalous fluctuations of $\mathcal{O}(1)$ as compared with the initial or final fluctuations of $\mathcal{O}(\sqrt{\epsilon})$.

This sort of stochastic approximation for calculating transient phenomena in terms of the first passage time statistics has shown good agreement in different physical problems. For example when it is compared with the Monte Carlo simulations of the marginal stochastic flux: $dX/dt = X^2 - X^3 + \sqrt{\epsilon}\,\xi(t)$ [29], or in supercritical bifurcations with color noise [8], or when a time dependent control parameter is swept through the instability at a finite rate [30, 31], and also in extended systems [11, 32]. In the next section we will present the stochastic path approach to tackle the escape time in nonlinear instabilities, and some comments on extended systems will be briefly presented at the end of the chapter.

9.2.5 Stochastic Paths Perturbation Approach for Nonlinear Instabilities

Generally speaking, we can distinguish three regimes in the decay process from an unstable state which is deterministically stationary but decays due to fluctuations. In the first regime the system is close to the initial state. In the second the system leaves that state, and in the third the final steady state is approached. Noise effects are obviously important in the first regime, while in the second they have a relatively small influence on the evolution of the system which is essentially deterministic. Fluctuations appear again around the final steady state, but they are often quite small.

A detailed description of a relaxation process depends on the nature of the initial state and the type of instability involved. A well-studied case is the one of a supercritical pitchfork bifurcation [24, 27] (this was summarized with a specific example in the previous section). In this situation an initially stable state becomes unstable when the instability is crossed. The initial stage of the relaxation process is associated with linear stochastic dynamics, so that an initial description in terms of Gaussian statistics is possible. The time scale separating the first and second regimes is the lifetime of the state calculated as a MFPT.

The relaxation process is much less understood when the initial state is one of marginal stability. These states appear in first-order-like instabilities at the end point of hysteresis cycles. Typical cases are those of a saddle-node instability and a subcritical pitchfork bifurcation. The essential difficulty of the description of a relaxation from a state of marginal stability is that there is no regime of interest in which a linear approximation is meaningful. The process requires a nonlinear stochastic description right from the beginning. As a consequence, no initial Gaussian regime exists [6].

Of particular importance is the case when the potential breaks the inversion symmetry $X \rightarrow -X$ and the unstable state is marginal at $X = 0$. Here we will focus on that special case in order to illustrate the methodology. Then, let us consider the stochastic differential equation

$$\frac{dX}{dt} = b + aX^2 + \mathcal{O}\left(X^2\right) + \sqrt{\epsilon}\,\xi(t); \; a > 0, \; b > 0, \; X(0) = 0, \qquad (9.65)$$

this dynamics can represent the critical behavior of a well stirred Semenov model for thermal explosive systems [33]. Note that higher order terms $\mathcal{O}\left(X^2\right)$ have not been written because in the small noise approximation, the dominant contribution to the MFPT is independent of the saturation. The marginal case corresponds to $b = 0$. In (9.65) $\xi(t)$ is as usual a Gaussian white noise (9.54), and X represents the order parameter of the system near the critical point. In what follows we always assume a small noise approximation: $\epsilon \to 0$.

The MFPT from $X = 0$ to reach some attractor of $\mathcal{O}(1)$ is given in terms of the lifetime of the marginal unstable state of the **sde** (9.65). In order to introduce a perturbation theory it is convenient to write the process $X(t)$ as the ratio of two stochastic processes

$$X(t) = \frac{H(t)}{Y(t)}. \tag{9.66}$$

Using this nonlinear transformation in (9.65) we obtain an equivalent set of coupled equations

$$\frac{dH(t)}{dt} = bY(t) + \sqrt{\epsilon}\,\xi(t), \; H(0) = 0 \tag{9.67}$$

$$\frac{dY(t)}{dt} = -aH(t); \; Y(0) = 1 \tag{9.68}$$

In the absence of noise ($\epsilon = 0$) from (9.67) and (9.68) we obtain

$$\frac{d^2}{dt^2} Y(t) = -abY(t),$$

which is in agreement with the dynamics of the deterministic system. For small ϵ an approximate solution of the coupled Eqs. (9.67) and (9.68) can be considered approaching suitable $Y(t)$ in Eq. (9.67). At the initial noise-diffusive regime in which $Y(t)$ is close to its initial value, $H(t)$ is essentially a Wiener process plus a drift. Hence, we obtain

$$H(t) \simeq bt + \sqrt{\epsilon}\,W(t), \; W(0) = 0 \tag{9.69}$$

where

$$W(t) = \int_0^t \xi(t')dt', \tag{9.70}$$

is the Wiener process. In order to find an iterative solution, starting with $Y(0) = 1$, we solve (9.68) with the approximate solution $H(t)$ given by (9.69)

$$Y(t) \simeq 1 - a \int_0^t \left[bt' + \sqrt{\epsilon}\,W(t') \right] dt' \tag{9.71}$$

$$= 1 - \frac{1}{2}abt^2 - a\sqrt{\epsilon}\,\Omega(t).$$

Where the stochastic process $\Omega(t)$ is defined by

$$\Omega(t) = \int_0^t W(t')dt',\tag{9.72}$$

thus $\Omega(t)$ is a renormalized Gaussian process (see Sect. 3.17.2).

Exercise. Show that the pdf of $\Omega(t = 1) \equiv \Omega$ is

$$P(\Omega) = \sqrt{\frac{3}{2\pi}}\exp\left(-\frac{3}{2}\Omega^2\right)\tag{9.73}$$

and so the moments are given by

$$\langle \Omega^m \rangle = \frac{\Gamma\left(m + \frac{1}{2}\right)}{\sqrt{\pi}}\left(\frac{2}{3}\right)^{m/2}.$$

Hence, from (9.69) and (9.71), our first approximation for the stochastic path $X(t)$ is

$$X(t) \simeq \frac{bt + \sqrt{\epsilon}\,W(t)}{1 - \frac{1}{2}abt^2 - a\sqrt{\epsilon}\,\Omega(t)}.\tag{9.74}$$

At this stage the complicated mechanism of the escape process can be noticed. The numerator is a Wiener process, of $\mathcal{O}\left(\sqrt{\epsilon}\right)$, at the early initial stage if $b = 0$, otherwise there is competition between drift and diffusion. From this perturbation it is easy to observe the nonsymmetric fluctuations of the paths. The denominator gives the corrections to the statistics due to the nonlinear contribution in the unstable evolution (normal form (9.65)). Note that the numerator of Eq. (9.74) is bounded for $t \neq \infty$, since the Wiener process satisfies $W(t)/t \to 0$ for $t \to \infty$ with probability one (to prove this affirmation, see exercise in Sect. 3.8).

Rescaling time as $s = t'/t$ in the integral of the Wiener process, we obtain from (9.72)

$$\Omega(t) = t^{3/2}\Omega(1)$$

therefore the **rv** Ω is characterized by the pdf (9.73). The escape time, defined by $X(t_e) = \infty$, can then be obtained as the zero of the denominator of the stochastic path given in Eq. (9.74).

Up to this order $\mathcal{O}\left(\sqrt{\epsilon}\right)$, the stochastic path perturbation approach gives the random escape time, t_e, as a mapping with the random number Ω. The random escape time is found by inverting t_e as a function of Ω from:

$$1 = \frac{1}{2}abt_e^2 + a\sqrt{\epsilon}\,\Omega\,t_e^{3/2}\tag{9.75}$$

Note that $P(\Omega)$ is a symmetric pdf. Nevertheless, from Eq. (9.75) it is easily seen that there is a symmetry breaking in $t_e(\Omega)$ (i.e., under the transformation $\Omega \to -\Omega$) as can be appreciated from a simple graph of the solution of Eq. (9.75). This is a consequence of the symmetry breaking $X \to -X$ in the associated potential of the deterministic dynamics of (9.65).

The First Passage Time Density

If $b \neq 0$, Eq. (9.75) cannot be easily inverted as $t_e = t_e(a, b, \epsilon, \Omega)$ (but uniqueness of a positive t_e is easily seen from a hand graph of this equation). Therefore, we will work out the FPTD which, in fact, can be obtained analytically from the Jacobian of the transformation $|d\Omega/dt_e|$. Because the inversion is unique, using the theorem of the transformation of an **rv** we can write

$$\mathcal{P}(t_e) = P(\Omega) \, |d\Omega/dt_e| . \tag{9.76}$$

Defining a time-scale τ (a deterministic value) and a group parameter K in the form

$$\tau = \sqrt{\frac{2}{ab}}, \ K = \frac{b^3}{(a\epsilon^2)} ,$$

from (9.76) we can write the FPTD in the form

$$\mathcal{P}(t_e) = \sqrt{\frac{3}{2\pi}} \exp\left[-\frac{3}{2}\frac{\tau^3}{t_e^3} \sqrt{\frac{K}{8}}\left(1 - \frac{t_e^2}{\tau^2}\right)^2 \right] \left|\frac{d\Omega}{dt_e}\right| \tag{9.77}$$

with

$$\left|\frac{d\Omega}{dt_e}\right| = \frac{1}{2\tau}\left(\frac{K}{8}\right)^{1/4}\left[\left(\frac{\tau}{t_e}\right)^{1/2} + 3\left(\frac{\tau}{t_e}\right)^{5/2}\right] ,$$

this function is shown in Fig. 9.4.

The approximation described above (case $b \neq 0$) has a systematic underestimation of the escape time; this can be checked from a numerical simulation of the process. To correct this problem we can introduce a simple modification which has shown to be very effective in different problems [6, 9, 29]. The idea is to compare an exact mean-value with some value from the iterative solution, then by renormalizing the parameters we can match each result. From Eqs. (9.67) and (9.68) we can write

$$\frac{d^2}{dt^2}Y(t) = -abY(t) - a\sqrt{\epsilon}Y(t)\xi(t) \tag{9.78}$$

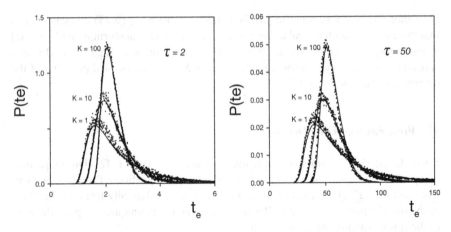

Fig. 9.4 Plot of the FPTD $P(t_e)$ as a function of t_e for three values of $K = \{100, 10, 1\}$ and two values of $\tau = \{2, 50\}$. The dots show the Monte Carlo simulations of the first passage time from $X(0) = 0$ to reach order $\mathcal{O}(1)$ using the SDE (9.65), see [34]

This equation can be identified with the Kubo oscillator [5]. Since $\xi(t)$ is a Gaussian white noise, it is possible to obtain an exact equation for the mean value of $Y(t)$

$$\frac{d^2}{dt^2} \langle Y(t) \rangle = -ab \langle Y(t) \rangle , \; \langle Y(0) \rangle = 1. \tag{9.79}$$

The solution of this equation is

$$\langle Y(t) \rangle = \cos \sqrt{ab}t$$

then the mean value vanishes when $t_m = \pi/\sqrt{4ab}$, independently of the noise parameter ϵ, while our approximation, Eq. (9.75) with $\epsilon = 0$ gives a different escape time $\tau = \vartheta \, t_m$, with $\vartheta = \sqrt{8}/\pi$. A simple way to improve our result is to force Eq. (9.75) to vanish at the correct time t_m for arbitrary ϵ. This is done by introducing the quantity ϑ by

$$1 = \frac{1}{2}ab \, (\vartheta \, t_e)^2 + a\sqrt{\epsilon} \, \Omega \, (\vartheta \, t_e)^{3/2} .$$

Thus, redefining the physical constants as $a' = \vartheta^{3/2}a$ and $b' = \vartheta^{1/2}b$, the formulae for $\mathcal{P}(t_e)$, K and τ remain valid. In this way we can improve the FPTD given by Eq. (9.77). The comparison with the numerical simulations can be seen in Fig. 9.4.

Exercise (**Marginal Case**). From Eq. (9.75) for the case $b = 0$ we get

$$t_e = \left(a^2\epsilon\Omega^2\right)^{-1/3}$$

for the escape time. Then the MFPT can be obtained (starting from the initial condition $X(0) = 0$) from the statistics of $\langle \Omega^{-2/3} \rangle$. Show that the MFPT can be written as

$$\langle t_e \rangle = \left(a^2 \epsilon \right)^{-1/3} \Gamma \left(1/6 \right) \left(3/2 \right)^{1/3} / \sqrt{\pi},$$

and compare with Suzuki's scaling time (9.59). In general, higher moments will diverge as can be seen from the mean values of $\langle \Omega^{-m} \rangle$ if $m > 1$, indicating the occurrence of an anomalous broad FPTD (which in fact can be calculated to first $\mathcal{O}\left(\sqrt{\epsilon} \right)$ from (9.75) with $b = 0$)

$$P(t_e) = \frac{3^{3/2}}{a\sqrt{2\pi\epsilon}t_e^{5/2}} \exp \left(-\frac{3}{2} \left(a^2\epsilon t_e^3 \right)^{-1} \right). \tag{9.80}$$

This is a consequence of the flatness at the unstable point $X = 0$ in the marginal case. However, this FPTD (9.80) does not give a good description of the random escape trajectories, but this issue can be overcome by going to the next order iteration, $\mathcal{O}\left(\epsilon \right)$, in the stochastic paths (9.74), see [29].

Exercise. Using Stratonovich's calculus and Novikov's Theorem obtain the exact evolution Eq. (9.79).

Exercise. From (9.77) obtain in the limit $b \to 0$ the FPTD (9.80).

Optional Exercise (**Subcritical Bifurcation**). Consider the relaxation from $X = 0$ in a subcritical bifurcation with additive Gaussian white noise [6]

$$\frac{dX}{dt} = bX^3 - cX^5 + \sqrt{\epsilon} \, \xi \left(t \right), \ b > 0, \ c \geq 0, \ X(0) = 0, \tag{9.81}$$

since we are only interested in the escape process we set $c = 0$ in (9.81). In order to approximate the individual paths of the relaxation process we write $X(t)$ as the ratio of two stochastic processes $X(t) = Z(t)/\sqrt{Y(t)}$, then (9.81) is equivalent to the set of equations

$$\frac{dZ(t)}{dt} = \sqrt{\epsilon}\sqrt{Y(t)}\xi \left(t \right)$$

$$\frac{dY}{dt} = -2bZ \left(t \right)^2,$$

with the initial conditions $Z(0) = X(0) = 0, \ Y(0) = 1$. As in the previous case, see (9.74), this set of equations can be solved iteratively from the initial conditions. Show that the zero-th order iteration is

$$X(t) \sim Z(t) = \sqrt{\epsilon} \, W \left(t \right),$$

where $W(t)$ is the Wiener Process. Show that in the first order iteration the stochastic paths can be approximated by

$$X(t) \simeq \frac{\sqrt{\epsilon}\, W(t)}{\sqrt{1 - 2b \int_0^t W(s)^2\, ds}}. \tag{9.82}$$

In this approximation the decomposition $X(t) = Z(t)/\sqrt{Y(t)}$ can be interpreted as follows. The numerator represents the first stage of evolution and there is no escape from $X = 0$. The nonlinearity introduced by the process $Y(t)$ is essential for the escape process to take place. There is no regime of escape process in which Gaussian statistics holds. The strong nonlinearity implies that the deterministic solution of (9.81) with $c = 0, X(0) \neq 0$ reaches $|X| = \infty$ in a finite time. Then, it is natural to identify the escape time from $X = 0$ as a random time t_e for which $|X(t_e)| = \infty$ in (9.82). Using the scaling of the Wiener process $W(\Lambda s)^2 = \Lambda W(s)^2$, show that we can write the random escape time as

$$t_e = \sqrt{\frac{1}{2b\epsilon\, \Xi}}, \tag{9.83}$$

with

$$\Xi \equiv \int_0^1 W(s)^2\, ds.$$

Therefore, (9.83) gives the statistics of t_e as a transformation of the random number Ξ (non-Gaussian variable). Using the generating function $G(\lambda) = \langle e^{-\lambda\, \Xi} \rangle = 1/\sqrt{\cosh(2\sqrt{\lambda})}$ [6], show that the characteristic scaling time is given by

$$\langle t_e \rangle = \left[\frac{\sqrt{2}}{16\sqrt{b\epsilon}} \right] \Gamma\left(\frac{1}{4}\right)^2.$$

Show that all the moments of the escape time can be calculated from the integral of the generating function as

$$\langle t_e^{2n} \rangle = \frac{(2b\epsilon)^{-n}}{\Gamma(n)} \int_0^\infty d\lambda\, \lambda^{n-1} G(\lambda),$$

and using the instanton-like approximation the transient behavior of the **sp** $X(t)$ can be studied.

Optional Exercise (**Semenov's Model**). The Semenov model describes a well-stirred gaseous combustion taking place in a vessel embedded in a heat bath maintained at constant temperature (the model does not allow for mass flow of reactants across the boundaries). In the limit when the thermal relaxation is much faster than the chemical one, we can neglect reactant consumption. Then, the energy balance gives the evolution for the temperature in the vessel. This equation takes into account the two essential mechanisms responsible for thermal effects: the chemical source and the cooling term describing heat exchange between the system and the reservoir. Introducing dimensionless variables the rate equation for the temperature becomes

$$\frac{dT}{d\tau} = \exp\left(\frac{-1}{ET}\right) - \gamma\,(T-1) \equiv f\,(T,\gamma) \tag{9.84}$$

Show that there exists a range of the parameter $\gamma \in [\gamma_c, \gamma_b]$ for which Eq. (9.84) admits two simultaneous stable stationary states separated by an intermediate unstable state provided that E is smaller than a value E^* independent of γ. Following the usual perturbation treatment in the form $T = T_c + \sqrt{\lambda}T_{(1)} + \mathcal{O}\,(\lambda)$, where $\lambda = (\gamma/\gamma_c - 1)$ (λ is a small parameter when dealing with spontaneous explosion), and introducing these expressions in Eq. (9.84) it is possible to isolate terms corresponding to the different orders in λ from

$$\frac{d}{d\tau}\left(T_c + \sqrt{\lambda}T_{(1)} + \mathcal{O}(\lambda)\right) \simeq \left.\frac{\partial^2 f(T,\gamma)}{\partial T^2}\right|_{T_c,\gamma_c} \frac{(T-T_c)^2}{2} + \left.\frac{\partial f(T,\gamma)}{\partial \gamma}\right|_{T_c,\gamma_c} (\gamma-\gamma_c) + \cdots$$

Therefore, we can get an amplitude equation for the correction $T_{(1)}$ of the form

$$\frac{dT_{(1)}}{dz} = b + aT_{(1)}^2, \tag{9.85}$$

where the slow time scale is defined as $z = \lambda\tau$ (compare with (9.65)). Find explicitly the expression for the constants a, b [35].

9.2.6 Genesis of Extended Systems: Relaxation from Unstable States*

Fischer's Equation

An extended system can be idealized as the continuous limit of a multidimensional process $X_i(t)$, therefore, we can work out a model where $\phi\,(r,t)$ is a stochastic process (field) with r a continuous index. For example the discrete operator that evolves a random walk in a lattice can be replaced by the Laplacian (diffusion operator) in a continuous representation, and so on. We can generalize the logistic equation considering the diffusion of the species in the form [36]:

$$\frac{\partial \phi}{\partial t} = a\phi - b\phi^2 + D\frac{\partial^2}{\partial x^2}\phi, \quad a > 0, b > 0, \phi\,(x,t) \geq 0. \tag{9.86}$$

This equation has two stationary (homogeneous) states: $\phi_{st} = 0$ and $\phi_{st} = a/b$. Linearizing the time evolution around ϕ_{st}, the stability analysis is quite simple. From Eq. (9.86) the linear dynamics close to $\phi_{st} = 0$, the unpopulated state, is

$$\partial_t \Delta\phi = \partial_x^2 \Delta\phi + a \, \Delta\phi, \qquad (9.87)$$

while close to $\phi_{st} = a/b$, the fully populated state, it is

$$\partial_t \Delta\phi = \partial_x^2 \Delta\phi - a\Delta\phi. \qquad (9.88)$$

To obtain a dispersion relation (spectrum) for the perturbation we take

$$\Delta\phi(x, t) \propto e^{\varphi t + ikx}$$

and substitute it in the time evolution equations above to get the relation connecting the Fourier wave number k and φ. We find that the spectrum near the unpopulated state is

$$\varphi(k) = -Dk^2 + a, \qquad (9.89)$$

and near the fully populated state it is

$$\varphi(k) = -Dk^2 - a. \qquad (9.90)$$

From (9.89) we can conclude that perturbations around the unpopulated state $\phi_{st} = 0$ are unstable for $-Dk^2 + a > 0$, i.e., to large wavelength perturbations. On the other hand, any perturbation around the fully populated state $\phi_{st} = a/b$ is always stable. Different results can be obtained depending on the type of nonlinearity and the *migration* operator that we use in the model. For example the diffusion operator and/or the interaction between species (nonlinear terms) might be nonlocal in space representing long-distance actions.[2] Such interactions can be mediated through vision, hearing, smelling, or other kinds of sensing in biological population models [36].

Traveling-wave (invading fronts) and monotonic solutions for extended systems is a fascinating subject that is related to instability analysis, but it is outside the scope of the present introduction. The noise-induced transition phenomena mediated by invading fronts is an interesting problem that can be tackled introducing the first passage time theory that we have discussed before (see for example [11, 37]).

---

[2]In the advanced Exercise 9.3.3 a simple model with a generalized migration (nonlocal effect) is solved. This approach allows to characterize the time-scale of the pattern formation. That is, when the fully populated homogeneous state changes the stability and a new nonhomogeneous stable state appears triggered by fluctuations.

To get a more complete idea of the instability analysis and the first passage time technique in an extended physical problem, we now present here the study of instability in a thermochemical explosion.

Frank–Kamenetskii's Model for a Thermochemical Explosion

The Frank–Kamenetskii model considers a nonhomogeneous chemical reactor, in mechanical equilibrium, closed to mass transfer but capable of exchanging energy with a thermal reservoir at constant temperature T_a. The chemical transformation taking place within the reactor is an irreversible process, which is assumed—in its simplest description—to be a unimolecular exothermic decomposition: *fuel + oxygen* $\rightarrow$ (k) $\rightarrow$ *oxide + heat*, where the rate constant $k \equiv k(T)$ is an increasing function of the temperature [38, 39]. If for simplicity one assumes that the concentration of the reactant varies on a scale that is much slower than heat transfer, this concentration can be taken as a constant c_0. Thus, the relevant variable is the temperature profile $T \equiv T(\vec{r}, t)$ (the order parameter of the system). From the energy balance in a continuous one-dimensional model, the temperature profile T satisfies the partial differential equation

$$\sigma c_v \frac{\partial}{\partial \tilde{t}} T = Q c_0 k(T) + \kappa \frac{\partial^2}{\partial \tilde{x}^2} T, \tag{9.91}$$

where κ is the thermal conductivity of the reactant, σ is the mass density of the mixture, c_v the specific heat at constant volume, Q the heat of reaction, and $k(T)$ gives the temperature dependence of the velocity of reaction. This is the Frank–Kamenetskii equation which gives rise to a propagating flame front. If the reactor is well stirred, the diffusion term is replaced by a Newton's cooling law and the balance equation turns out to be the Semenov model [35, 38, 39], which we have introduced before (see (9.84)) in the stochastic context of the MFPT. Equation (9.91) must be solved under the boundary conditions $T(\pm L, \tilde{t}) = T_a$ (here $2L$ is the length of the one-dimensional reactor), where T_a is the temperature of the reservoir. Introducing the dimensionless transformations

$$\theta = \frac{(T - T_a)\, U}{R T_a^2}, \ \rho = \tilde{x}/L, \ \tau = \frac{\kappa}{\sigma c_v L^2} \tilde{t}, \tag{9.92}$$

where R is the gas constant and U is an activation energy (for example, if we use the Arrhenius rate model we have $k(T) = k_0 e^{-U/RT}$, it is possible to rewrite (9.91) in a simpler form in terms of the dimensionless temperature profile $\theta(\rho, \tau)$

$$\frac{\partial}{\partial \tau} \theta = \delta f(\theta) + \kappa \frac{\partial^2}{\partial \rho^2} \theta, \ -1 \le \rho \le 1. \tag{9.93}$$

Thus, the boundary conditions are now $\theta(\pm 1, \tau) = 0$. In (9.93) $f(\theta)$ is an arbitrary function representing the dimensionless law for the rate constant $k(T)$, i.e.

$$
f(\theta) = \begin{cases} e^{\theta}, & \text{Exponential model} \\ \exp\left[\frac{\theta}{1+RT_a\theta/U}\right], & \text{Arrhenius model} \\ p + q\theta + r\theta^2, & \text{Quadratic model} \end{cases} .
$$

In (9.93) δ is the dimensionless Kamenetskii control parameter

$$
\delta = QUL^2 k_0 c_0 \frac{e^{-U/RT_a}}{\kappa RT_a^2} . \tag{9.94}
$$

From (9.93) and a typical $f(\theta)$ we see that depending on the Kamenetskii parameter δ there will coexist *nonhomogeneous* stable and unstable stationary (time independent) solutions. In particular, there exists a critical value of the control parameter, $\delta = \delta_c$, where the phase-space $\{\theta_{st}, \delta\}$ has a limit point. Thus, the value of the control parameter δ leads to a bifurcation scenario in the phase space of the stationary field $\theta_{st}(\rho)$.

To clarify this further we can analyze the stationary solutions of (9.93) for the particular case when $f(\theta)$ is approximated by a *piecewise* linear function:

$$
f(\theta) = A \Theta(\theta - \theta_c) + 1, \tag{9.95}
$$

here $\Theta(\theta - \theta_c)$ is the Heaviside function.

Excursus (**Stationary Profiles**). Using expression (9.95) for the nonlinear term in (9.93) the stationary equation is $0 = \delta f(\theta_{st}) + \kappa \frac{\partial^2}{\partial \rho^2}\theta_{st}$. There exist three stationary solutions θ_{st}, fulfilling $\theta_{st}(\pm 1) = 0$

$$
\theta_{st} = \begin{cases} \theta_{\text{cold}}, & \text{if } \delta \leq 2\theta_c \\ \theta_{\text{hot}}, & \text{if } \delta \geq \theta_c\left[\frac{(A-1)^2}{a(A-\frac{1}{2})} + \frac{1}{2}\right] \equiv \delta \\ \theta_{\text{unst}}, & \text{if } \delta_b \leq \delta \leq 2\theta_c \equiv \delta_c \end{cases} .
$$

Here the *hot* and the *unstable* profiles $\theta_{\text{hot}}, \theta_{\text{unst}}$ are even functions given by

$$
\theta_u^h(\rho) = \begin{cases} \frac{-A\delta}{2}\rho^2 - \left(p_u^h\right)^2 \delta \frac{(A-1)}{2} + \frac{\delta}{2} + (p_u^h)\delta(A-1), & \forall\, 0 \leq \rho \leq (p_u^h) \\ \frac{\delta}{2}(1 - \rho^2) + (p_u^h)\delta(A-1)(1-\rho), & \forall\, (p_u^h) \leq \rho \leq 1 \end{cases} ,
$$

where

$$
p_u^h = \frac{\delta(A-1) \pm \sqrt{\delta^2(A-1)^2 - 4\delta\left(A-\frac{1}{2}\right)\left(\theta_c - \frac{\delta}{2}\right)}}{2\delta\left(A-\frac{1}{2}\right)} ,
$$

Fig. 9.5 Portrait of the stationary field θ_{st} as a function of δ near the critical point

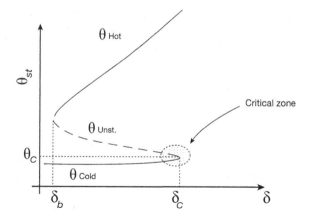

and the *cold* profile is characterized by the even function

$$\theta_{\text{cold}}\left(\rho\right) = -\frac{\delta}{2}\left(\rho - 1\right)\left(\rho + 1\right), \forall\, 0 \le \rho \le 1.$$

In general, due to the fact that the eigenvalue in the linear stability analysis of (9.93) is zero at θ_{st}, it is not possible to perform a perturbation analysis in order to study—analytically—which of the θ_{st} are asymptotically stable or unstable [32].

From the previous analysis (see Fig. 9.5) we see that for $\delta \in [\delta_b, \delta_c]$ one finds two branches of coexisting stationary states having opposite stability properties, in particular at the point δ_c the unstable and the cold branch *collide* at θ_c and subsequently they are annihilated. For this reason this point can be called a *limit point*. We should remark that a similar conclusion can also be obtained for a more realistic Arrhenius model of $f\left(\theta\right)$, but the present *piecewise linear* model gives us a simpler analytical description of the nonhomogeneous bifurcation scenario. In the next example we will be interested in the temporal evolution of a typical flame front coming from the vicinity of the critical point $\left(\theta_c, \delta_c\right)$, i.e., this is what is called the analysis of the *normal form*: the minimal dynamical expression that describes the time evolution of the order parameter near the critical point.

The results that follow are not based on any approximation concerning the nonlinear function $f\left(\theta\right)$. The stochastic evolution of the flame front can be studied on the basis of the normal form analysis, which of course is universal in the sense that its form is independent of the detailed structure of the underlying model $f\left(\theta\right)$. The structure of the function $f\left(\theta\right)$ only enters through renormalized coefficients in the normal form near the critical point $\left(\theta_c, \delta_c\right)$. At this point it is important to clarify what is meant by a *normal form*.

In order to study the relevant dynamics in the neighborhood of the critical point, it is necessary to introduce a multiple-scale transformation around the point $\left(\theta_c, \delta_c\right)$. A detailed analysis of the coalescence of the two stationary branches at the value δ_c is outside the scope of the present introduction (it can be seen in [32]). This technique is similar to the one we have introduced at the end of the previous section to solve the MFPT in higher dimension, and also when studying the dynamics near the critical point in the Semenov model (9.84). The idea is to expand the order parameter in the form

$$\theta = \theta_{(0)} + \sqrt{\lambda}\theta_{(1)} + \lambda\theta_{(2)} + \cdots ,$$

here each term satisfies the boundary conditions $\theta_{(j)} (\pm 1, t) = 0$, and $\lambda = \delta/\delta_c - 1 \geq 0$ measures the departure from the critical value δ_c. Expanding the space x and time τ variables in a consistent way, and introducing a time-dependent amplitude ϕ it is possible to derive a minimal dynamics that induces the perturbation to evolve around the critical point (θ_c, δ_c). The details of this sort of calculation can be found in many of the references on nonlinear dynamics, see for example [38].

The important point to note is that the universal saddle-node normal form that characterizes a limit point bifurcation in spatially distributed systems has the form (compare with (9.85))

$$\partial_t \phi = b + a\phi^2 + D\partial_x^2 \phi + \sqrt{\epsilon}\xi (x, t) , D > 0, \tag{9.96}$$

here we have added a space-time GWN $\xi (x, t)$ with mean value zero and correlation

$$\langle \xi (x, t) \xi (x', t') \rangle = \delta (x - x') \delta (t - t') , \tag{9.97}$$

in order to take fluctuations into account. This is the simplest model for the stochastic field $\xi (x, t)$, but to avoid possible divergences in some nonlinear cases it is necessary to smooth over the space correlation in some way, for example introducing a sharp function $f(x - x')$ instead of the delta function that appears in (9.97) [4]. In (9.96) we are facing the problem of a stochastic partial differential equation that, in principle, can be worked out using the Fourier transform technique. Therefore, the characterization of the flame explosion can be described by the analysis of the escape-times of Eq. (9.96). Nevertheless due to the fact that this equation is nonlinear, the Fourier modes of the amplitude field $\phi (x, t)$ are coupled to each other.

Optional Exercise. Using the Fourier expansions:

$$\phi (x, t) = \sum_{k=-\infty}^{+\infty} \phi_k(t) \exp (ik\pi x/L)$$

$$\xi (x, t) = \sum_{k=-\infty}^{+\infty} \xi_k(t) \exp (ik\pi x/L) ,$$

for periodic boundary conditions on $x \in [-L, L]$, here for a GWN in space-time the correlation is $\langle \xi_k (t) \xi_n (t') \rangle = \delta_{k+n,0}\delta (t - t')$. Show that at criticality the relaxation to the attractor $\theta_{\text{hot}}(x)$ is triggered by fluctuations, and the dominant equation under consideration in the Fourier space reads

$$\partial_t \phi_k = b\delta_{0,k} + a \sum_{n=-\infty}^{\infty} \phi_n \phi_{k-n} - \left(\frac{k\pi}{L}\right)^2 D\phi_k + \sqrt{\epsilon}\xi_k. \qquad (9.98)$$

The lifetime around the neighborhood of the unstable state $\phi(x,t) = 0$ is characterized by the FPTD for each Fourier mode ϕ_k to reach a macroscopic value $\phi_k \gg \sqrt{\epsilon}$. These probability distributions can be obtained (analytically) by analyzing the different stages of evolution of each Fourier mode ϕ_k, in the small noise approximation. Interestingly, it is possible to separate several stages of evolution, from which the dynamics of $\phi(x,t)$ toward the attractor $\theta_{\text{hot}}(x)$ of (9.91) can be inferred [32]. Therefore, the problem can be reduced to the first passage time analysis of a few Fourier modes.

Other models characterized by different instabilities can also be worked out, similarly, employing the stochastic path perturbation approach technique and the use of a minimum coupling approximation for the Fourier modes. This last approximation enables us to work out the escape process in terms of the most unstable Fourier mode and so to calculate the dominant contribution upon evaluating the MFPT and the anomalous transient behavior of the field, in the small noise approximation.

9.3   Additional Exercises with Their Solutions

9.3.1   Dynkin's Equation

In Sect. 9.1.3 we have presented the derivation of Dynkin's equation[3] starting from a small-time diffusion approach. Prove that Eq. (9.19) is a valid short-time evolution under the assumption that the **sp** $x(t)$ (defined on the state space $\mathcal{D}_x = [-\infty, +\infty]$) is Markovian and time-homogeneous. Hint: for the initial time $t_0 = 0$ the MFPT—in one dimension—to cross the level x_L is

$$\langle \tau(x_L | x_0) \rangle = \int_0^{\infty} dt \int_{-\infty}^{x_L} \bar{P}(x, t | x_0, 0) dx,$$

here $\bar{P}(x, t | x_0, 0)$ is a propagator that takes care of the absorbing boundary condition at $x = x_L$. Now use the Chapman–Kolmogorov equation for a short time Δt to obtain (9.19).

---

[3]E.B. Dynkin and A.A. Juschkewitz, *Satze und Aufgaben uber Markoffsche Prozesse*, Springer, Berlin 1969.

9.3.2 The Backward Equation and Boundary Conditions

It can be shown from the Chapman–Kolmogorov equation,[4] that the propagator, regarded as a function of $\vec{x}_0$ and t_0 satisfies the differential equation:

$$
\frac{\partial P(x_\nu, t \mid x_\nu^0, t_0)}{\partial t_0} = -\left[\sum_\mu K_\mu \left(x_\nu^0\right) \frac{\partial}{\partial x_\mu^0} + \sum_{\mu\lambda} D_{\mu\lambda} \left(x_\nu^0\right) \frac{\partial}{\partial x_\mu^0} \frac{\partial}{\partial x_\lambda^0} \right] P(x_\nu, t \mid x_\nu^0, t_0)
$$

$$
\equiv -\mathcal{L}^\dagger \left(\vec{x}_0, \frac{\partial}{\partial \vec{x}_0} \right) P(\vec{x}, t \mid \vec{x}_0, t_0).
$$

This is the *backward* equation in n-dimensions. If the **sp** $\vec{x}(t)$ is stationary $P(\vec{x}, t \mid \vec{x}_0, t_0) = P(\vec{x}, \tau \mid \vec{x}_0, 0)$, $\tau = t - t_0 \geq 0$. Show that the propagator satisfies also the equation: $\partial_\tau P(\vec{x}, \tau \mid \vec{x}_0, 0) = \mathcal{L}\left(\vec{x}_0, \frac{\partial}{\partial \vec{x}_0} \right) P(\vec{x}, \tau \mid \vec{x}_0, 0)$. Consider the **sp** $\vec{x}(t)$ to be confined to a region R with boundary ∂R, from the Chapman–Kolmogorov equation we can write

$$
0 = \frac{\partial}{\partial s} \int_R P(\vec{x}, t \mid \vec{q}, s) P(\vec{q}, s \mid \vec{x}_0, t_0) dV_q
$$

$$
= -\int_R \left[\mathcal{L}^\dagger \left(\vec{q}, \frac{\partial}{\partial \vec{q}} \right) P(\vec{x}, t \mid \vec{q}, s) \right] P(\vec{q}, s \mid \vec{x}_0, t_0) dV_q
$$

$$
+ \int_R P(\vec{x}, t \mid \vec{q}, s) \mathcal{L} \left(\vec{q}, \frac{\partial}{\partial \vec{q}} \right) P(\vec{q}, s \mid \vec{x}_0, t_0) dV_q.
$$

Then after some algebra using Gauss' theorem to integrate over a surface, we can infer which are the boundary conditions (on ∂R) associated with the *backward* equation, if there is a boundary condition imposed on the **F–P** equation [4].

For example, absorbing boundaries imply both $P(\vec{x}, t \mid \vec{x}_0, t_0) = 0$ for $\vec{x} \in \partial R$, and $P(\vec{x}, t \mid \vec{x}_0, t_0) = 0$ for $\vec{x}_0 \in \partial R$. A reflecting boundary on the *forward* equation demands that the current of probability be zero for $\vec{x} \in \partial R$; then, it implies

$$
\int_{\partial R} \sum_i dS_{qi} \, P(\vec{q}, s \mid \vec{x}_0, t_0) \left(\sum_j D_{ij} (q_\nu) \frac{\partial P(\vec{x}, t \mid \vec{q}, s)}{\partial q_j} \right) = 0.
$$

Therefore for all $P(\vec{q}, s \mid \vec{x}_0, t_0)$ this equation imposes a condition on the *backward* solution. In one-dimension this condition is: $\partial P(x, t \mid q, s)/\partial q = 0$ unless $D(q) = 0$ on the boundary ∂R.

---

[4] A.T. Bharuche-Reid, Elements of the Theory of Markov Process and their Applications, McGraw-Hill Book Co. Inc. New York (1960), and also [4]. In Sect. 3.18 we present an exercise on this topic.

9.3.3 Linear Stability Analysis and Generalized Migration Models

The dynamic model shown in the next equation accounts for the exponential growth of the population at short times, a competition leading to saturation, and a generalized migration term characterized by a kernel $G(x)$ [36]. We also model environmental/thermal fluctuations acting on such systems. To this end we introduce an additive fluctuating Gaussian field $\xi(x, t)$ in the dynamics, and we characterize the strength of the fluctuations with a small parameter ϵ. Then, the model is

$$\frac{\partial u(x, t)}{\partial t} = f[u(x, t)] + D \int_{\mathcal{D}} u(x - x', t) G(x') dx' + \sqrt{\epsilon} \xi(x, t) \tag{9.99}$$

$$1 = \int_{\mathcal{D}} G(x) dx, \quad u(x, t \geq 0) > 0, \quad \langle \xi(x, t) \xi(x', t') \rangle = \delta(x - x') \delta(t - t')$$

The homogeneous stationary state follows from the deterministic analysis:

$$0 = f[u_0] + u_0 D.$$

By means of a linear analysis around the (nontrivial) stable point u_0, we can use the perturbation

$$u(x, t) = u_0 + v(x, t),$$

then getting the following linear equation

$$\frac{\partial v(x, t)}{\partial t} = f'(u_0) v(x, t) + D \int_{\mathcal{D}} v(x - x', t) G(x') dx'. \tag{9.100}$$

Now, introducing a Fourier wave for the perturbation around u_0 in the form

$$v(x, t) \propto \exp(\varphi t + ikx),$$

the equation for the growth/decay rate of the Fourier modes (the so-called dispersion relation) is given by

$$\varphi = f'(u_0) + DG(k), \quad \text{where } G(k) = \int_{-\infty}^{+\infty} e^{ikx} G(x) dx. \tag{9.101}$$

We intend here to introduce the minimal model that shows a *generalized migration* dynamics. Thus, in what follows we will choose a logistic model for the interaction

$$f[u(x, t)] = u(x, t) - bu(x, t)^2.$$

Using this $f[u(x, t)]$ show that the dispersion relation is

$$\varphi = DG(k) - 2D - 1. \tag{9.102}$$

The Model for the Migration Kernel

In an ecological model the kernel must contain two important ingredients.[5] One is a positive term which accounts for the opportunities and lack of strong competition near the origin, and a negative term for competition effects beyond a certain range:

$$G(x) = \mathcal{N}\left(Be^{-Ax^2} - e^{-x^2}\right), \quad \mathcal{N}^{-1} = \sqrt{\pi}\left(\frac{B}{\sqrt{A}} - 1\right). \tag{9.103}$$

The Fourier transform of this kernel is

$$G(k) = \mathcal{N}\sqrt{\pi}\left(\frac{B\exp\left(-k^2/4A\right)}{\sqrt{A}} - e^{-k^2/4}\right). \tag{9.104}$$

With such characteristics the dispersion relation may present an instability as shown in Fig. 9.6.

Stochastic Pattern Formation

A nonhomogeneous pattern may be triggered by fluctuations due to the instability around the fully populated state u_0. To study this we now look at a finite system with periodic boundary conditions on $x \in [-1, 1]$. Thus from now on we will use a discrete Fourier transform ($k_n = n\pi$, $n = 0, \pm 1, \pm 2, \cdots$) characterized by the modes:

$$u(x, t) = \sum_{n=-\infty}^{\infty} A_n(t)\exp(in\pi x) \tag{9.105}$$

$$\xi(x, t) = \sum_{n=-\infty}^{\infty} \xi_n(t)\exp(in\pi x) \tag{9.106}$$

$$G(x) = \sum_{n=-\infty}^{\infty} G_n\exp(in\pi x) \tag{9.107}$$

---

[5]An application in a bio-physical scenario can be seen in: "Fairy circles and their non-local stochastic instability", M.A. Fuentes and M.O. Cáceres, Eur. Phys. J. (2017). doi:10.1140/epjst/e2016-60178-1.

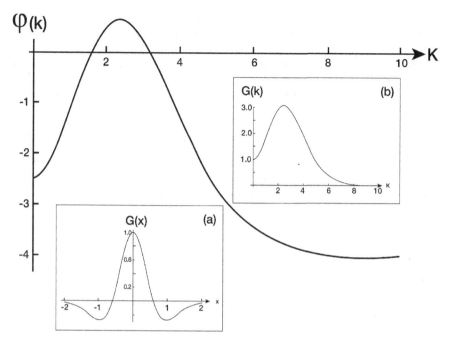

Fig. 9.6 Dispersion φ as a function of k. The instability occurs when $\varphi\,(k) > 0$. Here $D = 1.5$ in Eq. (9.102) and the function $G\,(k)$ can be read from Eq. (9.104) with parameters: $A = 3, B = 2$

$$\int_{-1}^{1} e^{i\pi\,(n-l)x}dx = 2\delta_{n,l}, \qquad G_n = \frac{1}{2}\int_{-1}^{1} G\,(x)\,e^{-in\pi x}dx,\ \text{etc.} \tag{9.108}$$

We note the relation between the discrete and continuous Fourier transform: $2G_n = G\,(k)$. Introducing (9.105)–(9.107) in the field equation (9.99) and using (9.108) show that the following set of stochastic coupled modes is found

$$\dot{A}_l = A_l\,(t)\,(1 + 2G_lD) - b\sum_{n=-\infty}^{\infty} A_n\,(t)\,A_{l-n}\,(t) + \xi_l\,(t)\,, \tag{9.109}$$

$$G_l = G_{-l},\ \langle\xi_l\rangle = 0,\ \langle\xi_l\,(t)\,\xi_n\,(t')\rangle = \delta_{n+l,0}\,\delta\,(t - t')\,. \tag{9.110}$$

In order to study the emergence of pattern formation (Fourier amplitude $A_{\pm e}$) from this set of coupled modes we introduce a minimum coupling approximation considering only the coupling between A_0 and $A_{\pm e}$. We thus get

$$\frac{dA_0}{dt} = A_0\,(1 + 2G_0D) - bA_0^2 - 2bA_e^2 + \sqrt{\epsilon}\xi_0\,(t) \tag{9.111}$$

$$\frac{dA_e}{dt} = A_e\,(1 + 2G_eD - 2bA_0) + \sqrt{\epsilon}\xi_e\,(t)\,, \tag{9.112}$$

in this approximation $A_e = A_{-e}$. This set of equations is a good approach during the initial stage of pattern formation and with them we can predict the random escape time.

The stochastic modes (9.111) and (9.112) can be analyzed introducing a multiple scale expansion. Since originally the initial state is the fully populated u_0, we take $A_n(0) = 0, \forall n > 0$. The proposed expansions for the homogeneous and the explosive modes are then

$$A_0(t) = H_0 + H_1\sqrt{\epsilon} + H_2\epsilon + \cdots, \qquad H_0 = \text{const.}, H_j(0) = 0, \forall j > 0 \qquad (9.113)$$

$$A_e(t) = 0 + E_1\sqrt{\epsilon} + E_2\epsilon + \cdots, \qquad E_j(0) = 0, \forall j > 0. \qquad (9.114)$$

We now introduce these expansions in (9.111) and (9.112). Show that to order $\mathcal{O}(\epsilon^0)$

$$0 = H_0(1+D) - bH_0^2 \Rightarrow H_0 = (1+D)/b,$$

and to order $\mathcal{O}(\epsilon^{1/2})$

$$\frac{dH_1}{dt} = H_1(1+D) - bH_0H_1 + \xi_0 \Rightarrow H_1(t) = \int_0^t \xi_0(s)ds \qquad (9.115)$$

$$\frac{dE_1}{dt} = E_1(1 + 2G_eD - 2(1+D)) + \xi_e \Rightarrow E_1(t) = \int_0^t e^{\varphi(t-s)}\xi_e(s)ds, \qquad (9.116)$$

here $\varphi = 2G_eD - 1 - 2D$ is the dispersion relation (9.102) in the discrete Fourier representation. The perturbation $H_1(t)$ is a Wiener process with mean value zero and it is statistically independent of the process $E_1(t)$, this last process can be rewritten in the form

$$H_1(t) = e^{\varphi t}h(t), \text{ with } \frac{dh}{dt} = e^{-\varphi t}\xi_e(t). \qquad (9.117)$$

Note that the process $h(t)$ is Gaussian with mean zero and dispersion

$$\left\langle h(t)^2 \right\rangle = \frac{1 - e^{-2\varphi t}}{2\varphi}.$$

Therefore for times $2\varphi t \gg 1$ the process $h(t)$ saturates to a Gaussian **rv** Ω which is characterized by the pdf

$$P(\Omega) = \frac{e^{-\Omega^2/2\sigma^2}}{\sqrt{2\pi\sigma^2}}, \quad \sigma^2 = \frac{1}{2\varphi}, \quad \varphi = 2G_eD - 1 - 2D.$$

With all this information we can calculate the lifetime around the unstable state u_0, and therefore study the pattern formation we mentioned earlier. In order to do this we need to find the FPTD, for the process $A_e(t)$, from the value 0 to some macroscopic (threshold) value A_e^*.

The First Passage Time Distribution

The solution of $H_1(t)$, see (9.117), can be used as a mapping to define the random escape times associated with the unstable mode $A_e(t)$ for the passage from $0 \to A_e^*$. If $t \gg 1/2\varphi$ we can approximate $h(t)$ by the **rv** Ω, then we get a map for the random escape times t_e as function of the **rv** Ω

$$H_1(t_e) = \exp(\varphi t_e) \, \Omega \;\Rightarrow\; t_e = \frac{1}{2\varphi} \ln\left(\frac{H_1^*}{\Omega}\right)^2. \tag{9.118}$$

Now using the scaling (9.114) we can write

$$t_e = \frac{1}{2\varphi} \ln\left(\frac{A_e^*}{\sqrt{\epsilon}\Omega}\right)^2, \tag{9.119}$$

which can be used to calculate the lifetime as the **mv** $\langle t_e \rangle$. In addition, the FPTD can also be calculated:

$$P(t_e) = \int_{\mathcal{D}} \delta\left(t_e - \frac{1}{2\varphi} \ln\left(\frac{A_e^*}{\sqrt{\epsilon}\Omega}\right)^2\right) P(\Omega) \, d\Omega, \tag{9.120}$$

here the domain $\mathcal{D}$ has to be chosen so as to ensure that $t_e \geq 0$. Using a dimensionless unit of time, show that

$$P(\tau_e) = \frac{2K}{\mathrm{erf}(K)\sqrt{\pi}} \exp\left(-\tau_e - K^2 \exp(-2\tau_e)\right) \tag{9.121}$$

$$K = A_e^* \sqrt{\frac{\varphi}{\epsilon}}, \quad \tau_e = \varphi t_e = (2G_e D - 1 - 2D)\, t_e,$$

here $2G_e = G(k_e)$ is given in (9.104) with $k_e = n_e \pi$, and n_e indicates the most unstable modes. Then the MFPT can be calculated as

$$\langle \tau_e \rangle = \int_0^\infty P(\tau_e) \, d\tau_e \simeq \ln(K) + \frac{E + \ln 4}{2\,\mathrm{erf}(K)}, \quad K \gg 1, \; E = \text{Euler constant}. \tag{9.122}$$

Interestingly we have reached from first principles to the Gumbel pdf (9.121), which is used to model the distribution of the maximum of a number of samples of various distributions. It is useful, for example, in predicting the chance that an extreme earthquake, flood or other natural disaster will occur in nature.[6] The applicability of Gumbel's pdf to represent the distribution of maxima relates to extreme value theory, as we have done here characterizing a threshold level crossing from the dynamics of the most unstable Fourier's mode of the harmonic analysis of the concentration field $u(x, t)$.

References

1. H.A. Kramers, Brownian motion in a field of force and the diffusion model of chemical reactions. Phys. **7**(4), 284–304 (1940)
2. H. Haken, *Synergetic*, Chap. 7 (Springer, Berlin, 1977)
3. W.F. Brown, Thermal fluctuations of a single-domain particle. Phys. Rev. **130**, 1677 (1963)
4. C.W. Gardiner, *Handbook of Stochastic Methods* (Springer, Berlin, 1983)
5. N.G. van Kampen, *Stochastic Process in Physics and Chemistry*, 2nd edn. (North-Holland, Amsterdam, 1992)
6. P. Colet, F. de Pasquale, M.O. Cáceres, M. San Miguel, Theory for relaxation at a subcritical pitchfork bifurcation. Phys. Rev. A **41**, 1901, (1990); P. Colet, F. de Pasquale, M. San Miguel, Relaxation in the subcritical pitchfork bifurcation: from critical to Gaussian scaling. Phys. Rev. A **43**, 5296 (1990)
7. M. San Miguel, H. Hernandez-Garcia, P. Colet, M.O. Cáceres, F. De Pasquale, in *Instabilities and Nonequilibrium Structures III*, ed. by E. Tirapegui, W. Zeller (Kluwer, Dordrecht, 1991), pp. 143–155
8. J.M. Sancho, M. San Miguel, Passage times for the decay of an unstable state triggered by colored noise. Phys. Rev. A **39**, 2722 (1989)
9. M.O. Cáceres, Passage time statistics in a stochastic verhulst model. J. Stat. Phys. **132**, 487–500 (2008)
10. M.O. Cáceres, J. Stat. Phys. **156**, 94–118 (2014)
11. M.O. Cáceres, M.A. Fuentes, First passage times for pattern formation in non-local partial differential equations. Phys. Rev. E **92**, 042122 (2015)
12. N.G. van Kampen, in *Instabilities and Nonequilibrium Structures III*, ed. by E. Tirapegui, W. Zeller (Kluwer, Dordrecht, 1987), pp. 241–270
13. R.L. Stratonovich, *Topics in the Theory of Random Noise*, Vols. 1 and 2 (Gordon and Breach, New York, 1963)
14. N.S. Goel, N. Richter-Dyn, *Stochastic Models in Biology* (Academic Press, New York, 1974)
15. J. Mathews, R.L. Walker, *Mathematical Methods of Physics*, 2nd edn., (W.A. Benjamin, INC., California, 1973)
16. H. Brand, A. Schenzlew, G. Schroeder, Lower and upper bounds for the eigenvalues of the Fokker-Planck equation in detailed balance. Phys. Rev. A **25**, 2324 (1982)
17. G.S. Agarwal, S.R. Shenoy, Observability of hysteresis in first-order equilibrium and nonequilibrium phase transitions. Phys. Rev. A **23**, 2719 (1981)
18. G.S. Agarwal, S. Dattagupta, K.P.N. Murthy, Relaxation behaviour of single-domain magnetic particles. J. Phys. C Solid State Phys. **17**, 6869 (1984)
19. R. Landauer, J.A. Swanson, Frequency factors in the thermally activated process. Phys. Rev. **121**, 1668 (1961)
20. J.S. Langer, Statistical theory of the decay of metastable states. Ann. Phys. (NY) **54**, 258 (1969)

---

[6]See, D. Sornette in: *Critical Phenomena in Natural Science*, Springer-Verlag, Berlin, 2000.

21. Z. Schuss, B.J. Matkowsky, The exit problem: a new approach to diffusion across potential barriers. SIAM J. Appl. Math. **36**(3), 604–623 (1979). doi:10.1137/0136043
22. Z. Schuss, Singular perturbation methods in stochastic differential equations of mathematical physics. SIAM Rev. **22**, 119–155 (1980)
23. P. Talkner, D. Ryter, Lifetime of a metastable state at low noise. Phys. Lett A **88**, 162 (1982)
24. M. Susuki, J. Stat. Phys. **16**, 11 (1977)
25. B. Caroli, C. Caroli, B. Roulet, Diffusion in a bistable potential: a systematic WKB treatment. J. Stat. Phys. **21**, 415 (1979)
26. F. Haake, Decay of unstable states. Phys. Rev. Lett. **41**, 1685 (1978)
27. F. de Pasquale, P. Tartaglia, P. Tombesi, Transient laser radiation as a stochastic process near an instability point. Phys. A. **99**, 581 (1979)
28. M.O. Cáceres, C.D. Rojas, Exponential distributed time-delay nonlinear models: Monte Carlo simulations. Phys. A **409**, 61 (2014)
29. M.O. Cáceres, M.A. Fuentes, C.E. Budde, Stochastic escape processes from a non-symmetric potential normal form II: the marginal case. J. Phys. A Math. Gen. **30**, 2287 (1997)
30. M.C. Torrent, M. San Miguel, Stochastic-dynamics characterization of delayed laser threshold instability with swept control parameter. Phys. Rev. A **38**, 245 (1988); M.C. Torrent, F. Sagues, M. San Miguel, Dynamics of sweeping through an instability: passage-time statistics for colored noise. Phys. Rev. A **40**, 6662 (1989)
31. M.O. Cáceres, A. Becker, Passage times for the decay of a time-dependent unstable state. Phys. Rev. A **42**, 696 (1990)
32. M.O. Cáceres, M.A. Fuentes, Stochastic escape processes from a non-symmetric potential normal form III: extended explosive systems. J. Phys. A Math. Gen. **32**, 3209 (1999)
33. P. Gray, W. Kordylewski, Dynamic responses to perturbations in the non-isothermal, continuous-flow stirred-tank reactor (CSTR)-II. Nonadiabatic operation: the general case. Chem. Eng. Sci. **40**, 1703 (1985)
34. M.O. Cáceres, C.E. Budde, G.J. Sibona, Stochastic escape processes from a nonsymmetric potential normal form. J. Phys. A Math. Gen. **28**, 3877 (1995)
35. M.O. Cáceres, G. Nicolis, C.E. Budde, About the shift between the advanced and delayed thermal explosion times. Chaos Solitons Fractals **6**, 51–57 (1995)
36. J.D. Murray, *Mathematical Biology*, vols. 1 and 2, 3rd edn. (Springer, Berlin, 2007)
37. M. San Miguel, R. Toral, in *Instabilities and Nonequilibrium Structures VI*, ed. by E. Tirapegui, W. Zeller (Kluwer, Dorchester, 1997)
38. G. Nicolis, *Introduction to Nonlinear Science* (Cambridge University Press, Cambridge, 1995)
39. G. Nicolis, Dissipative systems. Rep. Prog. Phys. **49**, 873 (1986); G. Nicolis, F. Baras, Intrinsic randomness and spontaneous symmetry-breaking in explosive systems. J. Stat. Phys. **48**, 1071 (1987)

Appendix A
Thermodynamic Variables in Statistical Mechanics

A.1 Boltzmann's Principle

The main pillar of equilibrium statistical mechanics is the hypothesis of equal probability a priori. From a quantum point of view this hypothesis states that a priori an isolated system[1] can be in any of its energy levels (with equal probability). The second pillar in the construction of equilibrium statistical mechanics is established by the ergodic hypothesis; that is, the time-average of an observable is equivalent to calculating the average over an *ensemble* of systems. Therefore, the theoretical task is to establish the relationship of the thermodynamic variables and state functions, with the dynamic (mechanical) variables of the system. The starting point for a statistical analysis (in ensemble) is established by Boltzmann's principle which defines entropy through the probability that an (isolated) system is in a given macroscopic state.[2]

Before analyzing in detail Boltzmann's principle we need to introduce some notation.[3] Suppose a given system has a certain energy E (nonnegative). Then, using the assumption of equal a priori probability we need only to count how many states there are in the energy range of interest. Representing the number of possible states in the interval $[0, E]$ by $j(E)$ and expressing this quantity in terms of the density of states $\Omega(E)$ we can write

$$j(E) = \int_0^E \Omega(E') \, dE'.$$

---

[1] A closed and isolated system does not exchange matter or energy with the rest of the universe. On an isolated system the energy is a constant of motion, so only time-independent external forces are acting on it.

[2] In the canonical ensemble the probability distribution at temperature T is given by Gibbs' distribution or ensemble distribution; see Appendix F.

[3] See [4] from Chap. 2.

© Springer International Publishing AG 2017
M.O. Cáceres, *Non-equilibrium Statistical Physics with Application to Disordered Systems*, DOI 10.1007/978-3-319-51553-3

From this relation we immediately see that in the differential element between E and $E + dE$ there will be $\Omega(E)dE$ microscopic states. So if we are interested in characterizing a macroscopic system, with energy between E and $E + dE$, we conclude that the quantity of interest is $W = \Omega(E)\,\delta E$, therefore from the principle of equal *a priori probability*, the probability that the macroscopic system has an energy between E and $E + \delta E$ is given by the number $P = 1/W$. We can then write Boltzmann's principle as follows:

$$S = -k \log P.$$

Now consider a case in which the (macroscopic) thermodynamic variables that define the state of the system are the energy E, the uncertainty δE, the volume V, and the number of particles N. It is possible to prove that the number of microscopic states is given, in the classical limit by the quantity

$$W(E, \delta E, N, V) = \int_{E < H(N,V) < E + \delta E} \frac{\mathcal{D}p\,\mathcal{D}q}{h^{3N}N!},$$

where h is the Planck constant, H the Hamiltonian of the system, and $\mathcal{D}p\,\mathcal{D}q$ the differential element in phase-space. In the quantum case the expression takes a simpler form:

$$W(E, \delta E, N, V) = \sum_{E < H(N,V) < E + \delta E} 1.$$

Then, Boltzmann's principle defines entropy of the macroscopic system as

$$S = k_B \log W(E, \delta E, N, V) = k_B \log\left(\Omega(E, N, V)\,\delta E\right), \tag{A.1}$$

that is: the logarithm of the number of microscopic states W compatible with the macroscopic state characterized by the variables $(E, \delta E, N, V)$.

A.1.1 Systems in Thermal Contact

From principle (A.1) and the assumption of *normal* thermodynamic systems[4] it is possible to prove that if two systems with different energies E_I and E_{II} (with $E_I + E_{II} = E$) come into thermal contact, then at equilibrium [i.e., with the most likely partition: $E_I^*, E_{II}^* = E - E_I^*$] their temperatures are equal. Then, the definition of temperature T coincides with the usual thermodynamic relation:

---

[4]In the sense that when the number of particles N or volume V of the system is large, the number of states $j(E, N, V)$ grows exponentially; see [4] Chap. 2.

$$T_{\mathrm{I}}(E_{\mathrm{I}}^*) = T_{\mathrm{II}}(E_{\mathrm{II}}^*), \quad \text{with} \quad \frac{\partial S(E)}{\partial E} = \frac{1}{T}, \quad E_{\mathrm{I}}^* + E_{\mathrm{II}}^* = E;$$

note that we have used the condition $S_{\mathrm{I}}(E_{\mathrm{I}}) + S_{\mathrm{II}}(E_{\mathrm{II}}) = \max$.

Similarly, if two systems in contact exchange matter (now the thermodynamic variables are E and N), in equilibrium their chemical potentials are equal [i.e., we use the condition $S_{\mathrm{I}}(E_{\mathrm{I}}, N_{\mathrm{I}}) + S_{\mathrm{II}}(E_{\mathrm{II}}, N_{\mathrm{II}}) = \max$]. Then, the definition of the chemical potential μ coincides with the usual thermodynamic relation:

$$\frac{\mu_{\mathrm{I}}}{T_{\mathrm{I}}}(E_{\mathrm{I}}^*, N_{\mathrm{I}}^*) = \frac{\mu_{\mathrm{II}}}{T_{\mathrm{II}}}(E_{\mathrm{II}}^*, N_{\mathrm{II}}^*), \quad \text{with} \quad \frac{\partial S(E, N)}{\partial N} = -\frac{\mu}{T}, \quad N_{\mathrm{I}}^* + N_{\mathrm{II}}^* = N.$$

Finally, if two systems come into contact and exert pressure on each other (i.e., the thermodynamic variables are E, N, V), in equilibrium their pressures are equal. Then, the definition of pressure p coincides with the usual thermodynamic relation:

$$\frac{p_{\mathrm{I}}}{T_{\mathrm{I}}}(E_{\mathrm{I}}^*, N_{\mathrm{I}}^*, V_{\mathrm{I}}^*) = \frac{p_{\mathrm{II}}}{T_{\mathrm{II}}}(E_{\mathrm{II}}^*, N_{\mathrm{II}}^*, V_{\mathrm{II}}^*), \quad \text{with} \quad \frac{\partial S(E, N, V)}{\partial V} = \frac{p}{T}, \quad V_{\mathrm{I}}^* + V_{\mathrm{II}}^* = V.$$

note that here we use the condition $S_{\mathrm{I}}(E_{\mathrm{I}}, N_{\mathrm{I}},, V_{\mathrm{I}}) + S_{\mathrm{II}}(E_{\mathrm{II}}, N_{\mathrm{II}}, V_{\mathrm{II}}) = \max$.

From all of this it follows that the entropy differential is given by the thermodynamic relation:

$$dS(E, N, V) = \frac{1}{T}(dE + p\, dV - \mu\, dN). \tag{A.2}$$

This differential represents the change in entropy of the system, from equilibrium states characterized by (E, V, N) and $(E + dE, V + dV, N + dN)$.

Finally, the increase in entropy can also be viewed from a probabilistic point of view. Consider two systems (originally in equilibrium) separated and characterized by the quantities $(E_{\mathrm{I}}^0, N_{\mathrm{I}}^0, V_{\mathrm{I}}^0)$ and $(E_{\mathrm{II}}^0, N_{\mathrm{II}}^0, V_{\mathrm{II}}^0)$. When these systems are put in thermodynamic contact, they reach a new equilibrium state (characterized by the thermodynamic variables E, N, V); then, we can relate the total entropy (for the whole system) and the new partition as follows:

$$\begin{aligned}
S_{\mathrm{I+II}}(E, N, V) &= S_{\mathrm{I}}(E_{\mathrm{I}}^*, N_{\mathrm{I}}^*, V_{\mathrm{I}}^*) + S_{\mathrm{II}}(E_{\mathrm{II}}^*, N_{\mathrm{II}}^*, V_{\mathrm{II}}^*) \\
&\geq S_{\mathrm{I}}(E_{\mathrm{I}}^0, N_{\mathrm{I}}^0, V_{\mathrm{I}}^0) + S_{\mathrm{II}}(E_{\mathrm{II}}^0, N_{\mathrm{II}}^0, V_{\mathrm{II}}^0) = S_{\mathrm{initial}}.
\end{aligned}$$

We have assumed that the most likely partition $\{(E_{\mathrm{I}}^*, N_{\mathrm{I}}^*, V_{\mathrm{I}}^*), (E_{\mathrm{II}}^*, N_{\mathrm{II}}^*, V_{\mathrm{II}}^*)\}$ is different from the initial partition. Now imagine that we separate the two systems, again. Almost certainly, they will keep the new partition $(E_{\mathrm{I}}^*, N_{\mathrm{I}}^*, V_{\mathrm{I}}^*); (E_{\mathrm{II}}^*, N_{\mathrm{II}}^*, V_{\mathrm{II}}^*)$, then we get

$$S_{\mathrm{final}} = S_{\mathrm{I}}(E_{\mathrm{I}}^*, N_{\mathrm{I}}^*, V_{\mathrm{I}}^*) + S_{\mathrm{II}}(E_{\mathrm{II}}^*, N_{\mathrm{II}}^*, V_{\mathrm{II}}^*) \geq S_{\mathrm{initial}}.$$

That is, the entropy has increased just from having put the two systems in contact. Entropy will stay constant only if the systems were originally in equilibrium with each other (the initial partition was equal to the *most likely partition* after contact). So when the systems are separated, after contact, the entropy of the two systems is almost certainly higher than they were before contact:

$$\Delta S = S_{\text{final}} - S_{\text{initial}} \geq 0.$$

A.2   First and Second Law of Thermodynamics

From the point of view of statistical mechanics these fundamental laws are derived from probabilistic concepts and mechanics. In particular, the first law of thermodynamics is a direct consequence of energy conservation. That is, if we consider the internal energy U as a thermodynamic variable, we can identify U with E as long as the latter does not include it in the mechanical energy of the system. From the differential (A.2) we can consider the change in energy (internal) when the system undergoes a process of *quasi-static* variation in the form:

$$dU = -p \, dV + \mu \, dN + T \, dS. \tag{A.3}$$

Then (A.3) represents the law of conservation of energy, i.e.: $-p \, dV$ represents the increase of internal energy due to mechanical work exerted on the system, $\mu \, dN$ explains the increase in internal energy when matter is transferred to the system, and finally $T \, dS = dQ$ represents other energy transfer (reversible) to the system (e.g., as heat).

The second law of thermodynamics can be formulated in many different ways.[5] Here we are particularly interested in emphasizing the concept of probability. In an adiabatic system entropy never decreases and *only* grows in irreversible processes. Note that Boltzmann's principle, that is, the very definition of statistical entropy, can generalize this concept for *nonequilibrium* states. The fact that there is a functional relationship between the entropy and the number of accessible microscopic states W ensures, in general, that if W increases (in an irreversible process) so does S (then the constant k_B has to be positive). This establishes the equality between thermodynamic entropy and statistical entropy.[6] On the other hand, the logarithmic functional

---

[5] See [2, 3] in Chap. 2. Here it is very appropriate to quote the words of R. Feynman: "Thermodynamics is a difficult subject because everybody uses a different approach. If only we could get together for once and decide what will be our variables and thus accept it always, thermodynamics would be quite easy."

[6] From the *pure* thermodynamics point of view, the second law (or Carnot's postulate) is stated as follows: There is no process whose *only* final result is taking heat from a reservoir and totally converting it into work W. From this postulate it follows that no heat-engine that extracts heat Q_1 at temperature T_1 and delivers heat Q_2 at temperature T_2 [regardless of the substance or matter

relationship (A.1) ensures that entropy is an additive property for independent systems. A simple way to characterize the second law of thermodynamics is showing that from probabilistic concept (A.1), heat flows in a certain direction. For example, let us study the contact of two bodies at different temperatures, $T_1 > T_2$. Here it is shown that heat flows from body 1 to body 2. To see this fact we consider two *normal* systems, in the statistical sense, characterized by entropies S_1 and S_2 before contact. Since both systems are normal, the number of microscopic states grows exponentially with the number of particles $N_i \gg 1$; that is, $j_i(E_i) \sim \exp(N_i \phi_i[E_i/N_i])$, where $\phi_i[z] > 0$, $\phi_i'[z] > 0$, $\phi_i''[z] < 0$. Then, on the grounds that $S(E_1, E_2) = k_B \log W(E_1, E_2, \delta E)$ [in general for all $\phi_i[E_i/N_i] \sim \mathcal{O}(1)$], it is possible to see that[7]

$$dS(E_1, E_2) = k_B \phi_1'[E_1/N_1] \, dE_1 + k_B \phi_2'[E_2/N_2] \, dE_2. \tag{A.4}$$

Denoting the variation of energy of the body 1 as $dE_1 = -dE_2 = -dQ$, it is clear that dQ represents the amount of heat delivered to the body 2. Then we can write (A.4) as follows

$$dS(E_1, E_2) = k_B \left\{ \phi_1'[E_1/N_1] - \phi_2'[E_2/N_2] \right\} dE_1. \tag{A.5}$$

Since $j_i(E_i) \sim \exp(N_i \phi_i[E_i/N_i])$, we conclude also that temperature [characterized by the relation $\partial S_i(E_i)/\partial E_i = \frac{1}{T_i}$] is $T_i(E_i) = 1/(k_B \phi_i')$. Finally, we can rewrite (A.5) as

$$dS(E_1, E_2) = (T_2^{-1} - T_1^{-1}) \, dQ. \tag{A.6}$$

Using that $T_1 > T_2$ and based on the second law of thermodynamics $(dS(E_1, E_2) > 0)$, it follows that $dQ > 0$, which is what we wanted to prove.

---

involved in its construction] can perform more work than a *reversible* machine (ideal) between the same values of Q and T, for which

$$\mathcal{W} = \frac{Q_1 (T_1 - T_2)}{T_1}.$$

In other words, the efficiency, $\mathcal{W}/Q_1$, of a reversible machine is proportional to the operating temperature difference divided by the highest temperature; whence it follows that efficiency is always less than unity.

[7] See [4] in Chap. 2.

Appendix B
Relaxation to the Stationary State

B.1 Temporal Evolution

Since the master Hamiltonian $\mathbf{H}$ is usually not a symmetric matrix, we cannot ensure its diagonalization. We can try to find a biorthonormal set of eigenvectors of the matrix $\mathbf{H}$ (dim $N \times N$) and its transpose $\mathbf{H}^T$. In other words, to solve the eigenvalue problem:

$$\mathbf{H}\phi_j = \lambda_j\phi_j , \; j = 1, \cdots, N \tag{B.1}$$

$$\mathbf{H}^T\psi_j = \lambda_j\psi_j \tag{B.2}$$

$$\sum_n \phi_j(n)\psi_l(n) = \delta_{jl}, \tag{B.3}$$

whence the following results are inferred:

- There is always an eigenvalue $\lambda_0 = 0$, which corresponds to the eigenvector $\psi_0 = (1, 1, \cdots, 1)$. This follows from the normalization of probability (i.e., the property $\sum_n \mathbf{H}_{nm} = 0$)

$$\mathbf{H}^T\psi_0 = \sum_n \mathbf{H}_{nm}\psi_0(n) = \sum_n \mathbf{H}_{nm} = 0.$$

- If the matrix $\mathbf{H}$ cannot be decomposed into sub-arrays of disconnected matrices $\mathbf{H}(a), \mathbf{H}(b)$, etc., that is, having states or groups of states isolated dynamically

© Springer International Publishing AG 2017
M.O. Cáceres, *Non-equilibrium Statistical Physics with Application to Disordered Systems*, DOI 10.1007/978-3-319-51553-3

from the rest,[1] then the eigenvalue λ_0 is not degenerate and the corresponding eigenvector ϕ_0 is the stationary state of the system $P_{st}(n)$. Under this condition the system approaches a single stationary probability $P(n, t \to \infty) = P_{st}(n)$. The eigenvectors with eigenvalue 0 are precisely the ones meeting that condition, but $P_{st}(n)$ is the only such

vector (save for a scalar factor).

- If the set $\{\phi_i, \psi_j\}$ is complete, we can expand the (conditional probability) solution of the ME in terms of this biorthonormal set of eigenvectors, i.e.:

$$P(n, t \mid n_0, t_0) = \sum_j \psi_j(n_0)\phi_j(n)e^{(t-t_0)\lambda_j}, \quad \forall t \geq t_0. \tag{B.4}$$

- The remaining eigenvalues have negative real part[2] (Perron-Frobenius' Theorem). Under the condition of non-reducible matrix $\mathbf{H}$, every solution of the ME asymptotically approaches $P_{st}(n)$ for $t \to \infty$; then, all other eigenvectors ϕ_j must vanish for $t \to \infty$, therefore, $\mathcal{R}_e[\lambda_j] < 0, \forall j \neq 0$.

Exercise. Show that properties: $\mathbf{H}_{n,m} \geq 0, \forall n \neq m$ and $\sum_n \mathbf{H}_{n,m} = 0, \forall m$ are preserved when permutations of rows ($n \leftrightarrow m$) and columns ($n \leftrightarrow m$) are applied simultaneously. Interpret the meaning of "simultaneously" applying permutations of rows and columns. Note that the above properties of $\mathbf{H}$ are not preserved if an (arbitrary) similarity transformation $\mathbf{H} \to S^{-1} \cdot \mathbf{H} \cdot S$ is applied.

Optional Exercise. Show that if $\mathbf{H}$ has two linearly independent eigenvectors (with zero eigenvalue) with positive components, then $\mathbf{H}$ is a reducible matrix.

Excursus (**Ergodicity in Markov Process**). We say that a Markov process is ergodic if for every pair of states $\{n, m\}$ there exists a *minimum* integer k such that $\left(W^k\right)_{n,m} > 0$. Then for sufficiently long times a Markov process reaches a steady state, where all states $\{n\}$ have nonzero probability of being occupied.

---

[1] Matrices that fulfill this condition and can, for example, be "rearranged" (redefining states) in the form:

$$\mathbf{H} = \begin{pmatrix} \mathbf{H}(a) & \mathbf{0} \\ \mathbf{0} & \mathbf{H}(b) \end{pmatrix},$$

are called "*completely reducible.*" Furthermore, matrices are called "*reducible*" if they can, for example, be rearranged in the form:

$$\mathbf{H} = \begin{pmatrix} \mathbf{H}(a) & \mathbf{0} & \mathbf{A} \\ \mathbf{0} & \mathbf{H}(b) & \mathbf{B} \\ \mathbf{0} & \mathbf{0} & \mathbf{C} \end{pmatrix},$$

where matrices $\mathbf{A}$ and $\mathbf{B}$ have nonnegative elements.

[2] An alternative proof in terms of the Lyapunov function will be presented in the next section; see also the advanced exercise in Sect. 4.9.7.

Example. In Sect. 6.3.1 we worked with a master Hamiltonian **H** (dim 3×3) that was not diagonalizable. In that example we explicitly constructed the solution $\mathbf{P}(t) = e^{t\mathbf{H}}$ from the theory of Jordan's matrices; note that this solution satisfies the initial condition $\mathbf{P}(t = 0) = \mathbf{1}$. Let us now look at the same example in the present context. It is easy to calculate the eigenvectors (right), i.e., to solve the problem $\mathbf{H}\phi_j = \lambda_j\phi_j$

$$\phi_0 = \frac{1}{6}(4, 1, 1), \quad \lambda_0 = 0$$

$$\phi_1 = (-2, 1, 1), \quad \lambda_1 = \lambda_2 = -3,$$

whence it is noted that precisely the eigenvalue -3 (degenerate) is the one introducing the difficulty to diagonalize the matrix. Furthermore, the set of attachments (left) eigenvectors, i.e., $\mathbf{H}^T\psi_j = \lambda_j\psi_j$ are given by

$$\psi_0 = (1, 1, 1), \quad \lambda_0 = 0$$
$$\psi_1 = (0, -1, 1), \quad \lambda_1 = \lambda_2 = -3.$$

From the set of eigenvectors $\{\phi_0, \phi_1, \psi_0, \psi_1\}$ it is easy to verify that one of the biorthonormal conditions (B.3) is not satisfied, so we cannot fulfill the initial condition $\mathbf{P}(t = 0) = \mathbf{1}$. Therefore, we cannot use formula (B.4). However, if we introduce a small quantity ϵ so that the eigenvalues are all different, we can use this approach and then study the solution in the limit $\epsilon \to 0$. For this purpose we modify the master Hamiltonian **H** in the form:

$$\mathbf{H} = \begin{pmatrix} -1 & 1 & 3 \\ 1/2 & -2 & \epsilon \\ 1/2 & 1 & -3 \end{pmatrix}.$$

In this case the eigenvalues can be approximated by

$$\lambda_0 = \frac{\epsilon}{6} + \mathcal{O}(\epsilon)^{3/2}$$

$$\lambda_1 = -3 - \sqrt{\frac{\epsilon}{2}} - \frac{\epsilon}{12} + \mathcal{O}(\epsilon)^{3/2}$$

$$\lambda_2 = -3 + \sqrt{\frac{\epsilon}{2}} - \frac{\epsilon}{12} + \mathcal{O}(\epsilon)^{3/2}.$$

The right eigenvectors are

$$\phi_0 = \frac{1}{\mathcal{N}_0}(4 - \frac{\epsilon}{3} + \mathcal{O}(\epsilon)^{3/2}, 1 + \frac{\epsilon}{3} + \mathcal{O}(\epsilon)^{3/2}, 1)$$

$$\phi_1 = \frac{1}{\mathcal{N}_1}(-2 + \sqrt{2\epsilon} - \frac{\epsilon}{2} + \mathcal{O}(\epsilon)^{3/2}, 1 - \sqrt{2\epsilon} + \frac{\epsilon}{6} - \frac{5\epsilon^{3/2}}{72\sqrt{2}} + \mathcal{O}(\epsilon)^2, 1)$$

$$\phi_2 = \frac{1}{\mathcal{N}_2}(-2 - \sqrt{2\epsilon} - \frac{\epsilon}{2} + \mathcal{O}(\epsilon)^{3/2}, 1 + \sqrt{2\epsilon} + \frac{\epsilon}{6} + \frac{5\epsilon^{3/2}}{72\sqrt{2}} + \mathcal{O}(\epsilon)^2, 1),$$

where $1/\mathcal{N}_j$ are factors that are introduced to meet normalization:

$$\mathcal{N}_0 = 6 - \frac{4\epsilon}{3} + \mathcal{O}(\epsilon)^{3/2}$$

$$\mathcal{N}_1 = 2\sqrt{2\epsilon} - \frac{10\epsilon}{3} + \mathcal{O}(\epsilon)^{3/2}$$

$$\mathcal{N}_2 = -2\sqrt{2\epsilon} - \frac{10\epsilon}{3} + \mathcal{O}(\epsilon)^{3/2}.$$

The left eigenvectors are

$$\psi_0 = (1 - \frac{5\epsilon}{18} + \mathcal{O}(\epsilon)^{3/2}, 1 - \frac{2\epsilon}{9} + \mathcal{O}(\epsilon)^{3/2}, 1)$$

$$\psi_1 = (-\frac{\sqrt{\epsilon}}{3\sqrt{2}} + \frac{11\epsilon}{36} + \mathcal{O}(\epsilon)^{3/2}, -1 + \frac{2}{3}\sqrt{2\epsilon} - \frac{8\epsilon}{9} + \mathcal{O}(\epsilon)^{3/2}, 1)$$

$$\psi_2 = (\frac{\sqrt{\epsilon}}{3\sqrt{2}} + \frac{11\epsilon}{36} + \mathcal{O}(\epsilon)^{3/2}, -1 - \frac{2}{3}\sqrt{2\epsilon} - \frac{8\epsilon}{9} + \mathcal{O}(\epsilon)^{3/2}, 1).$$

The explicit construction of $\mathbf{P}(t)$ is then obtained by multiplying the corresponding *perturbed* eigenvectors; for example, the first term of the series (B.4) is

$$\psi_0(n_0)\phi_0(n) = \frac{1}{\mathcal{N}_0} \begin{pmatrix} 4 - \frac{\epsilon}{3} \\ 1 + \frac{\epsilon}{3} \\ 1 \end{pmatrix} \left(1 - \frac{5\epsilon}{18}, 1 - \frac{2\epsilon}{9}, 1\right) + \mathcal{O}(\epsilon)^{3/2}$$

$$= \begin{pmatrix} 2/3 - 5\epsilon/54 & 2/3 - \epsilon/18 & 2/3 + 5\epsilon/54 \\ 1/6 + 5\epsilon/108 & 1/6 + \epsilon/18 & 1/6 + 5\epsilon/54 \\ 1/6 - 5\epsilon/108 & 1/6 & 1/6 + \epsilon/27 \end{pmatrix} + \mathcal{O}(\epsilon)^{3/2}.$$

From similar expressions it is possible to construct the (perturbed) general solution (B.4), in particular it is possible to see that in the limit $\epsilon \to 0$ the steady state, $\mathbf{P}(t \to \infty)$, coincides with the expression found in Chap. 6. On the other hand, it is also possible to observe that the initial condition $\mathbf{P}(t = 0) = \mathbf{1}$ holds for $\epsilon \sim 0$.

B.2   Lyapunov Function

Just as when we analyzed the irreversibility in the **F-P** equation, in the case of an ME it is also possible to define a positive and decreasing function over time. This allows us to study the steady-state relaxation of the equation $\dot{\mathbf{P}} = \mathbf{H} \cdot \mathbf{P}$.

The Lyapunov function is defined as follows:

$$\mathcal{H}(t) = \sum_n x_n \left(\log x_n - \log y_n \right), \tag{B.5}$$

where x_n, y_n are components of two arbitrary solutions of the ME.

To show that $\mathcal{H}(t)$ is a Lyapunov function we need to use the relationship:

$$\alpha \log \alpha - \alpha \log \beta - \alpha + \beta \geq 0, \quad \forall \alpha, \beta \text{ positive.} \tag{B.6}$$

For this reason first we prove the following result. If we divide (B.6) by $\beta > 0$ and define $R = \alpha/\beta$, we obtain

$$R \log R - R + 1 \geq 0, \quad \forall R > 0. \tag{B.7}$$

Furthermore, $\forall R > 0$ we have

$$\int_1^R \log x \, dx \geq 0,$$

equality is obtained when $R = 1$. Then, since $R \log R - R + 1 = \int_1^R \log x \, dx$, it follows (B.7) and thus (B.6).

To prove that $\mathcal{H}(t) \geq 0$, $\forall t$, by normalization we get $\sum_n x_n = \sum_n y_n = 1$; then, adding and subtracting 1 to (B.5), we obtain

$$\mathcal{H}(t) = \sum_n x_n \left(\log x_n - \log y_n \right) - x_n + y_n$$

$$= \sum_n y_n \left(R_n \log R_n - R_n + 1 \right),$$

where we have defined $R_n \equiv x_n/y_n > 0$. Assuming that $y_n \neq 0, x_n \neq 0$ and using the relationship (B.7) $\forall R_n$, we get that $\mathcal{H}(t) \geq 0$.

To prove that $d\mathcal{H}(t)/dt \leq 0$ we proceed as follows. First we differentiate with respect to time the function $\mathcal{H}(t)$, i.e.:

$$\frac{d}{dt} \mathcal{H}(t) = \sum_n \dot{x}_n \left(\log x_n - \log y_n \right) + \dot{x}_n - \frac{x_n}{y_n} \dot{y}_n$$

$$= \sum_n \dot{x}_n \log R_n - R_n \dot{y}_n,$$

where we have used $\sum_n \dot{x}_n = \frac{d}{dt} \sum_n x_n = 0$. If we now introduce the dynamics of the ME for each of the solutions (for example, $\dot{x}_n = \sum_m \mathbf{H}_{nm} x_m \equiv \sum_m W_{nm} x_m - x_n \sum_m W_{mn}$) we conclude that

$$\frac{d}{dt}\mathcal{H}(t) = \sum_n \left(\sum_m W_{nm} x_m - x_n \sum_m W_{mn} \right) \log R_n - R_n \left(\sum_m W_{nm} y_m - y_n \sum_m W_{mn} \right)$$

$$= \sum_n \sum_m W_{nm} (x_m \log R_n - y_m R_n) + \sum_n \sum_m W_{mn} (-x_n \log R_n + y_n R_n)$$

$$= \sum_n \sum_m W_{nm} (x_m \log R_n - y_m R_n - x_m \log R_m + y_m R_m)$$

$$= - \sum_n \sum_m W_{nm} y_m (R_m \log R_m - R_m \log R_n - R_m + R_n) \leq 0, \quad \forall t,$$

here we have used that $y_m > 0$, $W_{nm} \geq 0$ and relationship (B.6), $\forall \{R_n, R_m\}$ positive.

Since $\mathcal{H}(t)$ is positive and its derivative is negative, $\mathcal{H}(t)$ is a decreasing function in time and bounded from below. The lower limit $\mathcal{H}(t) = 0$ is attained when both (arbitrary) solutions $\{x_n\}$, $\{y_n\}$ approach each other; that is, when both solutions tend to the (unique) stationary state $P_{st}(n)$.

On the other hand, if there are several transition elements W_{nm} which are zero, so that for some solution $\{y_m\}$ of the system it is subdivided into groups having no dynamic coupling, the relaxation to a single steady state would not occur. But this case is precisely when the matrix $\mathbf{H}$ is reducible or completely reducible, which we have excluded from our analysis.

Appendix C
The Green Function of the Problem of an Impurity

C.1 Anisotropic and Asymmetric Case

The calculation of the Green function with an impurity may be performed in arbitrary number of dimensions. The perturbed (with *one* impurity) problem is defined by the master Hamiltonian: $\mathbf{H_i} = \mathbf{H_0} + \mathbf{V}$, where we assume that the homogeneous part $\mathbf{H_0}$ (with transitions to first neighbors) can be written generally in the form

$$\mathbf{H_0} = \sum_{s,s'(\neq)}^{N.N.} |s> W_{s,s'} <s'| - \sum_{s} |s> \Gamma_s <s|, \qquad (C.1)$$

here $\Gamma_s = \sum_{s'} W_{s',s}$ and s is a vector of arbitrary dimension. For example, in 2D we have

$$\mathbf{H_0} = \sum_{m,n} \{ B \mid m-1, n >< m, n \mid +A \mid m+1, n >< m, n \mid$$

$$+ D \mid m, n-1 >< m, n \mid +C \mid m, n+1 >< m, n \mid \}$$

$$- \sum_{m,n} \{ (A+B+C+D) \mid m, n >< m, n \mid \}.$$

In general, in nD dimensions, the operator $\mathbf{V}$ associated with bond impurity elements can be written as

$$\mathbf{V}_{\alpha\beta} = b_1 \delta_{\alpha a} \delta_{\beta a} + b_4 \delta_{\alpha b} \delta_{\beta b} + b_2 \delta_{\alpha a} \delta_{\beta b} + b_3 \delta_{\alpha b} \delta_{\beta a},$$

so that for each coordinate axis there will be a set of coefficients $\{b_j\}$. In Sect. 7.2.1 we display, with the help of a graphic, how to interpret the set $\{b_j\}$ for the 2D case.

© Springer International Publishing AG 2017
M.O. Cáceres, *Non-equilibrium Statistical Physics with Application to Disordered Systems*, DOI 10.1007/978-3-319-51553-3

The Green function of the problem of an impurity $G^i = (u\mathbf{1} - \mathbf{H}_i)^{-1}$ can be written in terms of the homogeneous Green function $G^0 = (u\mathbf{1} - \mathbf{H}_0)^{-1}$. As will be shown, this calculation can be done for any hypercubic network of dimension n, and for any location selected for the impure bond $a - b$.

If we define the operators $\mathbf{V}^1 = |\, a \rangle\, b_1\, \langle a\, |$, $\mathbf{H}_1 = \mathbf{H}_i - \mathbf{V}^1$, $G^1 = (u\mathbf{1} - \mathbf{H}_1)^{-1}$ and make a Dyson expansion[1]

$$G^i = G^1 + G^1\, \mathbf{V}^1\, G^1 + G^1\, \mathbf{V}^1\, G^1\, \mathbf{V}^1\, G^1 + \dots, \tag{C.2}$$

we can express G^i in terms of G^1; then

$$G^i_{\alpha\beta} = G^1_{\alpha\beta} + G^1_{\alpha a}\, b_1\, G^1_{a\beta}\, \left(1 - b_1\, G^1_{aa}\right)^{-1}. \tag{C.3}$$

The same scheme can be applied defining the operators $\mathbf{V}^2 = |\, a \rangle\, b_2\, \langle b\, |$, $\mathbf{H}_2 = \mathbf{H}_1 - \mathbf{V}^2$, and $G^2 = (u\mathbf{1} - \mathbf{H}_2)^{-1}$. This lets us write G^1 in terms of G^2:

$$G^1_{\alpha\beta} = G^2_{\alpha\beta} + G^2_{\alpha a}\, b_2\, G^2_{b\beta}\, \left(1 - b_2\, G^2_{ba}\right)^{-1}. \tag{C.4}$$

Then, defining $\mathbf{V}^3 = |\, b \rangle\, b_3\, \langle a\, |$, $\mathbf{H}_3 = \mathbf{H}_2 - \mathbf{V}^3$ and $G^3 = (u\mathbf{1} - \mathbf{H}_3)^{-1}$, we can express G^2 in terms of G^3:

$$G^2_{\alpha\beta} = G^3_{\alpha\beta} + G^3_{\alpha b}\, b_3\, G^3_{a\beta}\, \left(1 - b_3\, G^3_{ab}\right)^{-1}. \tag{C.5}$$

Defining $\mathbf{V}^4 = |\, b \rangle\, b_4\, \langle b\, |$ we obtain $\mathbf{H}_0 = \mathbf{H}_3 - \mathbf{V}^4$ and so we can express G^3 in terms of G^0 the homogeneous Green function in arbitrary dimension; that is:

$$G^3_{\alpha\beta} = G^0_{\alpha\beta} + G^0_{\alpha b}\, b_4\, G^0_{b\beta}\, \left(1 - b_4\, G^0_{bb}\right)^{-1}. \tag{C.6}$$

Finally, reversing the process, we obtain the desired expression:

$$G^i_{ab} = \frac{G^0_{ab} + b_4\left(G^0_{ab}G^0_{ba} - G^0_{aa}G^0_{bb}\right)}{1 + b_1\left(G^0_{ab} - G^0_{aa}\right) + b_4\left(G^0_{ba} - G^0_{bb}\right)}. \tag{C.7}$$

Note that by definition $b_3 = -b_1$, $b_2 = -b_4$. This (exact) expression is valid in the nD general case. Note that the anisotropy and asymmetry are covered by the structure of the homogeneous Green function G^0. The fact that the network is anisotropic involves breaking the discrete symmetry of rotation. On the other hand, if the ME transition elements are asymmetrical, this implies the existence of a drift (bias) in the transport

---

[1]See [10] in Chap. 7.

C.2   Anisotropic and Symmetrical Case

In the symmetric anisotropic case [i.e., no drift on any axis] we have that $W_{s,s'} = W_{s',s}$; then $b_2 = b_3$, whence, and using (C.7) we finally obtain:

$$
G_{ab}^i = \frac{G_{ab}^0 + b_1 \left(G_{ab}^0 G_{ba}^0 - G_{aa}^0 G_{bb}^0 \right)}{1 + b_1 \left(G_{ab}^0 + G_{ba}^0 - G_{bb}^0 - G_{aa}^0 \right)}. \tag{C.8}
$$

The symmetric isotropic case follows trivially from (C.8), whereas the Green function G^0 is the solution of an isotropic and homogeneous problem.

In the case of an asymmetric (or symmetrical) $1D$ network, in which there is a bond impurity $a - b$, the solution is given by (C.7) [or (C.8)] respectively, where G_{ab}^0 is the Green function corresponding to the asymmetric (or symmetric) $1D$ case.

Appendix D
The Waiting-Time Function $\psi(t)$ of the CTRW

The basic functions for the description of a (separable) CTRW are the jump structure in the network $\lambda(k)$ and the standby function (waiting-time) $\psi(t)$. This section presents the relationship between the CTRW waiting-time function $\psi(t)$ and the problem of the **mv**—over the disorder—of the Green function of the ME.

The starting point is again an ME in a nD network with transition elements characterized by the probability $W_{s,s'}$ ($\Gamma_{s'} \equiv \sum_s W_{s,s'}$), then the master Hamiltonian is

$$\mathbf{H} = \left[-\sum_s |s> \Gamma_s <s| + \sum_{s,s'(\neq)}^{N.N.} |s> W_{s,s'} <s'| \right] \equiv \mathbf{\Gamma} + \mathbf{W}, \qquad (D.1)$$

where we have defined two new operators: $\mathbf{\Gamma}$ and $\mathbf{W}$. Note that $\mathbf{\Gamma}$ is a diagonal operator in the basis $|s>$, while $\mathbf{W}$ has only nonzero off-diagonal elements. The Green function of the ME characterized by the master Hamiltonian (D.1) can be expanded in series as follows:

$$\begin{aligned}
G(u) &= (u\mathbf{1} - \mathbf{H})^{-1} = (u\mathbf{1} - \mathbf{\Gamma} - \mathbf{W})^{-1} \\
&= \left[(u\mathbf{1} - \mathbf{\Gamma}) \left(1 - (u\mathbf{1} - \mathbf{\Gamma})^{-1} \mathbf{W} \right) \right]^{-1} \\
&= \left(1 - (u\mathbf{1} - \mathbf{\Gamma})^{-1} \mathbf{W} \right)^{-1} (u\mathbf{1} - \mathbf{\Gamma})^{-1} = (u\mathbf{1} - \mathbf{\Gamma})^{-1} \left(1 - \mathbf{W}(u\mathbf{1} - \mathbf{\Gamma})^{-1} \right)^{-1} \\
&= (u\mathbf{1} - \mathbf{\Gamma})^{-1} \sum_{q=0}^{\infty} \left(\mathbf{W}(u\mathbf{1} - \mathbf{\Gamma})^{-1} \right)^q .
\end{aligned}$$

The CTRW approach involves opening the average over the disorder, when calculating $\langle G(u) \rangle$, in a certain way

© Springer International Publishing AG 2017
M.O. Cáceres, *Non-equilibrium Statistical Physics with Application to Disordered Systems*, DOI 10.1007/978-3-319-51553-3

$$\langle G(u) \rangle_{\text{Disorder}} \approx \left\langle (u\mathbf{1} - \mathbf{\Gamma})^{-1} \right\rangle \left[\sum_{q=0}^{\infty} \left\langle \left(\mathbf{W}(u\mathbf{1} - \mathbf{\Gamma})^{-1} \right)^q \right\rangle \right]$$
$$\approx \left\langle (u\mathbf{1} - \mathbf{\Gamma})^{-1} \right\rangle \left[\sum_{q=0}^{\infty} \left\langle \left(\mathbf{W}(u\mathbf{1} - \mathbf{\Gamma})^{-1} \right) \right\rangle^q \right].$$

If we define the operators (in Laplace's variable) in the form:

$$\hat{\phi} = (u\mathbf{1} - \mathbf{\Gamma})^{-1} \quad \text{and} \quad \hat{\psi} = \mathbf{W}(u\mathbf{1} - \mathbf{\Gamma})^{-1},$$

we see that the fundamental functions of a CTRW are closely related to the mean values of the operators $\hat{\phi}$ and $\hat{\psi}$.

In the general non-separable case we define

$$\psi(s - s', u) = \left\langle \hat{\psi}_{s,s'} \right\rangle = \left\langle \frac{W_{s,s'}}{u + \sum_s W_{s,s'}} \right\rangle, \tag{D.2}$$

an equation that expresses the fact that the **mv** of the operator $\hat{\psi}$ depends only on the distance $s - s'$ and the Laplace variable u. In other words, after taking the average over the disorder translational invariance is restored. Furthermore, the average of the operator $\hat{\phi}$, characterized by the function $(1 - \psi(u))/u$, also occurs naturally in CTRW theory:

$$\phi(u) = \left\langle \hat{\phi}_{s',s} \right\rangle = \left\langle \frac{1}{u + \sum_s W_{s,s'}} \right\rangle \delta_{s',s}. \tag{D.3}$$

This equation states that the probability of staying at site s of the lattice: $1 - \int_0^t \psi(t') \, dt'$ is independent of the site in question.

Exercise. Using the inverse Laplace transform verify that

$$\hat{\psi}_{s,s'}(t) = W_{s,s'} \exp \left(-t \sum_s W_{s,s'} \right).$$

From this result, show by direct integration that

$$\phi(t) = \left\langle 1 - \int_0^t \sum_s \hat{\psi}_{s,s'}(t') \, dt' \right\rangle = \left\langle \exp \left(-t \sum_s W_{s,s'} \right) \right\rangle,$$

which is in accordance with what one expects from (D.3).

In general, the function $\psi(s - s', t)$ of a CTRW is specified after the calculation of **mv**—over the disorder—of the quantity is done:

$$\psi(s - s', t) = \left\langle W_{s,s'} \exp \left(-t \sum_s W_{s,s'} \right) \right\rangle.$$

Depending on the probability density $\mathcal{P}(W_{s,s'})$ characterizing the random nature of the transition elements $W_{s,s'}$, the function $\psi(s - s', t)$ may or may not have a closed expression. On the other hand, in general, we are only interested in the long-time behavior of the waiting-time function $\psi(t)$. This fact simplifies considerably the calculation of $\psi(u)$ in the regime $u \sim 0$.

Excursus (**Controversy**). There is an interesting mathematical result due to J.K.E. Tunaley [Phys. Rev. Lett. 33, 1037 (1974); J. Stat. Phys. 14, 461 (1976)] which states that the CTRW theory should be extended considering the waiting-time function of the first step. This generalization leads to profound changes in the anomalous behavior of the CTRW. See the objections raised by M. Lax and H. Scher [Phys. Rev. Lett. 39, 781 (1977)] and A.A. Kumar and J. Heinrich [J. Phys. C 13, 2131 (1980)]; see exercises 7.5.1(d); 7.5.4., as well as the treatment given in [16] in Chap. 8.

Appendix E
Non-Markovian Effects Against Irreversibility

We have already mentioned that continuous or discrete Markov processes are adequate stochastic processes for describing the tendency to equilibrium or, more generally, the relaxation to the steady state, if it exists. This fact is closely related to the phenomenon of irreversibility, in fact, in order to verify the relaxation to the steady state, we have built a Lyapunov function showing, under certain conditions,[1] such irreversibility. Since a deterministic evolution, such as the Hamiltonian dynamics of a mechanical system, cannot present such behavior, the question arises of under what conditions will a non-Markov process cease to show tendency to irreversibility. To this end we shall study here how the memory effects of non-Markovian stochastic processes could destroy the Lyapunov function.

In a similar way as when we defined the Lyapunov function for a processes characterized by an ME, we define here the Lyapunov function for the CTRW process as

$$\mathcal{H}(t) = \sum_n x_n(t) \left(\log x_n(t) - \log y_n(t)\right), \tag{E.1}$$

where x_n, y_n are the components of two arbitrary solutions of a CTRW. That is, if we use the stochastic dynamics of a CTRW process—for each of the mentioned solutions—we have in real space[2]:

$$\frac{dx_n}{dt} = \int_0^t \sum_m \mathbf{H}_{nm}(t-\tau) \, x_m(\tau) \, d\tau \tag{E.2}$$

$$\equiv \int_0^t \Phi(t-\tau) \left(\sum_m W_{nm} x_m(\tau) - x_n \sum_m W_{mn}(\tau)\right) d\tau, \quad \forall t \geq 0.$$

---

[1] For example, in the case of the dynamics of a ME, the matrix $\mathbf{H}$ must not be "reducible."

[2] We are assuming that the CTRW process is separable; see Sect. 7.3.1.

© Springer International Publishing AG 2017
M.O. Cáceres, *Non-equilibrium Statistical Physics with Application to Disordered Systems*, DOI 10.1007/978-3-319-51553-3

To show that $\mathcal{H}(t)$ is a Lyapunov function we need to use the relationship (see Appendix B.2)

$$\alpha \log \alpha - \alpha \log \beta - \alpha + \beta \geq 0, \quad \forall \alpha, \beta \text{ positive}. \tag{E.3}$$

In particular, to show that $\mathcal{H}(t) \geq 0$ we use the normalization of the solutions, i.e.: $\sum_n x_n(t) = \sum_n y_n(t) = 1$. Then, adding and subtracting 1 to (E.1) we obtain

$$\mathcal{H}(t) = \sum_n x_n (\log x_n - \log y_n) - x_n + y_n \geq 0,$$

where we used (E.3).

To check whether or not $d\mathcal{H}(t)/dt \leq 0$, $\forall t \geq 0$, proceed as follows. First we differentiate with respect to time, that is,

$$\frac{d}{dt}\mathcal{H}(t) = \sum_n \dot{x}_n(t) (\log x_n(t) - \log y_n(t)) + \dot{x}_n(t) - \frac{x_n(t)}{y_n(t)} \dot{y}_n(t)$$

$$= \sum_n \dot{x}_n(t) \log R_n(t) - R_n(t) \dot{y}_n(t),$$

here we have defined $R_n(t) \equiv x_n(t)/y_n(t) > 0$ assuming that $x_n \neq 0$ and $y_n \neq 0$. If we use the CTRW dynamics for each of the solutions, it follows that

$$\frac{d}{dt}\mathcal{H}(t) = \sum_n \log R_n(t) \int_0^t \Phi(t-\tau) \left(\sum_m W_{nm} x_m(\tau) - x_n(\tau) \sum_m W_{mn} \right) d\tau$$

$$- \sum_n R_n(t) \int_0^t \Phi(t-\tau) \left(\sum_m W_{nm} y_m(\tau) - y_n(\tau) \sum_m W_{mn} \right) d\tau$$

$$= \int_0^t \Phi(t-\tau) \sum_n \sum_m W_{nm} (x_m(\tau) \log R_n(t) - y_m(\tau) R_n(t)) \, d\tau$$

$$+ \int_0^t \Phi(t-\tau) \sum_n \sum_m W_{nm} (-x_m(\tau) \log R_m(t) + y_m(\tau) R_m(t)) \, d\tau$$

$$= - \int_0^t \Phi(t-\tau) \sum_n \sum_m W_{nm} y_m(\tau)$$

$$\times [R_m(\tau) \log R_m(t) - R_m(\tau) \log R_n(t) - R_m(t) + R_n(t)] \, d\tau. \tag{E.4}$$

Note that W_{nm} and $y_m(t)$ are likely to be positive quantities, but non-Markovian effects introduce a memory kernel $\Phi(t-\tau) \neq \delta(t-\tau)$, which prevents us from using inequality (E.3) to prove the negativity of $d\mathcal{H}(t)/dt$ at all times. If the non-Markovian effect has a short correlation time, τ_c, the memory kernel $\Phi(t-\tau)$

approaches a very "narrow" function different from zero only for $t \sim \tau$; therefore, the function $\mathcal{H}(t)$ is decreasing for times t large enough (away from the initial condition). But if the time correlation scale τ_c is too long, or the function $\Phi(t - \tau)$ is very wide, the function $\mathcal{H}(t)$ is meaningless as a Lyapunov function.

Exercise (**Memory Kernel**). In Laplace's representation, the equation that characterizes the memory kernel of a generalized ME associated with a CTRW process, $\Phi(u)$, can be written in the form

$$\Phi(u) = \psi(u)/\phi(u), \tag{E.5}$$

where $\psi(t) = \mathcal{L}_u^{-1}[\psi(u)] \geq 0$ is the waiting-time function, and $\phi(t) = \mathcal{L}_u^{-1}[\phi(u)] \geq 0$ characterizes the probability of permanence at an arbitrary site during the time interval $[0, t]$ [see Sect. 7.3.1]. Show that $\mathcal{L}_u^{-1}[\Phi(u)]$ cannot, in general, be interpreted as a probability, because it can be negative. Note that we may also write the probability density $\psi(t)$ in the form

$$\psi(t) = \int_0^t \Phi(t - \tau)\phi(\tau)\, d\tau.$$

Using that $\phi(u) = (1 - \psi(u))/u$, we can invert Eq. (E.5) to get $\psi(u)$ as a function of $\Phi(u)$:

$$\psi(u) = \frac{1}{u/\Phi(u) + 1} \tag{E.6}$$

Now we may wonder on the *inverse* problem: given an arbitrary kernel $\Phi(u)$, Eq. (E.6) defines a positive waiting-time function on $t \in [0, \infty]$ if and only if $\psi(u)$ is a completely monotone (CM) function. That is, for all n, $\psi(u = 0) > 0$ and $(-1)^n d^n \psi/du^n \geq 0$. Using this concept show that in order to get a *bonafide* waiting-time function $\psi(t) \geq 0$, the conditions on the memory kernel are[3]

$$\frac{u}{\Phi(u)} \geq 0 \text{ and } \frac{d\,[u/\Phi(u)]}{du} \text{ to be a CM function.}$$

Examine under what conditions the memory kernel converges, asymptotically, for long times to a constant. Hint: see advanced exercises in Chap. 8, note also that only the choice $\psi(t) = \frac{1}{\langle t \rangle} e^{-t/\langle t \rangle}$ leads to a non-memory kernel, i.e., $\Phi(t - \tau) = \delta(t - \tau)/\langle t \rangle$.

Exercise. The Telegrapher's equation[4] is an equation that characterizes at long times a diffusive behavior; however, for short times it takes into account that the

---

[3] See I. Sokolov, Phys. Rev. E, 66, 041101 (2002).
[4] See advanced exercise 7.5.2, and [11, 16] in Chap. 7.

particle motion is performed with a finite speed c. Using the fact that this evolution equation can be written in the form of a generalized ME [with the initial condition $\frac{\partial}{\partial t}P(x,t)\big|_{t=0} = 0$; see Sect. 7.3.1], show that in the limit $D \to \infty$ the Telegrapher equation becomes the wave equation; i.e., deterministic. Check that at the same limit the scaling time τ_c, which appears in the memory kernel $\Phi(t - \tau)$ of the associated ME diverges. It should be emphasized that the limits $D \to \infty$ and $t \to \infty$ do not commute in the solution of Telegrapher's equation.

E.1 $\Phi(t)$ and the Generalized Differential Calculus

In Chap. 8 we noted that in the case of strong disorder, the memory kernel of a CTRW was related to a fractional differential structure [see optional exercise (*Fractional derivative*) in Sect. 8.6.3]. We do not intend here to make an introduction to study such a vast research field, just to give the minimum elements necessary for the reader to intuitively understand this fascinating topic.

Let us consider a generalized ME of the form (E.2), but here to simplify the analysis we assume this equation in a continuous nD space (we also introduce the change of variable $t_1 = t - \tau$), i.e.:

$$\frac{\partial P(x,t)}{\partial t} = \int_0^t \Phi(t_1) \, \nabla^2 P(x, t - t_1) \, dt_1. \tag{E.7}$$

First we introduce the Laplace transform of the memory kernel:

$$\int_0^\infty \Phi(\tau) e^{-u\tau} \, d\tau = \mathcal{L}_u\left[\Phi(t)\right] \equiv \hat{\Phi}(u).$$

Note that on this occasion only we have explicitly introduced the notation $\hat{\Phi}(u)$ to indicate the Laplace transform $\mathcal{L}_u\left[\Phi(t)\right]$, this will be useful for the analysis that we will present. Let us now define the following operator:

$$\hat{\Phi}\left(\frac{\partial}{\partial t}\right) \equiv \int_0^\infty \Phi(\tau) \exp\left(-\frac{\partial}{\partial t}\tau\right) d\tau.$$

Then, using the series expansion of $\exp\left(-\frac{\partial}{\partial t}\tau\right)$, it is clear that the application of the operator $\hat{\Phi}\left(\frac{\partial}{\partial t}\right)$ on an arbitrary "well-behaved" function $f(t)$ gives the following results:

$$\hat{\Phi}(\frac{\partial}{\partial t})f(t) = \left[\int_0^\infty \Phi(\tau) \exp\left(-\frac{\partial}{\partial t}\tau\right) d\tau\right] f(t) \tag{E.8}$$

$$= \int_0^\infty \Phi(\tau) \left(f(t) - \tau f'(t) + \frac{\tau^2}{2!}f''(t) - \frac{\tau^3}{3!}f'''(t) + \cdots\right) d\tau$$

$$= \int_0^\infty \Phi(\tau) f(t - \tau) \, d\tau.$$

When we say "well behaved," we mean continuous, differentiable, etc.; in general, an analytic function in the whole domain, i.e., the function can be expanded in Taylor's series at every point.

Thus, asymptotically, for long times and if the integrand in (E.7) decays fast enough, using (E.8), we can approximate (E.7) by

$$\frac{\partial P(x, t)}{\partial t} \simeq \hat{\Phi}(\frac{\partial}{\partial t}) \nabla^2 P(x, t). \tag{E.9}$$

Using this formula it is easy to obtain generalized (fractional) differential equations. For example, suppose the memory kernel $\Phi(t)$ represents a subdiffusive CTRW process, i.e.: $\hat{\Phi}(u) \sim u^{1-\theta}$ [see Sect. 7.3.3]. Then from (E.9) we get

$$\frac{\partial P(x, t)}{\partial t} \simeq \frac{\partial^{1-\theta}}{\partial t^{1-\theta}} \nabla^2 P(x, t).$$

Hence from this equation we immediately infer the result presented in Sect. 8.6.3.

Exercise. Consider the Telegrapher equation when it is associated with a generalized ME with memory kernel $\Phi(t) = c^2 \exp\left(-\frac{c^2}{D}t\right)$ [see exercise in Sect. 7.3.1]. Use (E.9) to show that in the limit $D \to \infty$ the wave equation is obtained. Compare with the exercise in the previous section.

Exercise. Consider a weak disorder model [see exercise in Sect. 8.8.2]. Use (E.9) to study the long time limit of the evolution of the generalized ME.

Appendix F
The Density Matrix

The density matrix definition appears naturally when calculating the expectation value of a quantum observable taking a thermal statistical average into account. Let B be an operator (Hermitian) representing an arbitrary physical observable. If $\psi(x, t)$ is the wave function representing the state of the system S, the expectation value of the observable B is given by[1]

$$\bar{B} = \int \psi^*(x, t) B \psi(x, t) \, dx. \tag{F.1}$$

Let $\{\varphi_n(x)\}$ be a complete set of eigenfunctions. Then we can expand the wave function $\psi(x, t)$, of the system S, in eigenfunctions $\varphi_n(x)$ as follows:

$$\psi(x, t) = \sum_n c_n(t) \varphi_n(x). \tag{F.2}$$

Then, the expectation value of the observable B can be written in the alternative form

$$\bar{B} = \int \sum_m c_m^*(t) \varphi_m^*(x) \, B \, \sum_n c_n(t) \varphi_n(x) \, dx \tag{F.3}$$

$$= \sum_{n,m} c_n(t) c_m^*(t) B_{mn}.$$

The elements of the operator B are defined in the usual manner:

$$B_{mn} = \int \varphi_m^*(x) \, B \, \varphi_n(x) \, dx.$$

---

[1] The symbol ψ^* means the complex conjugate of ψ.

© Springer International Publishing AG 2017
M.O. Cáceres, *Non-equilibrium Statistical Physics with Application to Disordered Systems*, DOI 10.1007/978-3-319-51553-3

If we now consider an ensemble of systems S (all with the same physical structure) and under the same macroscopic condition, we will have that the statistical expectation value of the observable B is given by

$$\langle B \rangle = \sum_{n,m} \left\langle c_n(t)c_m^*(t)\right\rangle_{ensemble} B_{mn} \equiv \sum_{n,m} \rho_{nm} B_{mn} = Tr\left[\rho B\right], \qquad \text{(F.4)}$$

where $Tr\left[\cdots\right]$ represents the trace, and here we have also defined the density matrix ρ denoting the mean value over the ensemble of the quantities $c_n(t)c_m^*(t)$, i.e.:

$$\rho_{nm} = \rho_{mn}^* = \left\langle c_n(t)c_m^*(t)\right\rangle_{ensemble}. \qquad \text{(F.5)}$$

Exercise. Show that if $B(x,x')$ and $\rho(x,x')$ are representations of operators B and ρ over the basis of the position operator, then the density matrix can be written as

$$\rho(x,x') = \left\langle \psi^*(x,t)\psi(x',t)\right\rangle_{ensemble},$$

where, as before, the average means taking the mean value in ensemble. Hint: use, for example, that $B(x,x') = \sum_m \sum_n \varphi_m(x) B_{m,n} \varphi_n^*(x')$, etc.

F.1   Properties of the Density Matrix

We can also define the density matrix considering different statistical weights associated with a set of normalized wave functions, i.e.:

$$\rho_{qq'} = \sum_n w_n \psi_n(q) \psi_n^*(q'). \qquad \text{(F.6)}$$

Note the equivalence in the notation: $\rho(q,q') = <q\mid\rho\mid q'> = \rho_{qq'}$. In general the density matrix provides a complete description of the system in the sense that all statistical expectation values can be obtained from ρ.

Note, however, that this definition does not determine uniquely the set of eigenfunctions $\{\psi_n\}$ nor the (thermal) statistical weights w_n for building such an *ensemble*.

Exercise. From (F.6), and using that $\langle B \rangle = Tr\left[\rho B\right]$, the following properties are easy to prove

$$\rho^\dagger = \rho, \quad \text{(Hermitian)}$$

$$Tr\left[\rho\right] = 1, \quad \text{(Normalization)}$$

$$\rho_{nn} \in [0,1] \quad \text{(positive for all } \psi_n).$$

On the other hand, in the case of having a discrete representation of dimension $(M \times M)$, and using that ρ is Hermitian, it is easy to realize that there are only $M^2 - 1$ real parameters that determine the density matrix.

Guided Exercise. Since ρ is Hermitian there is a basis in which it is diagonal, then $\rho_{nm} = \rho_{nn}\, \delta_{n,m}$; since $\rho_{nn} \in [0, 1]$, it is easy to see that

$$Tr\left[\rho^2\right] = \sum_n \rho_{nn}^2 \leq \left(\sum_n \rho_{nn}\right)^2 = 1,$$

whence it follows that $Tr\left[\rho^2\right] \leq 1$. On the other hand, in any other basis we will have

$$Tr\left[\rho^2\right] = \sum_m \sum_n \rho_{nm}\, \rho_{mn} = \sum_{n,m} \mid \rho_{nm} \mid^2 \leq 1,$$

thus it follows that all the elements of the density matrix are bounded.

F.2   The von Neumann Equation

From the very definition of density matrix, it is natural to consider what is its evolution equation. If H is the Hamiltonian of a closed system $\mathcal{S}$, and if the temporal variation of the wave function ψ is described by the Schrödinger equation[2]

$$i\hbar \frac{\partial \psi}{\partial t} = H\psi, \tag{F.7}$$

expanding the wave function ψ in eigenfunctions φ_n we see that the coefficients $c_n(t)$ satisfy the equation

$$i\hbar \frac{\partial c_n}{\partial t} = \sum_l H_{nl}\, c_l(t),$$

then the product $c_n(t)c_m^*(t)$ satisfies the following evolution equation:

$$i\hbar \frac{\partial}{\partial t} c_n(t)c_m^*(t) = \sum_l H_{nl}c_l(t)c_m^*(t) - c_n(t)c_l^*(t)H_{lm}. \tag{F.8}$$

---

[2] A closed system is one that can exchange energy with the rest of the universe—in the form of heat or work—but no matter. Thus the number of degrees of freedom of the system is constant. If H is independent of time, we say that (F.7) describes the time evolution of a closed and isolated system.

Here we have used Hermiticity of H (i.e., $H_{lm} = H^*_{ml}$). The evolution equation of the density matrix is obtained by taking the average value over the ensemble, from Eq. (F.8) it follows:

$$ i\hbar \frac{\partial}{\partial t} \rho_{nm}(t) = \sum_l H_{nl}\, \rho_{lm}(t) - \rho_{nl}(t)\, H_{lm}. \tag{F.9} $$

In matrix notation, and using the definition of a commutator, the von Neumann equation takes the form

$$ i\hbar \frac{\partial \rho}{\partial t} = H\rho - \rho H \equiv [H, \rho], \tag{F.10} $$

this is the quantum version of Liouville's equation.[3] Unlike a kinetic equation (Boltzmann, Fokker-Planck, ME, etc.), von Neumann's equation describes a unitary evolution for the density matrix ρ, i.e., time reversible.

Excursus (**Completely Positive Map**). A completely positive (CP) map is an important mathematical concept that describes a class of temporal evolution more general than the unitary dynamics (F.10). Physically, the condition of CP map is not an additional assumption if one assumes that the system together with its environment is described by a Hamiltonian dynamics; in that sense, it is possible to see that the reduced dynamic (eliminating the degrees of freedom of the *environment*) forms a CP map. The CP maps are the only candidates to describe an irreversible dynamics of an open quantum system. A particular case of CP map is the mean value of an unitary dynamic map generated by a von Neumann evolution when the Hamiltonian is stochastic. Another case of interest is the map generated by applying the trace operator over the variables of the thermal bath. However introducing a perturbation theory, in this case it is not evident that the map will be CP.[4] Here the important point is to note that under the condition of CP map, it is possible to generalize the second law of thermodynamics; i.e., the concept of irreversibility.[5] Finally, it is noteworthy that noncommutative semigroups which generalize Markovian processes in quantum mechanics are also CP maps, they are called Kossakowski-Lindblad semigroups; see also exercises in Appendix Sect. I.3.

---

[3]In a classical system a point in phase-space represents a microscopic state of the system S, then a set of identical dynamical systems, or ensemble of S, can be represented by a set of points in phase-space. When the system S is closed, this ensemble of points may be considered an incompressible fluid in phase-space. Liouville's theorem states that the density of representative points (or microscopic states) does not change if we follow the moving fluid; see Eq. (G.5) in next appendix.

[4]See: N.G. van Kampen and I. Oppenheim, J. Stat. Phys. 87, 1325 (1997); and N.G. van Kampen, J. Stat. Phys. 115, 1057 (2004).

[5]See [1] in Chap. 8.

Exercise. Show that an ensemble is stationary (that is, $d\rho/dt = 0$) if the density matrix is only a function of energy, i.e., $\rho = \rho(H)$. Then, for a stationary ensemble, the density matrix is diagonal in the basis where the operator H is diagonal. On the other hand, defining the probabilities $w_n = \langle |c_n|^2 \rangle$, and using the notation $<bra|$ and $|ket>$, it follows that the density matrix operator can be written as:

$$\rho = \sum_n |n> w_n <n|. \tag{F.11}$$

From this representation it is easily observed that

$$\rho(x, x') = \sum_n <\varphi|n> w_n <n|\varphi> = \sum_n \varphi_n(x) w_n \varphi_n^*(x').$$

Optional Exercise. In the microcanonical ensemble (the energy value is set-in as the system is isolated) the density matrix is defined from the principle of equal probability a priori. If φ_n is a state with energy in the range $(E, E + dE)$ and if δ is the total number of eigenstates in this range, show that the density matrix in the microcanonical ensemble is written in the form

$$\rho(x, x') = w \sum_{n=1}^{\delta} \varphi_n(x)\varphi_n^*(x'). \tag{F.12}$$

It is possible to prove that if the basis $\{\varphi_n\}$ is changed by an unitary transformation, the density matrix depends only on the δ-dimensional space and is independent of the basis[6]

$$\rho_E = \sum_{E \leq E' \leq E+\delta E} \frac{|E' >< E'|}{W(E, \delta E)}.$$

Exercise. In the microcanonical ensemble assume that the operator ρ is diagonal independently of the basis that we use, then we get:

$$\rho_{n,m} = \left\langle \sum_q \frac{C_n^q C_m^{q*}}{W(E, \delta E)} \right\rangle_{\text{ensemble}}$$

$$= \frac{|C|}{W(E, \delta E)} \sum_q \langle \exp[i(\varphi_n^q - \varphi_m^q)] \rangle_{\text{ensemble}} \propto \delta_{n,m}.$$

This means that phases among (point elements) systems of the ensemble are non-correlated.

---

[6]See [4] from Chap. 2.

Optional Exercise. In the canonical ensemble (the temperature is prescribed as the system is in contact with a thermal bath) the density matrix is defined as follows

$$\rho = \mathcal{Z}^{-1} e^{-\beta H} = \mathcal{Z}^{-1} \sum_{E'} e^{-\beta E'} \mid E' \rangle\langle E' \mid,$$

where, as always, $\beta = 1/k_B T$ and the vector $\mid E' \rangle$ is an eigenstate with energy E'; that is, $H \mid E' \rangle = E' \mid E' \rangle$ and $\mathcal{Z} = Tr\,[e^{-\beta H}] \equiv \sum_{E'} e^{-\beta E'}$ is the partition function of the system. In this case the energy of the system is an average value. Then, using the canonical distribution, the statistical average of any observable A is written as

$$\langle A \rangle = \frac{Tr\,[Ae^{-\beta H}]}{Tr\,[e^{-\beta H}]},$$

and its classical limit takes the form

$$\langle A \rangle = \frac{\int \mathcal{D}q\,\mathcal{D}p\,A(p,q)e^{-\beta H(p,q)}}{\int \mathcal{D}q\,\mathcal{D}p\,e^{-\beta H(p,q)}},$$

where $\mathcal{D}q\,\mathcal{D}p$ represents the (multidimensional) associated differential in phase-space. The proof of this result can be inferred considering the limit $\hbar \to 0$ and replacing commutators with Poisson's brackets.[7]

Similarly, in the grand canonical ensemble (the temperature and chemical potential are prescribed, as the system is in contact with a thermal bath and a source of particles) the density matrix is defined by

$$\rho = \mathcal{N}^{-1} e^{-\beta(H_N - \mu N)}, \quad \text{with} \quad \mathcal{N} = Tr\,[e^{-\beta(H_N - \mu N)}] \equiv \sum_{N,l} e^{-\beta(E_{N,l} - \mu N)},$$

where μ is the chemical potential, H_N is the Hamiltonian of N-body, and N is the number operator. In this case the energy and number of particles in the system are average values.

Optional Exercise. Using the canonical ensemble show that energy fluctuations are characterized by $\langle \Delta E^2 \rangle = Tr\,[\rho H^2] - Tr\,[\rho H]^2 = k_B T^2 C_V$. On the other hand, in the grand canonical ensemble the fluctuations in the number of particles are given by $\langle \Delta N^2 \rangle = k_B T^2\,\partial \langle N \rangle / \partial \mu$; compare with the results of Chap. 2.

---

[7]See, for example, the [4] in Chap. 2.

F.3   Information Entropy

In the particular case when all elements of the ensemble are in the same quantum state ψ_n the corresponding ρ is called a *pure* ensemble. In that case we have

$$\rho_{\text{pure}} = | \psi_n >< \psi_n | . \qquad (F.13)$$

Then the expectation value of any observable B matches the value of statistical expectation:

$$\bar{B} = < \psi_n | B | \psi_n > = \langle B \rangle = Tr\left[B\rho_{\text{pure}} \right] .$$

Optional Exercise. Show that a necessary and sufficient condition for an ensemble to be pure is that ρ is idempotent, i.e.: $\rho^2 = \rho$.

Information entropy (or von Neumann's entropy) is defined as the statistical average of the operator $\eta = -\ln\rho$, i.e.:

$$S^{(i)} \equiv \langle \eta \rangle = Tr\left[\rho\,\eta \right] .$$

Note that in a representation in which the density matrix is diagonal, the information entropy, $S^{(i)} = -\sum_n \rho_{nn} \ln \rho_{nn}$, is analogous to thermodynamic entropy.

Exercise. Show that $S^{(i)}$ is not negative; also it is additive when applied to describe the combined effect of independent ensembles [direct product of density matrices]. Also show that the operator η evolves according to von Neumann's equation.

Exercise. In the case of considering a pure ensemble, show that $S^{(i)} = 0$ is obtained; i.e., the minimum value of entropy implies having the maximum information about the system status. On the other hand, show that the maximum entropy $S^{(i)}$ is obtained when $\rho_{nn} =$const. (subjected to the constraint $\sum_n \rho_{nn} = 1$); that is, when we have the minimum information on system status. Then, information entropy $S^{(i)}$ is associated with the statistical imprecision of the system.

F.3.1   Quantum Correlations

Correlations between two classical random variables A, B are in information theory quantified by the mutual information $I(A:B) = S(A) + S(B) - S(A,B)$. Here $S(\cdot)$ stands for the Shannon entropy of a given variable; while $S(A,B)$ is the Shannon entropy of the joint probability distribution of both variables, see [5] in Chap. 1. For quantum systems A and B the function $S(\cdot)$ denotes von Neumann's entropy. In the classical case we can use Bayes' rule to find an equivalent expression for the mutual information $I(A:B) = S(A) - S(A|B)$ where $S(A|B)$ is the Shannon entropy of A

conditioned on the measurement outcome on B. For quantum systems, this quantity is different from the first expression for the mutual information and the difference define a *quantum correlation*.

Excursus. Consider a quantum composite system defined by the Hilbert space $\mathcal{H}_{AB} = \mathcal{H}_A \otimes \mathcal{H}_B$. Let the dimension of the local Hilbert space be dim $\mathcal{H}_A = d_A$ and dim $\mathcal{H}_B = d_B$, while for the joint system $d = $ dim $\mathcal{H}_{AB} = d_A d_B$. If ρ denotes the density matrix of the composite system, the total amount of correlations is quantified by $I(\rho) = S(\rho_A) + S(\rho_B) - S(\rho)$, where $S(\rho)$ is the von Neumann entropy and $\rho_A = Tr_B(\rho)$, $\rho_B = Tr_A(\rho)$ are reduced density matrices. A generalization of the *classical* conditional entropy is $S(\rho_{B|A})$, where $\rho_{B|A}$ is the state of B given a measurement on A. By optimizing over all possible measurements on A, and alternative version of the mutual information can be made

$$Q_A(\rho) = S(\rho_B) - \min_{E_k} \sum_k p_k S(\rho_{B|k}),$$

where $\rho_{B|k} = Tr_A(\mathbf{E}_k \otimes \mathbf{I}_B \rho) / Tr(\mathbf{E}_k \otimes \mathbf{I}_B \rho)$ is the state of B conditioned on outcome k in system A, and $\{\mathbf{E}_k\}$ represent the set of positive valued operator elements. Then the discrepancy between the two measures of information $D_A(\rho) = I(\rho) - Q_A(\rho)$ defines the "quantum discord."[8] Interestingly a lower bound for this quantity has been found, which facilitates the characterization of quantum correlations.[9]

---

[8] See: L. Henderson and V. Vedral, J. Phys. A **34**, 6899 (2001); H. Ollivier and W.H. Zurek, Phys. Rev. Lett. **88**, 017901 (2001).

[9] See B. Dakic, V. Vedral, and C. Brukner, Phys. Rev. Lett. **105**, 190502 (2010).

Appendix G
Kubo's Formula for the Susceptibility

The main idea in this theorem is to find the average variation of a physical quantity, represented by operator B, when the system of interest S (initially at equilibrium) is perturbed by the application of a time-dependent external force. This interaction is represented by an additional term in the total Hamiltonian, proportional to a linear operator A, which acts on the wave vectors of the Hilbert space $\mathcal{H}_S$ of the system S. Then the total Hamiltonian of the system S can be written in the form

$$H_T = H_0 + H_{\text{ext}}. \tag{G.1}$$

where H_0 is the unperturbed Hamiltonian and H_{ext} represents the external interaction on S through a given observable A; that is:

$$H_{\text{ext}} = -A\, F(t), \quad A = -\partial_F H_T. \tag{G.2}$$

Here F is a function representing the external force of a mechanical nature (but time-dependent), then A represents the corresponding thermodynamic displacement.

The instantaneous state of the system is characterized by the density matrix $\rho(t)$. The evolution equation that governs $\rho(t)$ is the Liouville equation[1]

$$\frac{\partial \rho}{\partial t} = i\left(\mathcal{L}_0 + \mathcal{L}_{\text{ext}}(t)\right)[\rho(t)]. \tag{G.3}$$

The Liouvillian is a super-operator characterized by Poisson's bracket (classically) or by a commutator in the quantum case (von Neumann's equation, see Appendix F); for example, without perturbation in the Hamiltonian we get:

---

[1]For a closed system.

© Springer International Publishing AG 2017
M.O. Cáceres, *Non-equilibrium Statistical Physics with Application to Disordered Systems*, DOI 10.1007/978-3-319-51553-3

$$i\mathcal{L}_0\left[\rho(t)\right] = \{H_0, \rho\} \equiv \frac{1}{i\hbar}\left[H_0, \rho\right]. \tag{G.4}$$

Then $\{\cdot, \cdot\}$ represents the commutator divided by $i\hbar$ (in the quantum case), or the Poisson bracket in the classical case:

$$i\mathcal{L}_0\left[f(p,q)\right] = \{H_0, f\} \equiv \sum_{p,q}\left(\frac{\partial H_0}{\partial q}\frac{\partial}{\partial p} - \frac{\partial H_0}{\partial p}\frac{\partial}{\partial q}\right)f(p,q,t). \tag{G.5}$$

The evolution equation for the density matrix $\rho(t)$ can be written as an integral equation in the form[2]

$$\rho(t) = \exp\left[i(t - t_0)\mathcal{L}_0\right]\rho(t_0) + \int_{t_0}^{t} dt_1 \, \exp\left[i(t - t_1)\mathcal{L}_0\right]i\mathcal{L}_{\text{ext}}(t_1)\rho(t_1), \tag{G.6}$$

where

$$\exp\left(it\mathcal{L}_0\right)\, g = \exp\left(\frac{t}{i\hbar}H_0\right)\, g \, \exp\left(-\frac{t}{i\hbar}H_0\right). \tag{G.7}$$

Suppose now that the system $\mathcal{S}$ was in thermodynamic equilibrium at the bath temperature T when the external disturbance began. That is, in the absence of disturbance it had the initial condition

$$\rho(t_0 = -\infty) = \rho^{\text{eq}} = \frac{\exp(-\beta H_0)}{Tr\left[\exp(-\beta H_0)\right]}, \quad \text{with} \quad \beta = \frac{1}{k_B T}. \tag{G.8}$$

To first order in the perturbation, $\mathcal{O}(\mathcal{L}_{\text{ext}})$, the integral Eq. (G.6) takes the form

$$\rho(t) \simeq \rho^{\text{eq}} + \int_{-\infty}^{t} dt_1 \, \exp\left[i(t - t_1)\mathcal{L}_0\right]i\mathcal{L}_{\text{ext}}(t_1)\rho^{\text{eq}}. \tag{G.9}$$

Substituting (G.4) and (G.7) in (G.9) we can write

$$\rho(t) \simeq \rho^{\text{eq}} + \int_{-\infty}^{t} dt_1 \, \exp\left[(t - t_1)H_0/i\hbar\right]\{H_{\text{ext}}(t_1), \rho^{\text{eq}}\}\exp\left[-(t - t_1)H_0/i\hbar\right]. \tag{G.10}$$

Defining the statistical variation of operator B as

$$\langle\delta B(t)\rangle = \langle B(t)\rangle - \langle B\rangle_{\text{eq}} \equiv Tr\left[\rho(t)B\right] - Tr\left[\rho^{\text{eq}}B\right], \tag{G.11}$$

---

[2]This is the formal solution of Liouville-Neumann's Eq. (G.3).

and given that

$$Tr\left[\rho(t)\Delta B\right] - Tr\left[\rho^{eq}\Delta B\right] = Tr\left[\rho(t)\left(B - \langle B\rangle_{eq}\right)\right] - Tr\left[\rho^{eq}\left(B - \langle B\rangle_{eq}\right)\right]$$
$$= \langle\delta B(t)\rangle\,,$$

we get

$$\langle\delta B(t)\rangle \simeq \int_{-\infty}^{t} dt_1\; Tr\left[\exp\left[(t-t_1)H_0/i\hbar\right]\{H_{ext}(t_1),\rho^{eq}\}\exp\left[-(t-t_1)H_0/i\hbar\right]\Delta B\right].$$
$$(G.12)$$

Using the fact that within the trace we can cyclically permute operators, we finally obtain to $\mathcal{O}(\mathcal{L}_{ext})$

$$\delta\langle B(t)\rangle = \int_{-\infty}^{t} dt_1\; Tr\left[\{H_{ext}(t_1),\rho^{eq}\}\exp\left(-(t-t_1)H_0/i\hbar\right)\Delta B\exp\left(+(t-t_1)H_0/i\hbar\right)\right].$$
$$(G.13)$$

Within the trace, the last part represents the operator ΔB evolved over time $(t-t_1)$; since in general $\Delta\dot{B} = +i\left[H,\Delta B\right]/\hbar$. Then we can rewrite Eq. (G.13) in the form

$$\langle\delta B(t)\rangle = \int_{-\infty}^{t} dt_1\; Tr\left[\{H_{ext}(t_1),\rho^{eq}\}\Delta B(t-t_1)\right].$$
$$(G.14)$$

Again, using the properties of cyclic permutation within the trace, we obtain

$$\langle\delta B(t)\rangle = \int_{-\infty}^{t} dt_1\; Tr\left[\rho^{eq}\{\Delta B(t-t_1),H_{ext}(t_1)\}\right],$$
$$(G.15)$$

and using (G.2) we can write

$$\langle\delta B(t)\rangle = \int_{-\infty}^{t} dt_1\, Tr\left[\rho^{eq}\{\Delta B(t-t_1),-A\}F(t_1)\right].$$
$$(G.16)$$

On the other hand, since $\{\Delta B(t-t_1),\langle A\rangle_{eq}F(t_1)\} = 0$ we can also write (G.16) in the form

$$\langle\delta B(t)\rangle = \int_{-\infty}^{t} dt_1\, Tr\left[\rho^{eq}\{\Delta B(t-t_1),-\Delta A\}F(t_1)\right]$$
$$(G.17)$$
$$= \int_{-\infty}^{t} dt_1\, Tr\left[\rho^{eq}\{\Delta A,\Delta B(t-t_1)\}F(t_1)\right].$$

Finally, this formula may be written as follows (using the shift $\tau = t - t_1$)

$$\langle\delta B(t)\rangle = \int_{-\infty}^{t} dt_1\;\phi_{BA}(t-t_1)F(t_1) = \int_{0}^{\infty} d\tau\;\phi_{BA}(\tau)F(t-\tau),$$
$$(G.18)$$

where $\phi_{BA}(\tau)$ is the response function of the system $\mathcal{S}$; that is[3]:

$$\phi_{BA}(t) = Tr\left[\rho^{eq}\{\Delta A, \Delta B(t)\}\right]. \tag{G.19}$$

This function defines the *canonical* Kubo's correlation.[4] The Fourier transform of the time convolution (G.18) is then[5]

$$\mathcal{F}_{\omega}\left[\langle\delta B(t)\rangle\right] = \phi_{BA}(\omega)F(\omega). \tag{G.20}$$

This relationship shows that the Fourier transform of the response function is the susceptibility. Note that this equality has the same structure as shown in linear response theory (see Chap. 5). Thus,

$$\phi_{BA}(\omega) = \int_0^{\infty} dt \, \exp(i\omega t)\phi_{BA}(t), \tag{G.21}$$

is the desired expression.

G.1   Alternative Derivation of Kubo's Formula

We will now present an alternative derivation of Kubo's formula, concluding with the fluctuation-dissipation theorem without using the concept of Liouville-Neumann's temporal evolution for the density matrix.[6]

The starting point is to consider that the Hamiltonian of system $\mathcal{S}$ can be divided into two terms

$$H_T = H_0 + V(t), \tag{G.22}$$

one of which is the time-independent Hamiltonian at the steady state H_0, while the other is a small perturbation $V(t)$ that begins in the distant past, and has the particular form

$$V(t) = -A \, F(t). \tag{G.23}$$

---

[3] The function $\phi_{BA}(\tau)$ represents the system response by measuring the variation of the observable B against a disturbance *force* in the Hamiltonian characterized in terms of the observable A.

[4] Sometimes it is called *The after effect function*; see [19] in Chap. 8.

[5] Where an extension of the integral to negative times has to be made, but for causality the response function is zero.

[6] That is to say, without using the evolution in phase-space to represent the dynamics of $\rho(t)$, while this proof is based on time-dependent perturbation theory. See [9] in Chap. 8.

Using time-dependent perturbation theory[7] the eigenvectors $|\psi_\alpha(t)\rangle$ of the total Hamiltonian H_T can be written [to first order in the perturbation $V(t)$] in the form

$$|\psi_\alpha(t)\rangle = |\alpha\rangle\, e^{-i\varepsilon_\alpha t/\hbar} + \sum_\beta |\beta\rangle\, a_{\alpha\beta}\, e^{-i\varepsilon_\beta t/\hbar} + \mathcal{O}(V^2), \qquad \text{(G.24)}$$

where $|\alpha\rangle$ are the eigenvectors of the unperturbed Hamiltonian, i.e.:

$$H_0\,|\alpha\rangle = \varepsilon_\alpha\,|\alpha\rangle.$$

Here $\{|\alpha\rangle\}$ is a basis of the Hilbert space $\mathcal{H}_S$. Time-dependent coefficients $a_{\alpha\beta}$ are given by

$$a_{\alpha\beta} = \frac{1}{i\hbar} \int_{-\infty}^{t} dt'\, e^{-i(\varepsilon_\alpha - \varepsilon_\beta)t'/\hbar} \langle\beta\,|\,V(t')\,|\alpha\rangle. \qquad \text{(G.25)}$$

The thermal mean value of any observable B [in Schrödinger's representation] is given by

$$\langle B(t)\rangle = \sum_\alpha p_\alpha\, \langle\psi_\alpha(t)\,|\,B\,|\psi_\alpha(t)\rangle, \qquad \text{(G.26)}$$

where p_α is the quantum occupation probability for the system S to be in the state described by the perturbed eigenvector $|\psi_\alpha(t)\rangle$.

The crucial point of the theory is how to assign a "specific" probability p_α for each state described by the perturbed state $|\psi_\alpha(t)\rangle$. So here we make the important thermal assumption[8]

$$p_\alpha \approx p_\alpha^{\text{eq}}. \qquad \text{(G.27)}$$

That is, we assume that the external perturbation $V(t)$ is small enough to leave unchanged the thermal statistical distribution of each eigenstate [i.e., the corresponding canonical distribution of fermions, bosons, etc., depending on the system under study]. With this condition we can now calculate thermal mean values of any observable B. Using (G.24), (G.26), and (G.27) we obtain

$$\langle B(t)\rangle = \sum_\alpha p_\alpha^{\text{eq}} \langle\psi_\alpha(t)\,|\,B\,|\psi_\alpha(t)\rangle \qquad \text{(G.28)}$$

$$= \sum_\alpha p_\alpha^{\text{eq}} \langle\alpha\,|\,B\,|\alpha\rangle$$

---

[7]See for example [14] in Chap. 8, or in the book: *Principles of Magnetic Resonance*, by C.P. Slichter, Springer-Verlag, Berlin, 1978.

[8]This assignment clearly places where the probabilistic hypothesis (random approach) or molecular chaos (Stosszahl-Anzatz) is introduced in the theory. That is, the canonical equilibrium distribution is the key ingredient that relates the response of a nonequilibrium system with the temporal correlations at equilibrium.

$$+ \sum_\alpha \sum_\beta p_\alpha^{eq} \left[\langle \beta \mid B \mid \alpha \rangle \, a_{\alpha\beta}^* \, e^{i\varepsilon_\beta t/\hbar} \, e^{-i\varepsilon_\alpha t/\hbar} + \langle \alpha \mid B \mid \beta \rangle \, a_{\alpha\beta} \, e^{i\varepsilon_\alpha t/\hbar} \, e^{-i\varepsilon_\beta t/\hbar} \right]$$

$$+ \; \mathcal{O}(V^2).$$

Noting that the first term of Eq. (G.28) is the average value over the unperturbed state $\langle B_{eq} \rangle$, and restricting ourselves only to a linear response, we can write:

$$\langle \delta B(t) \rangle = \sum_{\alpha\beta} p_\alpha^{eq} \left[\langle \beta \mid B \mid \alpha \rangle \, a_{\alpha\beta}^* \, e^{i\varepsilon_\beta t/\hbar} \, e^{-i\varepsilon_\alpha t/\hbar} + \langle \alpha \mid B \mid \beta \rangle \, a_{\alpha\beta} \, e^{i\varepsilon_\alpha t/\hbar} \, e^{-i\varepsilon_\beta t/\hbar} \right].$$

$$\text{(G.29)}$$

If now we replace the coefficients $a_{\alpha\beta}$ by their definition (G.25), and group the exponential terms together we get

$$\langle \delta B(t) \rangle = \sum_{\alpha\beta} p_\alpha^{eq} \left[\frac{-1}{i\hbar} \int_{-\infty}^t dt' \, \left[\langle \alpha \mid V(t') \mid \beta \rangle \, e^{i(t-t')\varepsilon_\beta/\hbar} \langle \beta \mid B \mid \alpha \rangle \, e^{-i(t-t')\varepsilon_\alpha/\hbar} \right. \right.$$

$$\left. \left. - \langle \beta \mid V(t') \mid \alpha \rangle \, e^{i(t-t')\varepsilon_\alpha/\hbar} \langle \alpha \mid B \mid \beta \rangle \, e^{-i(t-t')\varepsilon_\beta/\hbar} \right] \right].$$

The exponential can be placed within the "bra-kets" $\langle \alpha \mid \cdots \mid \beta \rangle$ and consequently we can eliminate the sum β (using the closure condition of the set $\{\mid\alpha\rangle\}$), to finally obtain

$$\langle \delta B(t) \rangle = \sum_\alpha \left[\frac{-1}{i\hbar} \int_{-\infty}^t dt' \, p_\alpha^{eq} \left[\langle \alpha \mid V(t') \, e^{i(t-t')H_0/\hbar} B \, e^{-i(t-t')H_0/\hbar} \mid \alpha \rangle \right. \right.$$

$$\left. \left. - \langle \alpha \mid \left(e^{i(t-t')H_0/\hbar} B \, e^{-i(t-t')H_0/\hbar} \right) V(t') \mid \alpha \rangle \right] \right].$$

Here we can identify the equilibrium evolution operator $e^{-i(t-t')H_0/\hbar}$ and also its conjugate acting on B. Then $e^{i(t-t')H_0/\hbar} B \, e^{-i(t-t')H_0/\hbar} \equiv B(t - t')$ is the operator B in the Heisenberg representation (i.e., the equilibrium temporal evolution of the observable B). Consequently this expression becomes

$$\langle \delta B(t) \rangle = \sum_\alpha \left[-\frac{1}{i\hbar} \int_{-\infty}^t dt' \, p_\alpha^{eq} \langle \alpha \mid V(t')B(t - t') - B(t - t')V(t') \mid \alpha \rangle \right]$$

$$= \int_{-\infty}^t dt' \, Tr \left[\rho^{eq} \{ -V(t'), B(t - t') \} \right]_0$$

$$= \int_{-\infty}^{t} dt'\ Tr\left[\rho^{eq}\{A, B(t - t')\}\right]_0 F(t')$$

$$= \int_{0}^{\infty} d\tau\ Tr\left[\rho^{eq}\{A, B(\tau)\}\right]_0 F(t - \tau). \tag{G.30}$$

As usual, here we used the notation

$$\{A, B\} \equiv \frac{1}{i\hbar}[A, B].$$

Note that $V(t)$ has been replaced by its explicit expression, Eq. (G.23). In Eq. (G.30) the subscript 0 has been placed to emphasize that the trace must be calculated using the equilibrium statistical weights $\{p_\alpha^{eq}\}$, which are written in a compact form using equilibrium density matrix notation

$$\rho^{eq} = \mathcal{Z}^{-1} e^{-\beta H_0} \equiv \sum_\alpha |\alpha\rangle p_\alpha^{eq} \langle\alpha|.$$

Then, comparing (G.30) with the formulation of linear response theory:

$$\langle \delta B(t) \rangle = \int_{0}^{\infty} d\tau\ \phi_{BA}(\tau)\ F(t - \tau),$$

we can conclude

$$\boxed{\phi_{BA}(t) = Tr\left[\rho^{eq}\{A, B(t)\}\right]_0,}$$

which is Kubo's theorem.

If the perturbation is a periodic force $F(t) = \mathcal{R}_e\left[F_0 \exp(i\omega t)\right]$, the linear response of the system may be written in the form

$$\langle \delta B(t) \rangle = \mathcal{R}_e\left[\phi_{BA}(\omega) F_0 \exp(-i\omega t)\right].$$

Using the causality principle and the Fourier convolution theorem, the susceptibility (or admittance) $\phi_{AB}(\omega)$ is just the half-Fourier transform of $\phi_{BA}(t)$; then we can identify $\phi_{BA}(\omega)$ with $\chi(\omega)$ in the notation of Chap. 5, (on the half-Fourier transform see advanced exercise 5.5.6)

$$\chi(\omega) = \int_{0}^{\infty} e^{i\omega t} \phi_{BA}(t)\ dt.$$

This is all that is needed to study nonequilibrium systems in the linear response approximation. If instead of using $Tr\left[\rho^{\text{eq}}\{A, B(t)\}\right]_0$ we use a symmetric correlation function (replacing commutators by anti-commutators), the admittance $\phi_{BA}(\omega)$ can be written by a temporal integration from $-\infty$ to ∞. From this expression one finds a relationship between the dissipated energy and the spectrum of thermal fluctuations, i.e., the fluctuation-dissipation theorem.[9]

[9]See, for example, the original paper: R. Kubo: *Statistical Mechanical Theory of Irreversible Processes I*, published in J. Phys. Soc. of Japan, Vol **12**, 570 (1957).

Appendix H
Fractals

H.1 Self-Similar Objects

The monumental work of B.B. Mandelbrot[1] generated countless research papers on the concept of fractal dimension. We do not intend to give here an introduction to the study of such a vast and complex research topic, but only to give the minimum elements necessary to motivate the reader in the study of this fascinating subject.[2]

If an object is self-similar it consists of parts that are all similar to each other. A self-similar object can be "built" by iterative rules, whence successive generations are produced with the characteristic that upon amplification of part of the k-th generation graph one recovers the graph of the $k - 1$ generation. The simplest example is the Cantor set. In Fig. H.1 four generations of the *object* are shown. For example, if in the $k = 2$ generation we amplify the right-part of the Cantor set by a factor of 3, we see that the same object is obtained exactly as in the generation $k = 1$. This is a typical deterministic fractal.[3]

Moreover, in nature there are random objects having self-similar characteristics. These types of objects are called random fractals and will be described briefly in the next section.

To characterize a self-similar object it is necessary first to give some definitions. d_E is called the minimum Euclidean dimension[4]—of the space—where we can "embed" this "object" or fractal. In an Euclidean space the hypervolume $V(l)$ of

---

[1]See for example, the book: *The Fractal Geometry of Nature*, New York, W.H. Freeman (1982).

[2]Other interesting books also available to study experiments and numerical simulations in the context of fractal geometry are: *Fractals,* J. Feder, N.Y., Plenun Press (1988); *The Science of Fractal Images,* Eds. Heinz-Otto Peitgen and Dietmar Saupe, Berlin, Springer-Verlag (1988).

[3]Often these objects are called "Monsters," perhaps because of their intrinsic mathematical beauty.

[4]The *Embedding dimension d_E.*

© Springer International Publishing AG 2017
M.O. Cáceres, *Non-equilibrium Statistical Physics with Application to Disordered Systems,* DOI 10.1007/978-3-319-51553-3

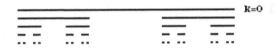

Fig. H.1 The Cantor set; its fractal dimension is $d_f = 0.639\cdots$. *Solid line* corresponds to the generation $k = 0$, the following is $k = 1$ and so on

an arbitrary object can be "measured" by covering it with linear "boxes" of size l. That is, we need $N(l)$ "boxes" to cover a given object in order to measure $V(l)$, i.e.:

$$V(l) = N(l)\, l^{d_E}.$$

Since the hypervolume cannot depend on a scale change in general we can expect $N(l) \sim l^{-d_E}$. However, for fractal objects it is proposed:

$$N(l) \sim l^{-d_f}. \tag{H.1}$$

If it holds that $d_f < d_E$, we say that this object is a fractal.[5] Then from (H.1)—after taking the logarithm—we conclude that

$$d_f = \lim_{l \to 0} \frac{\ln N(l)}{\ln(1/l)}$$

corresponds to the critical Hausdorff-Besicovitch's dimension.[6]

Example (**Cantor's Set**). Consider the fractal characterized in Fig. H.1. It is noted that to pass from generation k to generation $k + 1$ the scale factor of an arbitrary segment is $l = 1/3$ and hence the number of self-similar objects is $N(l) = 2$; then $d_f = \ln 2/\ln 3 = 0.639\cdots$. Furthermore, since $d_f < 1$, it follows that d_f is the fractal dimension.

Exercise (**Koch's Triangular Generator**). Consider the fractal in Fig. H.2. That is to say, at each iteration a segment is replaced with the "generator" appearing in the iteration $k = 1$. It is noted that in generation k the scale is $l_{(k)} = 3^{-k}$ and the total number of self-similar objects is $N(l_{(k)}) = 4^k$. Then, the total "length" of the curve in the generation k can be calculated using $k = -\ln l_{(k)}/\ln 3$, that is:

---

[5]In general we have $d_T < d_f < d_E$, where d_T is the topological dimension for the object in study. That is, $d_T = 0$ for a set of disconnected points; $d_T = 1$ for a curve; $d_T = 2$ for a surface; $d_T = 3$ for a solid body.

[6]See reference in the notes (1) and (2).

Fig. H.2 Koch's curve with triangular generator; its fractal dimension is $d_f = 1.2628\cdots$

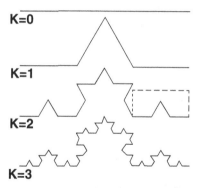

Fig. H.3 Koch's curve with square generator; in this case $d_f = 2$

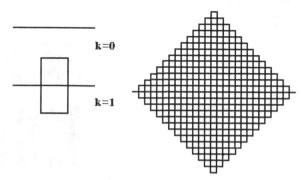

$$L(l_{(k)}) = N(l_{(k)})\, l_{(k)} = \left(\frac{4}{3}\right)^k = \left(\frac{4}{3}\right)^{-\ln l_{(k)}/\ln 3} = \exp\left(\ln\left[\left(\frac{4}{3}\right)^{-\ln l_{(k)}/\ln 3}\right]\right)$$

$$= \exp\left(\left[\frac{-\ln l_{(k)} \ln(4/3)}{\ln 3}\right]\right) = \exp\left(\left[\ln\left(l_{(k)}\right)^{-\ln(4/3)/\ln 3}\right]\right) = \left(l_{(k)}\right)^{1-d_f}.$$

where $d_f = \ln 4/\ln 3 = 1.2628\cdots < 2$ is its fractal dimension. Clearly, in the limit $l_{(k)} \to 0$ of the total length: $L(l_{(k)}) = N(l_{(k)})\, l_{(k)}$, is not a useful quantity. Show—in an analogous way as we did before—that $N(l_{(k)}) = l_{(k)}^{-d_f}$.

Example (**Koch's Square Generator**). Consider the fractal of Fig. H.3. That is to say, at each iteration a segment is replaced with the "generator" appearing in the iteration $k = 1$. It is noted that the scaling factor is $l = 1/3$ and $N(l) = 9$, then $d_f = \ln 9/\ln 3 = 2$. In this case we do not get a fractal dimension, but d_f is greater than the topological $d_T = 1$.

Let us now show examples of fractals where the topological dimension of the object under study is a surface or a solid body. Sierpinski's "carpets," whose generators are holed surfaces are typical examples of fractal surfaces.

Example (**Triangular Sierpinski's Carpet**). Consider the fractal in Fig. H.4. That is, in each iteration one triangular "surface" is replaced by the "generator" appearing

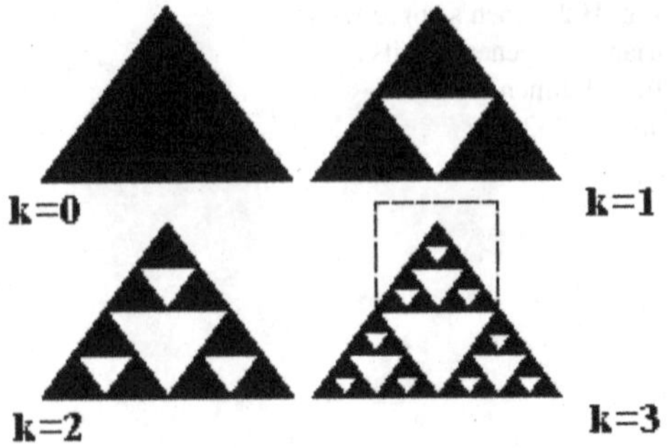

Fig. H.4 Sierpinski's carpet, its fractal dimension is $d_f = 1.585\cdots$

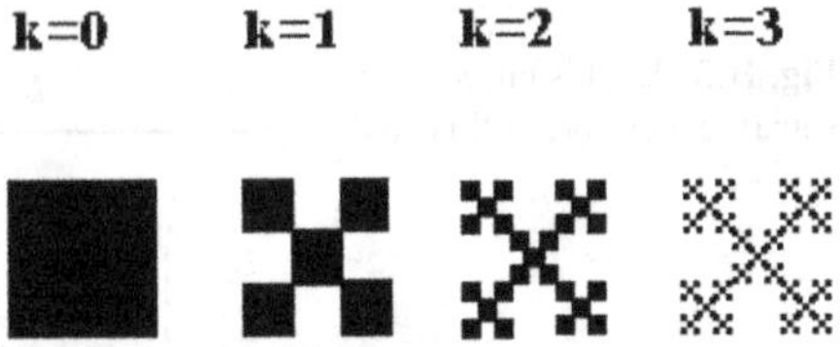

k=0   k=1   k=2   k=3

Fig. H.5 Vicsek's carpet; its fractal dimension is $d_f = 1.4649\cdots$

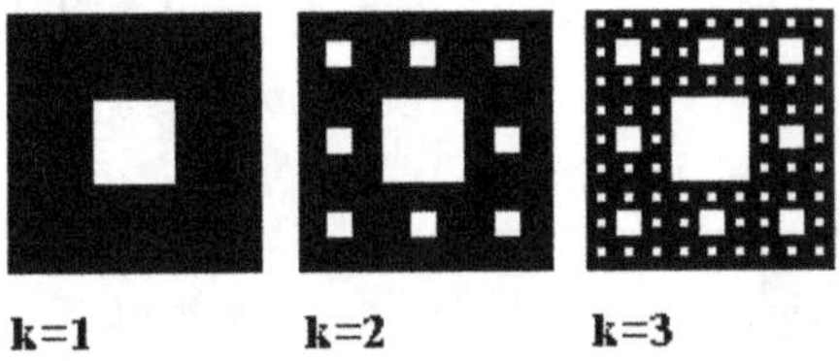

k=0   k=1   k=2   k=3

Fig. H.6 Sierpinski's square carpet; its fractal dimension is $d_f = 1.8927\cdots$

k=1   k=2   k=3

in the iteration $k = 1$. It is noted that going from generation k to generation $k + 1$ the scaling factor is $l = 1/2$ and $N(l) = 3$, then $d_f = \ln 3/\ln 2 = 1.585\cdots < 2$ is its fractal dimension .

Example (**Vicsek's Carpet**). Consider the fractal surface in Fig. H.5. In each iteration one square "surface" is replaced by the "generator" appearing in the iteration $k = 1$. It is noted that to pass from generation k to generation $k + 1$ the scale factor is $l = 1/3$ and $N(l) = 5$. Then $d_f = \ln 5/\ln 3 = 1.4649\cdots < 2$ is its fractal dimension.

Exercise (**Sierpinski's Solid Body**). Consider the square Sierpinski carpet shown in Fig. H.6. It is noted that to pass from generation k to generation $k + 1$ the scale factor is $l = 1/3$ and $N(l) = 8$; then $d_f = \ln 8/\ln 3 = 1.8927\cdots$. Now, from this "surface" fractal build a generalization for a fractal *solid body*, in this case show that the fractal dimension of the object generated is greater than 2 and it is given by $d_f = \ln 10/\ln 3 < 3$.

H.2   Statistically Self-Similar Objects

We have mentioned that there exist in nature random self-similar objects. Rough interfaces, advancing flame fronts, the outlines of a coast [see the map of southern Chile, for example], the outline of the clouds on a sunny day, the ramifications of rivers, etc. are some of the many examples of self-similar objects with random characteristics.

In general, if we focus on stochastic situations, i.e., if we study a **sp**, we can easily introduce the concept of random fractals for sorting the "roughness" of a given stochastic realization.

In Sect. 3.11, when the power spectrum of a **sp** was studied, we introduced the definition of scaled **sp**; let us specify this concept again. For example, consider the conditional probability of a diffusive Wiener process $P[X(t) \mid X(t_0)]$. In this case it is easy to verify the scaling relation

$$P\left[\Lambda^{1/2} X(\Lambda t) \mid \Lambda^{1/2} X(\Lambda t_0)\right] = \Lambda^{-1/2} P\left[X(t) \mid X(t_0)\right] , \ \forall \Lambda > 0.$$

As already mentioned in Chap. 3, this property can also be analyzed in terms of its corresponding characteristic function:

$$G_X(\frac{k}{\Lambda^{1/2}}, \Lambda t) = G_X(k, t).$$

That is, stochastic realizations are scale invariant in their distribution (then, we say that "in distribution" a realization of a **sp** fulfills scale invariance). We have denoted this fact by saying that the **sp** $X(t)$ satisfies the scale relationship:

$$X(\Lambda t) / \Lambda^{1/2} \doteq X(t).$$

This gives a procedure for calculating "mechanically" the fractal dimension associated with the *record* of a stochastic realization. However, by the very nature of this random fractal, the scale transformation on the space X-axis is different from the scale transformation on the time t-axis; as a result, the definition of fractal dimension of a statistically *self-affine* object is not unique. The difference with the self-similar concept is that in a self-affine transformation a point $\mathbf{z}$ in a E-dimensional "abstract-space" goes into a new point $\mathbf{z}'$, where the scaling ratios are *not* all equal for each component.[7] To show this fact we proceed as follows. First suppose that for an arbitrary scale ratio of the process it holds that

$$X(\Lambda t) / \Lambda^{H} \doteq X(t). \tag{H.2}$$

---

[7] In what follows of the text we do not make use of this rigorous (distinctive) classification. More details on self-affine and self-similar sets are found in Mandelbrot's book, see note (1); or in B.B. Mandelbrot, Physica Scripta **32**, 257, (1985).

We calculate the fractal dimension associated with the operation of "covering" the realization of **sp** $X(t)$ satisfying the scale relationship (H.2). One possible way to calculate this quantity is "counting" the number of parallelograms [minimum height a, and minimum width τ in the direction of the X-axis, and t-axis respectively] needed to fully cover the realization. Suppose $\mathcal{T}$ is the duration of the realization, then we need $\mathcal{T}/\Lambda\tau$ segments to cover the entire time axis. On the other hand, using (H.2) we see that the range in the time interval $\Lambda\tau$ (taking $X(0) = 0$) is of the order

$$\Delta X(\Lambda\tau) \equiv X(\Lambda\tau) - X(0) \doteq \Lambda^H \Delta X(\tau). \tag{H.3}$$

That is, in each $\Lambda\tau$ segment the range of variation in the X-axis is characterized by the quantity $\Delta X(\Lambda\tau) \doteq \Lambda^H \Delta X(\tau)$, from which it follows that we need to "stack" $\Lambda^H \Delta X(\tau)/\Lambda a$ "boxes" of height Λa to cover such variation. Then the total number of "boxes" needed to cover this object is given by

$$\mathcal{N}(\Lambda, a, \tau) = \frac{\Lambda^H \Delta X(\tau)}{\Lambda a} \times \frac{\mathcal{T}}{\Lambda\tau} \sim \Lambda^{H-2}.$$

Then, from (H.1) it follows that the fractal dimension is

$$D_B = 2 - H < 2. \tag{H.4}$$

Note that in this analysis we have used the argument that the boxes of minimum size are small compared with duration $\mathcal{T}$ and the range of the realization; that is, $\tau \ll \mathcal{T}$ and $a \ll \Delta X(\tau)$, respectively. If this is not the case, that is, if $a \sim \left(\Delta X(\tau)^2\right)^{1/2}$ it is only necessary to use *one* box to cover the range in a $\Lambda\tau$ segment, whence it follows that $\mathcal{N}(\Lambda, a, \tau) = 1 \times \mathcal{T}/(\Lambda\tau) \sim \Lambda^{-1}$, then $D = 1$. That is, in this case a fractal dimension is not obtained. Then this calculation only makes sense in the limit of high resolution, i.e., the fractal dimension D_B is a local quantity.[8]

Another possibility is to calculate along the curve, instead of computing the coverage of the realization. For this purpose we proceed as follows. For random self-similar objects, such as coastlines on a map, usually a "measuring unit" of *length* δ is used to calculate the total length of a stochastic realization. That is, the contribution to the total length measured on the realization of the **sp** $X(t)$ is performed with a "measuring unit" of length δ to cover a portion characterized by the time segment $\Lambda\tau$; in this case we have—using the Pythagorean theorem—that

$$\delta = \sqrt{(\Lambda\tau)^2 + \left(\frac{\Delta X(\Lambda\tau)}{a}\right)^2} = \sqrt{(\Lambda\tau)^2 + \Lambda^{2H}\left(\frac{\Delta X(\tau)}{a}\right)^2}. \tag{H.5}$$

---

[8]This fractal dimension is called *Box dimension* .

Note that in the second term of this expression we used (H.3); here a measures the unit on the X-axis. Then, in the limit of a very small a we infer that the dominant contribution is given by

$$\delta \sim \Lambda^H. \tag{H.6}$$

Note that in this case, when we put the measuring unit[9] of length δ *on* the realization, this is almost *parallel* to the X-axis. Then the number of measuring unit of length δ necessary to measure the object is given by

$$\mathcal{N}(\delta) = \frac{\mathcal{T}}{\Lambda \tau} \sim \Lambda^{-1} \sim \delta^{-1/H},$$

where we have used (H.6). Again, using (H.1) it follows that the fractal dimension of the set is[10]

$$D_D = \frac{1}{H}. \tag{H.7}$$

If the scale on the X-axis is not small, that is, $\Delta \mathbf{X}(\Lambda \tau) \ll a$, the second term in (H.5) is not dominant; then $\delta \sim \Lambda^1$, which implies that $\mathcal{N}(\delta) \sim \Lambda^{-1} \sim \delta^{-1}$, from which it follows that the apparent dimension is $D = 1$. That is, for the same random self-affine object, here we have presented at least three different dimensions: $D_B = 2 - H$, $D_D = 1/H$, and $D = 1$. In general, given a statistically self-similar object and if we set the mechanism to "measure" it, the interest is to plot $\ln N(l)$ versus $\ln(l)$ to be able to calculate its slope (the fractal dimension).

Example (**Brownian Motion**). The Wiener process has a scale ratio of the form (H.2) with $H = 1/2$; then, using (H.4) and (H.7) we obtain the following fractal dimensions: $D_B = 3/2$, $D_D = 2$.

Excursus (**Fractional Brownian Motion.**) In Sect. 3.6.1 we mentioned that B.B. Mandelbrot introduced a generalization of the Wiener process, which he called *fractional Brownian motion* (fBm). The advantage of this process is that it allows to vary the roughness of the realization by controlling its correlation function. In particular, Mandelbrot coined the use of the exponent $H \in (0, 1)$ to characterize the scale relationship (H.2); when $H \sim 0$ the realization is extremely rough [the process resembles white noise], while for $H \sim 1$ the realization is soft, the case $H = 0.5$ corresponds to the normal diffusive process. In Fig. H.7 realizations of different fBm processes are shown for three values of H, thus obtaining different fractal dimensions. From the definition of the fBm process (which is given in terms of a Wiener integral convoluted with a power-law kernel) it is simple to see that it

---

[9]Sometimes called the "rule."

[10]The fractal dimension D_D is called *Divider dimension* or *Compass dimension*.

is Gaussian and its dispersion is characterized by the behavior $\langle X(t)^2 \rangle \propto t^{2H}$. Also from the very definition of the fBm, it is possible to calculate the correlation of the fBm process[11]:

$$\langle X(t) X(s) \rangle = \frac{1}{2} \left(|t|^{2H} + |s|^{2H} - |t-s|^{2H} \right). \qquad (H.8)$$

Note that this is all that we need to know because the process is Gaussian, this function is homogeneous of order $2H$ and so it can be considered as a "fractal" property. From this expression it is simple to see that in the case $H = 1/2$, the Wiener correlation function $\langle X(t) X(s) \rangle = \min\{t, s\}$ is recovered. Using (H.8) calculate the increments $X(t) - X(s)$ in an interval $[s, t]$; then, show that they are Normal distributed with zero mean and the even moments are

$$\left\langle (X(t) - X(s))^{2k} \right\rangle = \frac{(2k)!}{k! 2^k} |t-s|^{2Hk}, \quad k = 1, 2, \cdots.$$

On the other hand, from (H.8) it is also simple to see that the persistent-antipersistent correlation is characterized by

$$\langle -X(-t) X(t) \rangle / \langle X(t)^2 \rangle = 2^{2H-1} - 1.$$

To close these comments note that knowing the statistical properties of the *increments* of the fBm $X(t) - X(s)$ it is possible to write a stochastic differential equation where the Wiener differential $dW(t)$ is replaced by the fBm differential: $dX_H(t) \equiv X(t) - X(t + \Delta t)$, allowing in this way to study a new type of stochastic differential equation; for example the *fractional* Ornstein–Uhlenbeck process:

$$dY = -Y dt + \sigma \, dX_H(t).$$

Optional Exercise (**Lévy's Process**). The conditional probability distribution of Lévy's process is characterized by the characteristic function[12]

$$G_X(k, t) \propto \exp\left(-Ct \, | \, k \, |^\mu \right), \quad \forall t > 0,$$

where $C > 0$ is a constant and $\mu \in]0, 2]$. The case $\mu = 2$ corresponds to a Gaussian process, $\mu = 1$ corresponds to Cauchy's distribution [see Sects. 1.4.2 and 6.2.2]. In general, this process obeys a scaling relation of the form (H.2) with $H = 1/\mu$. Show that we can define the following fractal dimension for Lévy's process:

---

[11]The calculation of this correlation has some complexity in its proof, it can be seen in: Mandelbrot, B. B., Van Ness, J. W., *Fractional Brownian motions, fractional noises and applications*. SIAM Review 10, 422–437 (1968).

[12]See *excursus* in Sects. 3.14 and 6.2.3.

Fig. H.7 Realization of the fBm process for three different values of H; (**a**) realization with $H = 0.2$; (**b**) $H = 0.5$ and (**c**) $H = 0.8$. The corresponding values for the "Box" fractal dimensions are (**a**) $D_B = 1.8$; (**b**) $D_B = 1.5$ and (**c**) $D_B = 1.2$

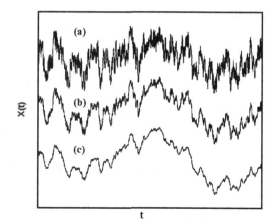

$$D_B = 2 - 1/\mu, \quad \forall \mu \in [\frac{1}{2}, 2[$$

$$D_B = 0, \quad \forall \mu \in (0, \frac{1}{2}].$$

Note that depending on the value of μ, moments of the **sp** are not defined; however, fractional moments can be calculated. Unfortunately, this calculation is not local, this is why its analysis is more difficult. For example, assuming a positive domain for X use

$$\frac{1}{X^{1-\beta}} = \frac{1}{\Gamma(1-\beta)} \int_0^\infty \xi^{-\beta} e^{-\xi X} \, d\xi, \quad 0 < \beta < 1,$$

to show that the fractional moments can be studied in terms of the characteristic function $G_X(k) \equiv \langle e^{ikX} \rangle$, using the formula

$$\langle X^\beta \rangle \equiv \int_0^\infty X^\beta P(X) \, dX = \frac{-1}{\Gamma(1-\beta)} \int_0^\infty \xi^{-\beta} \frac{dG_X(k = i\xi)}{d\xi} \, d\xi.$$

Guided Exercise (**Roughness of a Realization**). Consider a diffusion process $\mathbf{Z}(t)$ for times $t \in [0, T]$ which is not necessarily stationary, assuming that we do not know its *complete* statistics. If we only know the first and second moment, which are just given by $\langle \mathbf{Z}(t) \rangle = 0$, $\langle \mathbf{Z}(t_1) \mathbf{Z}(t_2) \rangle = g(t_1, t_2)$, we can introduce a sort of Gaussian approximation for studying the realization of the process in terms of the function $g(t_1, t_2)$, which in fact "signs" the roughness of realizations. In Sect. 3.16 we introduced an expansion in eigenfunctions for the stochastic realizations $\mathbf{Z}(t)$ solving the integral eigenvalue problem:

$$\int_0^T \langle \mathbf{Z}(t) \mathbf{Z}(s) \rangle \phi_j(s) ds = \lambda_j \phi_j(t) \tag{H.9}$$

$$\int_0^T \phi_l(s)\phi_j(s)ds = \delta_{lj} \tag{H.10}$$

$$\sum_j \phi_j(s)\phi_j(s') = \delta\left(s - s'\right) \tag{H.11}$$

In particular using the correlation function $g\left(t_1, t_2\right) = \left(\min\left(t_1, t_2\right)\right)^{2H}$, with $H \in$ [0, 1] in Eq. (H.9) we get

$$\int_0^t g\left(s\right)\phi(s)ds + \int_t^T g\left(t\right)\phi(s)ds = \lambda\phi(t), \tag{H.12}$$

as $g\left(0\right) = 0$, we find the boundary condition $\phi(0) = 0$. Taking the time derivative in (H.12) we get

$$g'\left(t\right)\int_t^T g\left(t\right)\phi(s)ds = \lambda\phi(t), \tag{H.13}$$

and considering that $g'\left(T\right) < \infty$ we get the second boundary condition $\phi'(T) = 0$. Taking again the time derivative in (H.13) we can write

$$\frac{g''\left(t\right)}{g'\left(t\right)}\lambda\phi'\left(t\right) - g'\left(t\right)\phi\left(t\right) = \lambda\phi''\left(t\right). \tag{H.14}$$

Then using the model $g(t) = t^{2H}$ in (H.14) we get the following differential equation:

$$t^2\phi''\left(t\right) - \left(2H - 1\right)t\phi'\left(t\right) + \frac{2H}{\lambda}t^{2H+1}\phi\left(t\right) = 0. \tag{H.15}$$

This equation admits two solutions (Bessel's functions of fractional order)

$$\phi\left(t\right) = \mathcal{N} J_{\pm v}\left(ut^{H+1/2}\right)t^H,$$

where

$$v = \frac{2H}{\left(1 + 2H\right)}, \quad u = \frac{\sqrt{2H}}{\left(H + 1/2\right)\sqrt{\lambda}}$$

and $\mathcal{N}$ is the normalization constant. Using the boundary condition $\phi(0) = 0$ and taking into account that

$$\lim_{t \to 0} J_{-v}\left(u\, t^{H+1/2}\right)t^H = \text{constant} \tag{H.16}$$

$$\lim_{t \to 0} J_{+v}\left(u\, t^{H+1/2}\right)t^H = 0, \tag{H.17}$$

we conclude that we must use only solution (H.17). The eigenvalues can be obtained from the second boundary condition $\phi'(T) = 0$. Then after some algebra we get

$$J_{\nu-1}\left(u_j\, T^{H+1/2}\right) = 0, \quad \text{with} \quad u_j = \frac{\sqrt{2H}}{(H+1/2)\sqrt{\lambda_j}}. \tag{H.18}$$

The normalization constant can be calculated using the orthogonality condition of boundary value problem (H.10), then

$$\mathcal{N} = \frac{\sqrt{1+2H}}{\left|J_{+\nu}\left(u_j\, T^{H+1/2}\right)\right|}, \quad u_j = \text{solution of (H.18)}.$$

Then, the eigenfunction of the problem (H.9) with $\langle \mathbf{Z}(t)\, \mathbf{Z}(s)\rangle = [\min(t,s)]^{2H}$ can be written in the form

$$\phi_j(t) = \frac{\sqrt{1+2H}}{\left|J_{+\nu}\left(u_j\, T^{H+1/2}\right)\right|}J_{+\nu}\left(u_j\, t^{H+1/2}\right)t^H,$$

and thus *one* realization of the **sp** $\mathbf{Z}(t)$ can be written as the expansion:

$$Z(t) = \sum_j z_j\phi_j(t), \quad \text{in terms of Gaussian } \mathbf{rvs}\ z_j,$$

where z_j for all j are characterized by moments:

$$\langle z_j\rangle = 0, \quad \langle z_j z_l\rangle = \frac{2H\,\delta_{jl}}{\left((H+1/2)\,u_j\right)^2}.$$

We should remark, however, that unfortunately the nontrivial part of this analysis is to verify or not the completeness condition (H.11).

Excursus (**Anomalous Diffusion on Fractal Networks**).   From the recognition that fractal networks are good candidates for modeling disordered systems, the study of the dynamics of an RW on a fractal structure has been an important technique in the analysis of anomalous diffusion. Often it is characterized the asymptotic behavior of the second moment on a non-Euclidean RW network (embedded in dimension d_E) by introducing an exponent d_ω in the form:

$$\langle r(t)^2\rangle \sim t^{2/d_\omega}, \quad \text{with} \quad d_E > d_\omega > 2,$$

in the Euclidean case $d_\omega = 2$. The reason why subdiffusive transport is obtained is that the RW is restricted to a reduced "volume" in a fractal structure. It is customary to relate the fractal dimension d_ω with the topology and dynamics information of the RW (d_ω represents the fractal dimension of RW in a disordered system).

Fig. H.8 Representation of a
Bethe's lattice (Cayley's tree)
for $z = 4$

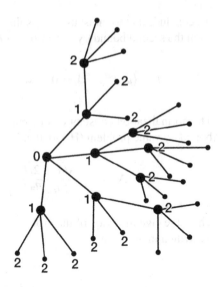

There are interesting conjectures among the different exponents that appear in the
theory of anomalous diffusion on non-Euclidean networks. Once again, Einstein's
relationship allows to establish the link between the static and dynamic critical
exponents [see, for example, the treatise on percolation and disorder by: A. Bunde
and S. Havlin in: *Fractals and Disordered Systems*, Berlin, Springer-Verlag (1991)].

Optional Exercise (**Bethe's Lattice**). The Bethe lattice or Cayley's "tree" is a
peculiar *net* structure making it possible to study analytically nontrivial percolation
problems depending on the coordination number z. This network is not a fractal,[13]
but consists of a net structure—without loops—in which there is only one way to go
from one level j to p.[14] To generate this network we proceed as follows. Consider
an arbitrary level, which we call 0, from which emanate z "branches" [of unit
length], then the end point of each branch is a new level [say 1] and so we get z
levels constituting the first "shell" of Cayley's tree. From each of these levels again
generate $z - 1$ levels [say 2], thus in the second shell we have $z(z - 1)$ equivalent
levels, and so on. The case $z = 2$ corresponds to the Euclidean lattice, that is a semi
infinite 1D lattice. Consider an RW on a Bethe lattice with coordination number z
and obtain the Green function of the ME associated with this homogeneous model.
Hint: The RW can be described as a biased one-dimensional lattice, where the origin
acts as a reflecting boundary condition.[15]

---

[13]It is possible to check that the *dimension* of this "object" is infinite $\forall z > 2$.

[14]It is important to remark that in this network Euclidean "distance" is meaningless, hence we
speak now about "levels," or chemical distance $j - p$; see Fig. H.8.

[15]The discrete-time representation of this problem and many interesting results concerning RW
Polya's problem on a Cayley's tree are shown in: B.D. Hughes and M. Sahimi, J. Stat. Phys. 29,
781 (1982).

Optional Exercise (**The "comb" Lattice**). A *comb* lattice is a two-dimensional structure that allows to study analytically nontrivial problems of anomalous diffusion. Obviously this is not a fractal network, but it is a peculiar structure in the form of "comb" in which, for example, there is only diffusion along the x-axis when $y = 0$. To generate an RW Markov dynamics on this comb network define the following master Hamiltonian (transition matrix, see Chap. 6)

$$\mathbf{H}_{\text{comb}} = \mu_x\, \delta_{y,0} \left(E_x^+ + E_x^- - 2 \right) + \mu_y \left(E_y^+ + E_y^- - 2 \right)$$
$$\equiv \mu_x\, \delta_{y,0}\, \mathbf{H}_x^0 + \mu_y\, \mathbf{H}_y^0.$$

Now consider the following ME

$$\dot{\mathbf{P}} = \mathbf{H}_{\text{comb}} \cdot \mathbf{P}.$$

Using the Green function of the problem in the $\hat{y}$ direction:

$$\mathbf{G} = \left[u\mathbf{1} - \mu_y\, \mathbf{H}_y^0 \right]^{-1},$$

and the term $\mu_x\, \delta_{y,0}\, \mathbf{H}_x^0 \cdot \mathbf{P}$ as a "source" of the formal solution, obtain an expression for the diffusion along the x-axis; that is calculate the probability:

$$\mathbf{P}(x,0,t) \equiv\, < x,0 \mid \mathbf{P}(t) \mid 0,0 >.$$

It can be seen that asymptotically $\sum_x \mathbf{P}(x,0,t) \sim t^{-1/2}$, that is to say $\mathbf{P}(x,0,t)$ is not normalized. Calculate the second moment of the displacement $\langle x(t)^2 \rangle$. Generalize this model to the case when the diffusion in the lattice direction $\hat{y}$ is disordered, i.e., use an effective Green function $\mathbf{G}$.[16] Hint: using Laplace ($t \to u$) and Fourier ($x \to k$) variables, expressions are often easier to handle. A good reference where an RW is analyzed—for discrete time—in this type of network is: G. Weiss and S. Havlin, Physica A 134, 474 (1986). Could we generalize a "comb" network to be a fractal structure itself? What happens if the "teeth" in the comb lattice have different lengths (finite) and reflecting boundary conditions at the edge? Or if teeth are located at random positions in a regular lattice?

---

[16]The generalization to a network with untidy comb teeth is presented in: S. Havlin, J.E. Kleifer and G. Weiss, Phys Rev A **36**, 1403 (1987).

Appendix I
Quantum Open Systems

I.1 The Schrödinger-Langevin Equation

The Kossakowski-Lindblad mathematical structure for a Markov semigroup[1] (which moreover embodies irreversibility in quantum mechanics) can be obtained heuristically introducing the following approach due to van Kampen.[2] As was mentioned in Chap. 8 a dissipation term in the Schrödinger equation violates the conservation of probability, to avoid this loss it is possible to add artificially a suitable noise term in such a way that the average norm of the wave function would be conserved. First we write the Schrödinger equation ($\hbar = 1$) for the vector state ψ in the form

$$\dot{\psi} = -iH\psi - U\psi + l(t) V\psi, \tag{I.1}$$

where H is the Hamiltonian of the system of interest and U can be taken to be Hermitian; here U and V are time-independent operators to be specified later. The noise $l(t)$ is complex and can be of any statistics structure,[3] at this point we only

---

[1] Any differential equation for the density matrix must have the following properties for all times: (a) Trace of the matrix is constant and equal to one. (b) The matrix remains Hermitian. (c) The matrix is positive definite. The most general form for such differential equation was derived by Kossakowski and Lindblad independently [A. Kossakowski, Bull. Acad. Pol. Sci, Série Math. Astr. Phys. 20, 1021 (1971); 21, 649 (1973); Rep. Math. Phys. 3, 247 (1972); G. Lindblad, Comm. Math. Phys. 40, 147 (1975); V. Gorini, A. Kossakowski, and E.C.G Sudarshan, J. Math. Phys. 17, 821 (1976)].

[2] As far as I know, reference on the Schrödinger-Langevin equation was presented in 1992 in the second edition of van Kampen's book, see [31] in Chap. 8.

[3] In quantum mechanics there is no need to calculate higher moments of the vector state ψ, so it is not necessary to worry about the statistical Gaussian (or not) structure of the complex noise $l(t)$.

© Springer International Publishing AG 2017
M.O. Cáceres, *Non-equilibrium Statistical Physics with Application to Disordered Systems*, DOI 10.1007/978-3-319-51553-3

demand that this noise has white correlations, in particular 1st moment zero, and 2nd moment of the form[4]

$$\overline{l(t)} = 0, \ \overline{l(t) \, l^*(t')} = \delta(t - t').$$

Equation (I.1) can be solved (using ordinary differential calculus) for a small increasing time $t + \Delta t$ in the form

$$\psi(t + \Delta t) \simeq \left[1 - iH\Delta t - U\Delta t + \int_t^{t+\Delta t} l(t') \, dt' V \right] \psi(t),$$

and similarly from the adjoint Schrödinger equation

$$\psi^\dagger(t + \Delta t) \simeq \psi^\dagger(t) \left[1 + iH\Delta t - U\Delta t + V^\dagger \int_t^{t+\Delta t} l^*(t') \, dt' \right].$$

Taking the scalar product of the wave vector, and the average over the noise we get the final expression

$$\overline{\psi^\dagger(t + \Delta t) \, \psi(t + \Delta t)} = \overline{\psi^\dagger(t) \left(1 - 2U\Delta t + V^\dagger V \Delta t \right) \psi(t)} + \mathcal{O}(\Delta t),$$

from this equation we see that to conserve the norm we must have the operational relation:

$$2U = V^\dagger V.$$

We can carry out the same calculation for the "*averaged*" density matrix:

$$\rho(t)_{jl} = \overline{\psi_j^\dagger(t) \, \psi_l(t)}. \tag{I.2}$$

For the general case having several traceless non-Hermitian[5] operators V_α in the Schrödinger-Langevin equation we have

$$\dot{\psi} = -iH\psi - \frac{1}{2} \sum_\alpha V_\alpha^\dagger V_\alpha \psi + \sum_\alpha l_\alpha(t) V_\alpha \psi \tag{I.3}$$

$$\overline{l_\alpha(t)} = 0, \ \overline{l_\alpha(t) \, l_\beta^*(t')} = \delta(t - t') \delta_{\alpha,\beta}.$$

---

[4] Note that in this Section the average over the noise is denoted as $\overline{l(t_1) \cdots l(t_n)}$.

[5] If among the V_α there is a Hermitian one, V_0, its final contribution may be incorporated into an effective Hamiltonian. The benefit of the Schrödinger-Langevin representation is the possibility to simulate an stochastic wave function instead of solving the reduced density matrix, see note (7).

Then, the corresponding evolution equation for the "*averaged*" density matrix is:

$$\dot{\rho} = -i\,[H,\rho] + \frac{1}{2}\sum_{\alpha}\left\{\left[V_{\alpha}\rho, V_{\alpha}^{\dagger}\right] + \left[V_{\alpha}, \rho V_{\alpha}^{\dagger}\right]\right\}, \tag{I.4}$$

which is precisely the canonical Kossakowski-Lindblad mathematical structure for a completely positive infinitesimal semigroup; that is, a quantum Markov process, see [1, 31] in Chap. 8.

We can also repeat this procedure considering nonwhite noises that depend on the temperature of the bath through the correlations: $\overline{l_{\alpha}(t)\,l_{\beta}^{*}(t')}$ and $\overline{l_{\alpha}(t)\,l_{\beta}(t')}$, in this case it is possible to see that the evolution equation for the density matrix has the same *form* as in (I.4) but it is not possible to assure that the infinitesimal generator is a completely positive one.[6] It is possible to see starting from an open quantum model that tracing out the bath variables is equivalent to using a general Schrödinger-Langevin equation (up to the Born-Markov approximation); that is, the *reduced* density matrix is equivalent to the *noise averaged* density matrix. In this case, the nonwhite noise terms in (I.3) must obey some restrictions in order fulfill the Kossakowski-Lindblad structural theorem, and the general quantum master equation has the form

$$\dot{\rho} = -i\,[H_{\text{eff}},\rho] + \frac{1}{2}\sum_{\alpha,\beta}a_{\alpha\beta}\left\{\left[V_{\alpha}\rho, V_{\beta}^{\dagger}\right] + \left[V_{\alpha}, \rho V_{\beta}^{\dagger}\right]\right\}, \tag{I.5}$$

where the matrix $a_{\alpha\beta}$ is positive definite.

In the context of *tracing-out* bath variables technique, we consider a system $\mathcal{S}$, interacting with a bath $\mathcal{B}$, to be described by a total Hamiltonian: $H_{\text{T}} = H_S + H_B + \lambda H_I$, where the density matrix ρ^{T} operates in the total Hilbert space $\mathcal{H}_{\text{T}} = \mathcal{H}_S \otimes \mathcal{H}_B$. Let us now give an important result in terms of the interaction Hamiltonian:

$$H_I = \sum_{\beta=1}^{n} V_{\beta} \otimes B_{\beta},\ n \le N^2 - 1, \tag{I.6}$$

here V_{β} belongs to Hilbert space $\mathcal{H}_S$ ($\dim\mathcal{H}_S = N$), and B_{β} are bath operators on Hilbert space $\mathcal{H}_B$; denoting

$$\chi_{\alpha\beta}(-\tau) = \text{Tr}_{\beta}\left[\rho_B^{\text{eq}} B_{\alpha}^{\dagger} B_{\beta}(-\tau)\right]$$

$$\Gamma_{\alpha\beta}(-\tau) = \text{Tr}_{\beta}\left[\rho_B^{\text{eq}} B_{\alpha} B_{\beta}(-\tau)\right]$$

---

[6]A generalization of the Schrödinger-Langevin equation considering a non-white noise correlation can be seen in: M.O. Cáceres and A.K. Chattah, Physica A 234, 322 (1996); Erratun 242, 317 (1997). Also some non-Markovian cases can be worked out using this approach, see in: A.K. Chattah and M.O. Cáceres, Condensed Matter Physics, Vol. 3, 51 (1997); M.O. Cáceres, A.A. Budini and A.K. Chattah, Rev. Mex. Fis. 45, 217 (1999).

$$V_\beta\left(-\tau\right) = e^{-i\tau H_S} V_\beta e^{+i\tau H_S} = \sum_{\gamma=1}^{N^2-1} C_{\beta\gamma}\left(-\tau\right) V_\gamma,$$

we can conclude that the algebraic structure in (I.5) is:

$$a_{\alpha\gamma} = \sum_\beta \int_0^\infty d\tau \left(\chi_{\gamma\beta}\left(-\tau\right) C_{\beta\alpha}\left(-\tau\right) + \chi_{\alpha\beta}^*\left(-\tau\right) C_{\beta\gamma}^*\left(-\tau\right) \right),$$

and the effective Hamiltonian in (I.5) can be written in terms of $\Gamma_{\alpha\beta}\left(-\tau\right)$ and $C_{\beta\gamma}\left(-\tau\right)$ in the form:

$$H_{\text{eff}} = H_S - \frac{i}{2} \sum_{\alpha\beta\gamma} \int_0^\infty d\tau \left(\Gamma_{\alpha\beta}\left(-\tau\right) C_{\beta\gamma}\left(-\tau\right) V_\alpha V_\gamma - \Gamma_{\alpha\beta}^*\left(-\tau\right) C_{\beta\gamma}^*\left(-\tau\right) V_\gamma^\dagger V_\alpha^\dagger \right).$$

Proposition *A necessary condition to assure that the algebraic structure $a_{\alpha\beta}$ will be positive can be seen in the following way. Let us assume that the interaction Hamiltonian is written in a particular basis as in (I.6) with $n \le N^2 - 1$, and that the half-Fourier transforms of the correlation of the bath are not zero. Then the set $\{V_\beta\}_{\beta=1}^n$ ought to be closed under Heisenberg representation, that is,*

$$V_\beta\left(-\tau\right) \equiv e^{-i\tau H_S} V_\beta e^{+i\tau H_S} = \sum_{\gamma=1}^m C_{\beta\gamma}\left(-\tau\right) V_\gamma, \text{ with } m \le n, \tag{I.7}$$

otherwise the matrix $a_{\alpha\gamma}$ will not be positive definite.[7]

I.2   The Quantum Master-Like Equation

In quantum mechanics there are several methods for obtaining an approximate evolution equation of the reduced density matrix, to mention just a few: Feynman's path integral, quantum trace algebra, Nakajima-Zwanzig's projector operators, van Kampen's elimination of fast variables, etc. In this section we are going to use the projector method which in fact was already presented in the context of disordered random matrices in Chap. 1. Let ρ^T be the density matrix of the total system $S + B$. Then ρ^T obeys von Neumann's equation ($\hbar = 1$)

---

[7] Several interesting comparisons between the Schrödinger-Langevin equation and the elimination of bath variables techniques can be seen in: A.A. Budini, A. K. Chattah and M.O. Cáceres, J. Phys. A: Math. Gen. 32, 631 (1999).

$$\frac{d\rho^{\mathrm{T}}}{dt} = -i\left[H_{\mathrm{T}}, \rho^{\mathrm{T}}\right] \equiv \mathcal{L}\rho^{\mathrm{T}}$$

$$= \left(\mathcal{L}_S + \mathcal{L}_B + \lambda\mathcal{L}_I\right)\rho^{\mathrm{T}}, \tag{I.8}$$

where $H_{\mathrm{T}} = H_S + H_B + \lambda H_I$, and $\mathcal{L}$ stands for a superoperator acting on ρ^{T}. We have introduced the parameter λ to keep track—in the future—of the perturbation order. We define the projector superoperator $\mathcal{P}$ by its action on any operator in the total Hilbert space of the system $\mathcal{S} + \mathcal{B}$:

$$\mathcal{P}\rho^{\mathrm{T}} = \rho_B^{\mathrm{eq}} \, Tr_B\left\{\rho^{\mathrm{T}}\right\} = \rho_B^{\mathrm{eq}} \, \rho_S$$

$$\mathcal{Q}\rho^{\mathrm{T}} = (1 - \mathcal{P})\rho^{\mathrm{T}}.$$

Before going ahead with the derivation of the evolution equation for the reduced density matrix ρ_S, it is important to consider the following identities:

$$\mathcal{P}\mathcal{L}_S = \mathcal{L}_S\mathcal{P} \tag{I.9}$$

$$\mathcal{P}\mathcal{L}_B = \mathcal{L}_B\mathcal{P} = 0 \tag{I.10}$$

$$\mathcal{P}\mathcal{L}_I\mathcal{P} = 0. \tag{I.11}$$

To prove these properties we apply each of these operators on ρ^{T}.

We start with (I.9)

$$\mathcal{P}\mathcal{L}_S\rho^{\mathrm{T}} = \rho_B^{\mathrm{eq}} \, Tr_B\left\{\mathcal{L}_S\rho^{\mathrm{T}}\right\} = \rho_B^{\mathrm{eq}} \, \mathcal{L}_S Tr_B\left\{\rho^{\mathrm{T}}\right\}$$

$$= \mathcal{L}_S\rho_B^{\mathrm{eq}} \, Tr_B\left\{\rho^{\mathrm{T}}\right\} = \mathcal{L}_S\mathcal{P}\rho^{\mathrm{T}}.$$

The second identity (I.10) follows because

$$\mathcal{P}\mathcal{L}_B\rho^{\mathrm{T}} = \rho_B^{\mathrm{eq}} \, Tr_B\left\{\mathcal{L}_B\rho^{\mathrm{T}}\right\} = -i\rho_B^{\mathrm{eq}} \, Tr_B\left\{H_B\rho^{\mathrm{T}} - \rho^{\mathrm{T}}H_B\right\}$$

$$= -i\rho_B^{\mathrm{eq}} \left[Tr_B\left\{H_B\rho^{\mathrm{T}}\right\} - Tr_B\left\{\rho^{\mathrm{T}}H_B\right\}\right] = 0,$$

here we have used the invariance of the trace under cyclic permutations. On the other hand

$$\mathcal{L}_B\mathcal{P}\rho^{\mathrm{T}} = \mathcal{L}_B\rho_B^{\mathrm{eq}} \, Tr_B\left\{\rho^{\mathrm{T}}\right\} = -i\left[H_B, \rho_B^{\mathrm{eq}}\rho_S\right]$$

$$= -i\left(H_B\rho_B^{\mathrm{eq}}\rho_S - \rho_B^{\mathrm{eq}}\rho_SH_B\right) = -i\left(H_B\rho_B^{\mathrm{eq}} - \rho_B^{\mathrm{eq}}H_B\right)\rho_S$$

$$= -i\left[H_B, \rho_B^{\mathrm{eq}}\right]\rho_S = 0,$$

this follows because H_B and ρ_S commute, and $\left[H_B, \rho_B^{\mathrm{eq}}\right] = 0$ because ρ_B^{eq} is invariant.

The identity (I.11) follows if

$$\mathcal{P}\mathcal{L}_I \mathcal{P}\rho^{\mathrm{T}} = \mathcal{P}\mathcal{L}_I \rho_B^{\mathrm{eq}} \, Tr_B \{\rho^{\mathrm{T}}\} = \mathcal{P}\mathcal{L}_I \rho_B^{\mathrm{eq}} \rho_S$$
$$= -i\mathcal{P}\left[H_I, \rho_B^{\mathrm{eq}}\rho_S\right] = -i\rho_B^{\mathrm{eq}} \, Tr_B \{[H_I, \rho_B^{\mathrm{eq}}\rho_S]\}$$
$$= -i\rho_B^{\mathrm{eq}}\left[Tr_B \{H_I\rho_B^{\mathrm{eq}}\} \rho_S - \rho_S Tr_B \{\rho_B^{\mathrm{eq}}H_I\}\right]$$
$$= -i\rho_B^{\mathrm{eq}}\left[Tr_B \{H_I\rho_B^{\mathrm{eq}}\}, \rho_S\right] \stackrel{?}{=} 0,$$

this would mean that the *bath average* over the interaction evolution vanishes. If this is not the case this term would give a von Neumann's contribution, which is not relevant for the irreversibility analysis.

In Chap. 1 (Sect. 1.14) we used projector operators in the Laplace representation, here we will do the same but in the time representation; that is, we apply separately $\mathcal{P}$ and $\mathcal{Q}$ to Eq. (I.8); then, we first get

$$\frac{d}{dt}\mathcal{P}\rho^{\mathrm{T}} = \mathcal{P}\left(\mathcal{L}_S + \mathcal{L}_B + \lambda\mathcal{L}_I\right)\rho^{\mathrm{T}}$$
$$= \mathcal{P}\mathcal{L}_S\rho^{\mathrm{T}} + \lambda\mathcal{P}\mathcal{L}_I\rho^{\mathrm{T}} = \mathcal{P}\mathcal{L}_S\rho^{\mathrm{T}} + \lambda\mathcal{P}\mathcal{L}_I\left(\mathcal{P} + \mathcal{Q}\right)\rho^{\mathrm{T}}$$
$$= \mathcal{L}_S\mathcal{P}\rho^{\mathrm{T}} + \lambda\mathcal{P}\mathcal{L}_I\mathcal{Q}\rho^{\mathrm{T}}, \tag{I.12}$$

here we have used (I.9)–(I.11) and the identity $\mathcal{P} + \mathcal{Q} = \mathbf{1}$.

Now applying $\mathcal{Q}$ to Eq. (I.8) we get

$$\frac{d}{dt}\mathcal{Q}\rho^{\mathrm{T}} = \mathcal{Q}\mathcal{L}\left(\mathcal{P} + \mathcal{Q}\right)\rho^{\mathrm{T}}$$
$$= \mathcal{Q}\left(\mathcal{L}_S + \mathcal{L}_B + \lambda\mathcal{L}_I\right)\mathcal{P}\rho^{\mathrm{T}} + \mathcal{Q}\mathcal{L}\mathcal{Q}\rho^{\mathrm{T}}$$
$$= \mathcal{Q}\lambda\mathcal{L}_I\mathcal{P}\rho^{\mathrm{T}} + \mathcal{Q}\mathcal{L}\mathcal{Q}\rho^{\mathrm{T}}, \tag{I.13}$$

here we have used (I.9) and (I.10). The solution of (I.13) is

$$\mathcal{Q}\rho^{\mathrm{T}}(t) = e^{\mathcal{Q}\mathcal{L}t}\mathcal{Q}\rho^{\mathrm{T}}(0) + \int_0^t e^{\mathcal{Q}\mathcal{L}(t-t')}\mathcal{Q}\lambda\mathcal{L}_I\mathcal{P}\rho^{\mathrm{T}}(t') \, dt'$$
$$= \int_0^t e^{\mathcal{Q}\mathcal{L}\tau}\mathcal{Q}\lambda\mathcal{L}_I\rho_B^{\mathrm{eq}}\rho_S(t-\tau) \, d\tau, \tag{I.14}$$

here we have used a change of variable $\tau = t - t'$, and a *separable* initial condition[8]

---

[8] Actually the system $\mathcal{S}$ and the bath $\mathcal{B}$ are never uncorrelated as they are constantly interacting. The glib excuse that the interaction is switched on at $t = 0$ is unrealistic and leads to some problems in trying to build up a correct semigroup, The issue on the assumption of a separable initial condition has been remarked recently, so a new proposal for studying fluctuations and relaxation in quantum open systems has been made; see N. van Kampen, J. Stat. Phys. 115, 1057 (2004).

$$\rho^{\mathrm{T}}(0) = \rho_S(0) \otimes \rho_B^{\mathrm{eq}},$$

then $\mathcal{Q}\rho^{\mathrm{T}}(0) = 0$. We can now introduce the solution of $\mathcal{Q}\rho^{\mathrm{T}}(t)$ in (I.12) to get an exact evolution equation for the mean value:

$$\frac{d}{dt}\mathcal{P}\rho^{\mathrm{T}} = \mathcal{L}_S\mathcal{P}\rho^{\mathrm{T}} + \lambda\mathcal{P}\mathcal{L}_I \int_0^t e^{\mathcal{Q}\mathcal{L}\tau}\mathcal{Q}\lambda\mathcal{L}_I\rho_B^{\mathrm{eq}}\rho_S(t-\tau)\,d\tau. \tag{I.15}$$

This equation is exactly equivalent to the original von Neumann's equation (I.8) after bath variables have been eliminated. Sometimes it is called *generalized quantum master equation*, but to call it in this way is not correct. In all this text we have preserved the name *generalized* ME to characterize a well-defined (classic) stochastic process (for example in the context of the CTRW with memory). In quantum mechanics if we want to get a semigroup (Markov process) the evolution equation should obey the structural theorem, which is represented by the Kossakowski-Lindblad completely positive infinitesimal generator (I.4). It is possible to see that even introducing some approximations in (I.15), in general this equation does not satisfy the structural theorem.[9] The quantum master-like Eq. (I.15) is just the starting point to introduce different approximations in order to get a Kossakowski-Lindblad semigroup.

I.2.1   Approximations to Get a Quantum ME

From (I.15) we see that the second term is multiplied by λ^2 but this parameter also appears (inside the integral) through the exponential $e^{\mathcal{Q}\mathcal{L}\tau}$. Then to maintain a second order perturbation we neglect the contribution $\lambda\mathcal{L}_I$ in this exponential. Then, we can write (I.15) in the form

$$\frac{d}{dt}\rho_S(t) \simeq \mathcal{L}_S\rho_S(t) + \lambda^2 \int_0^t \mathcal{G}(\tau)\rho_S(t-\tau)\,d\tau. \tag{I.16}$$

Using $\mathcal{P}\mathcal{L}_I\mathcal{P} = 0$ in the integrand we have

$$\mathcal{G}(\tau)\rho_S(t-\tau) = Tr_B\left\{\mathcal{L}_I\, e^{\mathcal{Q}(\mathcal{L}_S+\mathcal{L}_B)\tau}\mathcal{Q}\,\mathcal{L}_I\mathcal{P}\rho^{\mathrm{T}}(t-\tau)\right\}$$

$$= Tr_B\left\{\mathcal{L}_I\, e^{\mathcal{Q}(\mathcal{L}_S+\mathcal{L}_B)\tau}\mathcal{L}_I\mathcal{P}\rho^{\mathrm{T}}(t-\tau)\right\}$$

$$= Tr_B\left\{\mathcal{L}_I\, e^{(\mathcal{L}_S+\mathcal{L}_B)\tau}\mathcal{L}_I\rho_B^{\mathrm{eq}}\right\}\rho_S(t-\tau).$$

---

[9]See N.G. van Kampen and I Oppenheim, J. Stat. Phys. 87, 1325 (1997).

Then, the superoperator acting on $\rho_S(t - \tau)$ is

$$\mathcal{G}(\tau) = Tr_B\left\{\mathcal{L}_I e^{(\mathcal{L}_S + \mathcal{L}_B)\tau}\mathcal{L}_I\rho_B^{eq}\right\}. \tag{I.17}$$

Noting that $\dot{\rho} = -i[H, \rho] \equiv \mathcal{L}\rho$, this implies that $e^{\mathcal{L}t}$ produces an evolution forward in time: $\rho(t) = e^{\mathcal{L}t}\rho(0) = e^{-iHt}\rho e^{+iHt}$, but backward for any observable, so we can work out the integrand $\mathcal{G}(\tau)\rho_S(t - \tau)$ using (I.17) as:

$$\begin{aligned}
\mathcal{G}(\tau)\rho_S(t - \tau) &= Tr_B\left\{\mathcal{L}_I e^{(\mathcal{L}_S + \mathcal{L}_B)\tau}\mathcal{L}_I\rho_B^{eq}\rho_S(t - \tau)\right\} \\
&= -iTr_B\left\{\mathcal{L}_I e^{(\mathcal{L}_S + \mathcal{L}_B)\tau}\left[H_I, \rho_B^{eq}\rho_S(t - \tau)\right]\right\} \\
&= -iTr_B\left\{\mathcal{L}_I\left[H_I(-\tau), \rho_B^{eq}\rho_S(t)\right]\right\} \\
&= -Tr_B\left\{\left[H_I, \left[H_I(-\tau), \rho_B^{eq}\rho_S(t)\right]\right]\right\}.
\end{aligned}$$

Note that all evolutions have been done according to the unperturbed Liouville operator $\mathcal{L}_S + \mathcal{L}_B$. In order to be able to continue with this program we need to introduce a model for the interaction and for the bath.

To summarize, under the present approximations, the quantum master-like equation has the form:

$$\boxed{\frac{d}{dt}\rho_S \simeq \mathcal{L}_S\rho_S - \lambda^2\int_0^t Tr_B\left\{\left[H_I, \left[H_I(-\tau), \rho_B^{eq}\rho_S(t)\right]\right]\right\}d\tau.} \tag{I.18}$$

If there is a characteristic time scale τ_c for which $\mathcal{G}(\tau)$ is negligible for $\tau > \tau_c$, we can extend the upper limit in the integral to ∞ (this is the second step in order to get a master-like equation), but still in order to transform this equation into a *bonafide* Kossakowski-Lindblad semigroup more approximations must be introduced to fulfill the structural theorem (I.5), these approximations depend on the type of system we are working with, i.e., H_S and interaction Hamiltonian H_I; see next section.

I.3 Dissipative Quantum Random Walk

A dissipative quantum RW is an example of a quantum Markov semigroup. This model can be defined starting from a free particle in the lattice interacting with a thermal phonon bath. Then, eliminating the variables of the bath we can arrive to a Kossakowski-Lindblad infinitesimal semigroup I.5.[10] Let us present here the quantum ME in terms of the parameters of the system $S + B$.

---

[10]This model was originally introduced by: N.G. van Kampen, J. Stat. Phys. 78, 299 (1995). A slight difference in the interaction Hamiltonian—between the system and the bath—and considering the Schrödinger-Langevin approach can be seen in: M.O. Cáceres and A.K. Chattah, J. Mol. Liquids, 71, 187 (1997).

I.3.1   The Tight-Binding Quantum Open Model

The total Hamiltonian for *one* quantum random walk can be written in the form

$$H_T = \left(E_0 \mathbf{1} - \Omega \frac{a + a^\dagger}{2} \right) + \sum_{v=1}^{2} V_v \otimes B_v + H_\mathcal{B}. \tag{I.19}$$

The first term corresponds to the *tight-binding* Hamiltonian H_S where a and $a^\dagger$ are translational operators in the Wannier bases $|s\rangle$ (it is possible to write these operators in second quantization[11]) and $\mathbf{1}$ is the identity operator, i.e.:

$$a\,|s\rangle = |s - 1\rangle\,, \ a^\dagger\,|s\rangle = |s + 1\rangle\,, \ \mathbf{1} = \sum_s |s\rangle\,\langle s|\,,$$

note that $[a, a^\dagger] = 0$, here E_0 is the *tight-binding* energy of site and Ω the next neighbor hopping energy. The second term in (I.19) is the interaction Hamiltonian and corresponds to a linear coupling between phonon operators $B_1 = \sum_k v_k \mathcal{B}_k = B_2^\dagger$ and system operators $V_1 = \hbar \Gamma a = V_2^\dagger$, here $\Gamma > 0$ is the coupling parameter. The third term is the phonon bath Hamiltonian written in terms of boson operators $\sum_k \hbar \omega_k \mathcal{B}_k^\dagger \mathcal{B}_k$.

The quantum ME can be obtained eliminating the variables of thermal bath and assuming for the initial condition of the total density matrix:

$$\rho^T(0) = \rho(0) \otimes \rho_\mathcal{B}^{eq},$$

where $\rho_\mathcal{B}^{eq}$ is the equilibrium density matrix of the bath $\mathcal{B}$. Then, in the Born-Markov approximation, the evolution equation for the reduced density matrix ρ can be obtained from (I.18), and we get[12]:

$$\dot{\rho} \equiv \frac{d\rho}{dt} = \frac{-i}{\hbar} \left[H_{eff}, \rho \right] + D \left(a \rho a^\dagger + a^\dagger \rho a - 2 a^\dagger a \rho \right), \tag{I.20}$$

---

[11]For more details see Appendix A in: M. Nizama and M.O. Cáceres, J. Phys. A: Math. Theor. **45**, 335303 (2012). Also this model can be extended to work out many fermions or bosons interacting with a quantum thermal bath.

[12]For model (I.19) condition (I.7) is fulfilled. Bath thermal correlations $\chi_{\alpha\beta}(-\tau)$ can be calculated from equilibrium statistical mechanics. The derivation of (I.20) follows from the application of (I.18) with the use of (I.19), adopting a model for the spectral function of the bath: $g(\omega) = \sum_k |v_k|^2 \delta(\omega - \omega_k)$, and using Hilbert's transform to connect the half-Fourier of the correlation: $\int_0^\infty c(-\tau) \exp(-i\omega\tau)\, d\tau = \frac{1}{2} h(\omega) + i s(\omega)$ with the Fourier transform $h(\omega)$ and the function $s(\omega) = \frac{1}{2\pi} \mathcal{P} \int_{-\infty}^{+\infty} du\, h(u)(u - \omega)^{-1}$ (using symmetrization $g(\omega) = g(-\omega)$); see advanced exercise 5.5.6, and in M.O. Cáceres and N. Nizama, J. Phys. A Math. Theor. 43, 455306 (2010).

with a trivial effective Hamiltonian: $H_{eff} = H_S - \hbar\omega_c a^\dagger a$. The diffusion constant is given in terms of bath temperature T and the coupling constant: $D \propto \Gamma^2 k_B T/\hbar$, the additive energy $\hbar\omega_c$ is related to the Ohmic approximation.[13] For simplicity we can add an additive constant to the *tight-binding* Hamiltonian $-E_0 + \omega_c\hbar + \Omega$. This assumption does not change the general results and finally we can write (using that $a^\dagger a = 1$):

$$H_{eff} = \Omega \left(1 - \frac{a + a^\dagger}{2} \right), \tag{I.21}$$

as was presented originally in van Kampen's paper.[14] It can be noted from Eq. (I.20) that as $D \to 0$ (or temperature $T \to 0$), von Neumann's equation is recovered (unitary evolution).

Second Moment of a Dissipative Quantum Random Walk

From Eq. (I.20), we can obtain the thermal dynamics of any operator, in particular for the position operator $\mathbf{q}$, which in the Wannier basis has the matrix elements:

$$\langle s| \mathbf{q} |s'\rangle = s\, \delta_{s,s'}, \tag{I.22}$$

here s is an eigenvalue of $\mathbf{q}$. Note that $\mathbf{q}$ is defined as a dimensionless position operator with lattice parameter $\epsilon = 1$.

Time-derivative of the position operator is straightforward to calculate in the Heisenberg representation[15]

$$\frac{d\mathbf{q}}{dt} = \frac{-i}{\hbar} [\mathbf{q}, H] = \frac{-i\Omega}{2\hbar} \left(a - a^\dagger \right).$$

Then, from the fact that $a(t) = a(0), a^\dagger(t) = a^\dagger(0)$ (they do not evolve in time), we get for the position operator:

$$\mathbf{q}(t) = \frac{-i\Omega}{2\hbar} \left(a - a^\dagger \right) t + \mathbf{q}(0).$$

---

[13]This is called Caldeira and Legett's frequency cut-off for the spectral function of the bath $g(\omega) \propto \omega$, $0 \leq \omega \leq \omega_c$ and $g(\omega) = 0$ elsewhere; see: A.O. Caldeira and A.J. Legget; Ann. Phys. (USA), **149**, 374 (1983).

[14]Accordingly, the energy spectrum of this Hamiltonian contains a single energy band of width 2Ω and the motion of the particle would be purely ballistic without coupling to a fluctuations environment. Interestingly considering generalized translational operators that allow long range hopping for the RW, the associated Hamiltonian can present a spectrum of eigenenergies with a fractal structure, see reference in note (12).

[15]This result follows calculating the matrix elements in the Wannier basis $\langle s| [\mathbf{q}, H] |s'\rangle$; see also advanced exercise 8.8.1.

On the other hand, the quantum thermal statistical average—of any operator $\mathbf{A}$—in the Heisenberg picture can be written as $\langle \mathbf{A}(t) \rangle = Tr[\mathbf{A}(t)\rho(0)] = Tr[\mathbf{A}\rho(t)]$. Therefore from (I.20) we get for the dispersion of the dissipative quantum RW:

$$\sigma(t)^2 \equiv \langle \mathbf{q}(t)^2 \rangle - \langle \mathbf{q}(t) \rangle^2 = \frac{1}{2}\left(\frac{\Omega t}{\hbar}\right)^2 + 2Dt. \qquad (I.23)$$

From this result, it is possible to see that von Neumann's term gives the contribution $\propto t^2$, this is a well-known quantum result. In fact for the zero dissipation case, $D = 0$, we get that Anderson's boundaries (ballistic peak) move controlled by the linear law for the dispersion of the wave-packet:

$$\sigma(t) = \sqrt{\langle \mathbf{q}(t)^2 \rangle - \langle \mathbf{q}(t) \rangle^2} = \frac{1}{\sqrt{2}}\frac{\Omega t}{\hbar}, \text{ if } D = 0.$$

Therefore, we can associate the quantity $V_A = \frac{1}{\sqrt{2}}\frac{\Omega}{\hbar}$ to the velocity of Anderson's boundaries in a one-dimensional lattice. In addition, we can see that the quantum to classical transition is controlled by the dissipative coefficient D.

The Continuous Limit of the Quantum Random Walk

The continuous limit can be studied considering the lattice parameter $\epsilon \to 0$. Noting that shift operators can be written in the form:

$$a = \exp(i\hat{p}\epsilon/\hbar), \quad a^\dagger = \exp(-i\hat{p}\epsilon/\hbar), \quad \text{with} \quad \hat{p} = -i\hbar\partial_x,$$

and using the additional prescription:

$$\mathbf{q} \sim x/\epsilon, \quad \Pi \equiv \frac{-i\hbar}{2}(a - a^\dagger) = \hbar \sin\left(\frac{\epsilon\hat{p}}{\hbar}\right).$$

We conclude that Π can be called the pseudo-momentum operator. Now, it is simple to show, in the limit $\epsilon \to 0$, that the commutation relation between operators $\mathbf{q}$ and Π goes to

$$[\mathbf{q}, \Pi] = \frac{i\hbar}{2}(a + a^\dagger) \to \left[\frac{x}{\epsilon}, \epsilon\hat{p}\right] = i\hbar.$$

I.3.2 Quantum Decoherence, Dissipation, and Disordered Lattices

The extended analysis for a dissipative quantum semigroup, of the RW type, is a promising model. For example, from Eq. (I.20) the probability profile for this quantum open system ($D \neq 0$); that is, for a *dissipative* tight-binding free particle, can analytically be studied in terms of the reduced density matrix ρ[16]; then, we can characterize decoherence phenomena and solve several correlation functions associated with the coherent superposition feature for *one* or many particles, using the Fock representation approach.[17]

Interestingly the analysis of a dissipative quantum RW in a disordered lattice can be made in the context of the *Renewal* theory with the application of a completely positive map at discrete random times n (the map represents the interaction of the system of interest S with the bath B). This model of evolution leads to a sort of continuous-time quantum RW approach. That is, the (dissipative) quantum RW remaining (without bath disruptive intervention) at an arbitrary lattice site s during a random waiting-time interval $[0, \tau]$ until another completely positive map is applied again, and so on time and again in continuous time. Then, the resulting evolution equation for the density matrix is a generalized Kossakowski-Lindblad completely positive infinitesimal generator that has a time-convolution structure, reminiscent of non-Markov CTRW theory, and that we have explained in this text; see advanced exercise 7.5.4. Therefore, short and long times asymptotic results can analytically be studied in terms of the waiting-time model that we may have used to represent the disorder (weak or strong) for the trap-dynamics of a dissipative quantum RW particle, see advanced exercises 8.8.2–4. The important point is that introducing the disorder at this quantum level produces an evolution equation which is a generalized (non-Markov) completely positive infinitesimal generator of the form:

$$\dot{\rho}(t) = \frac{-i}{\hbar} \left[H_{eff}, \rho \right] + \int_0^t \Phi(t - \tau) e^{(t-\tau)\mathcal{L}_0} \left\{ a\rho(\tau) a^\dagger + a^\dagger \rho(\tau) a - 2aa^\dagger \rho(\tau) \right\} d\tau,$$
(I.24)

compare with (I.20), and with the classical CTRW approach of Sect. 7.3.1.

The present discussion can also be extended to tackle many fermionic (or bosonic) particles; then, pointing out the interplay between particle-particle and

---

[16] This density matrix can be solved in terms of Bessel functions, also Wigner's pseudo-density (in the lattice) can be calculated. This allows to study the quantum nature of the system (negativity of Wigner's function) in terms of the physical parameters and the temperature of the bath, see: M. Nizama and M.O. Cáceres, J. Phys. A: Math. Theor. **45**, 335303 (2012).

[17] Tools of quantum information theory can be used to study decoherence, correlations and entanglement bases on this exact solution, see M. Nizama and M.O. Cáceres, Physica A **392**, 6155, (2013); M. Nizama and M.O. Cáceres, Physica A **400**, 31 (2014). A model with two distinguishable particles can be seen in: M. Nizama and M.O. Cáceres, arXiv:1512.08700v1.

bath-particle interaction in the model. Using quantum information theory we can also perform analysis on the bath-induced quantum correlations for a many-body dissipative system.

Optional Exercise (**Proof of I.24**). Consider a completely positive (CP) super-operator acting on σ (density matrix in the interaction representation with respect to Hamiltonian)

$$\mathcal{E}\left[\bullet\right] = \sum_j V_j\left[\bullet\right] V_j^{\dagger}, \text{ with } \sum_j V_j V_j^{\dagger} = \mathbf{1}, \tag{I.25}$$

here V_j are any set of operators determining the irreversible, dissipative evolution of the density matrix of the system $\mathcal{S}$. In order to represent the intervention of a thermal bath $\mathcal{B}$ on the system $\mathcal{S}$, we assume a discrete time model (similar to a classical discrete recurrence relation for a Markov chain, see Chap. 6); then, apart from the unitary evolution we propose the following CP map on the density matrix

$$\sigma\left(n + 1\right) = \mathcal{E}\left[\sigma\left(n\right)\right], \text{ with initial condition } \sigma\left(n = 0\right).$$

The solution of this map can be found using the generating function technique

$$\sigma\left(z\right) = \sum_{n=0}^{\infty} z^n \sigma\left(n\right) = \left(1 - z\mathcal{E}\left[\bullet\right]\right)^{-1} \sigma\left(n = 0\right). \tag{I.26}$$

Now, consider a *renewal* process characterizing the number of events n in a continuous-time representation, see advanced exercise 7.5.1(d). The continuous time evolution of σ can be written in terms of the renewal probability density $\psi_{(n)}\left(t\right)$ in the form:

$$Q\left(t\right) = \sum_{n=0}^{\infty} \psi_{(n)}\left(t\right) \sigma\left(n\right), \tag{I.27}$$

here the quantity $Q\left(t\right) dt$ gives the density matrix *just* at time $t + dt$, and the renewal process is characterized by a waiting-time density function $\psi\left(t\right) \geq 0, \forall t \geq 0$ and the hierarchy $\left(n \in \mathcal{N}\right)$

$$\psi_{(n)}\left(t\right) = \int_0^t \psi_{(n-1)}\left(\tau\right) \psi\left(t - \tau\right) d\tau, \quad \psi_{(0)}\left(t\right) = \delta\left(t - 0^+\right).$$

Thus, from (I.27), in the Laplace representation we have

$$Q\left(u\right) = \left(1 - \psi\left(u\right) \mathcal{E}\left[\bullet\right]\right)^{-1} \sigma\left(n = 0\right). \tag{I.28}$$

In order to calculate the density matrix at time $t > 0$ we have to consider also the possibility of remaining without any disruptive intervention from the thermal bath. This effect is taken into account considering the probability of no-intervention during the time interval $[0, t]$:

$$\phi(t) = 1 - \int_0^t \psi(\tau) d\tau, \quad \int_0^\infty \psi(\tau) d\tau = 1. \tag{I.29}$$

From the solution $Q(t)$ and (I.29) we can write the density matrix in continuous time as:

$$\sigma(t) = \int_0^t \phi(t - \tau) Q(\tau) d\tau, \quad t > 0. \tag{I.30}$$

Introducing the Laplace transform in (I.30) we get

$$\sigma(u) = \frac{1 - \psi(u)}{u} (1 - \psi(u) \mathcal{E}[\bullet])^{-1} \sigma(n = 0).$$

From this equation, after some algebra, we can show the identity:

$$u\sigma(u) - \sigma(n = 0) = \frac{u\psi(u)}{1 - \psi(u)} (-1 + \mathcal{E}[\bullet]) \sigma(u).$$

Then, using the inverse Laplace transform we can write an evolution equation for the density matrix in the interaction representation $\sigma(t) = e^{-t\mathcal{L}_0} \rho(t)$, in the form $(-i[H, \rho] \equiv \mathcal{L}_0 \rho)$:

$$\partial_t \sigma(t) = \int_0^t \Phi(t - \tau) e^{-t\mathcal{L}_0} \mathcal{L}[\bullet] e^{+t\mathcal{L}_0} \sigma(\tau) d\tau, \quad \text{with} \quad \mathcal{L}[\bullet] = (\mathcal{E}[\bullet] - 1)$$
$$\tag{I.31}$$

where $\Phi(t)$ is the memory kernel[18] of the *generalized* Kossakowski-Lindblad completely positive infinitesimal generator $\mathcal{L}[\bullet]$. Note that from the inverse Laplace transform the kernel is given by

$$\Phi(t) = \frac{1}{2\pi i} \int_{c-i\infty}^{c+i\infty} \frac{u\psi(u)}{1 - \psi(u)} e^{ut} du.$$

Thus if $\psi(t)$ is exponential we recover the Markovian case for an ordered lattice: $\Phi(t) \propto \delta(t)$. Different random trap models can be introduced in terms of the waiting-time function $\psi(t)$.

Exercise. Prove that $\int_0^\infty \psi_{(n)}(\tau) d\tau = 1$, $\forall n \in \mathcal{N}$.

---

[18] Note that $\Phi(t)$ is not a pdf.

Exercise. Take the Laplace transform of $Q(t)$ and use $\mathcal{L}_u\left[\psi_{(n)}(t)\right] = \psi(u)^n$ to prove (I.28).

Exercise. Use the CP map (I.25) to show that Kossakowski-Lindblad completely positive infinitesimal generator is

$$\mathcal{L}[\bullet] = (\mathcal{E}[\bullet] - 1) = \frac{1}{2}\sum_j \left[V_j\bullet, V_j^\dagger\right] + \left[V_j, \bullet V_j^\dagger\right].$$

Note that the inverse (in general) is not true.

Exercise. Use a CP map with $V_j \propto \{a, a^\dagger\}$, $j = 1, 2$, in the quantum RW model (I.21) (*tight-binding* Hamiltonian) to obtain, in the original representation, the non-unitary evolution term of equation (I.24), interpret the meaning of the waiting-time $\psi(t)$, (see, M.O. Cáceres in: Euro. Phys. Jour. B; Topical issue 50 year of CRTW, 2017).

Exercise. Show that for any CP map $\mathcal{E}[\bullet]$ and waiting-time $\psi(t) \geq 0$, by construction (I.31) is a well-defined CP infinitesimal generator even when the kernel $\Phi(t)$ is not necessarily positive at all times (see advanced exercise 8.8.2–4, and Appendix E).

Index

© Springer International Publishing AG 2017
M.O. Cáceres, *Non-equilibrium Statistical Physics with Application to Disordered Systems*, DOI 10.1007/978-3-319-51553-3

Printed in the United States
By Bookmasters